New Comprehensive Biochemistry

Volume 15

General Editors

A. NEUBERGER
London

L.L.M. van DEENEN
Utrecht

ELSEVIER
AMSTERDAM · NEW YORK · OXFORD

Photosynthesis

Editor

J. AMESZ
Leiden

1987
ELSEVIER
AMSTERDAM · NEW YORK · OXFORD

ISBN 0-444-80864-7 (volume)
ISBN 0-444-80303-3 (series)

Published by:
Elsevier Science Publishers B.V. (Biomedical Division)
P.O. Box 211
1000 AE Amsterdam
The Netherlands

Sole distributors for the USA and Canada:
Elsevier Science Publishing Company, Inc.
52 Vanderbilt Avenue
New York, NY 10017
USA

Library of Congress Cataloging-in-Publication Data
Main entry under title:

Photosynthesis.

(New comprehensive biochemistry ; v. 15)
Includes bibliographical references and index.
1. Photosynthesis. I. Amesz, Jan. II. Series.
QD415.N48 vol. 15 574.19'2 s [581.1'3342] 87-9229
[QK882]
ISBN 0-444-80864-7

Printed in The Netherlands

Introduction

In the early 17th century Van Helmont (1577–1644) performed one of the first modern experiments in plant physiology. He planted a willow branch in a tub of soil and watered it regularly until it had developed into a reasonably large tree. After 5 years Van Helmont terminated the experiment and found that the tree had accumulated a considerable amount of dry material (164 pounds to be precise) whereas the weight of the soil had decreased by only a few ounces during the same period. From this he concluded that plants do not feed on soil, as postulated by the then prevailing theory, but on the only substance supplied to the tree: water. Van Helmont's experiment was probably the first to show that plants have a special form of metabolism that distinguishes them from animals, but it took approximately one and a half centuries before the discoveries of Priestley, Ingen-Housz and others established the existence of the process we now call photosynthesis. Although the importance of this process was immediately realized (the reader should consult Rabinowitch's monograph* for a vivid description of the early years of photosynthesis research), it took another 150 years before some insight into the molecular mechanisms of photosynthesis began to evolve. The post-war years, which showed such a rapid development of biochemical and physical techniques, also witnessed an unprecedented expansion of photosynthesis research, based on the application of these very techniques.

Due to the work of Calvin, Benson and associates in the forties and fifties it became clear that carbon dioxide fixation, once supposed to be the basic photosynthetic reaction, occurs by an intricate sequence of enzymatic processes that can in principle function in the dark if fueled by the products of photosynthesis. Duysens' studies established the role of pigments in harvesting and transferring the energy of light, and gradually it became clear that the primary energy conversion steps consist of electron transfer reactions that take place in an entity called the reaction center. Around 1960 the basic difference between plant and bacterial photosynthesis became known: bacteria have only one type of reaction center, whereas plants have two, one of which produces a strong oxidant able to oxidize water to oxygen.

During the last five or ten years many important developments have taken place in photosynthesis research. The combined efforts of biochemists and (bio)physicists have now provided a picture of the mechanisms of the photosynthetic reactions and of the structure of the various components of the photosynthetic membrane which is vastly more detailed than might have been envisaged a few years ago. The application of advanced optical instrumentation, both in the visible region (e.g. by laser spectroscopy) and by use of electron spin resonance, has provided a wealth of information concerning the primary reactions of photosynthesis and the inter-

* E.I. Rabinowitch, Photosynthesis and Related Processes, Vol. I. Interscience Publishers, New York, 1945, 599 pp.

actions between the primary reactants. On the other hand, the work of protein chemists and molecular biologists and the recent X-ray analysis of the bacterial reaction center together with optical measurements have given increasingly detailed information on the structure and organization of the protein complexes which are embedded in the photosynthetic membrane and are involved in energy conversion and electron transport. Also the mechanism of oxygen evolution and the role of manganese in this reaction, for a long time a 'black box' in the gradually emerging picture of the electron transfer scheme, are now beginning to reveal their secrets.

Although these recent developments have not basically altered our concepts of the mechanism of photosynthesis, they have certainly clarified the picture to a considerable extent, and altogether they signify an important leap forward to a better understanding of the intricacies of the molecular processes of photosynthesis. Many points that used to be blurred have now come into focus, and many questions can now be asked with more precision and are now amenable to further experimentation.

It is hoped that this book conveys some of the excitement of the recent discoveries. The first two chapters give an introduction to photosynthesis in plants and bacteria, while the other chapters give a discussion of more specialized topics in the areas of primary charge separation, electron transport, the secondary products of photosynthesis, structure and genetics of protein complexes, and, finally, evolution. Together they should present a comprehensive overview of the current state of knowledge of the molecular processes of photosynthesis, which have fascinated so many investigators of various disciplines and scientific backgrounds during the last decades.

In a book written by specialists in the various areas of photosynthesis research, there are bound to be some overlaps and some gaps. One area that may not have been adequately covered, although its impact can be discerned in various chapters, is the wealth of information regarding energy and electron transfer and structure derived from studies of prompt and delayed fluorescence of chlorophyll and bacteriochlorophyll. However, the reader interested in this area should find enough information in this book for further literature on the subject.

At this point the editor wishes to express his thanks to the authors of this volume, both for their willingness to write a chapter and for the quality of their contributions. Due to their efforts to keep to the projected time scheme, this book can be published with minimal delay, and give an up-to-date account of research into the molecular aspects of the most fundamental life process on earth.

J. Amesz

Department of Biophysics
Huygens Laboratory
University of Leiden
The Netherlands

Contents

Introduction, by J. Amesz .... *V*

Non-standard abbreviations used in this volume.......... *XV*

Chapter 1. Energy conversion in higher plants and algae, by G. Forti.......... *1*

1. Introduction 1
2. Electron transport from water to NADP: an overview 2
3. Photosynthetic phosphorylation 8
4. Molecular and supramolecular structure of thylakoids 12
   4.1. Lateral heterogeneity, fluorescence and electron transport 12
   4.2. Excitation energy distribution between the photosystems 14

References 17

Chapter 2. Photosynthetic bacteria, by B.K. Pierson and J.M. Olson.......... *21*

1. Introduction 21
2. Filamentous photosynthetic bacteria 23
   2.1. General characteristics 23
   2.2. Light-harvesting, reaction center and electron transport 24
3. Green sulfur bacteria 26
   3.1. General characteristics 26
   3.2. Light-harvesting, reaction center and electron transport 27
4. *Heliobacterium chlorum* – the gram-positive line 28
   4.1. General characteristics 28
   4.2. Light-harvesting, reaction center and electron transport 28
5. Purple bacteria 29
   5.1. General characteristics 29
   5.2. Light-harvesting, reaction center and electron transport 32
6. Bacteriochlorophyll *a*-containing non-phototrophic bacteria 34
7. Phylogeny 35
8. Halobacteria 37
9. Summary 38

References 39

Chapter 3. The bacterial reaction center, by W.W. Parson.......... *43*

1. Introduction 43

2. Purification and crystallization of reaction centers . . . 46
3. Protein structure . . . 47
4. BChl, BPh and other prosthetic groups . . . 51
5. Spectroscopic properties and the distinction between BPh_L and BPh_M . . . 53
6. Electron transfer kinetics and mechanisms . . . 55
Acknowledgements . . . 57
References . . . 57

Chapter 4. The primary reactions of photosystems I and II of algae and higher plants, by P. Mathis and A.W. Rutherford . . . *63*

1. Introduction . . . 63
2. Photosystem I reactions . . . 64
   2.1. The primary donor P-700 . . . 65
      2.1.1. Basic properties of P-700 . . . 65
      2.1.2. P-700: a chlorophyll species . . . 65
      2.1.3. P-700: probaby a dimer of chlorophyll . . . 66
   2.2. Sequence of electron acceptors . . . 67
      2.2.1. Terminal acceptors . . . 67
      2.2.2. Centre X, an intermediate acceptor? . . . 69
      2.2.3. Primary acceptors: A_0, A_1 . . . 70
      2.2.4. Overview of primary reactions and of electron acceptors . . . 72
   2.3. Electron donation to P-700 . . . 72
   2.4. Structure of the PS I reaction centre . . . 73
      2.4.1. Polypeptides and redox centres . . . 73
      2.4.2. Photosystem I light-harvesting antenna . . . 74
      2.4.3. Organization of the reaction centre in the membrane . . . 75
3. Photosystem II reactions . . . 75
   3.1. Introduction . . . 75
   3.2. PS II photochemistry – a comparative view . . . 76
   3.3. The electron acceptor side . . . 76
      3.3.1. The quinone-iron complex . . . 76
      3.3.2. Pheophytin – the intermediate electron acceptor . . . 81
      3.3.3. Other possible acceptors and heterogeneity . . . 82
   3.4. The electron donor side of PS II . . . 84
      3.4.1. P-680, the primary donor . . . 84
      3.4.2. Z, the electron donor to P-680$^+$ . . . 85
      3.4.3. D, the component associated with Signal II slow . . . 86
      3.4.4. Other electron donors in PS II . . . 87
   3.5. Photochemical electron transfer in PS II – an overview . . . 88
   3.6. Structural aspects . . . 89
References . . . 91

Chapter 5. Electron paramagnetic resonance in photosynthesis, by A.J. Hoff *97*

1. Introduction . . . 97
2. Magnetic resonance for the layman . . . 97

3. Physics of EPR . . . 99
3.1. Basic principles . . . 99
3.2. The EPR spectrum . . . 102
3.3. Electron nuclear double resonance, ENDOR . . . 105
4. EPR of primary reactants in photosynthesis . . . 106
4.1. The primary electron donor . . . 107
4.1.1. Bacterial photosynthesis . . . 107
4.1.2. Photosystem I . . . 108
4.1.3. Photosystem II . . . 108
4.2. The primary acceptor . . . 109
4.2.1. Purple bacteria . . . 109
4.2.2. Green bacteria . . . 110
4.2.3. Photosystem I . . . 110
4.2.4. Photosystem II . . . 111
4.3. The intermediary acceptor . . . 111
4.3.1. Bacterial photosynthesis . . . 111
4.3.2. Photosystem I . . . 112
4.3.3. Photosystem II . . . 113
5. The triplet state . . . 113
6. The oxygen-evolving complex . . . 115
6.1. Manganese . . . 115
6.2. Signal II . . . 115
7. Electron spin polarization . . . 116
8. New techniques: ESE and RYDMAR . . . 117
9. Conclusions and prospects . . . 119
Acknowledgements . . . 119
References . . . 120

Chapter 6. The photosynthetic oxygen-evolving process, by G.T. Babcock . . . 125

1. Introduction . . . 125
2. Oxygen evolution – the minimal unit . . . 126
2.1. Polypeptide composition and function in the PS II/OEC . . . 129
2.1.1. Intrinsic polypeptides . . . 129
2.1.2. Extrinsic polypeptides . . . 131
2.2. Electron transfer components . . . 132
2.2.1. P-680 and Z . . . 132
2.2.2. Manganese . . . 134
2.3. Cofactor requirements . . . 138
3. Electron transfer in the oxygen-evolving unit . . . 139
3.1. Electron transfer in the untreated PS II/OEC . . . 139
3.2. Electron transfer in the PS II/OEC following inhibition . . . 143
4. Water oxidation in the oxygen-evolving unit . . . 146
4.1. Substrate and substrate analogue binding . . . 147
4.2. The occurrence of water chemistry . . . 148
4.3. Representative models of oxygen evolution . . . 149
5. Conclusions . . . 151
Acknowledgements . . . 152
References . . . 152

Chapter 7. Photophosphorylation in chloroplasts, by M. Avron *159*

1. History 159
2. General characteristics 160
   2.1. Relation to electron transport 160
   2.2. Coupling sites 161
   2.3. Uncouplers and energy-transfer inhibitors 162
3. Partial reactions 162
   3.1. ATPase 162
   3.2. ATP-P_i exchange 163
   3.3. ^{18}O exchange 163
   3.4. Post-illumination phosphorylation 164
   3.5. Acid-base phosphorylation 164
   3.6. Electric-field phosphorylation 165
4. Mechanism 165
   4.1. The electrochemical potential hypothesis 165
   4.2. ΔpH generation and utilization 165
   4.3. $\Delta\Psi$ generation and utilization 166
   4.4. The threshold 166
   4.5. Bulk vs. local electrochemical potentials 167
5. The ATP synthase 167
   5.1. CF_1 – isolation, properties and reconstitution 167
   5.2. CF_0-CF_1 – isolation, properties and reconstitution 169
6. Reverse reactions 169
   6.1. ATP-driven reactions 169
   6.2. Reactions driven by an electrochemical potential 170
7. Conclusion 170

References 171

Chapter 8. Carbon dioxide assimilation, by F.D. Macdonald and B.B. Buchanan *175*

1. Introduction 175
2. The reductive pentose phosphate cycle 176
3. The C_4 pathway 178
4. Crassulacean acid metabolism 180
5. Regulation of the reductive pentose phosphate cycle 183
   5.1. Identification of the sites of regulation 183
   5.2. Mechanisms of regulation 184
      5.2.1. Regulation of ribulose-1,5-bisphosphate carboxylase oxygenase 184
      5.2.2. The ferredoxin/thioredoxin system 185
      5.2.3. Coordinate regulation of photosynthetic enzymes 186
6. Compartmentation and triose phosphate transport 187
7. Coordination of CO_2 fixation and sucrose synthesis 188
   7.1. Fructose 2,6-bisphosphate 189
      7.1.1. Relationship to carbon partitioning 190
8. Regulation of C_4 photosynthesis 191
9. Regulation of Crassulacean acid metabolism 193

10. Concluding comments 194
Acknowledgements 195
References 196
Note added in proof 197

Chapter 9. Substrate oxidation and NAD$^+$ reduction by phototrophic bacteria, by D.B. Knaff and C. Kämpf *199*

1. Introduction 199
2. Energy-dependent vs. direct reduction of NAD(P)$^+$ 201
    2.1. Purple bacteria 201
    2.2. Green sulfur bacteria 203
3. Succinate oxidation 203
4. Sulfide oxidation 204
5. Thiosulfate oxidation 207

Acknowledgements 208
References 208

Chapter 10. Structure and function of protein complexes in the photosynthetic membrane, by N. Nelson *213*

1. Introduction 213
2. Cytochrome b_6-f complex 214
    2.1. Structure and function of the isolated complex 214
    2.2. Biogenesis of cytochrome b_6-f complex 215
3. The proton-ATPase complex 216
    3.1. Structure and function 216
    3.2. Biogenesis of the proton-ATPase complex 218
4. Photosystem I reaction center 219
    4.1. Structure and function 219
    4.2. Biogenesis of photosystem I reaction center 222
5. Photosystem II 223
    5.1. Structure and function 223
    5.2. Biogenesis of photosystem II 225

References 227

Chapter 11. Structure and function of light-harvesting pigment-protein complexes, by H. Zuber, R. Brunisholz and W. Sidler *233*

1. Introduction 233
2. Light-harvesting antennae of photosynthetic bacteria 236
    2.1. Purple photosynthetic bacteria 238
        2.1.1. Purple bacteria with one type of antenna system 238
        2.1.2. Purple bacteria with two types of antenna systems 243
        2.1.3. Purple bacteria with three or more types of antenna systems 244

2.2. Green photosynthetic bacteria: intramembrane antenna complexes, baseplate systems and the accessory antenna systems (chlorosomes) 246
2.2.1. The antenna system of *Chloroflexus aurantiacus* 246
2.2.2. The BChl *a*-protein of *Prosthecochloris aestuarii* 246
3. Accessory light-harvesting antenna systems (phycobilisomes) of cyanobacteria, red algae and of cryptomonads 247
3.1. Pigment structure and absorption properties of phycobilin chromophores 249
3.2. Classification, occurrence and distribution of phycobiliproteins 249
3.3. Linker polypeptides 255
3.4. The architecture of the phycobilisome 256
3.5. The three-dimensional structure and the function of phycobiliproteins 256
3.6. Cryptomonad phycobiliproteins 259
4. Light-harvesting antennae of algae and higher plants 261
4.1. General features 261
4.2. Antenna complexes of photosystem I 262
4.3. Antenna complexes of photosystem II 263
References 266

Chapter 12. Molecular organization of thylakoid membranes, by J.M. Anderson 273

1. Introduction 273
2. Transverse organization of thylakoid membranes 274
2.1. Transverse asymmetry of thylakoid lipids 274
2.2. Transverse asymmetry of thylakoid proteins 275
2.2.1. Hydropathy index plots 276
2.2.2. Topology of the Cyt *b/f* complex 277
2.2.3. Transverse organization of the Chl-proteins 279
2.2.4. Intrinsic proteins of the PS II complex 280
2.2.5. Extrinsic proteins of the PS II complex 280
2.2.6. Transverse organization of the PS I complex 281
3. Lateral distribution of thylakoid components 281
3.1. Lateral asymmetry of acyl lipid distribution 281
3.2. Lateral heterogeneity in the location of thylakoid intrinsic complexes 283
3.2.1. Electron microscopic studies 283
3.2.2. Biochemical studies 283
3.2.3. 'Seeing is believing' 287
4. Consequences of lateral heterogeneity 288
4.1. Light-harvesting strategies 288
4.1.1. Protein phosphorylation 289
4.2. Electron transport strategies 290
4.3. Adaptation of photosynthetic capacity 291
5. Thylakoid stacking 292
5.1. Mechanisms of thylakoid stacking 292
5.2. Significance of thylakoid stacking 293
6. Epilogue 294
References 295

Chapter 13. Structure and exciton effects in photosynthesis, by R.M. Pearlstein *299*

1. Introduction 299
2. Theoretical concepts 299
3. Purple bacterial antennas 301
   3.1. Scherz-Parson model 303
   3.2. 'Structure-first' models 304
4. Chl *a*/*b*-protein complex 306
5. BChl *a*-protein from *P. aestuarii* 308
6. Purple bacterial reaction centers 311
7. C-phycocyanin 314

References 315

Chapter 14. Genetics and synthesis of chloroplast membrane proteins, by J.C. Gray *319*

1. Introduction 319
2. Photosystem II 319
   2.1. Polypeptides of PS II 320
   2.2. Genes for PS II components 321
   2.3. Synthesis of PS II 327
3. Cytochrome *b-f* complex 329
   3.1. Polypeptides of the cytochrome *b-f* complex 330
   3.2. Genes for components of the cytochrome *b-f* complex 330
   3.3. Synthesis of the cytochrome *b-f* complex 331
4. Photosystem I 332
   4.1. Polypeptides of PS I 332
   4.2. Genes for PS I components 333
   4.3. Synthesis of PS I 334
5. ATP synthase 335
   5.1. Polypeptides of ATP synthase 335
   5.2. Genes for ATP synthase polypeptides 336
   5.3. Synthesis of ATP synthase 337
6. Conclusions and future directions 338

Acknowledgements 338

References 339

Chapter 15. Evolution of photosynthesis, by H.J. van Gorkom *343*

1. Introduction 343
2. The origin of chloroplasts 343
3. The origin of photosynthesis 345
4. Reaction center structure 345
5. A minimal model 348
6. Photosynthesis 349

References 350

Non-standard abbreviations used in this volume

A_1, A_2	electron acceptors of photosystem I
AP, APC	allophycocyanin
APB	allophycocyanin B
B880	antenna complex absorbing at 880 nm
B800-850	antenna complex absorbing at 800 nm and 850 nm
B800-820	antenna complex absorbing at 800 nm and 820 nm
B1015	antenna complex absorbing at 1015 nm
BChl	bacteriochlorophyll
BPh	bacteriopheophytin
CAM	crassulacean acid metabolism
CC	core complex
CF_0	the membrane-embedded part of the ATP synthase
CF_0-CF_1	the complete ATP synthase
CF_1	the stroma-facing part of the ATP synthase
CH_2O	carbohydrate
Chl	chlorophyll
C-PC	C-phycocyanin
C-PE	C-phycoerythrin
Cyt	cytochrome
DCCD	*N,N′*-dicyclohexylcarbodiimide
DCMU	3-(3′,4′-dichlorophenyl)-1,1-dimethylurea
ΔpH	the transmembrane proton concentration gradient
$\Delta\Psi$	the transmembrane electrical gradient
$\Delta\bar{\mu}_{H^+}$	the transmembrane proton electrochemical gradient
DHAP	dihydroxyacetone phosphate
DGDG	digalactosyldiacylglycerol
DiPGA	1,3-diphosphoglycerate
DTT	dithiothreitol
ENDOR	electron nuclear double resonance
E_m	midpoint potential
EPR	electron paramagnetic resonance
FCCP	carbonylcyanide *p*-trifluoromethoxyphenylhydrazone
Fd	ferredoxin
FBPase	fructose 1,6-bisphosphatase
FTR	ferredoxin-thioredoxin reductase
Fru-6-P	fructose 6-phosphate
Fru-1,6-P_2	fructose 1,6-bisphosphate
Fru-2,6-P_2	fructose 2,6-bisphosphate

Fru-6-P,2K	fructose 6-phosphate,2-kinase
Fru-2,6-P_2ase	fructose 2,6-bisphosphatase
G3P	glyceraldehyde 3-phosphate
G3PDH	glyceraldehyde 3-phosphate dehydrogenase
G6PDH	glucose 6-phosphate dehydrogenase
Glu-6-P	glucose 6-phosphate
I	primary electron acceptor
L_C	linker polypeptide, core
LHC	light-harvesting complex
L_R	linker polypeptide, rod
MDH	malate dehydrogenase
MGDG	monogalactosyldiacylglycerol
NMR	nuclear magnetic resonance
OAA	oxaloacetate
OEC	oxygen-evolving complex
PBS	phycobilisome
P, P-700, P-680, P-870, P-840, etc.	primary electron donors
PC	phycocyanin
PCB	phycocyanobilin chromophore
PE	phycoerythrin
PEB	phycoerythrobilin
PEC	phycoerythrocyanin
PEP	phospho*enol*pyruvate
PFK	phosphofructokinase
PFP	pyrophosphate,fructose 6-phosphate,1-phosphotransferase
PG	phosphatidylglycerol
3-PGA	3-phosphoglycerate
Pheo	pheophytin
PPDK	pyruvate,phosphate dikinase
PQ, PQH_2	plastoquinone, plastoquinol
PRK	phosphoribulokinase
PS	photosystem
PUB	phycourobilin chromophore
PVB	phycobiliviolin chromophore
Q_A, Q_B	electron acceptors (quinones) of photosystem II and of purple and green filamentous bacteria
Rbu-5-P	ribulose 5-phosphate
Rbu-1,5-P_2	ribulose 1,5-bisphosphate
RC	reaction center
RPP	reductive pentose phosphate pathway
rubisco	ribulose 1,5-bisphosphate carboxylase oxygenase
SBPase	sedoheptulose 1,7-bisphosphatase
SG	sulphoquinovosylglycerol
Td	thioredoxin

J. Amesz (ed.) *Photosynthesis*

CHAPTER 1

Energy conversion in higher plants and algae

GIORGIO FORTI

Dipartimento di Biologia dell'Università di Milano – Centro CNR sulla Biologia Cellulare e Molecolare delle piante, Via Celoria 26, Milano, Italy

1. Introduction

Energy conversion in oxygenic photosynthesis of higher plants and algae is the process which converts the energy of electromagnetic radiation, in the visible region of the solar spectrum, into chemical energy in the form of NADPH and ATP, which are subsequently utilized by a sequence of enzymatic reactions to convert CO_2 into organic molecules.

This review will deal with the events involved in the generation of NADPH and ATP, while the assimilation of CO_2 will be dealt with elsewhere in this volume (Chapter 8). These two parts of the photosynthetic process can be considered separately, since it is now generally recognized that while CO_2 has a regulatory (or, possibly, catalytic) role in photosynthetic electron transport [1,2], its assimilation into organic molecules is a separate process occurring in the stroma of chloroplasts.

Hill's hypothesis on photosynthetic electron transport from water to NADP has been a landmark in photosynthesis research [3], and has inspired all subsequent work in the field. The 'Z scheme' originally proposed by Hill, in its present version (Fig. 1) has received experimental support from a very large number of differently conceived experiments performed with a variety of techniques and approaches, so as to be generally accepted by most scientists. However, Arnon et al. [4] have proposed a different hypothesis which will be briefly discussed in Section 2.

Recent research has made relevant progress in several directions; this chapter will present a synthesis of these, whereas the reader is referred to other chapters of this volume for detailed discussions of the individual topics.

An overview of electron transport from water to NADP will be presented, and a discussion of photophosphorylation. This will include an appraisal of the recent observations and controversies about the localized versus delocalized nature of the proton pool(s) contributing to the proton electrochemical gradient involved in the mitchellian coupling of electron transport to ATP synthesis [5,6].

Finally, the importance of molecular and supramolecular organization of the photosynthetic membranes (namely, the distribution of the Chl-protein and electron transport complexes in the different regions of the membranes when they are appressed to form grana) will be discussed in relation to its influence on light energy distribution between the two photosystems and on electron transport.

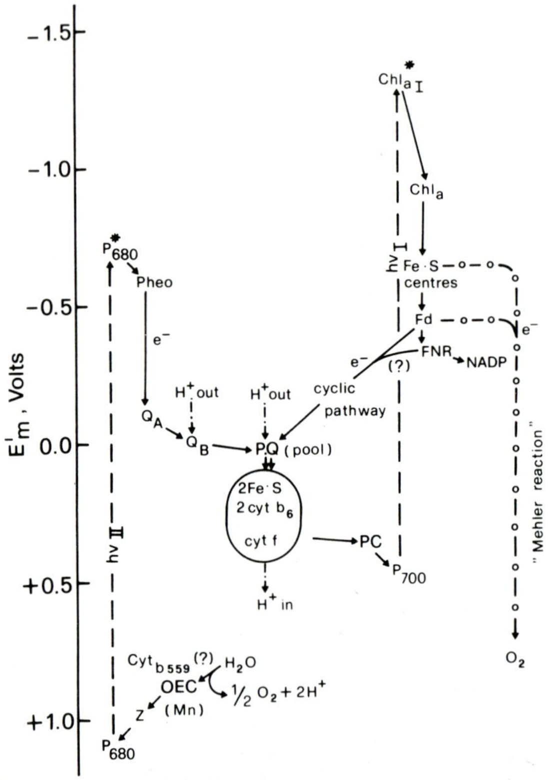

Fig. 1. Scheme of electron transport in oxygenic photosynthesis. The solid arrows (→) indicate the direction of electron transport; -·→, proton transport across the thylakoid membrane; — →, light reactions; —o→, electron transport to 0_2 at the reducing side of PS I (Mehler reaction).

2. *Electron transport from water to NADP: an overview*

The photosynthetic apparatus of green plants and cyanobacteria oxidizes water and transfers electrons to NADP, with a net gain in electrochemical potential of 1.13 eV (at pH 7), utilizing the energy of two light quanta per electron. The complete system is contained in the chloroplasts, and is localized within the thylakoid membranes, with the exception of the electron carrier ferredoxin, which is in solution in the stroma, and serves to transfer electrons from the reducing end of photosystem I (PS I) to a membrane-bound flavoprotein which then reduces NADP, and of the copper protein plastocyanin (PC, the electron donor to PS I), which is in solution in the internal phase of thylakoids.

The two photochemical reactions are performed by two photosystems. Each photosystem consists of a so-called reaction centre, where the primary energy conversion takes place, associated with a few hundred pigment molecules (chlorophylls and carotenoids; see Fig. 2) serving as light-harvesting antennas, which transfer the absorbed energy as electronic excitation energy to the reaction centres.

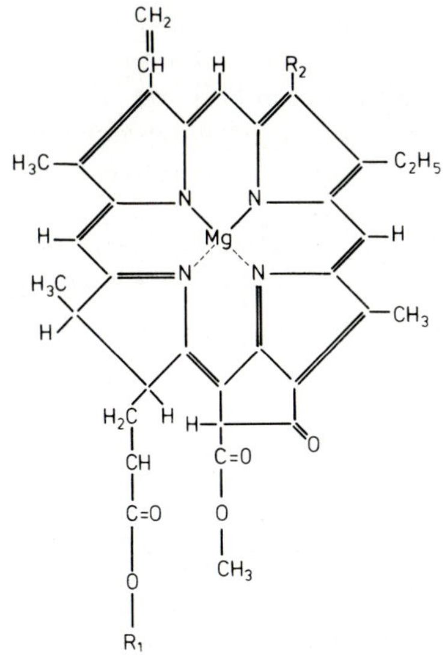

Fig. 2. Structures of chlorophylls *a* and *b*. R_1: phytyl; R_2 is either $-CH_3$ (Chl *a*) or -CHO (Chl *b*).

PS II is responsible for the oxidation of water and the reduction of a stable acceptor at the potential of ca. 0.0 to −0.2 V, while PS I transfers electrons from a donor of E'_m = 0.45 V to an acceptor of E'_m of ca. −0.65 V. An electron transport chain connects the reducing side of PS II to the oxidizing side of PS I, down the electrochemical gradient. At the reducing side of PS I NADP is reduced, while at the oxidizing side of PS II water is oxidized and O_2 is evolved.

The evolution of O_2 from water has been shown to occur every 4th flash, when flashes of saturating intensity, short enough to allow only one turnover of the PS II reaction centres, are fired, separated by a dark period long enough to permit the reoxidation of the electron acceptors on the reducing side of PS II [7]. This observation has been the basis of the 'S states' model. Each flash promotes the transition from the state S_n to S_{n+1}, in the sequence [8,9]:

$$S_0 \xrightarrow{h\nu} S_1 \xrightarrow{h\nu} S_2 \xrightarrow{h\nu} S_3 \xrightarrow{h\nu} S_4 + 2H_2O \longrightarrow S_0 + O_2 + 4H^+$$

The S states represent the accumulation of positive charges on the oxygen-evolving complex (OEC), and O_2 is evolved only when 4 charges are accumulated. Starting with dark-adapted chloroplasts (or intact photosynthetic cells), O_2 evolution is maximal at the third flash, then proceeds with a periodicity of 4, because the state S_1 is the most abundant at equilibrium in the dark. After a number of cycles of the system, the periodicity tends to disappear due to 'misses' and 'double hits', which finally randomize the PS II units into the 4 S states [8,9].

The oxidation of H_2O by PS II and the OEC has been until recently the least understood step of photosynthesis. Only recently a number of components have been discovered and hypotheses on mechanisms proposed (see recent reviews: Refs. 10–14), but the mechanism of the reaction is still unknown. The primary electron donor of PS II, discovered by Döring et al [15] and called P-680 or Chl a_{II}, transfers an electron in the excited state to a pheophytin molecule (Pheo) of $E'_m \cong -0.6$ V [16,17] in a few picoseconds. The subsequent step is the transfer of the electron to a one-electron acceptor bound quinone, Q_A, which was discovered as a quencher of PS II fluorescence [18] and was later identified as a molecule of plastoquinone [19] bound to the PS II reaction centre complex. Q_A behaves as a quencher when it is in the oxidized state, not when reduced. This is interpreted to indicate that fluorescence quenching occurs when electron transfer from the excited state of P-680 competes successfully with fluorescence emission and other pathways of energy dissipation (such as thermal decay).

The oxidation of P-680 generates a strong oxidant, P^+-680, which oxidizes a primary electron donor (Y or Z) which has been proposed to be a semiquinone cation PQH^+ bound to a protein [20]. The oxidation of Z is coupled to the re-reduction of P^+-680 in a very fast reaction [21–23]. Z^+ oxidizes the Mn-containing OEC, which accumulates the four oxidation equivalents needed to oxidize water. The participation of Mn in the O_2 evolution reaction is firmly established [24] and is theoretically well founded on the fact that the thermodynamic equilibrium of the [25]. Several schemes of reaction mechanisms for H_2O oxidation by the Mn-con- $(OH^-)+H_3O^+$ is much more favorable than with any other transition metal ion [25]. Several schemes of reaction mechanisms for H_2O oxidation by the Mn-containing OEC complex have been presented, which will not be discussed here (see, for a review, Ref. 10). Dekker et al. [26] have presented evidence that all the S-state transitions are accompanied by the same absorption spectrum changes in the ultraviolet, which they have suggested to be due to the oxidation of Mn^{3+} to Mn^{4+}. This is in contrast to other hypotheses on the mechanism of Mn participation [10]. Participation of cytochrome *b*-559 in the oxidation of water is indicated by experiments with mutants lacking this component: a mutant lacking *only* Cyt *b*-559 is unable to oxidize water, while it can use diphenyl carbazide as an artificial electron donor to PS II, and the rest of the electron transport chain is normally functioning [27].

The requirement for chloride ion of O_2 evolution has been known for a long time [28]; however, its mode of action is a matter of speculation: a catalytic role and an allosteric one have been suggested [10].

The pattern of proton release during the S-state transitions has been shown to be 1:0:1:2 [6,23,29,30]. It is therefore well established that, unlike O_2, protons are released during at least three of the S-state transitions. This indicates that water must be oxidized step-wise, while bound to OEC, probably through manganese.

Several polypeptide components of PS II and OEC have been isolated from thylakoids and PS II preparations capable of O_2 evolution, after the initial isolation by Kuwabara and Murata [31] of a 33–34 kDa polypeptide (see, for a review, Ref. 10). On the basis of several criteria, such as the extraction by different reagents and the accessibility to antibodies in thylakoids or in inside-out vesicles prepared from thylakoids, a tentative and certainly incomplete picture has been proposed

[10]. Polypeptides of 43 and 47 kDa are thought to be the components of the reaction centre and antenna chlorophylls complex, also binding pheophytin and Q_A, with Q_A and the Fe centre on the outer side of the membrane. Polypeptides of 24, 18 and 33 kDa seem to be on the internal side (exposed to the lumen of thylakoids), while a 34 kDa polypeptide which was co-isolated with a 31 kDa component [32] seems to bind Mn in a cleft facing the thylakoid lumen (see model in Ref. 10).

The evidence that water is split on the inner side of thylakoids is convincing: early experiments by Fowler and Kok [33] and more recent ones [6] have shown that the protons generated by water splitting are detected inside the thylakoid lumen. Furthermore, it has been shown that the 24 and 18 kDa polypeptides are accessible to antibodies only in so-called inside-out preparations; these polypeptides can be extracted in salt solutions from the inside-out vesicles, and subsequently rebound to them [34,35].

On the reducing side of PS II, the 'primary' acceptor Q_A (Q_A had been considered the primary acceptor until pheophytin was discovered to precede it) is reduced in less than 400 ps by $Pheo^-$. The reduction of Q_A is conveniently monitored by the increase of PS II fluorescence from an initial value, F_0, to a maximal level, F_m, indicative of the steady-state level of Q_A^-/Q_A. If reoxidation of Q_A^- is prevented by the specific inhibitor DCMU (or other herbicides having the same effect), the fluorescence yield of PS II increases sharply, because Q_A becomes fully reduced. The reduced form is an anion semiquinone (see the review by Cramer and Crofts [36]), and the absorption spectrum of this compound with a maximum at 326 nm serves for its identification [19] and offers an alternative method for kinetic studies of Q_A redox reactions (see Ref. 37 for review).

Q_A^- is reoxidized in 0.1–0.6 ms by a two-electron acceptor, Q_B [38,39]. Q_B has been identified as a plastoquinone molecule bound to a 32 kDa protein partially exposed on the outer surface of thylakoids [40,41]. At this level the light quantum-activated one-electron process converts to a two-electron one (Q_B is the 'two-electron gate'). Q_B^{2-} becomes protonated by protons from the outer aqueous phase, then released into the plastoquinone pool, and substituted on the 32 kDa polypeptide site by a molecule of PQ from the pool (for a detailed model of this sequence, see Ref. 36). The reoxidation of Q_BH_2 has been shown to be strongly dependent on the presence of HCO_3^- (or CO_2), which has been proposed to accelerate the plastoquinone–plastoquinol exchange at the two-electron gate of electron transport from PS II to PS I [1]. It should be mentioned, however, that others support the idea of a participation of CO_2 as a catalyst on the oxidizing side of PS II [2,42]. Plastoquinone, in the reduced as well as the oxidized form, diffuses freely within the thylakoid membrane; it has been shown that PQ is present at similar concentrations in the granal as well as the stromal regions of the thylakoids on the basis of its functional activity [43] and chemical analysis after fractionation of the membranes [44].

PQH_2 is reoxidized by the Cyt f-b_6-Rieske protein complex [45]. This has been known for a long time to be the rate-limiting step of photosynthetic electron transport, with a half-time of ca. 15–20 ms (see Refs. 29 and 37 for reviews). The reoxidation of PQH_2 releases protons into the lumen of thylakoids [29,37]; this is the second protolytic reaction contributing to the generation of the electrochemical proton gradient across the thylakoid membrane which is the driving force for ATP-synthesis coupled to electron transport.

Recent work has introduced the idea that the reoxidation of PQH_2 might be coupled to the transfer of three protons, rather than two, into the lumen of thylakoids (see Refs. 6, 29 and 36 for reviews). This concept is based on the operation of a 'Q-Cyt *b* cycle', similar to the one operating in photosynthetic bacteria [46]. Though several versions of the Q cycle have been proposed (which will not be discussed here), the general scheme implies that PQH_2 is oxidized to the semiquinone level when one of the two Fe^{3+} of the Rieske-Fe-S protein present in the Cyt *f*-Cyt b_6-Rieske protein complex is reduced at the inner aqueous surface of the thylakoids, releasing two protons into the lumen. The semiquinone is then oxidized by one of the two Cyt b_6 molecules of the complex. A second molecule of PQH_2 is oxidized in the same manner, and the two reduced Cyt b_6 are then reoxidized by PQ. The PQ^{2-} generated in this way is bound near the outer surface of the membrane and becomes protonated; its reoxidation by the Fe-S centre will then discharge 2 H^+ into the thylakoid lumen.

The result of such a process would be that two electrons are cycled twice through the PQ, and the ratio of H^+/e^- between PS II and PS I would be higher than one. This, if definitively confirmed, would be of great importance from the point of view of understanding the coupling of electron transport to the synthesis of ATP, and of the quantum yield of photosynthesis (see discussion under photophosphorylation).

Cyt *f* (E_m = 340–365 mV) is present in the complex in the ratio of one mole per two moles of Cyt b_6 and two Fe-S centres; it is reduced by the Fe^{2+}-S, then reoxidized by plastocyanin (E_m = 380 mV), which is dissolved in the lumen of the thylakoids. Reduced PC is oxidized directly by PS I, with a half-time of ca. 20 μs [29], corresponding to the half-time of the reduction of the oxidized reaction centre of PS I, Chl a_I, also called P-700 [47]. The kinetics of P-700 oxidation is very fast: a rise time of 30–50 ps has been reported (see for review Ref. 48 and Chapter 4) and a redox potential, E_m, of 450 mV.

The primary electron acceptors of PS I have been extensively studied spectroscopically [29,48]. The formation of a Chl *a* anion radical has been proposed, of midpoint potential as low as −900 mV.

Three bound Fe-S centres have been proposed to be the next acceptors (see Refs. 29, 48 and 49 for reviews), on the basis of optical and EPR spectroscopy and Mössbauer studies. The stable, one-electron acceptor of PS I is a soluble Fe-S protein, ferredoxin (Fd) [50], of molecular weight of 10 kDa and E_m = −440 mV. So, PS I transfers electrons against an apparent electrochemical gradient of ca. 0.9 V.

Ferredoxin has been shown to interact with the thylakoids at two distinct sites [51]: it accepts electrons from the reducing side of PS I, then is reoxidized by the thylakoid-bound FAD-flavoprotein, ferredoxin-NADP reductase (FNR) [50]. It has been shown that FNR forms a one-to-one complex with Fd when the two proteins are in solution [52] as well as when FNR is membrane-bound [53], with a dissociation constant of ca. 5 μM under the conditions prevailing in the chloroplasts. The binding of NADP to FNR has also been shown [52,54] and spectroscopic evidence has suggested that the flavoprotein might be reduced to the level of the semiquinone of FAD, then reoxidized by NADP [55,56]. The flavoprotein seems then to function as the 'two-electron gate' at the reducing end of the photosynthetic electron transport chain. Though the detailed mechanism of NADP reduction is still unknown, a reasonable hypothesis emerging from the available data may be summarized as follows: (1) Fd is reduced by one electron at the reducing end of PS I; (2) reduced Fd diffuses to the site where the FNR-NADP complex is bound to the membrane, in the stroma-exposed regions [57], and binds to form the ternary complex $Fd^- \cdot FNR \cdot NADP$ (alternatively, one molecule of Fd is bound to FNR on the thylakoids, and the ternary complex receives one electron from Fd^- in solution; (3) NADP is reduced (in a two-step process) then released into the medium.

Reduced Fd is known to be the electron donor for a number of different acceptors, both artificial, such as mammalian Cyt *c* [58], and of physiological importance, such as nitrite reductase [59]. It is also known to be a catalyst of cyclic electron transport around PS I [59], a process coupled to the synthesis of ATP (cyclic photophosphorylation), in which electrons are transferred back to a component (Cyt b_6 ?) of the chain between the two photochemical reactions.

The participation of FNR in cyclic photophosphorylation has been suggested on the basis of inhibition of cyclic phosphorylation by antibodies raised against FNR [60,61] and more recently on the basis of inhibitor studies [62]. Studies on isolated FNR have shown that this enzyme can reduce Cyt *f* [63] and the enzyme has recently been extracted from thylakoids together with Cyt *f* and Cyt b_6 by a procedure involving the use of detergents [64]. Whether the catalytic activity of FNR as Cyt *f* reductase and its possible association with the Cyt f-b_6 complex have any relation to its participation in cyclic photophosphorylation remains to be established.

The rates of cyclic photophosphorylation around PS I catalysed by the natural catalysts are rather low, about one order of magnitude lower than those of linear electron transport [59], while they are very high when artificial electron carriers, such as phenazine methosulfate, are added to the system. Cyclic photophosphorylation has been shown to occur in intact leaves [65] and algae [66].

At variance with Hill's scheme [3], which has been discussed above in its recent developments, a three-light reaction scheme has been proposed by Arnon and coworkers [4,59]. According to this scheme, Fd and subsequently NADP would be reduced by PS II directly, and PS II would perform two different photoacts with two acceptors: Fd and Q (Q_A?) [4]. The role of PS I would be limited to the performance of cyclic photophosphorylation, catalysed by Fd as the electron carrier. Recent experiments showing that PS II-enriched, inside-out thylakoid vesicles are capable of low rates of NADP reduction upon addition of Fd, FNR and plastocyanin [67] have been designed to investigate the view that only PS II is required to transfer electrons from water to NADP. However, the presence of PS I in the preparations, though in low proportions, was not ruled out, and the cause of the absolute requirement for PC, which is known to be oxidized by P-700 [29], was unexplained.

3. *Photosynthetic phosphorylation*

The mechanism of ATP synthesis coupled to electron transport in thylakoids is discussed in Chapter 7 of this volume, and the reader is referred there. Some general aspects of photophosphorylation will be dealt with here in relation to the structure of thylakoids, their supramolecular organization and the overall efficiency of the process.

Mitchell's chemiosmotic theory [68–70] is generally accepted (see reviews in Refs. 5,37 and 71), though a large number of important details are still undefined, including the mechanism of action of the ATP synthase itself, and the ratio of ATP formed to electron transported.

Mitchell's theory holds that an electrochemical proton gradient across the membrane (which is only slightly permeable to many ionized species and particularly to H^+) is formed by the vectorial transport of H^+ into the thylakoid lumen coupled to electron transport, as a consequence of the alternate disposition across the membrane of electron carriers which can bind protons and others which cannot be protonated.

The experimental use of artificial electron acceptors and donors has demonstrated, in agreement with Mitchell's theory, that electron transport can be coupled to ATP synthesis only when the chemical structure and the lipophilicity of the electron carriers added is such as to allow vectorial proton transport across the membrane [72].

In this way, the loss of redox free energy occurring during electron transport is partially conserved as electrochemical potential energy of the proton gradient. The synthesis of ATP occurs when the protons accumulated inside the thylakoid lumen are transported out into the external water phase by an anisotropic, proton-translocating ATP synthase–ATPase (the complex CF_o-CF_1), which catalyses the reaction

$$\mathrm{ADP} + \mathrm{P_i} + n\mathrm{H^+_{in}} \rightarrow \mathrm{ATP} + \mathrm{H_2O} + n\mathrm{H^+_{out}} \qquad (1)$$

The free energy change of ATP synthesis is given by

$$\Delta G_\mathrm{p} = \Delta G^0_\mathrm{p} + RT \ln \frac{[\mathrm{ATP}]}{[\mathrm{ADP}]\,[\mathrm{P_i}]} \qquad (2)$$

and the free energy change of H^+ efflux is

$$\Delta G_\mathrm{H} = RT \ln \frac{[\mathrm{H^+_{out}}]}{[\mathrm{H^+_{in}}]} + F\,(\Psi_\mathrm{out} - \Psi_\mathrm{in}) \qquad (3)$$

(where Ψ is the electric potential and F is Faraday's constant). ΔG^0_p is dependent upon pH, Mg^{2+} concentration, H_2O concentration, ionic strength and temperature. At pH 8, $[Mg^{2+}]$ 1 mM, ionic strength 0.1 M and 25°C, ΔG^0_p = 32.2 kJ/mol [73].

Synthesis of ATP can only occur when $\Delta G_p + \Delta G_H < 0$, while ATP hydrolysis occurs when the reverse is true, provided that the ATPase is activated.

As shown by Eq. 3, ΔG_H (which is also called $\Delta\mu_H +$) can be separated into two components: the chemical potential of protons and the electric field across the membrane, to which all charged species present are contributing. The source of free energy for ATP synthesis is therefore the 'protonmotive force',

$$\Delta p = \frac{\Delta\mu_{H^+}}{F}$$

$$\text{or} \quad \Delta p = \Delta\Psi - 2.3\frac{RT}{F}\Delta\text{pH}$$

(where $\Delta\text{pH} = \text{pH}_{out} - \text{pH}_{in}$ $\Delta\Psi = \Psi_{out} - \Psi_{in}$). At 30°C, $\Delta p = \Delta\Psi - 60\ \Delta\text{pH}$ (if $\Delta\Psi$ is expressed in mV).

In isolated thylakoids, a large $\Delta\Psi$ is generated by each of the two primary photochemical reactions (negative on the outside), due to the fact that the primary electron acceptors are on the outside of the membrane, and the primary donors on the inside (see the review by Witt [37]. The rise time of the electric field generation is therefore very fast (ca. 1 ns), and is conveniently measured by the red-shifts of the absorption bands of endogenous pigments (Chl *b*, carotenoids) when subjected to a large electric field [37]. The formation of ΔpH across the membrane is a much slower process, linked to electron transport along the chain [5,6,37,71]. On the other hand, $\Delta\Psi$ decays rather rapidly in thylakoids [37] owing to the diffusion of counter-ions (such as Cl^-), so that while in the pre-steady-state period at the onset of illumination Δp is mainly made up by $\Delta\Psi$, in the steady-state regime $\Delta\Psi$ is vanishingly small and Δp is mostly due to ΔpH [71,74,75].

The synthesis of ATP starts 4–5 ms after the onset of illumination with saturating light intensity [74], which is the turnover time of the ATP synthase [71]. This means that $\Delta\Psi$ *or* ΔpH can fulfill the energy requirement for ATP synthesis.

ATP formation has been demonstrated in the absence of light, if ΔpH is imposed artificially across the thylakoid membrane [76], or by imposing a $\Delta\Psi$ large enough to supply the energy required [37,71]. In both cases, the activity of the ATP synthase complex is required and ATP synthesis is concomitant with the transfer of protons from the internal water space of the thylakoid lumen to the external bulk phase.

All available evidence indicates that the synthesis of ATP is not *directly* coupled to electron transport, but is dependent only on the protonmotive force. If an uncoupler (a substance which equilibrates H^+ across the membrane) is added in continuous light, ATP synthesis is decreased or abolished, while electron transport is accelerated, due to the release of the control exerted by $\Delta\mu_H{}^+$ on the rate of electron transport.

A large body of evidence indicates that the generation of the protonmotive force utilized for ATP synthesis is the cooperative result of the activity of a large number of electron transport chains. When the ionophore gramicidin (which opens channels for all monovalent cations, including protons) is added in the ratio of one gramicidin channel (two molecules of gramicidin form a conducting channel according to Bamberg and Laüger [77]) per 10^5 Chls (ca. 200 electron transport chains) the synthesis of ATP in a single turnover flash is inhibited by a factor of more than two [78]. Further evidence on this point has been provided recently by Hangarter and Ort [79]: ATP synthesis was measured in a series of single turnover flashes of saturating intensity under conditions where $\Delta\Psi$ was abolished by the presence of the K^+-specific ionophore nonactin and was therefore only dependent on ΔpH. The uptake of ca. 60 mmol H^+/mol of Chl was required before ATP synthesis could be observed at constant yield of ATP/flash, independently of which part of the electron transport chain is activated. These experiments are easily interpreted, according to Mitchell's theory, on the basis of a delocalized pool of protons, available to the ATP synthase complex. Furthermore, the delocalization of the charges generated by the primary photochemical reactions of PS II and PS I has been shown to occur in ca. 10 μs due to ionic conduction [80].

At variance with this strictly mitchellian view, some authors have proposed that protons generated by the protolytic reactions of electron transport into restricted 'domains' within the membrane might be utilized by the ATP synthase *before* being equilibrated with the H^+ pool of the internal phase (see Ref. 5 for review). Such a concept has some similarities with the hypothesis of Williams [81,82], according to which electron transport would produce high-potential protons within the membrane. The membrane would therefore be the reservoir of high-potential protons, and provide within its structure the proton-conducting link between the different enzyme complexes. Only two phases would be required, according to this hypothesis, to couple ATP synthesis to electron transport: the membrane and the external water phase.

Recent work by Hangarter and Ort [79] in agreement with previous results [5], has shown that the introduction of permeating buffers into the inner phase of the thylakoids does not alter the number of single turnover flashes required to produce the threshold value of Δp necessary to initiate ATP synthesis. This would indicate that the pooled protons utilized by the ATP synthase are not located in the internal water phase, nor are they rapidly equilibrated with it: the only alternative would be that the proton pool is located within the membrane, and a proton-conducting system must then exist capable of transferring protons from where they are generated to the ATP synthase complex, without allowing equilibration with the buffered internal phase. The intramembrane proton conduction could be imagined as being due to intrinsic proteins [81,82] and should span rather large distances within the thylakoids, as it is well known that only about one ATP synthase complex per two electron transport chains [83,84] is present, and this enzyme complex is confined to the non-appressed regions of the thylakoids [85]. So, the protons produced by the water-splitting reaction, which occurs mainly, if not exclusively, in the granal appressed regions, have to be pooled with those produced where plastoquinol is oxidized by the Cyt f-b_6 complex, and made available to the ATP synthase, which may be as far away as several hundred nanometers.

The stoichiometry of protons translocated per ATP formed (H^+/ATP), the related ratio of protons transported into the thylakoid per electron flowing through the chain (H^+/e^-) and the resulting ratio ATP/2 e^- are still controversial. As the assimilation of CO_2 requires 3 ATP and 2 NADPH per CO_2 assimilated in the plants utilizing the Calvin cycle (the ATP requirement is 5 molecules/CO_2 in 'C_4' plants), additional light quanta are necessary if the ratio ATP/NADPH (or ATP/2 e^-) is lower than 1.5, to produce ATP in the amount needed. This could be provided by cyclic photophosphorylation or the phosphorylation coupled to the Mehler reaction (reoxidation by oxygen of the reduced acceptors of PS I).

Most experimental results indicate that the ATP/2 e^- ratio observed with isolated chloroplasts (or washed thylakoids) ranges between 1 and 1.3 (see review by Ort and Melandri [5]). Only a few reports of higher ratios have appeared in the literature [86,87] related to 'class I' chloroplasts; in spite of the correction applied [87] it is difficult to rule out completely the possibility that some cyclic electron transport occurring together with NADP reduction might have contributed to ATP synthesis under the conditions of the experiments.

Photosynthetic electron transport occurs at appreciable rates ('basal rate') in the absence of ADP and P_i and under these non-phosphorylating conditions the highest values of ΔpH are observed. The addition of ADP and P_i accelerates electron transport, and decreases the steady-state level of ΔpH, as expected [88]. The rate of proton efflux in the absence of ADP and P_i is proportional to ΔpH (see Ref. 79 for review), and apparently occurs mostly through the ATP synthase (CF_0-CF_1 complex). Upon addition of ADP and P_i, the phosphorylating proton efflux through the ATP synthase proceeds at rates at least an order of magnitude higher than in the absence of phosphorylation (when light intensity and electron acceptor concentration are saturating), and in these conditions a ratio of 2.4 H^+/ATP formed has been measured [89]. A ratio of 3 has been reported by Portis and McCarty [90].

If one assumes that the ratio H^+/e^- is 2, then an ATP/2 e^- of 1.66 can be calculated. Izawa and Good [91] (see also Ref. 92) have established that a ratio ATP/2 e^- of 2 can be calculated for 'phosphorylating' electron transport, if one subtracts the independently occuring electron transport rate observed in the absence of ADP-P_i: obviously, such a correction cannot be made if the purpose is to evaluate the quantum requirement of photosynthesis, but it is important in the investigation of the efficiency of energy coupling.

The number of protons translocated into the thylakoids per electron transported (H^+/e^- ratio) is still controversial (see Refs. 5 and 6 for reviews). The reason for the controversy lies in the still unclear mechanism of the PQ-Cyt *b* cycle and its role in cyclic and/or non-cyclic electron transport (see Section 2), and in the differences in the methods used. However, most authors find a ratio H^+/e^- of ≈ 2 in isolated thylakoids, while a few reports of higher values have appeared [5,6].

Unfortunately, the H^+/e^- ratio measurement in intact chloroplasts (where the outer envelope of the organelle is intact and the stroma is integrally conserved around the thylakoids) is impossible, and no data are available based on direct observations with these chloroplasts.

It has been reported that CO_2 assimilation by intact isolated chloroplasts is always accompanied by O_2 uptake due to the Mehler reaction [93,94]. This is likely to indicate that the linear electron transport from H_2O to NADP is unable to provide ATP in the amount required (1.5 ATP/NADPH) by the operation of the Calvin cycle, and that either phosphorylation coupled to the Mehler reaction [95] or cyclic phosphorylation must intervene to supply the extra ATP needed in the process. This is in agreement with observations on quantum requirement for CO_2 assimilation in intact isolated chloroplasts, which indicated that at least 10–12 quanta are required per molecule of CO_2 assimilated [96]. In intact chloroplasts electron flow at the reducing end of PS I is diverted to the cyclic pathway (or to the Mehler reaction) when ATP is deficient: lack of ATP (or a low ATP/ADP ratio) causes the drop in concentration of 1,3-bisphosphoglycerate (1,3-bisPGA), the final electron acceptor in the Calvin cycle (Chapter 8), the accumulation of NADPH (which is anyway in large excess with respect to NADP in steady-state photosynthesis) and the almost complete disappearance of NADP. Lack of NADP will prevent the reoxidation of Fd_{red} by FNR; therefore reoxidation of Fd_{red} can only occur through the cycle around PS I or the Mehler reaction (Fd_{red} is readily oxidized by O_2), until ATP regenerated by one of these processes allows the synthesis of 1,3-bisPGA, followed by reoxidation of NADPH and restoration of linear electron flow.

In agreement with this regulatory mechanism, Heber [96] has observed that, upon a sudden decrease of light intensity, steady-state CO_2 assimilation and ATP concentration were sharply decreased in intact chloroplasts, while NADPH concentration was unchanged or increased.

4. *Molecular and supramolecular structure of thylakoids*

4.1. *Lateral heterogeneity, fluorescence and electron transport*

Murata has discovered [97] that the addition of Mg^{2+} to isolated thylakoids increases the fluorescence yield of PS II and its photochemical activity, decreasing at the same time the photochemical activity of PS I. These observations received further experimental support and widespread acceptance (see review by W.P. Williams [98]), and have been interpreted to indicate that the presence of Mg ion (and other cations as well) interrupts the transfer of excitation energy to PS I (which is not fluorescent at room temperature). So the concentration of cations in the medium seems to regulate the distribution of excitation energy between the two photosystems. The addition of cations was also shown to cause the stacking of thylakoids and the formation of grana [99]; this process was found to be correlated to the fluorescence increase [100,101].

The effect of cations in inducing thylakoid appression and the correlated phenomena is independent of the cation present, but only depends on its charge: on the basis of this observation, grana formation was quantitatively explained by Barber and his associates as due to screening by cations of the negative charges present on the surface of thylakoids [102,103]. Screening of the surface charges abol-

ishes electrostatic repulsion between the membranes, then stacking occurs due to Van der Waals interaction between proteins of opposite membranes, and possibly interactions through oriented water [104]. In nature the thylakoids are found in the granal structure, as expected on the basis of the cation composition of the chloroplast stroma.

The localization in the appressed and non-appressed regions of thylakoids of the Chl-protein complexes of PS I, PS II and LHC (the light-harvesting Chl-protein complex) has been the object of controversy. The early fractionation experiments of Boardman and Anderson [105] revealed that PS II activity is concentrated in the 'heavy fraction', mostly the partitions of grana, while PS I activity is concentrated in the 'light fraction' consisting mostly of the stroma membranes (probably including the margins of grana and their end membranes). More recent investigations have concluded that most of the PS II activity is indeed found in the grana, while the non-appressed membranes contain very little PS II and about 15% of the total Chl [106,107]. It should be noted that the different methods of fractionation used do not allow it to be established in which fraction the margins of grana are found; this gives rise to uncertainty concerning the quantitative distribution of the activities and of the protein-Chl complexes (estimated by electrophoretic analysis) among the fractions separated. The granal membranes contain essentially all the large freeze-fracture particles thought to be PS II-LHC complexes, while the smaller particles supposed to be the PS I complexes have been found uniformly distributed between grana and stromal thylakoids [107].

Upon the advent of a method [108] which permitted the isolation of relatively pure fractions of partition zone membranes, isolated as inside-out vesicles, Andersson and Anderson found [109] that PS II and LHC are mainly in the partition zones, in agreement with the previous reports, but they also reported that PS I is confined to the stroma-exposed regions and practically absent in the partitions (see also Chapter 12). The fluorescence rise observed upon cation addition and grana formation is easily explained with this picture of thylakoid structure in mind: PS II-LHC is segregated far apart from PS I, and the excitation energy cannot be transferred to the latter, which is not fluorescent. The fluorescence rise can therefore be taken as a measure of the increase of energy remaining within PS II. Observations on the kinetics of fluorescence changes upon removal [110] or addition [111] of cations have indicated that the quenching of fluorescence can indeed be envisaged as the result of PS I collision with PS II when the complexes diffuse laterally in the membrane, and the kinetic heterogeneity of one of the complexes is suggested by the observed initial deviation from second-order kinetics [110].

If the model proposed by Andersson and Anderson [109] of total separation of PS I and PS II in the granal chloroplasts were to be accepted, electron transport from the PS II acceptors to P-700 would require a mobile electron carrier(s) which should diffuse laterally in the membrane fast enough to account for the observed electron transport rate. Plastoquinone [112] and plastocyanin are the candidates of choice for this role. The former has been shown to be present at approximately the same activity in the partitions and in the stroma-exposed membranes [43], while PC is known to be located in the intrathylakoid space [113].

could provide a picture of the effects of cations on energy distribution between the two photosystems [120].

In agreement with previous results [121,122], Jennings [120] reported that upon removal of Mg^{2+} from the medium fluorescence quenching (of F_m) is greater when excitation is at 475 nm than when 435 nm light is used. However, the decrease of Mg^{2+} from 2.5 to 0.5 mM produced no change in the F_{475}/F_{435} fluorescence ratio and a slight increase of F_0, while F_m decreased about 50%. Upon further decrease of the Mg ion concentration, the ratio F_{475}/F_{435} decreased sharply as well as F_m, while F_0 returned to the same level observed at high Mg^{2+}. These results have been interpreted to indicate that Mg^{2+} (and the other cations) prevents spillover of energy from PS II to PS I in connection with grana formation, but also increases energy coupling between LHC and PS II (the latter requiring a lower concentration for saturation), and increases the efficiency of energy transfer to P-680. Titration of fluorescence with exogenous quenchers competing with PS I for PS II-LHC excitations [43] has indicated that the LHC-PS II matrix is homogeneously quenched upon removal of Mg^{2+}.

It seems therefore that Butler's bipartite model is adequate to describe the interactions of the Chl-protein complexes when the concentration of cations is above a level which ensures tight coupling of LHC and PS II, whereas the tripartite model is needed when cation concentration is so low as to cause uncoupling of these two complexes. The mechanism of this regulatory effect is unknown.

It was recently reported that Mg^{2+} addition to thylakoids and grana formation decreases light absorption (corrected for light scattering) in the main absorption bands [123]. This was attributed to the 'sieve effect', which is due to inhomogeneous distribution of pigments. It is therefore expected to affect mainly absorption by the grana membranes, where PS II is concentrated. So, grana formation may influence the balance of PS II and PS I energy distribution merely by changing their relative light absorptions.

An important step towards the understanding of the regulation of excitation energy partition between the two photosystems has been the discovery of LHC phosphorylation by a thylakoid-bound protein kinase and its dephosphorylation by a phosphatase [124]. The kinase is activated when the plastoquinone pool is reduced, and inactivated when it becomes oxidized [125,126]. Phosphorylation of LHC leads to a decrease of PS II fluorescence of ca. 15–20%, and dephosphorylation to the opposite changes [127–129]. PS I photochemical activity is at the same time enhanced [130–133].

It has been proposed that LHC phosphorylation-dephosphorylation is a regulatory mechanism to adjust any imbalance between PS II and PS I photochemical activities. When PS II prevails, the PQ pool becomes over-reduced, the kinase is activated, then LHC is phosphorylated and more excitation energy flows to PS I. Oxidation of PQH_2 follows, then inactivation of the kinase: the phosphatase (for this enzyme no regulation has been reported) will then dephosphorylate LHC [127–129].

This mechanism could provide the adjustment of the photochemical apparatus to the prevailing illumination conditions, and would also respond to the redox state

ishes electrostatic repulsion between the membranes, then stacking occurs due to Van der Waals interaction between proteins of opposite membranes, and possibly interactions through oriented water [104]. In nature the thylakoids are found in the granal structure, as expected on the basis of the cation composition of the chloroplast stroma.

The localization in the appressed and non-appressed regions of thylakoids of the Chl-protein complexes of PS I, PS II and LHC (the light-harvesting Chl-protein complex) has been the object of controversy. The early fractionation experiments of Boardman and Anderson [105] revealed that PS II activity is concentrated in the 'heavy fraction', mostly the partitions of grana, while PS I activity is concentrated in the 'light fraction' consisting mostly of the stroma membranes (probably including the margins of grana and their end membranes). More recent investigations have concluded that most of the PS II activity is indeed found in the grana, while the non-appressed membranes contain very little PS II and about 15% of the total Chl [106,107]. It should be noted that the different methods of fractionation used do not allow it to be established in which fraction the margins of grana are found; this gives rise to uncertainty concerning the quantitative distribution of the activities and of the protein-Chl complexes (estimated by electrophoretic analysis) among the fractions separated. The granal membranes contain essentially all the large freeze-fracture particles thought to be PS II-LHC complexes, while the smaller particles supposed to be the PS I complexes have been found uniformly distributed between grana and stromal thylakoids [107].

Upon the advent of a method [108] which permitted the isolation of relatively pure fractions of partition zone membranes, isolated as inside-out vesicles, Andersson and Anderson found [109] that PS II and LHC are mainly in the partition zones, in agreement with the previous reports, but they also reported that PS I is confined to the stroma-exposed regions and practically absent in the partitions (see also Chapter 12). The fluorescence rise observed upon cation addition and grana formation is easily explained with this picture of thylakoid structure in mind: PS II-LHC is segregated far apart from PS I, and the excitation energy cannot be transferred to the latter, which is not fluorescent. The fluorescence rise can therefore be taken as a measure of the increase of energy remaining within PS II. Observations on the kinetics of fluorescence changes upon removal [110] or addition [111] of cations have indicated that the quenching of fluorescence can indeed be envisaged as the result of PS I collision with PS II when the complexes diffuse laterally in the membrane, and the kinetic heterogeneity of one of the complexes is suggested by the observed initial deviation from second-order kinetics [110].

If the model proposed by Andersson and Anderson [109] of total separation of PS I and PS II in the granal chloroplasts were to be accepted, electron transport from the PS II acceptors to P-700 would require a mobile electron carrier(s) which should diffuse laterally in the membrane fast enough to account for the observed electron transport rate. Plastoquinone [112] and plastocyanin are the candidates of choice for this role. The former has been shown to be present at approximately the same activity in the partitions and in the stroma-exposed membranes [43], while PC is known to be located in the intrathylakoid space [113].

could provide a picture of the effects of cations on energy distribution between the two photosystems [120].

In agreement with previous results [121,122], Jennings [120] reported that upon removal of Mg^{2+} from the medium fluorescence quenching (of F_m) is greater when excitation is at 475 nm than when 435 nm light is used. However, the decrease of Mg^{2+} from 2.5 to 0.5 mM produced no change in the F_{475}/F_{435} fluorescence ratio and a slight increase of F_0, while F_m decreased about 50%. Upon further decrease of the Mg ion concentration, the ratio F_{475}/F_{435} decreased sharply as well as F_m, while F_0 returned to the same level observed at high Mg^{2+}. These results have been interpreted to indicate that Mg^{2+} (and the other cations) prevents spillover of energy from PS II to PS I in connection with grana formation, but also increases energy coupling between LHC and PS II (the latter requiring a lower concentration for saturation), and increases the efficiency of energy transfer to P-680. Titration of fluorescence with exogenous quenchers competing with PS I for PS II-LHC excitations [43] has indicated that the LHC-PS II matrix is homogeneously quenched upon removal of Mg^{2+}.

It seems therefore that Butler's bipartite model is adequate to describe the interactions of the Chl-protein complexes when the concentration of cations is above a level which ensures tight coupling of LHC and PS II, whereas the tripartite model is needed when cation concentration is so low as to cause uncoupling of these two complexes. The mechanism of this regulatory effect is unknown.

It was recently reported that Mg^{2+} addition to thylakoids and grana formation decreases light absorption (corrected for light scattering) in the main absorption bands [123]. This was attributed to the 'sieve effect', which is due to inhomogeneous distribution of pigments. It is therefore expected to affect mainly absorption by the grana membranes, where PS II is concentrated. So, grana formation may influence the balance of PS II and PS I energy distribution merely by changing their relative light absorptions.

An important step towards the understanding of the regulation of excitation energy partition between the two photosystems has been the discovery of LHC phosphorylation by a thylakoid-bound protein kinase and its dephosphorylation by a phosphatase [124]. The kinase is activated when the plastoquinone pool is reduced, and inactivated when it becomes oxidized [125,126]. Phosphorylation of LHC leads to a decrease of PS II fluorescence of ca. 15–20%, and dephosphorylation to the opposite changes [127–129]. PS I photochemical activity is at the same time enhanced [130–133].

It has been proposed that LHC phosphorylation-dephosphorylation is a regulatory mechanism to adjust any imbalance between PS II and PS I photochemical activities. When PS II prevails, the PQ pool becomes over-reduced, the kinase is activated, then LHC is phosphorylated and more excitation energy flows to PS I. Oxidation of PQH_2 follows, then inactivation of the kinase: the phosphatase (for this enzyme no regulation has been reported) will then dephosphorylate LHC [127–129].

This mechanism could provide the adjustment of the photochemical apparatus to the prevailing illumination conditions, and would also respond to the redox state

could provide a picture of the effects of cations on energy distribution between the two photosystems [120].

In agreement with previous results [121,122], Jennings [120] reported that upon removal of Mg^{2+} from the medium fluorescence quenching (of F_m) is greater when excitation is at 475 nm than when 435 nm light is used. However, the decrease of Mg^{2+} from 2.5 to 0.5 mM produced no change in the F_{475}/F_{435} fluorescence ratio and a slight increase of F_0, while F_m decreased about 50%. Upon further decrease of the Mg ion concentration, the ratio F_{475}/F_{435} decreased sharply as well as F_m, while F_0 returned to the same level observed at high Mg^{2+}. These results have been interpreted to indicate that Mg^{2+} (and the other cations) prevents spillover of energy from PS II to PS I in connection with grana formation, but also increases energy coupling between LHC and PS II (the latter requiring a lower concentration for saturation), and increases the efficiency of energy transfer to P-680. Titration of fluorescence with exogenous quenchers competing with PS I for PS II-LHC excitations [43] has indicated that the LHC-PS II matrix is homogeneously quenched upon removal of Mg^{2+}.

It seems therefore that Butler's bipartite model is adequate to describe the interactions of the Chl-protein complexes when the concentration of cations is above a level which ensures tight coupling of LHC and PS II, whereas the tripartite model is needed when cation concentration is so low as to cause uncoupling of these two complexes. The mechanism of this regulatory effect is unknown.

It was recently reported that Mg^{2+} addition to thylakoids and grana formation decreases light absorption (corrected for light scattering) in the main absorption bands [123]. This was attributed to the 'sieve effect', which is due to inhomogeneous distribution of pigments. It is therefore expected to affect mainly absorption by the grana membranes, where PS II is concentrated. So, grana formation may influence the balance of PS II and PS I energy distribution merely by changing their relative light absorptions.

An important step towards the understanding of the regulation of excitation energy partition between the two photosystems has been the discovery of LHC phosphorylation by a thylakoid-bound protein kinase and its dephosphorylation by a phosphatase [124]. The kinase is activated when the plastoquinone pool is reduced, and inactivated when it becomes oxidized [125,126]. Phosphorylation of LHC leads to a decrease of PS II fluorescence of ca. 15–20%, and dephosphorylation to the opposite changes [127–129]. PS I photochemical activity is at the same time enhanced [130–133].

It has been proposed that LHC phosphorylation-dephosphorylation is a regulatory mechanism to adjust any imbalance between PS II and PS I photochemical activities. When PS II prevails, the PQ pool becomes over-reduced, the kinase is activated, then LHC is phosphorylated and more excitation energy flows to PS I. Oxidation of PQH_2 follows, then inactivation of the kinase: the phosphatase (for this enzyme no regulation has been reported) will then dephosphorylate LHC [127–129].

This mechanism could provide the adjustment of the photochemical apparatus to the prevailing illumination conditions, and would also respond to the redox state

and Chl *b* in a ratio of ca. 1.1, and accounting for ca. 50% of the total Chl, and the last one including the PS I reaction centre, its antenna comprising mostly Chl *a* and a minor amount of Chl *b*), one can conceive that they constitute either three separate entities absorbing light ('separate packages') endowed with definite probabilities of transferring excitation energy to each other, or PS II-LHC can be viewed as a single matrix (meaning by this that all the Chl molecules of the matrix have the same probability of transferring energy to the reaction centre P-680 or to PS I). The latter is a separate entity capable of receiving excitons from the PS II-LHC matrix. The reverse process is much less probable, owing to the lower energy of the PS I pigment molecules.

Butler and Kitajima [116] have developed a model, the 'bipartite' model for the latter situation, and Butler and Strasser [118] have provided a 'tripartite' model to analyse the former situation (see also the review by Butler [119]). From the bipartite model, the following equations can be derived to analyse the fluorescence yield at F_0 and F_m.

$$F_m = \frac{k_F}{k_F + k_x + k_{T21} + k_{T'}\Psi_d} \tag{4}$$

$$F_0 = \frac{k_F}{k_F + k_x + k_{T21} + k_T} \tag{5}$$

$$F_v/F_m = \frac{k_T - k_{T'}\Psi_d}{k_F + k_x + k_{T21} + k_T} \tag{6}$$

where k_F, k_{T21}, k_T, $k_{T'}$ and k_x are, respectively, the rate constants for fluorescence, energy transfer from PS II-LHC to PS I (the 'spillover'), energy transfer to open PS II traps, energy transfer to closed PS II traps, and other processes which compete for PS II excitons. Ψ_d is the probability of non-radiative decay at the closed reaction centre (reaction centre where Q_A is reduced).

The fluorescence yield in the tripartite model [118] is

$$\phi_F = \frac{\gamma\Psi F_3 + \beta\Psi T_{23}\Psi F_3 + \beta\Psi T_{32}\Psi F_2 + \beta\Psi F_2}{1 - \Psi T_{23}\Psi T_{32}} \tag{7}$$

where β and γ are the relative optical cross-sections of PS II and LHC respectively; ΨF_2 and ΨF_3 are the fluorescence probabilities of PS II and LHC, respectively; ΨT_{23} and ΨT_{32} are the transfer probabilities from PS II to LHC and from LHC to PS II, and define the degree of coupling between the two types of complexes. All the indicated probabilities are defined as the ratios of rate constants [118].

On the basis of these models, measurements of F_0 and F_m at different ionic compositions of the medium, and of fluorescence excited either at 475 nm (a wavelength absorbed mainly by LHC) or at 435 nm (absorbed mainly by PS II and LHC)

However, this extreme picture of spatial segregation of the two photosystems has been seriously challenged. The substantial presence of PS I activity in the grana partitions was reported by Vaughn et al. [114] on the basis of immunocytochemical and cytochemical evidence.

Furthermore, Atta-Asafo-Adjei and Dilley [115] reported a thorough investigation of the inside-out vesicles from grana partitions (prepared essentially according to Andersson et al. [108]), comparing their P-700 content and their photochemical and electron transport activities with those of intact thylakoids. P-700 was found in the ratio of 1/1500 Chl (mol/mol) in inside-out vesicles, while the ratio was 1/600 in the intact thylakoids, as is usually the case. Whole-chain electron transport rate in the inside-out vesicles was as high as in the intact thylakoids, provided PC was added (PC is lost from the inside-out vesicles, being located in the intrathylakoid space). PS I activity was very high when PC was present in saturating concentration. These observations demonstrate that PS I is present and active in the partition region and its activity is fully adequate to ensure maximal electron transport along the whole chain from water to NADP. Atta-Asafo-Adjei and Dilley [115] conclude that a more accurate representation of thylakoid structure may be one with moderate lateral heterogeneity.

4.2. Excitation energy distribution between the photosystems

Grana formation and the segregation, even 'moderate', of PS I from PS II has important effects on the partition of excitation energy between the photosystems. As the formation of grana can be observed reversibly in isolated thylakoids upon addition of cations (see Section 4.1.), the additional problem arises of discriminating the effect of cations per se, if any, from that of membrane stacking and the segregation of the Chl-protein complexes in different regions of the membranes.

Investigations of the excitation energy available to PS II or, alternatively, to PS I, take advantage of the fact that PS II is fluorescent and PS I is not fluorescent at room temperature. At cryogenic temperatures (77 K or below) the contribution of PS I and PS II to fluorescence can be measured, respectively at 730–740 nm and 690 nm [116]. The transfer of excitation energy from PS II to PS I can therefore be measured at room temperature by the decrease of fluorescence yield and PS I can be viewed as a quencher of PS II fluorescence. Furthermore, when dark-adapted isolated thylakoids (or photosynthetic cells) are illuminated, the increase in fluorescence from an initial low level, F_0, to a final high level, F_m, is due to the conversion of a strong photochemical quenching (due to trapping of excitons at the reaction centre by the very efficient primary photochemical reaction; see Section 2) to a weak non-photochemical quenching when Q_A is reduced [116,117]. The former situation is described as one of 'open traps', and the latter of 'closed traps'. F_v ('variable fluorescence') is the difference $F_m - F_0$, and its increase signals the progress of Q_A reduction (see Section 2).

If one considers the three major Chl-protein complexes of thylakoids (namely PS II, LHC and PS I, the former including the reaction centre and its Chl *a* antenna of the water-splitting photochemical reaction, the second containing Chl *a*

ishes electrostatic repulsion between the membranes, then stacking occurs due to Van der Waals interaction between proteins of opposite membranes, and possibly interactions through oriented water [104]. In nature the thylakoids are found in the granal structure, as expected on the basis of the cation composition of the chloroplast stroma.

The localization in the appressed and non-appressed regions of thylakoids of the Chl-protein complexes of PS I, PS II and LHC (the light-harvesting Chl-protein complex) has been the object of controversy. The early fractionation experiments of Boardman and Anderson [105] revealed that PS II activity is concentrated in the 'heavy fraction', mostly the partitions of grana, while PS I activity is concentrated in the 'light fraction' consisting mostly of the stroma membranes (probably including the margins of grana and their end membranes). More recent investigations have concluded that most of the PS II activity is indeed found in the grana, while the non-appressed membranes contain very little PS II and about 15% of the total Chl [106,107]. It should be noted that the different methods of fractionation used do not allow it to be established in which fraction the margins of grana are found; this gives rise to uncertainty concerning the quantitative distribution of the activities and of the protein-Chl complexes (estimated by electrophoretic analysis) among the fractions separated. The granal membranes contain essentially all the large freeze-fracture particles thought to be PS II-LHC complexes, while the smaller particles supposed to be the PS I complexes have been found uniformly distributed between grana and stromal thylakoids [107].

Upon the advent of a method [108] which permitted the isolation of relatively pure fractions of partition zone membranes, isolated as inside-out vesicles, Andersson and Anderson found [109] that PS II and LHC are mainly in the partition zones, in agreement with the previous reports, but they also reported that PS I is confined to the stroma-exposed regions and practically absent in the partitions (see also Chapter 12). The fluorescence rise observed upon cation addition and grana formation is easily explained with this picture of thylakoid structure in mind: PS II-LHC is segregated far apart from PS I, and the excitation energy cannot be transferred to the latter, which is not fluorescent. The fluorescence rise can therefore be taken as a measure of the increase of energy remaining within PS II. Observations on the kinetics of fluorescence changes upon removal [110] or addition [111] of cations have indicated that the quenching of fluorescence can indeed be envisaged as the result of PS I collision with PS II when the complexes diffuse laterally in the membrane, and the kinetic heterogeneity of one of the complexes is suggested by the observed initial deviation from second-order kinetics [110].

If the model proposed by Andersson and Anderson [109] of total separation of PS I and PS II in the granal chloroplasts were to be accepted, electron transport from the PS II acceptors to P-700 would require a mobile electron carrier(s) which should diffuse laterally in the membrane fast enough to account for the observed electron transport rate. Plastoquinone [112] and plastocyanin are the candidates of choice for this role. The former has been shown to be present at approximately the same activity in the partitions and in the stroma-exposed membranes [43], while PC is known to be located in the intrathylakoid space [113].

of the electron acceptor, NADP/NADPH, and therefore to the metabolic activity of the Calvin cycle.

It has been demonstrated that phosphorylation of LHC causes the detachment of a fraction of it from PS II and its lateral migration in the membrane to become incorporated into PS I [134–136]. It has indeed been shown that the fluorescence quenching caused by LHC phosphorylation is qualitatively different from spill-over, because only LHC is quenched, not PS II [136], and F_0 as well as F_m are quenched [136,137]. The phosphorylation of LHC and/or of other thylakoid polypeptides may have more complex effects, and their interactions are far from being understood. It has been reported that protein phosphorylation enhances PS I-dependent cyclic photophosphorylation even under light saturation conditions [133], which could not be explained merely on the basis of PS I antenna enlargement.

In conclusion, LHC phosphorylation influences the balance of excitation energy in the two photosystems by increasing PS I and decreasing PS II optical cross-section. The mechanism is different from the cation regulation, which involves changes of the rate constants of energy transfer to the PS II reaction centre and of transfer from the LHC-PS II matrix to PS I (see above).

A discussion of the role of LHC phosphorylation and/or cation effects as mechanisms of regulation of energy distribution between the two photosystems in vivo is beyond the scope of this review. However, it seems likely that both mechanisms might cooperate in vivo to achieve a fine regulation of energy distribution to the two light reactions of photosynthesis and therefore an adaptation to the prevailing illumination conditions.

References

1 Eaton Rye, F.F. and Govindjee (1984) Photobiochem. Photobiophys. 8, 279–288.
2 Fisher, K. and Metzner, H. (1981) Photobiochem. Photobiophys. 2, 133–140.
3 Hill, R. and Bendall, F. (1960) Nature (London) 186, 136–137.
4 Arnon, D.I., Tsujimoto, H.Y. and Tang, G.M.S. (1981) Proc. Natl. Acad. Sci. USA 78, 2942–2946.
5 Ort, D.R. and Melandri, B.A. (1982) in Photosynthesis (Govindjee, ed.) Vol. I, pp. 537–587, Academic Press, New York.
6 Junge, W. and Jackson, J.B. (1982) in Photosynthesis (Govindjee, ed.) Vol. I, pp. 589–646, Academic Press, New York.
7 Joliot, P., Barbieri, G. and Chabaud, R. (1969) Photochem. Photobiol. 10, 309–329.
8 Kok, B., Forbush, B. and McGloin, M. (1970) Photochem. Photobiol. 11, 457–475.
9 Joliot, P. and Kok, B. (1975) in Bioenergetics of Photosynthesis (Govindjee, ed.) pp. 387–412, Academic Press, New York.
10 Govindjee, Kambara, T. and Coleman, W. (1985) Photochem. Photobiol. 42, 187–210.
11 Renger, G. and Govindjee (1985) Photosynth. Res. 6, 33–55.
12 Amesz, J. (1983) Biochim. Biophys. Acta 726, 1–12.
13 Babcock, G.T., Buttner, W.J., Ghanotakis, D.F., O'Malley, P.J., Yerkes, C.T. and Yocum, C.F. (1984) in Advances in Photosynthesis Research (Sybesma, C. ed.) Vol. I, pp. 243–252, Martinus Nijhoff/Dr. W. Junk Publ., Den Haag.
14 van Gorkom, H.J. (1985) Photosynth. Res. 6, 97–112.
15 Döring, G., Stiehl, H.H. and Witt, H.T. (1967) Z. Naturforsch. 22b, 639–644.
16 Klimov, V.V., Klevanik, A.V., Shuvalov, V.A. and Krasnovskii, A.A. (1977) FEBS Lett. 82, 183–186.

17 Klimov V.V. (1984) in Advances in Photosynthesis Research (Sybesma, C., ed.) Vol. I, pp. 131–138, Martinus Nijhoff/Dr. W. Junk Publ., Den Haag.
18 Duysens, L.N.M. and Sweers, H.E. (1963) Microalgae and Photosynthetic Bacteria, pp. 353–372, Japanese Society of Plant Physiologists and The University of Tokyo Press.
19 van Gorkom, H.J. (1974) Biochim. Biophys. Acta 347, 439–442.
20 O'Malley, P.J. and Babcock, G.T. (1983) Biophys. J. 41, 351a.
21 Babcock, G.T., Ghanotakis, D.F., Ke, B. and Diner, B.A. (1983) Biochim. Biophys. Acta 723, 276–283.
22 Sauer, K., Boska, M., Casey, J.L. and Karukstis, K.K. (1984) in Advances in Photosynthesis Research (Sybesma, C., ed.) Vol. I, pp. 121–126, Martinus Nijhoff/Dr. W. Junk Publ., Den Haag.
23 Brettel, K., Schlödder, E. and Witt, H.T. (1984) Biochim. Biophys. Acta 766, 403–415.
24 Cheniae, G.N. (1980) Methods Enzymol. 69, 349–362.
25 Wells, C.F. (1965) Nature 205, 693–694.
26 Dekker, J.P., van Gorkom, H.J., Wensink, J. and Ouwehand, L. (1984) Biochim. Biophys. Acta 767, 1–9.
27 Bishop, N.J. (1984) in Advances in Photosynthesis Research (Sybesma, C., ed.) Vol. I, pp. 321–328, Martinus Nijhoff/Dr. W. Junk Publ., Den Haag.
28 Warburg, O. and Lüttgens, W. (1944) Naturwissenschaften 32, 161, 301.
29 Haehnel, W. (1984) Annu. Rev. Plant Physiol. 35, 659–693.
30 Saygin, Ö. and Witt, H.T. (1985) Photobiochem. Photobiophys. 10, 71–82.
31 Kuwabara, T. and Murata, N. (1979) Biochim. Biophys. Acta 581, 228–236.
32 Yamamoto, Y. and Nishimura, M. (1984) in Advances in Photosynthesis Research (Sybesma, C., ed.) Vol. I, pp. 333–336, Martinus Nijhoff/Dr. W. Junk Publ., Den Haag.
33 Fowler, C.F. and Kok, B. (1974) Biochim. Biophys. Acta 357, 299–307.
34 Åkerlund, H.E., Jansson, C., and Andersson, B. (1982) Biochim. Biophys. Acta 681, 1–10.
35 Larsson, C., Jansson, C., Ljunberg, H., Åkerlund, E. and Andersson, B. (1984) in Advances in Photosynthesis Research (Sybesma, C., ed.) Vol. I, pp. 363–366, Martinus Nijhoff/Dr. W. Junk Publ., Den Haag.
36 Cramer, W.A. and Crofts, A.R. (1982) in Photosynthesis (Govindjee, ed.) Vol. I, pp. 387–467, Academic Press, New York.
37 Witt, H.T. (1979) Biochim. Biophys. Acta 505, 355–427.
38 Velthuys, B.R. and Amesz, J. (1974) Biochim. Biophys. Acta 333, 85–94.
39 Bouges-Bocquet, B. (1973) Biochim. Biophys. Acta 314, 250–256.
40 Renger, G. (1976) Biochim. Biophys. Acta 440, 287–300.
41 Trebst, A. (1979) Z. Naturforsch. 34C, 986-991.
42 Stemler, A. (1982) in Photosynthesis (Govindjee, ed.) Vol. II, pp. 513–539, Academic Press, New York.
43 Jennings, R.C., Garlaschi, F.M. and Gerola, P.D. (1983) Biochim. Biophys. Acta 722, 144–149.
44 Melis, A. and Brown, J.S. (1980) Proc. Natl. Acad. Sci. USA 77, 4712–4716.
45 Hurt, E. and Hauska, G. (1981) Eur. J. Biochem. 37, 1–9.
46 Crofts, A.R., Meinhardt, S.W., Jones, K.R. and Snozzi, M. (1983) Biochim. Biophys. Acta 723, 202–218.
47 Kok, B. (1957) Acta Bot. Neerl. 6, 316–336.
48 Parson, W.W. and Ke, B. (1982) in Photosynthesis (Govindjee, ed.) Vol. I, pp. 332–385, Academic Press, New York.
49 Malkin, R. (1982) Annu. Rev. Plant Physiol. 33, 455–479.
50 Whatley, F.R., Tagawa, K. and Arnon, D.I. (1963) Proc. Natl. Acad. Sci. USA 49, 266–270.
51 Forti, G. and Grubas, P.M.G. (1985) FEBS Lett. 186, 149–152.
52 Foust, G.P., Mayhew, S.G. and Massey, V. (1969) J. Biol. Chem. 244, 964–970.
53 Forti, G. and Bracale, M. (1984) FEBS Lett. 166, 81–84.
54 Ricard, F., Nari, F. and Diamantidis, G. (1980) Eur. J. Biochem. 108, 55–66.
55 Forti, G., Melandri, B.A., San Pietro, A. and Ke, B. (1970) Arch. Biochem. Biophys. 140, 107–112.
56 Bouges-Bocquet, B. (1978) FEBS Lett. 94, 95–99.
57 Jennings, R.C., Garlaschi, F.M., Gerola, P.D. and Forti, G. (1979) Biochim. Biophys. Acta 546, 207–219.

58 Lazzarini, R.A. and San Pietro, A. (1962) Biochim. Biophys. Acta 62, 417–420.
59 Arnon, D.I. (1977) in Encyclopedia of Plant Physiol., New Series (Trebst, A. and Avron, M., eds.) Vol. 5, pp. 7–56, Springer-Verlag, Berlin.
60 Forti, G. and Zanetti, G. (1969) in Progress in Photosynthesis Research (Metzner, H., ed.) Vol. III, pp. 1213–1216, Laupp. Tübingen.
61 Forti, G. and Rosa, L. (1972) in Proceedings of the II International Congress on Photosynthesis Research (Forti, G., Avron, M. and Melandri, B.A., eds.) Vol. II, pp. 1261–1269, Dr. W. Junk N.V., The Hague.
62 Shahak, Y., Crowther, D. and Hind, G. (1981) Biochim. Biophys. Acta 636, 234–243.
63 Forti, G. and Sturani, E. (1968) Eur. J. Biochem. 3, 461–472.
64 Clark, R.D., Hawkesford, M.F., Coughlan, S.F., Bennett, J. and Hind, G. (1984) FEBS Lett. 174, 137–142.
65 Forti, G. and Parisi, B. (1963) Biochim. Biophys. Acta 71, 1–6.
66 Tanner, W. and Kandler, O. (1969) in Progress in Photosynthesis Research (Metzner, H., ed.) Vol. III, p. 1217–1223, Laupp Tübingen.
67 Albertsson, P.Å., Ban-Dar, Hsu, Tang, G.M.S. and Arnon, D.I. (1983) Proc. Natl. Acad. Sci. USA 80, 3971–3975.
68 Mitchell, P. (1961) Nature 191, 144–148.
69 Mitchell, P. (1966) Biol. Rev. 41, 445–502.
70 Mitchell, P. (1977) FEBS Lett. 78, 1–20.
71 Schlödder, E., Gräber, P. and Witt, H.T. (1982) in Electron Transport and Photophosphorylation (Barber, J., ed.) pp. 105–175, Elsevier Biomedical Press, Amsterdam.
72 Trebst, A. (1974) Annu. Rev. Plant Physiol. 25, 423–458.
73 Rosing, F. and Slater, E.C. (1972) Biochim. Biophys. Acta 267, 275–286.
74 Ort, D.R. and Dilley, R.A. (1976) Biochim. Biophys. Acta 449, 95–107.
75 Ort, D.R., Dilley, R. and Good, N. (1976) Biochim. Biophys. Acta 449, 95–107.
76 Jagendorf, A.T. and Uribe, E.G. (1966) Proc. Natl. Acad. Sci. USA 55, 170–177.
77 Bamberg, E. and Laüger, P. (1973) J. Biol. Chem. 11, 177–194.
78 Boeck, H. and Witt, H.T. (1972) Proc. 2nd Int. Congr. Photosynthesis Research, Stresa 1971 (Forti, G., Avron, M. and Melandri, B.A., eds.) pp. 903–911, Dr. W. Junk N.V., The Hague.
79 Hangarter, R. and Ort, D.R. (1985) Eur. J. Biochem. 149, 503–510.
80 Witt, H.T. and Zickler, A. (1973) FEBS Lett. 37, 307–310.
81 Williams, R. (1961) J. Theor. Biol. 1, 1–13.
82 Williams, R. (1978) Biochim. Biophys. Acta 505, 1–44.
83 Strotmann, H., Hesse, H. and Edelman, K. (1973) Biochim. Biophys. Acta 314, 202–210.
84 Moroney, J.V., Lopresti, L., McEwen, B.F., McCarty, R.E. and Hammes, G.G. (1983) FEBS Lett. 158, 58–62.
85 Miller, K. and Staehelin, L.A., (1976) J. Cell Biol. 68, 30–47.
86 Reeves, S. and Hall, D. (1973) Biochim. Biophys. Acta 314, 66–78.
87 Forti, G. (1968) Biochem. Biophys. Res. Commun. 32, 1020–1024.
88 Portis, A.R., Jr. and McCarty, R.E. (1976) J. Biol. Chem. 251, 1610–1617.
89 Gräber, P. and Witt, H.T. (1976) Biochim. Biophys. Acta 423, 141–163.
90 Portis, A.R., Jr. and McCarty, R.E. (1974) J. Biol. Chem. 249, 6250–6254.
91 Izawa, S. and Good, N.E. (1968) Biochim. Biophys. Acta 162, 380–391.
92 Winget, G. Izawa, S. and Good, N. (1969) Biochemistry 8, 2067–2074.
93 Egneus, H., Heber, U., Matthiesen, U. and Kirk, M. (1975) Biochim. Biophys. Acta 408, 252–268.
94 Forti, G. and Gerola, P.D. (1977) Plant Physiol. 59, 859–862.
95 Forti, G. and Jagendorf, A.T. (1961) Biochim. Biophys. Acta 54, 322–330.
96 Heber, U. (1973) Biochim. Biophys. Acta 305, 140–152.
97 Murata, N. (1969) Biochim. Biophys. Acta 189, 171–181.
98 Williams, W.P. (1977) in Primary Processes in Photosynthesis (Barber, J., ed.) pp. 99–147, Elsevier Biomedical Press, Amsterdam.
99 Izawa, S. and Good, N.E. (1966) Plant Physiol. 41, 544–552.
100 Murakami, S. and Packer, L. (1971) Arch. Biochem. Biophys. 146, 337–347.

101 Gross, E.L. and Prasher, S.H. (1974) Arch. Biochem. Biophys. 164, 460–468.
102 Barber, J., Mills, J.D. and Love, A. (1977) FEBS Lett. 74, 174–181.
103 Rubin, B.T. and Barber, J. (1980) Biochim. Biophys. Acta 592, 87–102.
104 Duniec, J.T., Sculley, M.F. and Thorne, S.W. (1979) J. Theor. Biol. 79, 473–484.
105 Boardman, N.K. and Anderson, J. (1964) Nature 203, 166–167.
106 Sane, P.V., Goodchild, D.J. and Park, R.B. (1970) Biochim. Biophys. Acta 216, 162–178.
107 Arntzen, C.J., Dilley, R.A., Peters, G.A. and Shaw, E.R. (1972) Biochim. Biophys. Acta 256, 85–107.
108 Andersson, B., Sundby, C. and Albertsson, P.A. (1980) Biochim. Biophys. Acta 599, 391–402.
109 Andersson, B. and Anderson, J.M. (1980) Biochim. Biophys. Acta 593, 427–440.
110 Jennings, R.C., Gerola, P.D., Garlaschi, F.M. and Forti, G. (1982) FEBS Lett. 142, 167–170.
111 Vernotte, C., Briantais, J.M. and Maison-Peteri, B. (1982) Biochim. Biophys. Acta 681, 11–14.
112 Anderson, J.M. (1982) FEBS. Lett. 138, 62–66.
113 Haehnel, W., Propper, A. and Krause, H. (1980) Biochim. Biophys. Acta 593, 383-399.
114 Vaughn, K.C., Vierling, E., Duke, S.O. and Alberte, R.S. (1983) Plant Physiol. 73, 203–207.
115 Atta-Asafo-Adjei, E. and Dilley, R.A. (1985) Arch. Biochem. Biophys. 243, 660–667.
116 Butler, W.L. and Kitajima, M. (1975) Biochim. Biophys. Acta 396, 72–85.
117 Duysens, L.N.M. (1979) in Chlorophyll Organization and Energy Transfer in Photosynthesis, Ciba Foundation Symp., New Series, pp. 323–340, Elsevier-North Holland, Amsterdam.
118 Butler, W.L. and Strasser, R.J. (1977) Proc. Natl. Acad. Sci. USA 74, 3382–3385.
119 Butler, W.L. (1978) Annu. Rev. Plant Physiol. 29, 345–378.
120 Jennings, R.C. (1984) Biochim. Biophys. Acta 766, 303–309.
121 Loos, E. (1976) Biochim. Biophys. Acta 440, 314–321.
122 Wong, D. and Govindjee (1981) Photochem. Photobiol. 33, 103–108.
123 Jennings, R.C. and Zucchelli, G. (1985) Photobiochem. Photobiophys. 9, 215–221.
124 Bennett, J. (1977) Nature 269, 344–346.
125 Horton, P. and Black, M.T. (1980) FEBS Lett. 119, 141–144.
126 Horton, P., Allen, J.F., Black, M.T. and Bennett, J. (1981) FEBS Lett. 125, 193–196.
127 Bennett, J. (1983) Biochem. J. 212, 1–13.
128 Horton, P. (1983) FEBS Lett. 152, 47–52.
129 Staehelin, L.A. and Arntzen, C.J. (1983) J. Cell Biol. 97, 1327–1337.
130 Telfer, A., Bottin, H., Barber, J. and Mathis, P. (1984) Biochim. Biophys. Acta 764, 324–330.
131 Farchaus, J.W., Widger, W.R., Cramer, W.A. and Dilley, R.A. (1982) Arch. Biochem. Biophys. 217, 362–367.
132 Jennings, R.C. and Zucchelli, G. (1986) Arch. Biochem. Biophys. in the press.
133 Forti, G. and Grubas, P.M.G. (1986) Photosynth. Res. 10,
134 Haworth, P., Kyle, D.J. Horton, P. and Arntzen, C.J. (1982) Photochem. Photobiol. 36, 743–748.
135 Kyle, D.J., Haworth, P. and Arntzen, C.J. (1982) Biochim. Biophys. Acta 680, 336–342.
136 Torti, F., Gerola, P.D. and Jennings, R.C. (1984) Biochim. Biophys. Acta 767, 321–325.
137 Horton, P. and Black, M.T. (1981) Biochim. Biophys. Acta 635, 53–62.

J. Amesz (ed.) *Photosynthesis*

CHAPTER 2

Photosynthetic bacteria

BEVERLY K. PIERSON[a] and JOHN M. OLSON[b]

[a]*Biology Department, University of Puget Sound, Tacoma, WA, U.S.A. and* [b]*Institute of Biochemistry, Odense University, Odense M, Denmark*

1. Introduction

The high level of diversity among photosynthetic bacteria (including cyanobacteria) stands in contrast to the relative uniformity found among the chloroplasts of photosynthetic eukaryotes. Part of the excitement of working with photosynthetic bacteria today stems from the realization that this group is not only diverse but that we are only beginning to recognize the extent of this diversity. Those of us who work with the photosynthetic bacteria realize that many species are still waiting to be discovered.

Recent advances in understanding the phylogeny of photosynthetic prokaryotes have been made by comparing oligonucleotides derived from 16S rRNA [1]. Results of these analyses have renewed general interest in the comparative biochemistry of photosynthetic bacteria, since it is now clear that they are closely related to many non-photosynthetic bacteria. Extensive phylogenetic analysis has indicated that the ancestry of most, if not all, eubacteria and even the ancestry of the cellular organelles of eukaryotes, the mitochondria and chloroplasts, lies deeply entrenched in the history of the photosynthetic bacteria.

Photosynthetic bacteria convert light energy into chemical free energy. Most of these bacteria belong to the five eubacterial groups shown in Fig. 1 [1,2], but some, the halobacteria, belong to the archaebacteria. Although cyanobacteria are certainly photosynthetic eubacteria, they are considered separately in Chapters 1, 4 and 6 because of their unique and important ability to evolve oxygen.

All photosynthetic eubacteria contain photochemical reaction centers (RCs) containing one or more chlorophyll molecules. Each reaction center consists of a primary electron donor P (bacteriochlorophyll), an initial electron acceptor I (bacteriochlorophyll or bacteriopheophytin), and one or more secondary acceptors (Fe-S centers, quinones). Sometimes a secondary electron donor D (Cyt *c*) is tightly bound to the RC.

When a quantum of excitation reaches the RC, the primary donor P is excited to a new state P^* in which it is a powerful reducing agent. P^* transfers an electron to the initial acceptor I. To prevent the electron from falling back to P^+, the secondary acceptor X 'takes' the electron from I and stabilizes the charge separation. To further stabilize this separation the secondary donor D gives an electron to P^+.

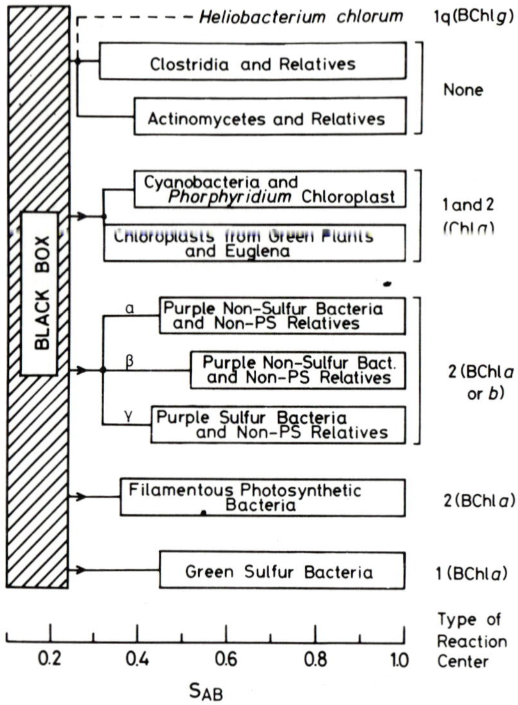

Fig. 1. Dendrogram of relationships among photosynthetic prokaryotes and their relatives, after Stackebrandt and Woese [2]. Five bacterial 'phyla' [1] containing photosynthetic members are shown. The exact relationship of *Heliobacterium chlorum* to the gram-positive bacteria is not yet known. Not shown are the other five 'phyla' without known photosynthetic members: peptidoglycan-less bacteria; bacteroids, cytophagas and flavobacteria; spirochaetes and leptospiras; bdellovibrios, myxococci, and certain S^0 and SO_4^{2-} reducers; and *Deinococcus*. PS = photosynthetic; for further explanation, see text.

These steps are summarized below:

$$\mathrm{DPIX} \xrightarrow{h\nu} \mathrm{DP^{*}IX} \longrightarrow \mathrm{DP^{+}I^{-}X} \longrightarrow \mathrm{DP^{+}IX^{-}} \longrightarrow \mathrm{D^{+}PIX^{-}}$$

These reactions taking place in the RC are the 'primary' chemical reactions of photosynthesis. (A detailed description of these primary chemical reactions in RCs of purple bacteria is given in Chapter 3.) The primary 'physical' processes of photosynthesis are light absorption and transfer of excitation energy. These processes take place mainly in the light-harvesting complexes (LHCs) described in Chapter 11.

The absorption cross-section of a single RC is so small that it cannot trap light fast enough to drive the organism's electron transport system up to capacity. LHCs exist to enlarge the effective absorption cross-section for each RC. The LHCs range in size from 50 BChl *a* molecules (Fig. 2) per RC in some purple bacteria to 2000 BChl *c* molecules per RC in some green sulfur bacteria. Some LHCs are integral

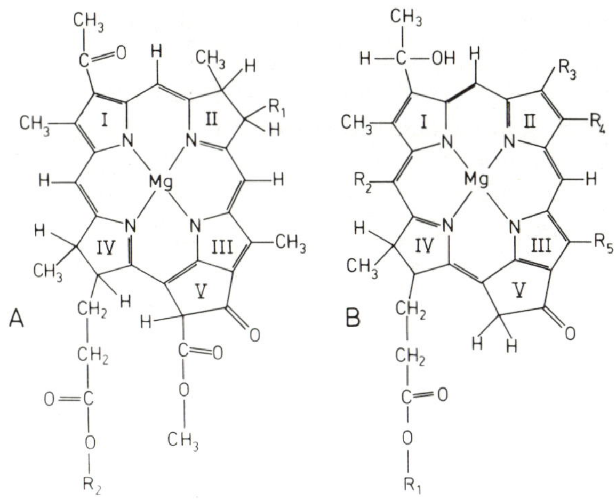

Fig. 2 (A) Structures of bacteriochlorophylls *a* and *b*. For BChl *a*, the substituents R_1 and R_2 are -H and either phytyl or geranyl-geranyl, respectively. For BChl *b*, R_1 is $=C\text{-}CH_3$ (with omission of the adjacent hydrogen on ring II) and R_2 is either phytyl (*Rhodopseudomonas viridis*) or 2,10-phytadienyl (*Ectothiorhodospira halochloris*). (B) Structures of bacteriochlorophylls *c, d* and *e*. R_1 is mostly farnesyl for green sulfur bacteria, mainly stearyl for *Chloroflexus aurantiacus*. R_2–R_5 are various substituents; each BChl exists as a series of homologs.

components of a membrane (purple bacteria), while others are housed in extramembranous bodies (e.g. green bacterial chlorosomes) which are attached to the membrane. In all cases the function of an LHC is to absorb light and to transfer the resulting excited state to the RC.

This chapter covers general characteristics of photosynthetic bacteria, with special emphasis on RCs, light-harvesting, electron transport and bacterial phylogeny.

2. *Filamentous photosynthetic bacteria*

2.1. *General characteristics*

Filamentous photosynthetic bacteria are anoxygenic phototrophic bacteria that are grouped together on the basis of their distinctive filamentous morphology [3]. All members of the group have been recently described, and most have not yet been isolated in pure culture. The most thoroughly studied genus is *Chloroflexus*, for which only one species, *Chloroflexus aurantiacus*, has been described [4]. Strains of *Chloroflexus* range in diameter from 0.5 to 1.5 μm. The cells form long septate filaments indeterminate in length. The filaments move by gliding and are major components of microbial mats in a variety of habitats. The only strains in pure culture are thermophiles isolated from hot spring mats, where they form conspicuous layers. While not successfully isolated in pure culture, several mesophilic strains have been observed from other habitats, including freshwater lakes, intertidal mats, and other hypersaline mat environments [3,5–8].

C. aurantiacus grows best as a photoheterotroph, but autotrophy has been successfully demonstrated in some strains [9]. Either hydrogen or sulfide can serve as electron donors for photoautotrophy. The pathway of CO_2 fixation is not known but does not appear to be either the reductive pentose phosphate pathway or the reductive tricarboxylic acid cycle [10].

Oxygen suppresses the synthesis of BChl *a* and *c* [11,12], and under aerobic conditions *C. aurantiacus* switches to hetero/-organotrophic respiration using an electron transfer system involving cytochromes of the *b, c* and probably *a* types [11,13,14]. Under these conditions synthesis of the chlorosomes is also repressed. [12].

A metabolically different type of *Chloroflexus* has recently been isolated from hot spring mats that are exposed to relatively high levels of sulfide and are totally devoid of cyanobacteria. This strain of *Chloroflexus*, designated GCF *Chloroflexus*, grows photoheterotrophically in culture but carries on sulfide-dependent CO_2 fixation in situ. It cannot be grown aerobically as a chemoorganotroph. While it appears to be an obligate phototroph, it can tolerate exposures to oxygen (Giovannoni, S. and Castenholz, R.W., personal communication).

Other filamentous gliding anoxygenic phototrophs containing bacteriochlorophylls have been observed in aquatic habitats. Many of these form septate filaments larger in diameter than *Chloroflexus*. *Heliothrix oregonensis* is a major mat-forming photoheterotroph found in certain hot springs in central Oregon and in Yellowstone National Park [15,16]. It lacks chlorosomes, contains only BChl *a* and carotenoids, and appears to be quite tolerant of oxygen. It is grown in co-culture with an obligate aerobic chemotroph, *Isosphaera pallida*. While quite different from *Chloroflexus* and definitely not a ‘green’ bacterium, *Heliothrix* is more closely related to *Chloroflexus* than to other bacteria tested on the basis of 5S rRNA sequence homologies [16].

Two other genera, *Oscillochloris* [17] and *Chloronema* [18], are known to contain BChl *a* and either BChl *c* or *d* (Fig. 2), apparently housed in chlorosomes.

2.2. Light-harvesting, reaction center and electron transport

Chloroflexus aurantiacus is metabolically versatile in being able to grow either anaerobically as a photoautotroph or photoheterotroph or aerobically as a chemoheterotroph using respiration. When grown phototrophically, it synthesizes two chlorophylls, BChl *a* and *c*. The BChl *c* functions entirely as a light-harvesting pigment and is associated with a 5.6-kDa polypeptide [19] in small sub-cellular structures called chlorosomes, adjacent to the cytoplasmic membrane [12,20,21]. The BChl *a* serves two functions. Most of the pigment is located in the cytoplasmic membrane in the B808-866 antenna complex and functions as an energy-transfer and light-harvesting pigment [22,23]. A very small fraction of the BChl *a* (BChl 790) is associated with the chlorosomes, probably at the site of attachment to the membrane, and functions as an energy-transfer pigment [21,22,24]. Some of the

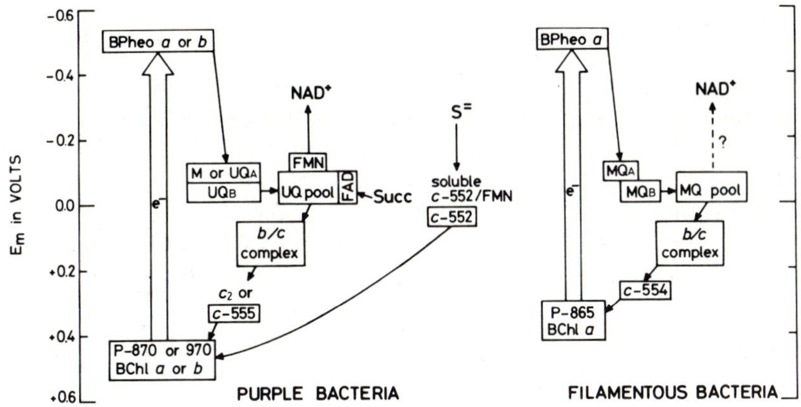

Fig. 3. Light-driven electron transport pathways in purple bacteria and in filamentous bacteria. Boxed-in components are membrane-bound. Secondary electron acceptors (MQ and UQ) in RCs are placed according to their effective midpoint potentials (without proton exchanges). Dashed pathways have not yet been demonstrated. The pathway from S^{2-} to P-870 or P-970 is found only in purple *sulfur* bacteria.

BChl *a* is found in the RC, where it is associated with two protein subunits (M_r = 28 000 and 30 000) [25,26]. As in purple bacteria (see Fig. 3) the primary electron donor P-865 ($E_{m,8}$ = + 0.36 V) is a BChl *a* dimer [22,27], and the initial acceptor is BPh *a* [28,29]. The secondary acceptor ($E_{m,8}$ = 0.05 V) is menaquinone (MQ) [30,31,32]. Unlike the RC in purple bacteria the RC in *C. aurantiacus* contains three BChl *a* and three BPh *a* molecules [33,34].

As far as we know, the cyclic electron transfer pathways in *Chloroflexus* are similar to those found in purple bacteria [13,35]. As shown in Fig. 3, there appear to be cyclic pathways involving *b*- and *c*-type cytochromes, Fe-S centers, and quinones in both kinds of bacteria grown phototrophically. However, *Chloroflexus* contains only MQ [32], while any given purple bacterium will always contain ubiquinone (UQ) with or without some MQ. On the electron acceptor side of the RC in *Chloroflexus* there are apparently two MQ molecules in series (MQ_A and MQ_B), which are assumed to feed electrons into an MQ pool [30,36]. Electrons presumably enter a Cyt *b*/*c* complex from the MQ pool and leave the complex via Cyt *c* ($E_{m,7}$ = 0.21 V) and Cyt *c*-554 ($E_{m,7}$ = 0.28 V) [35]. The membrane-bound Cyt *c*-554 ($E_{m,8}$ = +0.26 V, M_r = 43 000) [37] is the direct electron donor to P-865 in the RC [22,30,36]. It contains two hemes with redox potentials of +0.14 and +0.26 V respectively [36] and is absent from aerobically grown cells [38].

During electron flow the *b*/*c* complex functions as an energy transducer that converts a substantial fraction of the energy difference between the quinone pool and Cyt *c*-554 to potential energy in the form of a transmembrane proton gradient.

The evidence for a *b*/*c* complex in *Chloroflexus* is impressive. Membrane fragments from high-light-intensity cells contain cytochromes *c*-552 and *c*-554, cytochromes *b*-562 and multiple Fe-S centers [35]. Three different *c*-type cytochromes were found with $E_{m,7}$ values (n = 1) of +0.02, +0.21 and +0.28 V respectively.

Cyt *c* (+0.02 V) appeared to be analogous to the autooxidizable, CO-binding Cyt *c*-552 in *Chromatium vinosum*, and Cyt *c* (+0.21 V) was assigned the role of Cyt *c* in the proposed *b/c* complex. Cyt *c* (+0.28 V) was identified with Cyt *c*-554, the secondary donor to the RC. Four different *b*-type cytochromes were found with $E_{m,7}$ values (n = 1) of −0.07 V, +0.06, +0.16 and +0.24 V respectively. Cytochromes *b* (−0.07 and +0.06 V) were assigned to the proposed *b/c* complex. Six different Fe-S centers were distinguished, and two were characterized as HiPIP-type ($E_{m,7}$ = +0.06 V) and Rieske-type ($E_{m,7}$ = +0.10 V). The HiPIP center was tentatively assigned to succinic dehydrogenase, and the Rieske center to the proposed *b/c* complex.

Little is known about the electron flow pathways when *Chloroflexus* oxidizes S^{2-} to S^0 and fixes CO_2, but it seems reasonable to suppose that NAD^+ is reduced by reverse electron flow from the MQ pool in analogy to the mechanism used by the purple sulfur bacteria (Chapter 9).

While little is known of the nature of the phosphorylation mechanism, the presence of a thermostable ATP synthetase has been demonstrated [39].

3. Green sulfur bacteria

3.1. General characteristics

Green sulfur bacteria are small unicellular organisms [40]. Most cells are around 0.5 μm in diameter and from 1.0 to 2.0 μm in length. With one exception they are non-motile. The recently described species, *Chloroherpeton thalassium* [41], has exceptionally long cells (from 8 to 30 μm), is unicellular and moves by gliding. The green sulfur bacteria are found in strongly reducing habitats exposed to low levels of light such as the hypolimnion of lakes, at the surface of anaerobic sediments, and in the lower layers of microbial mats. While they may form dense planktonic blooms in the anaerobic zones of both freshwater and saline aquatic habitats, they rarely form the dense mat layers typical of *Chloroflexus*.

As a group the green sulfur bacteria have relatively little diversity. The five genera, *Ancalochloris, Chlorobium, Chloroherpeton, Pelodictyon,* and *Prosthecochloris*, appear to be very similar physiologically [40]. All are obligate anaerobic N_2-fixing [42] photolithoautotrophs using hydrogen or reduced sulfur compounds to reduce CO_2. Most physiological and biochemical studies have been done on species of *Chlorobium* and *Prosthecochloris*. Carbon dioxide is fixed via the reductive tricarboxylic acid cycle rather than the reductive pentose phosphate cycle [43]. Respiration does not occur.

Analyses of 1-alkyl-2,3,6-trimethylbenzenes from Silurian and Devonian oils in western Canada indicate the existence of green sulfur bacteria in restricted seas about 400 million years ago [44].

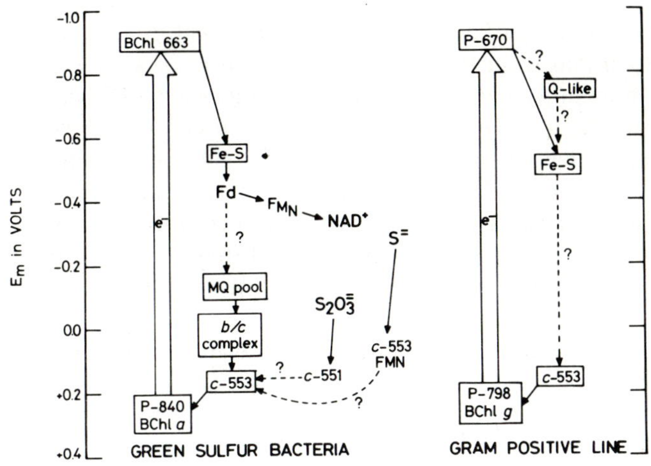

Fig. 4. Light-driven electron transport pathways in green sulfur bacteria and in the gram-positive line (*H. chlorum*). The true midpoint potential for the $S_2O_3^{2-}$ /HS^- + HSO_3^- couple is −0.40 V [62], not −0.10 V as shown in this figure. BChl 663 is a specialized lipophilic chlorophyll related to BChl *c* [52], and P-670 is a pigment related to BChl *c* [69] and/or Chl *a* [66]. See legend for Fig. 3.

3.2. Light-harvesting, reaction center and electron transport

All species contain two different chlorophylls, BChl *a* and an additional light-harvesting pigment, either BChl *c, d* or *e* [40]. As in *Chloroflexus* these light-harvesting pigments are housed in chlorosomes located adjacent to the cytoplasmic membrane [45,46]. In *Chlorobium* BChl *c* is associated with a 7.5-kDa polypeptide. (J.M. Olson and P. Roepstorff, unpublished). A small amount of energy-transferring BChl *a* is also found in the chlorosomes [47,48], but most of the BChl *a* is found in a water-soluble protein associated with the cytoplasmic membrane [45]. The RC (see Fig. 4 and Ref. 36) is similar to that of PS I in cyanobacteria and chloroplasts. The primary electron donor P-840 ($E_{m,7}$ = +0.25 V) is a BChl *a* dimer [49] and is associated with a 65-kDa polypeptide [50]. The initial electron acceptor is BChl 663, a special lipophilic form of BChl *c* [51,52], and the secondary acceptor is a −0.54 V Fe-S center capable of reducing ferredoxin directly [53,34].

The electron transfer system [36] contains cytochromes of the *b* and *c* types, a Rieske-type Fe-S protein [54] and MQ [55]. A membrane-bound Cyt *c*-553 ($E_{m,7}$ = +0.17 V) is the immediate electron donor to P-840$^+$ in both *Chlorobium* and *Prosthecochloris* [56,57,36]. Another membrane-bound Cyt *c* is Cyt *c*-550.5 ($E_{m,7}$ = +0.22 V), which may be part of a Cyt *b*/*c* complex along with a Rieske-type Fe-S center ($E_{m,7}$ = +0.16 V) [50]. There are three *b*-type cytochromes in the membrane: Cyt *b*-563, Cyt *b*-564 ($E_{m,7}$ = 0.09 V) and Cyt *b*-562 ($E_{m,AV}$ = +0.01 V) [36]. Cyt *b*-562 does not titrate as a single component and is thought to be part of a *b*/*c* complex [50]. There is indirect evidence for cyclic electron flow in green sulfur bacteria [58–60] and, in analogy to the path of cyclic electron flow in photosystem I, we suppose that electrons from the −0.54 V Fe-S center in the RC flow to soluble ferredoxin and from there to the presumed MQ pool. Electrons from the MQ pool are thought to enter the *b*/*c* complex and to return to the RC via Cyt *c*-553.

It is well established that green sulfur bacteria carry out linear electron flow from external donors such as S^{2-}, S^0, $S_2O_3{}^{2-}$ and $SO_3{}^{2-}$ to ferredoxin and NAD^+ (see Ref. 36 and Chapter 9). The use of O_2-uptake as a valid assay of linear flow in membrane preparations has also been demonstrated [61]. Sulfide oxidation is mediated by a soluble flavocytochrome *c*-553 ($E_m \approx +0.09$ V) [36,62], and thiosulfate oxidation seems to require the soluble Cyt *c*-551 ($E_m \approx 0.14$ V) [36]. Soluble Cyt *c*-555 ($E_{m,7} \approx 0.15$ V) is reduced by both S^{2-} and $S_2O_3{}^{2-}$, but its exact role is not yet known [36]. We suppose that linear electron flow via the soluble *c*-type cytochromes enters the RC via the membrane-bound Cyt *c*-553 and leaves the RC via ferredoxin. Soluble ferredoxin and flavoprotein ferredoxin-NAD(P) reductase have been found in several green sulfur bacteria [36].

A membrane-bound adenosine triphosphatase has been demonstrated in *Chlorobium* [58].

4. *Heliobacterium chlorum* – the gram-positive line

4.1. General characteristics

Heliobacterium chlorum is a recently described anoxygenic N_2-fixing photosynthetic bacterium [63]. It is a relatively large (1.0 × 7–10 μm) unicellular organism with gliding motility. The one culture that has been studied was isolated from the soil, and it is not yet known if this organism forms dense planktonic blooms or benthic mats. *H. chlorum* is a gram-negative bacterium like all the other photosynthetic bacteria that have been described. However, phylogenetic analysis using sequence data for 16S rRNA has revealed that this organism is more closely related to the gram-positive bacteria than to the other gram-negative bacteria (see Fig. 1) [64]. This is the first time that such a phylogenetic relationship has been found for a photosynthetic bacterium.

While *H. chlorum* tolerates exposure to high light intensities well, it is extremely sensitive to oxygen and grows only as a strict anaerobe [63]. Physiologically it is an obligate photoheterotroph, since no respiratory metabolism has been observed. Autotrophic CO_2 fixation is not sustained by either hydrogen or sulfide [63].

4.2. Light-harvesting, reaction center and electron transport

H. chlorum has one major chlorophyll, a unique pigment called BChl *g* [65] (see Fig. 5), which has a major absorption band at 788 nm in vivo [63,66]. The light-harvesting BChl *g*, which exists in at least three spectroscopically different forms [67], is apparently housed in the cytoplasmic membrane, as there are no intracytoplasmic membranes or chlorosomes [63,66]. The cytoplasmic membrane is unusually rich in protein – 70% on a dry weight basis [66]. *H. chlorum* also has a pigment absorbing at 670 nm and an unusually low content of a single carotenoid, neurosporene [63,67]. Both the 670-nm pigment and the carotenoid transfer excitation energy to BChl *g* with 70% efficiency [67].

In preliminary studies on isolated membranes Fuller et al. [66] and Prince et al. [68] observed a primary donor (P-798) with properties similar to those of P-840 in the RC of green sulfur bacteria, but it was not clear whether P-798 is a dimer or monomer of BChl *g*. Nuijs et al. [69] showed that the initial acceptor was a molecule absorbing at 670 nm and suggested that it might be related to BChl *c*. The evidence is also compatible with a Chl *a*-like acceptor [66]. The secondary electron acceptor appeared to be an Fe-S center ($E_{m,10} = -0.51$ V) as in the green sulfur bacteria, but Brok et al. [70] very recently obtained EPR evidence for a quinone-type secondary acceptor ($E_m < -0.62$ V) in addition to an Fe-S center $E_m > -0.42$ V). The difference between the data of Prince et al. [68] and those of Brok et al. [70] have not yet been resolved. This new RC may be unique in having both an Fe-S center and a quinone-like molecule as secondary acceptors as shown in Fig. 4. We will call this new type of reaction center RC-1q to indicate its similarity to RC-1 of green sulfur bacteria and the possibility of a quinone-like molecule (q) not found in RC-1. Nothing is yet known about the polypeptide(s) associated with RC-1q, but a 50-kDa protein may be associated with the membrane-bound cytochrome *c*-553 which is the electron donor to P-798 [66]. The other components of the electron transport system are as yet unknown.

5. *Purple bacteria*

5.1. *General characteristics*

Purple bacteria are the largest, most diverse, and most thoroughly studied group of the anoxygenic photosynthetic bacteria. Recent studies on the phylogenetic relationships among the various species of this group have greatly influenced contemporary notions regarding the evolutionary significance of photosynthetic bacteria in general. Oligonucleotide cataloging using 16S rRNA has revealed that many of the purple photosynthetic bacteria have closer phylogenetic associations with non-photosynthetic bacteria than with each other [2]. Some of these phylogenetic associations differ significantly from the classical taxonomic categories used to identify and describe members of this group [71]. For our purposes recognition of subdivisions within the group is not essential. To avoid confusion, however, we will summarize the various categories that have been used or have been proposed for use in either phylogenetic or taxonomic schemes for the purple bacteria.

Purple bacteria were initially divided into two taxonomic groups: *Thiorhodaceae*, which oxidize sulfide to sulfur and accumulate the latter inside the cells, and *Athiorhodaceae*, which do not [72]. Subsequently the names were changed to *Chromatiaceae and Rhodospirillaceae* respectively [73]. Recognition of these two major groups has rested mainly on physiological criteria. More recently the *Chromatiaceae*, were subdivided, creating a third family, *Ectothiorhodaceae*, species of which oxidize sulfide but deposit the resultant sulfur extracellularly [72]. While this division into three families is taxonomically useful, it does not reveal the interesting phylogenetic relationships among these bacteria and their non-photosynthetic relatives. On the basis of oligonucleotide catalogs of 16S rRNA, Woese et al. [74] have grouped the purple photosynthetic bacteria into three major subdivisions (see

TABLE 1

General properties of photosynthetic bacteria. Parentheses ( ) denote properties within the group but not major characteristics of the group. Question marks (?) denote properties not yet determined or controversial.

Group	Genera	Morphology
Filamentous photosynthetic bacteria	*Chloroflexus* *Chloronema* *Heliothrix* *Oscillochloris*	Filamentous rods
Green sulfur bacteria	*Ancalochloris* *Chlorobium* *Chloroherpeton* *Pelodictyon* *Prosthecochloris*	Unicellular rods, spheres, vibrios
Gram-positive line (*H. chlorum*)	*Heliobacterium*	Unicellular rods
Purple sulfur bacteria	*Amoebobacter* *Chromatium* *Ectothiorhodospira* *Lamprocystis* *Thiocapsa* *Thiocystis* *Thiodictyon* *Thiopedia* *Thiospirillum*	Unicellular rods, spheres, spirals
Purple non-sulfur bacteria	*Rhodobacter* *Rhodocyclus* *Rhodomicrobium* *Rhodopseudomonas* *Rhodopila* *Rhodospirillum*	Unicellular rods, spheres, spirals
BChl *a*-containing non-phototrophic bacteria	*Erythrobacter* *Protaminobacter* *Pseudomonas*	Unicellular rods
Halobacteria	*Halobacterium* *Halococcus*	Unicellular rods, spheres, discs

Fig. 1) referred to as alpha, beta and gamma and formerly designated groups I–III by Gibson et al. [75]. The alpha subdivision includes most of the species of the classical *Rhodospirillaceae* (e.g. *Rhodobacter* (formerly '*Rhodopseudomonas*') *sphaeroides*) and several non-photosynthetic genera as well [74]. The beta subdivision contains three other species of the *Rhodospirillaceae* (e.g. *Rhodocyclus gelatinosus*) and several species of non-photosynthetic bacteria [76]. The gamma subdivision includes all members of the *Chromatiaceae* and *Ectothiorhodaceae* and no members of the *Rhodospirillaceae* [77,78]. The significance of these phylogenies is that they clearly show that the several species of purple bacteria are very closely related to many different non-photosynthetic bacteria.

Habitat	Carbon source	External reductant	Respiration	N fixation
Mats (planktonic)	Organic (CO_2)	H_2, H_2S, organic	+	−?
Planktonic (mats)	CO_2	H_2, H_2S, S^0, $S_2O_3^{2-}$	−	+
Soil, but not known to form large accumulations	Organic	Organic	−	+
Mats, planktonic	CO_2 (organic)	H_2, H_2S, S^0, (organic), ($S_2O_3^{2-}$), (SO_3^{2-})	(+)	+
Soil or water, but not known to form large accumulations	Organic (CO_2)	Organic, H_2, (H_2S), ($S_2O_3^{2-}$)	+	+
Marine epiphytic, planktonic	Organic	Organic (not for photosynthesis)	+	?
Planktonic, mats	Organic	Organic (not for photosynthesis)	+	−?

For this discussion we will overlook the complexities of current taxonomic and phylogenetic organization in this group of organisms and lump them all under the simple and useful term, purple photosynthetic bacteria. The morphological diversity of the group is tremendous, including rods, cocci, vibrios, spirals and budding forms. All are unicellular, however, and no filaments or filamentous tendencies have been observed. Cell dimensions range from less than one to several μm. When motile, cells move by flagellar rotation. Gliding motility is unknown in this group [40].

Physiologically and ecologically this group can be subdivided into the purple sulfur and the purple non-sulfur bacteria, although the properties of these two groups are not mutually exclusive. As shown in Table 1, the purple sulfur bacteria include nine genera (taxonomically, all members of the *Chromatiaceae* and *Ectothiorhodaceae* or phylogenetically, all members of the gamma subdivision). They are mostly anaerobic photoautotrophs using hydrogen or reduced sulfur to fix CO_2 via the reductive pentose phosphate cycle [40]. Their tolerance for oxygen varies, and a few can grow aerobically in the dark [79]. They commonly form dense planktonic blooms and benthic mats in habitats rich in sulfide and exposed to light, such as the anaerobic zones of freshwater and saline aquatic environments, sulfur springs and intertidal or supratidal salt marshes.

The purple non-sulfur bacteria presently include several species in six genera (see Table 1) all of which were included taxonomically in the *Rhodospirillaceae* and now are included phylogenetically in the alpha and beta subdivisions [74,76]. The purple non-sulfur bacteria grow best as photoheterotrophs although many are capable of autotrophic CO_2 fixation via the reductive pentose phosphate cycle using hydrogen or reduced sulfur [40], and most can also fix nitrogen [80]. They vary in their tolerance to oxygen and many can grow facultatively using respiration [71]. Oxygen suppresses the synthesis of their pigments. They can also ferment and are known for their metabolic versatility. As expected from this versatility, members of the purple non-sulfur bacteria can be isolated from a large variety of habitats, including ponds, standing fresh or brackish water and soil [71]. They are rarely, if ever, observed in massive planktonic blooms, and are not known to form benthic mats.

5.2. Light-harvesting, reaction center and electron transport

Despite the physiological diversity of the purple bacteria the photosynthetic apparatus is much the same in all species. All purple bacteria contain only one type of chlorophyll, either BChl *a* or *b*. The light-harvesting and RC chromophores are all located in the cytoplasmic membrane or elaborate invaginations of it in the form of vesicles, tubules or lamellae [40].

The light-harvesting system of the purple bacteria containing BChl *a* is organized into various pigment-protein complexes [81]. Closely associated with the RC is the longer-wavelength-absorbing complex (B890-protein complex in *Rhodospirillum rubrum* and B875-protein complex in other purple bacteria). The complexes include two BChl *a* and one or two carotenoid molecules bound to two low molecular weight hydrophobic polypeptides. In addition to the longer-wave-length-absorbing complexes, the light-harvesting system includes a shorter-wavelength-absorbing complex (B800-850-protein complex) in multimeric units built up from subunits including two or three low molecular weight hydrophobic polypeptides associated with three BChl *a* molecules and carotenoid.

The light-harvesting system of purple bacteria containing BChl *b* is similarly organized around low molecular weight hydrophobic polypeptides. Two of these from the B1015-protein complex of *Rps. viridis* have been sequenced and show about a 50% homology with similar polypeptides from *R. rubrum* [82]. A B800-1020-protein complex isolated from *Ectothiorhodospira halochloris* contains five low-to-medium molecular weight polypeptides, BChl *b* and no carotenoids [83].

Comparisons of RCs show fundamental similarities throughout the group [81]. The RC (see Fig. 3) is composed of three subunits, L (31 kDa), M (34 kDa) and H ($\approx$ 28 kDa), and contains four BChl *a*/*b* molecules and two BPh *a*/*b* molecules [84,85]. Subunits L and M are homologous to each other and to the D-1 (32 kDa) and D-2 (34 kDa) proteins found in the chloroplast thylakoid [85,86,87]. The BChl *a* and BChl *b* RCs appear to function similarly with the primary electron donor ($+0.4$ V $\leqslant E_m \leqslant +0.5$ V) being a BChl dimer, and the initial acceptor (-0.6 V $\leqslant E_m \leqslant -0.4$ V) being a BPh *a* or *b* molecule [88]. The first secondary acceptor (-0.1 V $\leqslant E_m \leqslant 0.0$ V), may be either UQ or MQ, but the second acceptor is always UQ. These two quinones are bound to the M and L subunits [89,85].

The membrane-bound ATP synthetase couples phosphorylation to a proton gradient [90] which is generated by the cyclic electron transfer system (Fig. 3). This system includes the RC, a UQ pool [91], a Cyt *b*/*c* complex [92,93], and a specialized Cyt *c* ($E_{m,7} = +0.34$ V) for transferring electrons to the oxidized primary donor (P-870$^+$ or P-970$^+$) of the RC. In some bacteria such as *Chromatium vinosum* and *Rhodopseudomonas viridis* this specialized Cyt *c* is bound to the RC in the membrane [93,94], whereas in other bacteria such as *Rb. sphaeroides* and *Rhodospirillum rubrum* this cytochrome is a periplasmic protein (Cyt c_2) that binds to the membrane-bound RC [90].

In bacteria that fix CO_2, NAD^+ can be reduced by H_2 via ferredoxin [95] or by reverse electron flow from the UQ pool [96]. Succinate and malate are examples of exogenous reductants able to feed electrons directly into the UQ pool via membrane-bound dehydrogenases [97]. Although the redox potential of the H_2S/S^0 couple ($E_{m,7} = -0.28$ V) is low enough for H_2S to reduce the UQ pool directly in the dark, the evidence from purple sulfur bacteria indicates that reduced sulfur compounds (H_2S, S^0, $S_2O_3^{2-}$ and SO_3^{2-}) donate electrons to the UQ pool indirectly via the RC with a light-drive electron transfer from P-870 (or P-970) to BPh *a* or *b* required [62,97]. The low-potential ($E_{m,8} \approx 0.0$ V) membrane-bound Cyt *c*-552 (*Chr. vinosum*) and Cyt *c*-553 (*Rps. viridis*) may function in linear electron flow from exogenous electron donors directly to the oxidized primary donor (P-870$^+$ or P-970$^+$) of the RC [94]. Various soluble *c*-type cytochromes are thought to be involved in the oxidation of H_2S to S^0: Cyt *c*-550 and Cyt *c* in *Thiocapsa roseopersicina* and flavo-Cyt *c*-552 in *Chr. vinosum* as in the green sulfur bacteria [62]. The oxidation of S^0 to SO_3^{2-} may be catalysed by a siroheme-containing enzyme, and the oxidation of SO_3^{2-} to SO_4^{2-} takes place in many bacteria via adenosine 5'-phosphosulfate (APS) in an AMP-dependent reaction [62].

The purple non-sulfur bacteria are particularly well adapted to living by photoassimilation of dissolved organic compounds. Most can utilize certain intermediates of the tricarboxylic acid cycle (succinate, fumarate and malate) as well as pyruvate and lactate as the sole source of carbon and the sole source of reducing power [98]. Some, such as *R. rubrum* and *Rb. sphaeroides*, assimilate acetate or butyrate and store it internally as β-hydroxybutyrate [99]. Other purple bacteria, such as *Chr. vinosum, Rb. capsulatum* and *Rps. palustris*, contain a glyoxylate cycle which permits them to convert assimilated acetate to carbohydrate without fixing CO_2 [100]. All these examples of photoassimilation require a light-driven cyclic electron transport chain supplemented by a reverse electron flow pathway which can be driven by cyclic electron flow. Carbon dioxide fixation with H_2 as the exogenous electron donor uses the same cyclic electron transport system as does photoassimilation. In purple sulfur bacteria the picture changes when CO_2 fixation is carried out with reduced sulfur compounds as electron donors. In addition to cyclic electron flow for ATP production, linear electron flow from reduced sulfur compounds to NAD^+ via the UQ pool requires a light-driven step through the RC as shown in Fig. 3.

When purple non-sulfur bacteria switch from photosynthesis in the light to respiration in the dark, the content of BChl *a* or *b* is diminished, and the synthesis of cytochrome oxidase is increased [101]. In some bacteria, such as *Rb. sphaeroides*, two oxidases are formed, Cyt aa_3 [90] and Cyt *o*. In most others (e.g. *R. rubrum*) only Cyt *o* is formed [97]. Work during the past decade has revealed a strong similarity between the electron transfer pathways in purple non-sulfur bacteria and in the mitochondrial inner membrane [92].

6. *Bacteriochlorophyll a-containing non-phototrophic bacteria*

These bacteria appear to be related phylogenetically to the purple bacteria, and one species, *Erythrobacter longus*, has been grouped with the alpha subdivision (non-sulfur) [74]. However, unlike the classical purple bacteria, these bacteria are unable to synthesize BChl *a* anaerobically, and are unable to grow phototrophically under anaerobic conditions. They require oxygen for growth and for BChl *a* synthesis. The species in which these properties have been observed include the facultative methylotrophs *Protaminobacter ruber* and *Pseudomonas* AM1 [102] and *Erythrobacter longus*, an obligately aerobic marine bacterium found growing epiphytically on marine algae [103].

Cells of *Protaminobacter ruber* cannot be grown anaerobically and BChl *a* synthesis occurs under aerobic conditions only. However, as in the purple non-sulfur bacteria, the level of BChl *a* synthesis can be increased by lowering the level of oxygen as long as sufficient oxygen is present to sustain growth [104]. The membranes of this organism contain photochemically active RCs containing BChl *a*, and a cyclic electron transfer system including membrane-bound Cyt *c*-554 [105] appears to be functionally connected to a phosphorylation system [106]. It is clear, however, that photosynthesis cannot sustain growth. Although photosynthetic, these bacteria are not phototrophic.

The obligate aerobe *Erythrobacter longus* is unicellular (cells 0.5 by 1.5 μm) and motile. All strains of *Erythrobacter* isolated so far are marine and are found primarily as epiphytes. It has not been reported to form planktonic blooms or to grow in microbial mats. The synthesis of BChl *a* and the proliferation of photosynthetic membranes increases with increasing levels of oxygen in *Erythrobacter*, contrary to what occurs in the purple non-sulfur bacteria and *Protaminobacter ruber* [107]. However, the BChl *a*-protein complexes found in *E. longus* and *Erythrobacter* species OCH114 are similar to those found in purple bacteria [108]. Photochemical RC activity, electron transfer and light-dependent phosphorylation have all been observed in *Erythrobacter* [109,102]. Under aerobic conditions the growth rate is doubled when cells are grown in the light rather than in the dark. The simultaneous operation of both photosynthesis and respiration therefore appears to be advantageous to these cells [102]. Photochemical activity is stopped completely in cells placed under anaerobic conditions, and hence oxygen is required for growth. The dependence of photochemistry on the presence of oxygen and/or respiratory activity appears to be due at least in part to the relatively high potential of the secondary acceptor, a quinone, which remains reduced under anaerobic conditions, thereby preventing photochemical oxidation of the reaction center BChl *a* [110]. Thus while these organisms are indeed photosynthetic and growth may be enhanced in the light over levels attained using respiration alone, and survival in the absence of exogenous substrate may be sustained by photosynthetic activity, they are not phototrophic [102,110].

7. *Phylogeny*

As shown in Fig. 1, the phylogeny of photosynthetic eubacteria based on 16S rRNA catalogs is inseparable from the phylogeny of eubacteria in general. There are photosynthetic organisms in five (out of ten) eubacterial 'phyla' [1,77]: purple bacteria and relatives, green sulfur bacteria, filamentous photosynthetic bacteria, cyanobacteria and chloroplasts, and gram-positive bacteria [111,64]. This wide distribution of photosynthetic organisms among half the branches of the eubacterial tree strongly suggests a common photosynthetic ancestor for all these branches and probably for all other eubacteria as well.

The rRNA catalogs for the various eubacterial 'phyla' do not permit us to infer a branching order, because the S_{AB} (association coefficient) values between the major branches are too low to be significant. For example, between members of the green sulfur bacteria and members of the filamentous photosynthetic bacteria S_{AB} values range from 0.12 to 0.20 [111]. These values are too low to establish branching orders with respect to other 'phyla', for example *Escherichia coli* (purple bacteria and relatives, S_{AB} = 0.17–0.28) and *Bacillus pumilus* (gram-positive bacteria, S_{AB} = 0.18–0.24). As shown in Fig. 1, the various 'phyla' disappear into a 'black box' corresponding to S_{AB} values between 0.1 and ≈ 0.25.

In addition to the molecular data in the form of oligonucleotide catalogs, support for the notion of a common ancestor for these highly diverse photosynthetic bacteria is found in the comparative biochemistry of the RCs. All RCs function by similar mechanisms even though the specific redox components vary (see Table 2 and Figs. 3 and 4). The RC chlorophylls are closely related as shown in Fig. 5, and the RC polypeptides sequenced so far (including those of RC-2 in cyanobacteria and chloroplasts) show some degree of homology. This thread of commonality extends to the known cyclic electron transport chains, all of which contain a quinone pool and a Cyt *b*/*c* complex.

All known photosynthetic eubacteria (including cyanobacteria) contain Chl *a*, BChl *g*, BChl *a* or BChl *b* in their RCs. Chl *a* is restricted to the cyanobacteria and BChl *g* to the gram-positive line (as far as we know). As shown in Fig. 5 chlorophyllide *a* and BChl-ide *g* are isomers of one another [65]. BChl *a* is found in the RCs of green bacteria, purple bacteria and filamentous bacteria. So far BChl *b* has been found only in purple bacteria. We have suggested that all photosynthetic eubacteria are descendents of a common ancestor containing Chl *a* in an RC-1 type RC [112], and we suppose that the gram-positive line of bacteria (BChl *g*)

TABLE 2

Reaction center and light-harvesting pigments of photosynthetic bacteria.
Q = quinone; UQ = ubiquinone; MQ = menaquinone. Question marks (?) denote uncertainty or lack of information.

Group	Reaction center			Non-cytochrome polypeptides	
	Primary donor	Initial acceptor	Secondary acceptor(s)	*M* (kDa)	$M_r \times 10^{-3}$
Filamentous photosynthetic bacteria	BChl *a* (P-865)	BPh *a*	MQ		28 30
Green sulfur bacteria	BChl *a* (P-840)	BChl 663	Fe-S		65
Gram-positive line, *H. chlorum*	BChl *g* (P-798)	P-670	Q-like, Fe-S		?
Purple sulfur and non-sulfur bacteria	BChl *a* (P-870) or BChl *b* (P-960)	BPh *a* or *b*	UQ, MQ	31 (L) 34 (M) 28 (H)	19 22 28
BChl *a*-containing non-phototrophic bacteria	BChl *a* (P-870)	?	Q	?	
Halobacteria	No redox reactions, cyclic protonation/deprotonation			Bacterioopsin	

may be a direct off-shoot from the common ancestor. The cyanobacteria and chloroplast line (Chl *a* in RC-1 and RC-2) is thought to be another direct off-shoot of the common ancestor, while the purple bacteria (BChl *a* or *b* in RC-2), the filamentous bacteria (BChl *a* in RC-2), and the green sulfur bacteria (BChl *a* in RC-1) are thought to have arisen from a hypothetical common ancestor (BChl *a* in RC-1 and RC-2) derived from the cyanobacteria and chloroplast line.

More detailed speculations about the origin and evolution of photosynthesis may be found in Chapter 15.

8. *Halobacteria*

The *Halobacteriaceae*, commonly referred to as the halobacteria, are a family of extremely halophilic archaebacteria [113]. As in other archaebacteria, their membranes contain ether-linked lipids. The primary lipids present are diphytanyl phospholipids [113]. Their cell walls are also unique in structure and lack muramic acid. There are several species of halobacteria that vary considerably in their physiological characteristics. The halobacteria are unicellular rods or cocci. More recently flat, square and box-shaped cells have been described. Halobacteria are found growing in salterns or natural salt lakes and on the surface of salted fish. They often form dense planktonic blooms and can form massive accumulations on solid substrates. They may be involved in mat communities in hypersaline environments.

Electron transport components	Light-harvesting pigments		Carotenoids
	Chls	Location	(derivatives not specified)
MQ, Cyts *b* and *c*	BChl *a*, BChl *c* or *d*	Plasma membrane chlorosome	β, γ-Carotene
MQ, oxoMQ, Cyts *b* and *c*	BChl *a*, BChl *c*, *d* or *e*	Plasma membrane chlorosome	Chlorobactene, isorenieratene, γ-carotene
Cyt *c* ?	BChl *g*	Plasma membrane	Neurosporene
UQ, Cyts *b* and *c*	BChl *a* or *b*	Intracytoplasmic membrane	Spirilloxanthin, okenone, (β-carotene), spheroidene, rhodopin, lycopene, neurosporene
Q, Cyts *b* and *c*	BChl *a*	Intracytoplasmic membrane ?	?
None associated with photochemistry	None	–	Bacterioruberins (C_{50}); retinal – direct photochemistry (not involved in light-harvesting)

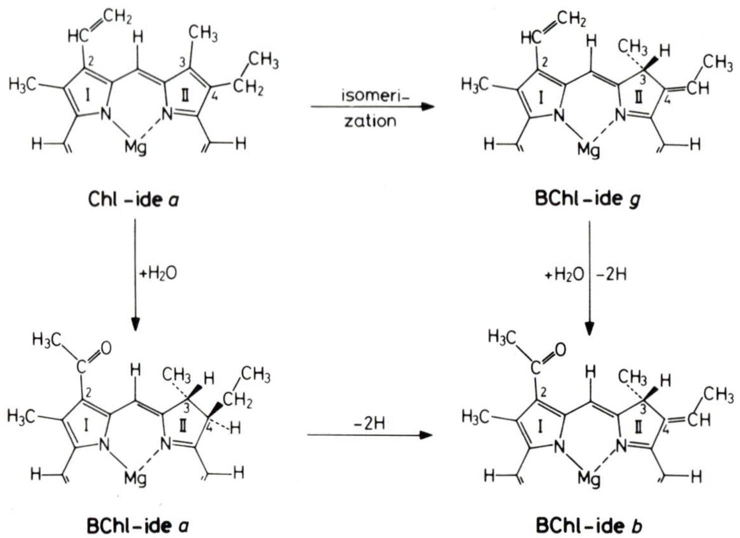

Fig. 5. Comparison of RC chlorophyllides: Chl-ide *a*, BChl-ide *g* [65], BChl-ide *a* and BChl-ide *b*.

Halobacteria grow primarily as aerobic chemoheterotrophs using an electron transfer chain containing cytochromes to generate a proton gradient which drives the synthesis of ATP. Many species of halobacteria also synthesize a membrane-bound pigment-protein complex, bacteriorhodopsin, which contains a retinal chromophore. Retinal is synthesized by the oxygen-requiring cleavage of β-carotene [114]. Bacteriorhodopsin forms crystalline arrays in the membranes of halobacteria grown in the presence of light and low levels of oxygen. These purple membrane patches mediate a light-driven extrusion of protons from the cell which can then drive the synthesis of ATP. These halobacteria are therefore photosynthetic. When grown aerobically in the light, halobacteria have higher growth rates and yields than when grown in the dark [115]. Light cannot sustain growth anaerobically indefinitely, however, because of the oxygen requirement for the synthesis of retinal [114]. The halobacteria are therefore limited in their phototrophic capabilities.

Porphyrin-based photosynthesis has not yet been observed in any archaebacteria, although the capacity for porphyrin biosynthesis is widely distributed in this group [112].

9. *Summary*

Photosynthetic bacteria are found both among the eubacteria and the archaebacteria. Those found among the eubacteria all contain (B)Chl, while those found among the archaebacteria, i.e. the halobacteria, contain the carotenoid retinal, but no (B)Chl.

Photosynthetic eubacteria are classified as filamentous, green sulfur, gram-positive linked, purple, and cyanobacteria. All contain membrane-bound RCs in which (B)Chl serves as the primary electron donor. The RCs may be divided into two main types: RC-1, in which the initial electron acceptor is a (B)Chl molecule and the secondary acceptor is an Fe-S center, and RC-2, in which the initial acceptor is a (B)Ph molecule and the secondary acceptor is a quinone. RC-1 centers are found in green sulfur and gram-positive linked bacteria, while RC-2 centers are found in filamentous bacteria and purple bacteria. Cyanobacteria contain both RC-1 and RC-2 centers in which the chlorophyll is Chl *a*. BChl *a* is found in filamentous, green sulfur and purple bacteria, while BChl *g* is characteristic of the gram-positive line. BChl *b* is found in certain purple bacteria instead of BChl *a*.

All photosynthetic eubacteria contain LHCs for delivering light energy to the RC(s). In some filamentous bacteria and in all green sulfur bacteria the LHCs contain BChl *c* or a related pigment and are found in chlorosomes appressed to the inner surface of the cytoplasmic membrane where the RCs are located. In purple bacteria the LHCs are found in the intracytoplasmic membranes along with the RCs. These LHCs contain BChl *a* or *b* and highly colored carotenoids. In the gram-positive line (*H. chlorum*) the LHCs (BChl *g*) are found only in the cytoplasmic membrane along with the RCs.

Most photosynthetic eubacteria appear to contain cyclic electron transfer pathways driven by the RCs. Electrons from the secondary acceptor of the RC are transferrred first to a quinone pool and then to the secondary donor (Cyt *c*) via a Cyt *b*/*c* complex which stores some of the electron redox energy as potential energy in the form of a transmembrane proton gradient. Evidence for cyclic electron flow in the gram-positive line has not yet been found, but it would be surprising not to find it.

Those photosynthetic eubacteria with RC-2 centers (filamentous and purple bacteria) reduce NAD^+ for CO_2 fixation by reverse electron flow from the quinone pool, whereas the green sulfur bacteria (RC-1 center) reduce ferredoxin and NAD^+ directly from the secondary acceptor (Fe-S center) of the RC. In both cases an external reductant such as H_2S is required. The mechanism of NAD^+ reduction in the gram-positive line has not yet been investigated, but *H. chlorum* is a heterotroph rather than an autotroph, and may not need to fix CO_2.

Many photosynthetic purple bacteria are closely related phylogenetically to non-photosynthetic respiring eubacteria. Some photosynthetic eubacteria are autotrophic (e.g. green and purple sulfur bacteria), while others are mainly heterotrophic (e.g. filamentous bacteria, purple non-sulfur bacteria and *H. chlorum*). All convert light energy into chemical free energy.

References

1 Stackebrandt, E. and Woese, C.R. (1984) Microbiol. Sci. 1, 117–122.

2 Stackebrandt, E. and Woese, C.R. (1981) in Molecular and Cellular Aspects of Microbial Evolution (Carlile, M.J., Collins, J.F. and Moseley, B.E.B., eds), pp. 1–31, Cambridge University Press, Cambridge, U.K.

3 Castenholz, R.W. and Pierson, B.K. (1981) in The Prokaryotes (Starr, M.P., Stolp, H., Trüper, H.G., Balows, A. and Schlegel, H.G., eds.), pp. 290–298, Springer-Verlag, Berlin.
4 Pierson, B.K. and Castenholz, R.W. (1974) Arch. Microbiol. 100, 5–24.
5 Cohen, Y (1984) in Microbial Mats: Stromatolites, (Cohen, Y., Castenholz, R.W. and Halvorson, H.O., eds.), pp. 133–148, Alan R. Liss, New York.
6 Gorlenko, V.M. (1975) Microbiology (Eng. transl. of Mikrobiologiya) 44, 682–684.
7 Stolz, J.F. (1984) in Microbial Mats: Stromatolites, (Cohen, Y., Castenholz, R.W. and Halvorson, H.O., eds.), pp. 23–38, Alan R. Liss, New York.
8 Pivovarova, T.A. and Gorlenko, V.M. (1977) Microbiology (Eng. transl. of Mikrobiologiya) 46, 276–282.
9 Madigan, M.T. and Brock, T.D. (1977) FEMS Microbiol. Lett. 1, 301–304.
10 Holo, H. and Sirevåg, R. (1986) Arch. Microbiol. 145, 173–180.
11 Pierson, B.K. and Castenholz, R.W. (1974) Arch. Microbiol. 100, 283–305.
12 Sprague, S.G., Staehelin, L.A., DiBartolomeis, J. and Fuller R.C. (1981) J. Bacteriol. 147, 1021-1031.
13 Pierson, B.K. (1985) Arch. Microbiol. 143, 260–265.
14 Zannoni, D. (1986) FEBS Lett. 198, 119–124.
15 Pierson, B.K., Giovannoni, S.J. and Castenholz, R.W. (1984) Appl. Environ. Microbiol. 47, 576–584.
16 Pierson, B.K., Giovannoni, S.J., Stahl, D.A. and Castenholz, R.W. (1985) Arch. Microbiol. 142, 164–167.
17 Gorlenko, V.M. and Pivovarova, T.A. (1977) (In Russian with English summary) Izvestiya Akademii Nauk SSSR, Seriya Biologicheskaya 3, 396–409.
18 Dubinina, G.A. and Gorlenko, V.M. (1975) Microbiology (Eng. transl. of Mikrobiologiya) 44, 452–458.
19 Wechsler, T., Suter, F., Fuller, R.C. and Zuber, H. (1985) FEBS Lett. 181, 173–178.
20 Staehelin, L.A., Golecki, J.R., Fuller, R.C. and Drews, G. (1978) Arch. Microbiol. 119, 269–277.
21 Schmidt, K., Maarzahl, M. and Mayer, F. (1980) Arch. Microbiol. 127, 87–97.
22 Bruce, B.D., Fuller, R.C. and Blankenship, R.E. (1982) Proc. Natl. Acad. Sci. USA 79, 6532–6536.
23 Wechsler, T., Brunisholz, R., Suter, F., Fuller, R.C. and Zuber, H. (1985) FEBS Lett. 191, 34–38.
24 Feick, R.G. and Fuller, R.C. (1984) Biochemistry 23, 3693–3700.
25 Pierson, B.K., Thornber, J.P. and Seftor, R.E.B. (1983) Biochim. Biophys. Acta 723, 322–326.
26 Mancino, L.J., Hansen, P.L., Stark, R.E. and Blankenship, R.E. (1985) Biophys. J. 47, 2a.
27 den Blanken, H.J., Vasmel, H., Jongenelis, A.P.J.M., Hoff, A.J. and Amesz, J. (1983) FEBS Lett. 161, 185–189.
28 Kirmaier, C., Holten, D., Feick, R. and Blankenship, R.E. (1983) FEBS Lett. 158, 73–78.
29 Parot, P., Delmas, N., Garcia, D. and Vermeglio, A. (1985) Biochim. Biophys. Acta 809, 137–140.
30 Blankenship, R.E., Feick, R., Bruce, B.D., Kirmaier, C., Holten, D. and Fuller, R.C. (1983) J. Cell. Biochem. 22, 251–261.
31 Vasmel, H. and Amesz, J. (1983) Biochim. Biophys. Acta 724, 118–122.
32 Hale, M.B., Blankenship, R.E. and Fuller, R.C. (1983) Biochim. Biophys. Acta 723, 376–382.
33 Pierson, B.K. and Thornber, J.P. (1983) Proc. Natl. Acad. Sci. USA 80, 80–84.
34 Blankenship, R.E. (1984) Photochem. Photobiol. 40, 801–806.
35 Zannoni, D. and Ingledew, W.J. (1985) FEBS Lett. 193, 93–98.
36 Blankenship, R.E. (1985) Photosynth. Res. 6, 317–333.
37 Blankenship, R.E., Huynh, P., Gabrielson, H. and Mancino, L.J. (1985) Biophys. J. 47, 2a.
38 Wynn, R., Fuller, R., Redlinger, T., Foster, J., Blankenship, R., Shaw, R. and Knaff, D. (1986) Biophys. J. 49, 23a.
39 Oelze, J. and Fuller, R.C. (1983) J. Bacteriol. 155, 90–96.
40 Trüper, H.G. and Pfennig, N. (1981) in The Prokaryotes (Starr, M.P., Stolp, H., Trüper, H.G., Balows, A. and Schlegel, H.G., eds.) pp.299–312, Springer-Verlag, Berlin.
41 Gibson, J., Pfennig, N. and Waterbury, J.B. (1984) Arch. Microbiol. 138, 96–101.
42 Heda, G.D. and Madigan, M.T. (1986) Arch. Microbiol. 143, 330–336.
43 Ormerod, J.G. and Sirevåg, R. (1983) in The Phototrophic Bacteria (Ormerod, J.G., ed.), pp. 100–119, University of California Press, Berkeley.

44 Summons, R.E. and Powell, T.G. (1986) Nature (London) 319, 763–765.
45 Olson, J.M. (1980) Biochim. Biophys. Acta 594, 33–51.
46 Staehelin, L.A., Golecki, J.R. and Drews, G. (1980) Biochim. Biophys. Acta 589, 30–45.
47 Gerola, P.D. and Olson, J.M. (1986) Biochim. Biophys. Acta 848, 69–76.
48 van Dorssen, R.J., Gerola, P.D., Olson, J.M. and Amesz, J. (1986) Biochim. Biophys. Acta 848, 77–82.
49 Wasielewski, M.R., Smith, U.H. and Norris, J.R. (1982) FEBS Lett. 149, 138–140.
50 Hurt, E.C. and Hauska, G. (1984) FEBS Lett. 168, 149–154.
51 Nuijs, A.M., Vasmel, H., Joppe, H.L.P., Duysens, L.N.M. and Amesz, J. (1985) Biochim. Biophys. Acta 807, 24–34.
52 Braumann, T., Vasmel, H., Grimme, L.H. and Amesz, J. (1986) Biochim. Biophys. Acta 848, 83–91.
53 Olson, J.M. and Thornber, J.P. (1979) in (Capaldi, R., ed.) Membrane Proteins in Energy Transduction, pp. 279–340, Marcel Dekker, New York.
54 Knaff, D.B., Olson, J.M. and Prince, R.C. (1979) FEBS Lett. 98, 285–289.
55 Takamiya, K.-I. (1970) Biochim. Biophys. Acta. 234, 390–398.
56 Prince, R.C. and Olson, J.M. (1976) Biochim. Biophys. Acta 423, 357–362.
57 Swarthoff, T., van der Veek-Horsley, K.M. and Amesz, J. (1981) Biochim. Biophys. Acta 635, 1–12.
58 Burns, D.D. and Midgley, M. (1976) Eur. J. Biochem. 67, 323–333.
59 Sissons, A. and Midgley, M. (1981) J. Gen. Microbiol. 122, 211–216.
60 Olson, J.M. and Gromet-Elhanan, Z. (1979) Biophys. J. 25, 54a.
61 Shill, D.A. and Wood, P.M. (1985) Arch. Microbiol. 143, 82–87.
62 Hansen, T.A. (1983) in The Phototrophic Bacteria (Ormerod, J.G., ed.), pp. 76–99, Blackwell, Oxford.
63 Gest, H. and Favinger, J.L. (1983) Arch. Microbiol. 136, 11–16.
64 Woese, C.R., Debrunner-Vossbrinck, B.A., Oyaizu, H., Stackebrandt, E. and Ludwig, W. (1985) Science 229, 762–765.
65 Brockmann, H., jr. and Lipinski, A. (1983) Arch. Microbiol. 136, 17–19.
66 Fuller, R.C., Sprague, S.G., Gest, H. and Blankenship, R.E. (1985) FEBS Lett. 182, 345–349.
67 van Dorssen, R.J., Vasmel, H. and Amesz, J. (1985) Biochim. Biophys. Acta 809, 199–203.
68 Prince, R.G., Gest, H. and Blankenship, R.E. (1985) Biochim. Biophys. Acta 810, 377–384.
69 Nuijs, A.M., van Dorssen, R.J., Duysens, L.N.M. and Amesz, J. (1985) Proc. Natl. Acad. Sci. USA 82, 6865–6868.
70 Brok, M., Vasmel, H., Horikx, J.T.G. and Hoff, A.J. (1986) FEBS Lett. 194, 322–326.
71 Imhoff, J.F., Trüper, H.G. and Pfennig, N. (1984) Int. J. Syst. Bacteriol. 34, 340–343.
72 Imhoff, J.F. (1984) Int. J. Syst. Bacteriol. 34, 338–339.
73 Pfennig, N. and Trüper, H.G. (1974) in Bergey's Manual of Determinative Bacteriology, 8th edition (Buchanan, R.E. and Gibbons, N.E., eds.) pp. 24–64, Williams and Wilkins, Baltimore.
74 Woese, C.R., Stackebrandt, E., Weisburg, W.G., Paster, B.J., Madigan, M.T., Fowler, V.J., Hahn, C.M., Blanz, P., Gupta, R., Nealson, K.H. and Fox, G.E. (1984) Syst. Appl. Microbiol. 5, 315–326.
75 Gibson, J., Stackebrandt, E., Zablen, L.B., Gupta, R. and Woese, C.R. (1979) Curr. Microbiol. 3, 59–64.
76 Woese, C.R., Weisburg, W.G., Paster, B.J., Hahn, C.M., Tanner, R.S., Krieg, N.R., Koops, H.-P. Harms, H. and Stackebrandt, E. (1984) Syst. Appl. Microbiol. 5, 327–336.
77 Woese, C.R., Weisburg, W.G., Hahn, C.M., Paster, B.J., Zablen, L.B., Lewis, B.J., Macke, T.J., Ludwig, W. and Stackebrandt, E. (1985) Syst. Appl. Microbiol. 6, 25–33.
78 Fowler, V.J., Pfennig, N., Schubert, W. and Stackebrandt, E. (1984) Arch. Microbiol. 139, 382–387.
79 Wynn, R.M., Kämpf, C., Gaul, D.F., Choi, W.-K., Shaw, R.W. and Knaff, D.B. (1985) Biochim. Biophys. Acta 808, 85–93.
80 Yoch, D.C. (1978) in The Photosynthetic Bacteria (Clayton, R.K. and Sistrom, W.R., eds.) pp. 657–676, Plenum, New York.
81 Thornber, J.P., Gogdell, R.J., Pierson, K.B. and Seftor, R.E.B. (1983) J. Cell. Biochem. 23, 159–169.
82 Brunisholz, R.A., Jay, F., Suter, F. and Zuber, H. (1985) Biol. Chem. Hoppe-Seyler, 366, 87–98.
83 Steiner, R. and Scheer, H. (1985) Biochim. Biophys. Acta 807, 278–284.

84 Deisenhofer, J., Epp, O., Miki, K., Huber, R. and Michel, H. (1984) J. Mol. Biol. 180, 385–398.

85 Deisenhofer, J., Epp, O., Miki, K., Huber, R. and Michel, H. (1985) Nature (London) 318, 618–624.

86 Williams, J.C., Steiner, L.A., Feher, G. and Simon, M.I. (1984) Proc. Natl. Acad. Sci. USA 81, 7303–7307.

87 Youvan, D.C., Bylina, E.J., Alberti, M., Begusch, H. and Hearst, J.E. (1984) Cell 37, 949–957.

88 Olson, J.M. (1981) BioSystems 14, 89–94.

89 de Vitry, C. and Diner, B.A. (1984) FEBS Lett. 167, 327–331.

90 Baccarini-Melandri, A., Casadio, R. and Melandri, B.A. (1981) Curr. Topics Bioenerg. 12, 197–258.

91 Cramer, W.A. and Crofts, A.R. (1982) in Photosynthesis: Energy Conversion by Plants and Bacteria (Govindjee, ed.) Vol. I, pp. 387–467, Academic Press, New York.

92 Hauska, G., Hurt, E., Gabellini, N. and Lokau, W. (1983) Biochim. Biophys. Acta 726, 97–133.

93 Coremans, J.M.C.C., van der Wal, H.N., van Grondelle, R., Amesz, J. and Knaff, D.B. (1985) Biochim. Biophys. Acta 807, 134–142.

94 Thornber, J.P. and Olson, J.M. (1971) Photochem. Photobiol. 14, 329–341.

95 Knaff, D.B. (1978) in The Photosynthetic Bacteria (Clayton, R.K. and Sistrom, W.R., eds.), pp. 629–640, Plenum, New York.

96 Dutton, P.L. and Prince, R.C. (1978) in The Photosynthetic Bacteria (Clayton, R.K. and Sistrom, W.R., eds.), pp. 525–570, Plenum, New York.

97 Jones, C.W. (1982) Bacterial Respiration and Photosynthesis, Nelson, Walton-on-Thames, U.K.

98 Trüper, H.G. and Pfennig, N. (1978) in The Photosynthetic Bacteria (Clayton, R.K. and Sistrom, W.R., eds.), pp. 19–27, Plenum, New York.

99 Merrick, J.M. (1978) in The Photosynthetic Bacteria, (Clayton, R.K. and Sistrom, W.R., eds.), pp. 199–219, Plenum, New York.

100 Sojka, G.A. (1978) in The Photosynthetic Bacteria (Clayton, R.K. and Sistrom, W.R., eds.), pp. 707–718, Plenum, New York.

101 Hüdig, H. and Drews, G. (1985) J. Bacteriol. 162, 897–901.

102 Shiba, T. (1984) J. Gen. Appl. Microbiol. 30, 239–244.

103 Shiba, T. and Simidu, U. (1982) Int. J. Syst. Bacteriol. 32, 211–217.

104 Sato, K., Kazuhiko, H. and Shoichi, S. (1985) Agric. Biol. Chem. 49, 1–5.

105 Takamiya, K.-I. (1985) Arch. Microbiol. 143, 15–19.

106 Takamiya, K.-I. and Okamura, K. (1984) Arch. Microbiol. 140, 21–26.

107 Harashima, K., Hayasaki, J., Ikari, T. and Shiba, T. (1980) Plant Cell Physiol. 21, 1283–1294.

108 Shimada, K., Hayashi, H. and Tasumi, M. (1985) Arch. Microbiol. 143, 244–247.

109 Harashima, K., Nakagawa, M. and Murata, N. (1982) Plant Cell Physiol. 23, 185–193.

110 Okamura, K., Takamiya, K.-I. and Nishimura, M. (1985) Arch. Microbiol. 142, 12–17.

111 Gibson, J., Ludwig, W., Stackebrandt, E. and Woese, C.R. (1985) Syst. Appl. Microbiol. 6, 152–156.

112 Olson, J.M. and Pierson, B.K. (1987) Int. Rev. Cytol. 108, in the press.

113 Kushner, D.J. (1985) in The Bacteria (Gunsalus, I.C., ed.) Vol. VIII, pp. 171–214, Academic Press, New York.

114 Hartmann, R., Sickinger, H.-D. and Oesterhelt, D. (1980) Proc. Natl. Acad. Sci. USA 77, 3821–3825.

115 Rodriguez-Valera, F., Nieto, J.J. and Ruriz-Berraquero, F. (1983) Appl. Environ. Microbiol. 45, 868–871.

J. Amesz (ed.) *Photosynthesis*

CHAPTER 3

The bacterial reaction center

WILLIAM W. PARSON

Department of Biochemistry, University of Washington, Seattle, WA, 98195 U.S.A.

1. *Introduction*

The idea that the initial photochemical reaction in bacterial photosynthesis is the oxidation of a bacteriochlorophyll (BChl) complex in a special site, or 'reaction center', developed from pioneering studies by L.N.M. Duysens and R.K. Clayton. Duysens [1,2] discovered that illuminating cell suspensions of *Rhodospirillum rubrum* or other species of purple photosynthetic bacteria caused optical absorption changes indicative of the oxidation of *c*-type cytochromes. In addition, there were small absorption changes in the region of 870 nm that suggested the oxidation of BChl. Clayton [3] and Kuntz et al. [4] showed that the BChl that underwent photooxidation (P-870, or more simply, P) was somehow distinct from the much more abundant BChl making up the light-harvesting antenna system in the chromatophore membrane. Much of the antenna BChl could be destroyed by chemical oxidants, or by exposure to strong light in the presence of O_2, without causing any permanent damage to P-870. When chromatophores of *Chromatium vinosum* were excited with a short flash of light, P-870 was oxidized with a high quantum yield in less than 1 μs, and returned to the reduced state with a time constant of about 2 μs [5]. Cyt *c*-556 became oxidized as the BChl complex regained an electron.

EPR and ENDOR spectroscopic studies [6–12] indicate that the oxidized BChl complex (P^+) is a π-radical cation in which the spin of the unpaired electron is delocalized over the π electron systems of two BChl molecules. Chemical redox titrations of P-870 follow a one-electron Nernst curve. The midpoint potential at pH 7 ($E_{m,7}$) ranges from +0.36 to +0.50 V depending on the species of bacteria, but is typically about +0.45 V [13–17].

Attempts to identify the 'primary' electron acceptor that takes an electron from P-870 focused first on nonheme iron. Illumination of chromatophores at cryogenic temperatures gave rise to a broad EPR signal at g = 1.8, in addition to the sharp signal at g = 2.0025 characteristic of P^+ [18–21]. However, the photooxidation of P-870 still occurred in preparations which were depleted of Fe, and in these the reduced electron acceptor gave a sharp EPR signal consistent with an organic semiquinone [22]. Purified reaction centers from *Rhodobacter sphaeroides*** were

** *Rhodobacter sphaeroides* and *Rhodobacter capsulatus* were formerly classified in the genus *Rhodopseudomonas*. They have recently been reclassified [23]. *Rhodopseudomonas gelatinosa* is now classified as *Rhodocyclus gelatinosus*.

found to contain two molecules of ubiquinone, one of which appeared to be essential for photochemical activity [24,25]. These observations suggest that one of the quinones (Q_A) serves as the initial electron acceptor, and that the EPR spectrum of the semiquinone is normally broadened by magnetic interactions with a nearby nonheme iron atom. The other quinone (Q_B) acts as a secondary electron acceptor that removes an electron from Q_A [26–32].

Redox titrations of Q_A in *Rb. sphaeroides* reaction centers give an $E_{m,7}$ value of about −0.05 V [33]. Subtracting the $E_{m,7}$ of Q_A from that of P (+0.45 V) gives a standard free energy change of about 11.8 kcal/mole, or 0.50 eV, for the formation of $P^+Q_A^-$ from P Q_A. As a point of reference, the lowest excited singlet state of P (P*) in *Rb. sphaeroides* reaction centers is about 1.38 eV above the ground state. The charge-separation reaction thus captures about 36% of the free energy of P* (Fig. 1). Fluorescence measurements of the amount of P* that is in equilibrium with $P^+Q_A^-$ lead to a similar conclusion [34].

In chromatophores, the E_m of Q_A decreases by 59 mV/pH unit as the pH is raised, up to an apparent pK_A that is between 7.8 and 9.8, depending on the species [16,30,35,36]. The pK_A probably reflects the binding of a proton to a group other than the quinone itself, because the absorption spectrum and EPR spectrum of Q_A^- match those expected for an anionic semiquinone [31,37–40]. The ENDOR spectrum of Q_A^- suggests that the quinone is hydrogen-bonded to a histidine residue of the protein [41]. The E_m value of about −0.18 V measured above the pK_A may be the most relevant value when Q_A is photoreduced, because Q_A^- probably transfers an electron to Q_B before proton uptake occurs. In isolated reaction centers of *Rb. sphaeroides*, the $E_{m,7}$ of Q_B is about 0.07 V more positive than that of Q_A [29,34,42–44]. The difference between the two E_m values appears

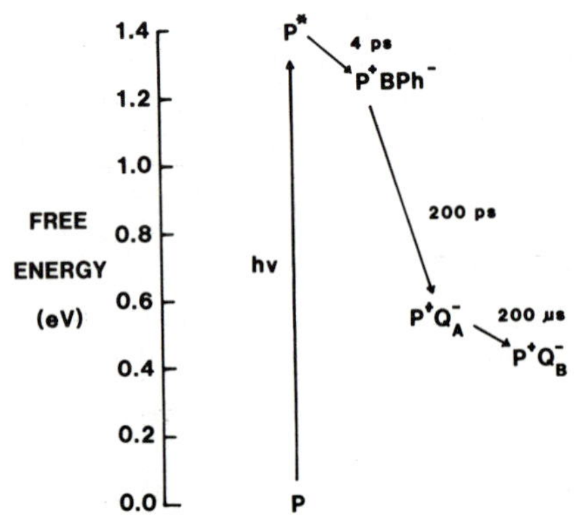

Fig. 1. Kinetics and standard free energy changes of electron transfer steps in reaction centers isolated from *Rb. sphaeroides*. In the chromatophore membrane, a *c*-type cytochrome (Cyt c_2) normally reduces P^+ before an electron moves from Q_A^- to Q_B. The cytochrome oxidation has a time constant of about 20 μs in *Rb. sphaeroides*, and 0.5 to 2 μs in reaction centers of *Rp. viridis* and *Ch. vinosum*, which have bound cytochromes. When the reaction center is excited a second time, Q_B^- is reduced to Q_BH_2.

to be larger in intact chromatophores (about 0.12 V at pH 8) [45].

In *Ch. vinosum* and *Rhodopseudomonas viridis*, Q_B is ubiquinone, but Q_A is a menaquinone [25,32]. In *Chloroflexus aurantiacus*, both quinones probably are menaquinones [46,47]. The $E_{m,7}$ of Q_A is about 100 mV more negative in these species than it is in species that contain ubiquinone.

The quantum yield of charge separation in the reaction center is remarkably high. If *Rb. sphaeroides* reaction centers are excited in the 870-nm absorption band of P, $P^+Q_A^-$ is formed with a quantum yield of 1.02 ± 0.04 [48]. In accord with the high yield of photochemical products, the yield of fluorescence from P* is only about 1×10^{-4} [49,50].

An electron moves from Q_A^- to Q_B in about 200 μs [28–31,51]. Excitation of the reaction center by a second photon sends another electron from P* to Q_A, and then on to Q_B with similar kinetics. The fully reduced Q_B now probably picks up two protons from the solvent, dissociates from the reaction center as the quinol (Q_BH_2), and is replaced by a fresh molecule of ubiquinone. Electrons from Q_BH_2 return to P^+ via a Cyt bc_1 complex and a high-potential, *c*-type cytochrome. This cyclic electron flow drives proton translocation across the chromatophore membrane, and is coupled to the formation of ATP.

Indications that there might be another electron acceptor prior to Q_A emerged from studies in which the photoreduction of Q_A was blocked by reducing the quinone chemically [51–54]. Excitation of purified reaction centers with a short flash under these conditions resulted in the formation of a transient state (P^F or P^+I^-), in which an electron appeared to have moved from P to a bacteriopheophytin (BPh) or BChl. (BPh differs from BChl only in having two hydrogens in the center, in place of Mg.) Model studies on the π-radical anions of BChl and BPh in solution indicate that BPh is thermodynamically the easier molecule to reduce [9,55,56], but the optical absorption changes associated with P^+I^- suggest that the electron acceptor (I) might be a complex of BChl and BPh [57].

The lifetime of P^+I^- in reaction centers that have Q_A reduced before the excitation is about 12 ns. The radical-pair decays partly by back reactions which return the reaction center to its original state (P I), and partly by the formation of an excited triplet state of P [53,58–66]. P^+I^- is formed initially in a singlet state, in which the spins of the unpaired electrons on P^+ and I^- are antiparallel. With time, the relationship between the two spins changes as a result of interactions with nuclear spins on the two molecules, or with the electronic spins on Q_A^- and the nonheme iron atom. Back reactions that occur when the radical-pair is predominantly triplet in character lead to the triplet state of P.

The initial electron acceptor can be made to accumulate in the reduced state (I^-) if reaction centers which have bound (or added) cytochromes are illuminated continuously after the reduction of Q_A [56,67–69]. Each time the radical-pair state P^+I^- is formed, P^+ has a brief opportunity to oxidize the cytochrome instead of recovering an electron from I^-. The probability of electron transfer from the cytochrome is low, because the back reactions between P^+ and I^- are much faster than the cytochrome oxidation. After many turnovers, however, essentially all of the reaction centers may be left with I reduced, particularly if the return of electrons to the cytochrome is prevented by lowering the temperature. Again, the optical absorption changes that accompany the reduction of I suggest that the electron acceptor consists of a complex of BPh and BChl. However, the EPR and ENDOR spectra favor the view that, at least at low temperatures, the odd electron is localized mainly on a single molecule, which seems most likely to be the BPh [9,56].

The radical-pair state P^+I^- is also formed if unreduced reaction centers are excited with a short flash, but it then decays with a time constant of about 200 ps [54,70–74]. The rapid decay of the transient state presumably reflects electron transfer from I^- to Q_A (Fig. 1), because it is prevented if the quinone is already reduced or is extracted from the reaction centers. The transient absorption changes suggest that I^- is a BPh^- π-radical anion, which interacts with a nearby BChl [75–77]. The absorbance changes in a band associated with the BChl decay with somewhat different kinetics from those in bands associated with the BPh or BPh^-, perhaps because they reflect nuclear motions in the electron carriers or the surrounding protein [75]. The possible role of the BChl in the initial transfer of an electron from P* to the BPh will be discussed below.

The free energy gap between P* and P^+I^- can be calculated from measurements of the fluorescence that occurs during the lifetime of the radical-pair in reaction centers that have electron transfer to Q_A blocked by the reduction or extraction of the quinone [65,78–81]. The fluorescence emitted by P* at any given time is a measure of the amount of the excited singlet state that is in equilibrium with the radical-pair. By this measure, the earliest form of P^+I^- that can be resolved lies about 0.17 eV below P* in free energy, both in chromatophores and in isolated reaction centers (Fig. 1). The amplitude of the fluorescence decays in several steps, possibly because of nuclear relaxations in the radical-pair.

Like the purple bacterial species mentioned above, *Prostheocochloris aestuarii* and other members of the Chlorobiaceae subgroup of the green photosynthetic bacteria appear to use a BChl dimer as an initial electron donor, but they evidently use BChl *c* istead of BPh as an initial electron acceptor [82–85]. The Chlorobiaceae also differ in using iron-sulfur proteins as the next electron carriers, instead of quinones. Their electron acceptor system appears to resemble that found in PS I of plants and cyanobacteria more than it does that of other groups of photosynthetic bacteria.

2. *Purification and crystallization of reaction centers*

Reaction centers of the purple, nonsulfur photosynthetic bacteria (*Rhodospirillaceae*) have proved easier to isolate than those of other photosynthetic organisms. In a typical procedure, chromatophore membranes are first collected from broken cells by differential centrifugation. Mild disruption of the membranes with lauryldimethylamine-*N*-oxide or another non-ionic detergent solublizes the reaction centers, leaving most of the antenna BChl in a particulate fraction that is removed by centrifugation. Sucrose-gradient centrifugation or fractionation with ammonium sulfate is sometimes used at this point. The reaction centers are then purified by column chromatography on a cationic resin such as DEAE-Sephacryl in the continuing presence of a low concentration of detergent, and are concentrated by pressure dialysis. Purified reaction centers generally retain their spectroscopic properties and photochemical activity for months if they are stored frozen or at 4°C.

Among the *Rhodospirillaceae* whose reaction centers have been purified in this way are *Rhodobacter sphaeroides* (formerly called *Rhodopseudomonas sphaeroides*) strains 2.4.1 (a wild-type strain), GA (a mutant with altered carotenoids) and R-26 (a carotenoidless mutant), *Rhodobacter capsulatus, Rhodospirillum rubrum, Rhodocyclus gelatinosus* (formerly called *Rhodopseudomonas gelatinosa*), and *Rhodopseudomonas viridis* [16,21,29,86–92]. A similar preparation has been obtained from the green, filamentous thermophile *Chloroflexus aurantiacus*, a member of the Chloroflexaceae subgroup of the green photosynthetic bacteria [46,93–95]. The purple sulfur (*Chromatiaceae*) species *Chromatium vinosum* has yielded somewhat less satisfactory preparations, from which persistent antenna pigments can be removed by extraction with aqueous acetone [67,68,96]. Attempts to isolate reaction centers from the Chlorobiaceae have not been successful, although some purification has been achieved from *Prosthecochloris aestuarii* [97].

The first reaction centers to be crystallized were those of *Rhodopseudomonas viridis* [92]. The key to this dramatic breakthrough, which led quickly to the first high-resolution structural model for a hydrophobic, integral membrane protein, was to include small amphiphilic molecules such as heptane-1,2,3-triol in the crystallization solution. The rationale was that small amphiphiles would aid in filling the spaces around hydrophobic regions of the protein, much as water fills the extra space in crystals of water-soluble proteins. The same technique, and another method based on phase separation in solutions containing octyl glucoside and polyethylene glycol, subsequently yielded crystals of reaction centers from *Rhodobacter sphaeroides* [98–100]. A variety of crystal forms has been obtained, depending on the conditions of the crystallization. The current structural model of the *Rp. viridis* reaction center [101,102,172] is based on X-ray diffraction data to 3.0 Å and is still undergoing refinement; the crystals diffract to a resolution of at least 2.5 Å [92].

3. *Protein structure*

Reaction centers isolated from *Rb. sphaeroides*, *Rb. capsulatus* and *Rs. rubrum* contain three polypeptides in 1:1:1 stoichiometry, with a total molecular weight of about 10^5. The polypeptides are generally called 'L, M and H', signifying their apparent relative molecular weights ('light, middle and heavy') as judged from SDS-polyacrylamide gel electrophoresis. However, subunits L and M are more hydrophobic than H and their electrophoretic mobilities are misleading: H is actually the smallest subunit, and M the largest. The molecular weights of the *Rb. capsulatus* subunits are 28 534 (H), 31 565 (L) and 34 440 (M) [103]. Reaction centers obtained from *Cf. aurantiacus* have only two subunits, which are probably homologous to L and M [46]. The H subunit can be removed from *Rb. sphaeroides* reaction centers without much effect on photochemical electron transfer between P and Q_A [21,104].

In addition to the L, M and H subunits, reaction centers isolated from *Rp. vir-*

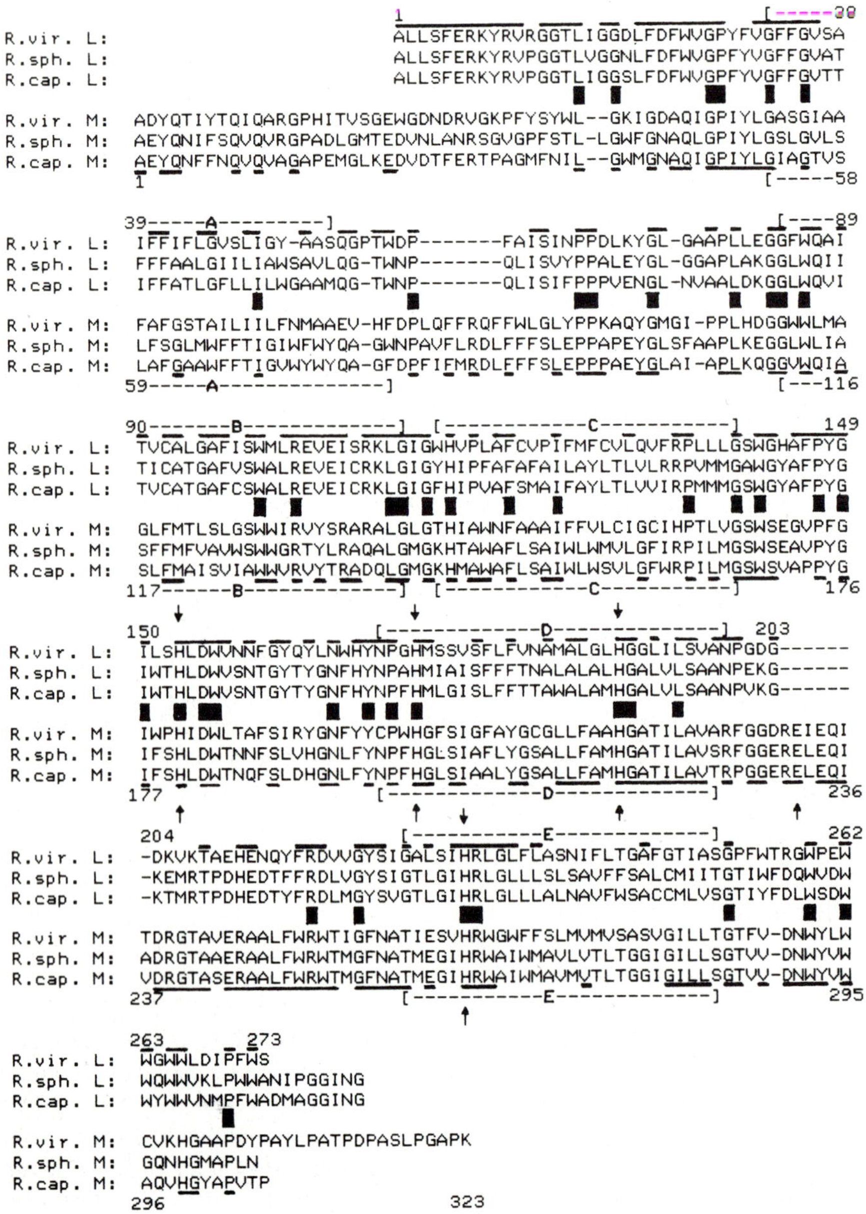

idis contain a bound cytochrome that is similar to the M subunit in size and has four *c*-type hemes [101,102]. Those of *Ch. vinosum* have a similar four-heme cytochrome [14,105,106]. Of the four heme groups in these cytochromes, two have $E_{m,7}$ values of about +0.30 V, and two have $E_{m,7}$ values of about +0.01 V. All four heme groups are capable of reducing P^+. Reaction centers isolated from *Rb. sphaeroides* or *Rb. capsulatus* do not have tightly bound cytochromes. In chromatophores of the *Rhodobacter* species, P^+ is reduced by the soluble Cyt c_2, which has an $E_{m,7}$ of about +0.3 V [107].

The amino acid sequences of the L and M subunits from *Rb. capsulatus* [103], *Rb. sphaeroides* [108,109] and *Rp. viridis* [110] are highly conserved from species to species (Fig. 2). The L and M sequences are also homologous to each other. Both proteins are strongly hydrophobic: about 70% of their amino acids are nonpolar. In each subunit, there are five regions that could form transmembrane helices, because they contain stretches of 19 or more hydrophobic amino acids. The H subunit has one such region [103,111].

The crystal structure of the *Rp. viridis* reaction centers [102] bears out the main structural predictions based on the amino acid sequences (Fig. 3). Subunits L and M have homologous secondary and tertiary structures. They both contain five helices that are more or less parallel and are likely to traverse the chromatophore membrane, in addition to several shorter helices that run approximately parallel to the plane of the membrane. The putative transmembrane helices are labeled A, B, C, D and E in Figs. 2 and 3. In the intact reaction center, subunits L and M pack together side-by-side, with helices D and E of both subunits cooperating to form the iron-binding site (see below). An axis of 2-fold rotational pseudosymmetry runs through the L-M complex in a direction perpendicular to the plane of the membrane. Rotation of the M subunit by 180° about this axis superimposes approximately 2/3 of its C^{α} carbon atoms on the corresponding atoms of the L subunit.

The conclusion that helices A–D traverse the chromatophore membrane is based primarily on the lengths of the helical regions and on the hydrophobic nature of their amino acid residues, but is in accord with the accessibility of the *Rb. sphaeroides* M and H subunits to labeling from either side of the membrane [112,113].

Fig. 2. Amino acid sequences of the L and M subunits from *Rb. sphaeroides* strain R-26 [108,109], *Rb. capsulatus* [103] and *Rp. viridis* [110]. The sequences are aligned as shown by Michel et al. [110]. The numbering of the amino acids starting at the amino-terminal end of the *Rp. viridis* L subunit is given above the sequences, and that of the *Rp. viridis* M subunit below. Residues that are identical in all six sequences are marked with black boxes between the L and M subunits. Residues that are conserved in the L subunits are indicated with bars above the *Rp. viridis* L sequence; those conserved in the M subunits are indicated similarly below the *Rp. capsulatus* M sequence. The transmembrane helical regions identified in the *Rp. viridis* crystal structure [102] are indicated with bracketed dashed lines and the letters A–E above and below the sequences. The helical regions predicted on the basis of hydropathy plots are similar, but terminate at somewhat different positions [103,108–110]. Arrows mark the histidines that ligate the two BChls of P (His L173, His M200) and the other two BChls (His L153, His M180), and the ligands of the nonheme Fe atom (His L190, His L230, His M217, His M264 and Glu M232).

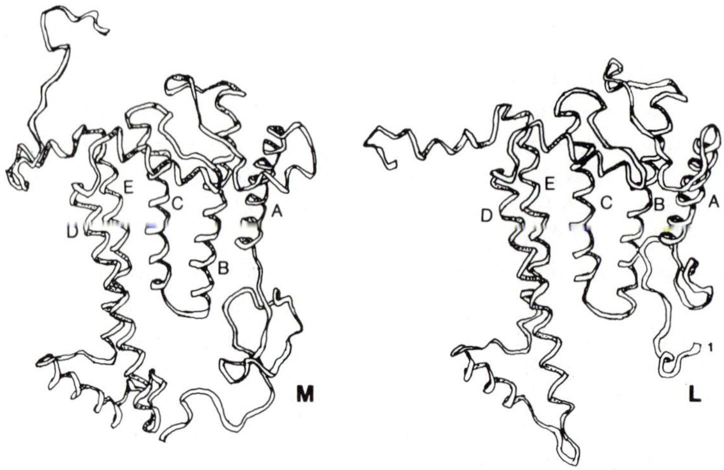

Fig. 3. Ribbon drawings of the polypeptide chains in the M and L subunits of the *Rp. viridis* reaction center, redrawn from Deisenhofer et al. [102]. The drawings are oriented so that the normal to the chromatophore membrane is approximately vertical, with the periplasmic side of the membrane at the top and the cytoplasmic side at the bottom. The amino-terminal ends of the chains are on the cytoplasmic side of the membrane; that of the L subunit is labeled 1. The five transmembrane helices are labeled A–E. In each subunit, the histidine residue that ligates one of the BChls of P is located near the top of helix D, on the periplasmic side of the hydrophobic region. The L and M subunits are closely appressed in the reaction center complex, so that the two BChls of P overlap (Fig. 4). The histidine ligands of the nonheme Fe are located toward the cytoplasmic ends of helices D and E in each subunit; the glutamyl ligand in the M subunit is in the connecting region between D and E.

The CD and IR spectra of oriented samples also agree with the view that the reaction center contains considerable α-helical structure, and that the helices are oriented (on the average) approximately normal to the chromatophore membrane [114]. The observation that the amino-terminal end of the L subunit can be labeled from the cytoplasmic side of the membrane [115] identifies the bottom of Fig. 3 with this side of the membrane, and the top of the figure with the periplasmic side.

The other two subunits of the *Rp. viridis* reaction center are more globular in shape [102]. The amino-terminal end of the H subunit has a hydrophobic (and presumably transmembrane) helix that runs parallel to the contact region of helices D and E of subunit M. Most of the rest of H forms a large globular domain at the cytoplasmic end of the L-M complex. The cytochrome subunit sits on the relatively flat surface at the periplasmic end of the L-M complex, in agreement with the observation that the cytochromes react with P^+ from this side of the membrane in chromatophores or whole cells. The cytochrome also has an internal axis of 2-fold rotational pseudosymmetry, which includes about 1/3 of its amino acid residues. Two of the four hemes lie on one side of this axis, and two on the other.

4. BChl, BPh and other prosthetic groups

Reaction centers isolated from the *Rhodospirillaceae* contain four molecules of BChl, two molecules of BPh, one or two quinones (depending on the isolation procedure), and one atom of nonheme Fe [21, 116]. As mentioned above, the quinones can be either ubiquinone or menaquinone, depending on the species. The Fe can be replaced by Mn, Zn or other metals with only minor effects on photochemical activity [42,117,118]. In reaction centers from *Rp. viridis* the BChl and

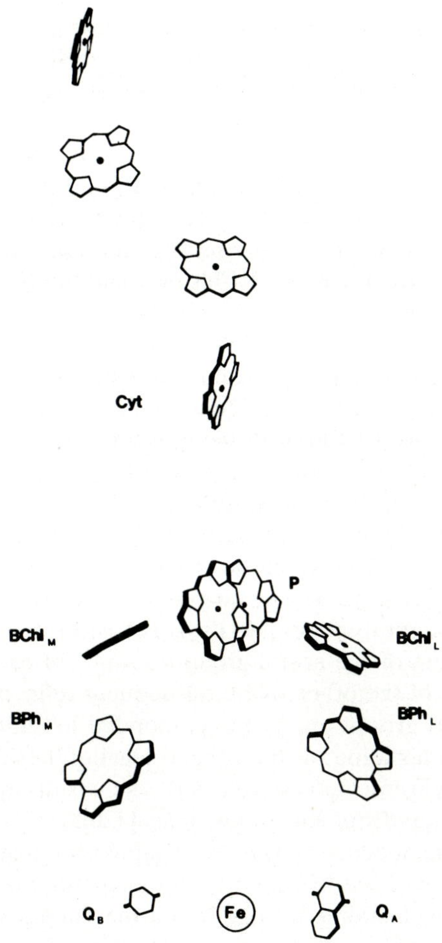

Fig. 4. Arrangement of the prosthetic groups in the *Rp. viridis*, reaction center, redrawn from Ref. 101. Q_B is shown at the site identified by Deisenhofer et al. [102], but the orientation of the quinone in this site is drawn arbitrarily; the exact orientation of Q_B in the crystal structure has not been described. The four hemes at the top of the figure are in the cytochrome subunit; the other components are in the L-M complex. As in Fig. 3, the normal to the chromatophore membrane is approximately vertical and the periplasmic side of the complex is at the top.

BPh are BChl *b* and BPh *b*; in most of the other species that have been characterized, they are BChl *a* and BPh *a*. (BChl *b* differs from BChl *a* in having a vinyl group on ring II in place of an ethyl group. *Thiocapsa pfennigii*, another bacterial species that contains BChl *b*, resembles *Rp.viridis* in its photochemical activities [119,171].) Reaction centers isolated from *Cf. aurantiacus* are unusual in having three molecules of BChl *a* and three of BPh *a*, instead of four BChls and two BPhs [46,93].

Fig. 4 shows the arrangement of the pigments and the iron, as seen in the crystal structure of *Rps. viridis* reaction centers [101,102]. The BChls, BPhs, quinones and nonheme Fe all reside in the central, hydrophobic domain of the L-M polypeptide complex. Near the periplasmic edge of the hydrophobic region are two BChl molecules that are particularly close together. Ring I of each of these BChls overlaps ring I of the other. The planes of the two molecules are approximately parallel (the angle between the normals is about 15°), and are about 3.2 Å apart where the molecules overlap; the molecular centers are about 7 Å apart. This pair of BChls can be identified as the electron-donating dimer (P) on the basis of the optical absorption spectrum of the reaction center (see below) and the EPR and ENDOR properties of P^+ (see above). The rotational symmetry axis that relates the L and M polypeptides passes between the two BChls of P and through the Fe atom, which is located near the cytoplasmic edge of the hydrophobic region. A 180° rotation about this axis interchanges the two BChls. The other two BChls, the two BPhs, and the two quinones also lie in an approximately symmetric arrangement on either side of the same axis. The center of each of the additional BChls is about 11 Å from the center of the nearest BChl of P; the shortest distance edge-to-edge is about 4 Å. The BPh on each side is about 10 Å (center-to-center) from the neighboring BChl, and about 12 Å from the nearest quinone.

The Mg atom in each of the four BChl molecules is attached to a histidine residue of either the L or the M polypeptide [102]. The pair of BChls that make up P are bound to His L173 and M200; the neighboring BChls to His L153 and M180 (Fig. 2). The initial structural model [101] was not accurate enough to determine whether the oxygen atom of the acetyl group on ring I of each of the BChls of P was attached to the Mg of the other BChl. Subsequent refinement of the structure indicated that the acetyl groups are hydrogen-bonded to amino acid residues (tyrosine on one side and histidine on the other), leaving the Mg atoms pentacoordinate [172]. This conclusion agrees well with the results of resonance Raman studies on reaction centers from *Rb. sphaeroides* [120].

The nonheme Fe atom appears to have five ligands: two histidine residues of the L subunit, two histidines of the M subunit, and a glutamyl residue of M (Fig. 2). The coordination to four histidine nitrogens and the finding that the Fe is not attached directly to either of the quinones are in accord with measurements of the EXAFS spectrum of the Fe [121,122].

Q_A, which is menaquinone-9 in *Rp. viridis*, is located near the BPh that is attached to the L subunit (BPh_L, Fig. 4), but the quinone itself is surrounded mainly by amino acid residues of subunit M. Near the headgroup of the quinone are His M217, which is one of the ligands of the Fe, Trp M250, and a peptide nitrogen. The second quinone, Q_B, tends to dissociate from reaction centers during purification, and it was not seen in the original crystal structure. By soaking the crystals with ubiquinone or with electron-transfer inhibitors that are known to displace Q_B from the reaction center, Deisenhofer et al. [102] showed that Q_B binds at a site which is related to the Q_A site by a 180° rotation about the pseudosymmetry axis (Fig. 4). The proximity of the Fe to both of the quinones is consistent with the broadening of the EPR spectra of both Q_A^- and Q_B^-. The binding pocket for Q_B differs from that for Q_A in being exposed to subunit H. As yet, the structural model does not explain why Q_A undergoes reduction only to the level of the semiquinone (Q_A^-), whereas Q_B can accept two electrons.

The center-to-center distance from either of the BChls of P to the nearest heme in the cytochrome subunit is about 21 Å. A tyrosine residue of the protein sits squarely in the path from the heme to P [102]. Because the complete amino acid sequence of the cytochrome subunit has not yet been fitted to the crystallographic map of the reaction center, it is not clear which two of the four hemes are the low-potential hemes, and which two the high-potential, but information on this point should be available shortly.

Reaction centers from all of the wild-type bacterial strains that have been examined contain a carotenoid [123–126]. The carotenoid serves as an antenna pigment in the reaction center, but it also plays a protective role by quenching the excited triplet state of P [53,127]. This prevents the triplet BChl complex from reacting with O_2 to generate singlet O_2. The carotenoid could not be identified in the initial electron-density map of the *Rp. viridis* reaction center [101,102].

5. *Spectroscopic properties and the distinction between BPh_L and BPh_M*

The optical absorption spectrum of reaction centers isolated from the carotenoidless strain R-26 of *Rb. sphaeroides* has major bands near 530, 545, 600, 760, 800 and 880 nm (Fig. 5). There also is a set of strong, overlapping absorption bands

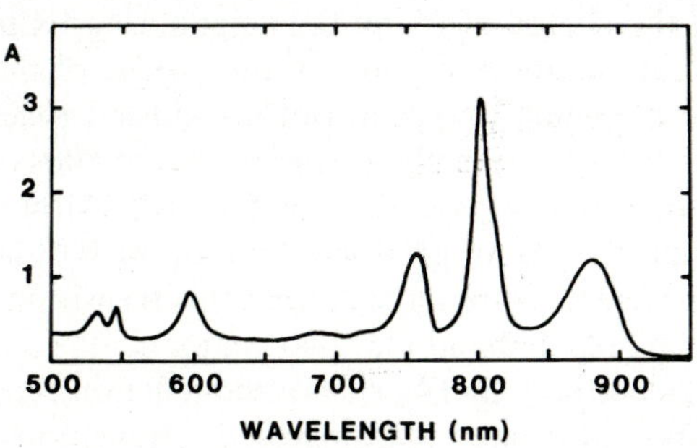

Fig. 5. Absorption spectrum of reaction centers from *Rb. sphaeroides* strain R-26, measured at 5 K with a film of reaction centers in polyvinylalcohol. Redrawn from Kirmaier et al. [76].

at shorter wavelengths, in the region from 350 to 410 nm. Reaction centers obtained from *Rp. viridis*, which contain BChl *b* instead of BChl *a*, have a qualitatively similar spectrum, except that the main bands in the near-IR region are near 800, 830 and 960 nm, and the carotenoid and cytochromes contribute additional bands in the 400–600-nm region. For comparison, monomeric BChl *a in vitro* has four main absorption bands, near 360, 390, 600 and 770 nm, depending somewhat on the solvent. The 600- and 770-nm bands are referred to as the Q_x and Q_y bands; the two bands at shorter wavelengths, as the Soret bands. Monomeric BPh *a* in solution has four corresponding absorption bands near 360, 380, 530 and 760 nm. In the reaction center, the BChls and BPhs are close enough together that their optical transitions are mixed, and each absorption band contains contributions from all six pigments [128–130]. However, one can reasonably attribute the bands near 530 and 545 nm primarily to the Q_x transitions of the two BPhs, that near 600 nm to overlapping Q_x transitions of the four BChls, that near 760 to the BPh Q_y transitions, and those near 800 and 880 nm to Q_y transitions of the BChls.

The reaction center's long-wavelength band near 880 (or 960 nm, in *Rp. viridis*) must be due principally to the two BChls that make up P, because it bleaches when P is oxidized to P^+, or is raised to an excited singlet state [131,132] or triplet state [52,53,133–135]. The linear dichroism of the long-wavelength band in oriented crystals of *Rp. viridis* reaction centers is consistent with this assignment [136,137]. The shift of the band to a much longer wavelength relative to the position of the Q_y band of monomeric BChl in solution can be explained partly by exciton interactions between the two BChls [128,130,138]. A spectral shift also could result from interactions of a single BChl with nearby charged groups [139], but in the *Rp. viridis* crystal structure there are no charged amino acids sufficiently close to the BChls to have such an effect [101,102,172]. Molecular orbital calculations based on the crystal structure indicate that a large part of the shift probably results from a mixing of the $\pi\pi^*$ excited states of the two BChls with charge-transfer states in which an electron is transferred from one of the BChls of P to the other [130].

The absorption band near 800 nm in *Rb. sphaeroides* or 830 nm in *Rp. viridis* is probably due largely to the two BChls that are not part of P, with some contributions of dipole strength from P and the two BPhs. This band moves about 5 nm to shorter wavelengths when P is oxidized. The shift to shorter wavelengths most likely reflects an effect of the charge of P^+ on the neighboring BChls (an 'electrochromic', or Stark effect), in addition to a loss of the mixing of the transitions of P with those of the other pigments. In spectra of unoxidized reaction centers at low temperatures, such as the spectrum shown in Fig. 5, the 800- or 830-mn band has a distinct shoulder on the long-wavelength side [76,140]. Although it has been suggested that the shoulder reflects an exciton transition of P [124,137,141], calculations using the *Rp. viridis* structure indicate that this transition would be very weak [128,130]. The splitting of the band into two components could reveal a distinction between the BChls on the L and M side of the reaction center ($BChl_L$ and $BChl_M$ in Fig. 4), or simply an exciton splitting of the transitions of two nearly identical molecules [76,142,143].

A low temperatures, the reaction center has two well-resolved absorption bands in the BPh Q_x region, at 530 and 545 nm (Fig. 5). The presence of two Q_x absorption bands indicates that the two BPh molecules in the reaction center are not equivalent. (There is no basis for interpreting the two bands as the exciton transitions of a strongly coupled pair of molecules, because they have essentially parallel linear dichroism in oriented reaction centers [137]. In addition, the BPhs in the *Rp. viridis* crystal structure are too far apart for their exciton interactions to give the observed splitting of the absorption bands.) Although the structural differences between the two BPhs are not yet fully clear, the C-9 keto group of the BPh on the L side of the reaction center (BPh_L) appears to be hydrogen-bonded to a glutamic acid residue, whereas the keto group of the BPh on the M side (BPh_M is next to a hydrophobic amino acid side chain [172].

The BPh (I) which is reduced transiently when the reaction center is excited with a short flash of light (or is reduced semipermanently if reaction centers are illuminated continuously at low temperature in the presence of Q_A^- and reduced cytochromes) is the one that absorbs at longer wavelengths, 545 and 760 nm in *Rb. sphaeroides* [52,70,71,75,76]. Measurements of the linear dichroism of oriented *Rp. viridis* crystals indicate that this is BPh_L [129,136]. This assignment agrees with the identification of Q_A as menaquinone and Q_B as ubiquinone in *Rp. viridis* [32], since menaquinone is on the same side of the reaction center as BPh_L (Fig. 4).

Whether $BChl_M$ and BPh_M play any role in the photochemical electron transfer reactions is unclear. Prolonged illumination under special conditions can result in the reduction of BPh_M [125,144], but this process has a low quantum yield and does not occur readily even if BPh_L and Q_A are already reduced [54,145, 146].

6. *Electron transfer kinetics and mechanisms*

If isolated reaction centers from *Rb. sphaeroides* or *Rp. viridis* are excited with a subpicosecond flash, the transfer of an electron from P* to BPh_L occurs with a time constant of 3 to 4 ps [72,131,132,147,148]. The kinetics can be measured by following the bleaching of the BPh's absorption bands at 545 and 760 (or 800 nm for the latter in *Rp. viridis*) and the appearance of broad absorption bands due to BPh^- and P^+ at 760 and 1250 nm (1325 nm in *Rp. viridis*). Prior to the reduction of the BPh, P* can be detected by its broad absorption bands in the visible and near-IR regions of the spectrum, and by its stimulated emission (fluorescence) at 920 or 1000 nm. The stimulated emission from P* decays with kinetics that match the formation of BPh^-.

The electron transfer reaction between P* and BPh is slightly *faster* at 80 K than it is at 295 K, indicating that it does not require a thermal activation energy [131]. This agrees with earlier observations that charge separation in the reaction center occurs with a high quantum yield at temperatures as low as 1 K [149].

Measurements with flashes lasting 10 to 40 ps have suggested that a transient P^+BChl^- state precedes the formation of P^+BPh^- [74,150,151]. However, the evidence for this conclusion has been criticized [77,152], and recent studies with higher time resolution do not support it [131,132,133,148]. Because $BChl_L$ is located almost in between P and BPh_L in the *Rp. viridis* crystal structure (Fig. 4), the BChl does seem likely to play a role in the electron transfer reaction and, as discussed above, the reduction of the BPh evidently perturbs the absorption spectrum in bands that are associated with the BChl. But P^+BChl^- appears not to be a kinetically resolvable intermediate in the electron transfer process. This could mean that P^+BChl^- is generated from P*, but decays too quickly to be detected. However, molecular orbital calculations, together with an analysis of the reaction center's absorption spectrum, indicate that the P^+BChl^- charge-transfer state probably lies significantly *above* P* in energy [130]. It is thus not likely to be populated to a significant extent, particularly at cryogenic temperatures.

Even if the $P^+BChl_L^-$ charge-transfer state is not formed as a distinct intermediate, it probably mixes quantum mechanically with P* and with $P^+BPh_L^-$. This mixing could facilitate electron tunneling from P* to the BPh by the process known as 'superexchange' [131,153]. Mixing of the excited states of $BChl_L$ with those of P could also play an important role in the reaction [130].

Spectroscopic hole-burning experiments at temperatures below 2 K [154,155] indicate that P* may undergo relaxations on the time scale of 20 fs, which is considerably faster than the movement of an electron from P* to $BChl_L$ as judged from the lifetime of stimulated emission. Although a conversion from a $\pi\pi^*$ state to a charge-transfer state has been suggested, this interpretation is not (in its simplest form) in accord with considerations that place the lowest charge-transfer state of P above the lowest $\pi\pi^*$ state in energy [130]. An alternative interpretation is that P* undergoes rapid changes in nuclear geometry. Such motions do not occur in monomeric BChl at these temperatures [154], but they might be expected in the reaction center if the excited state has substantial charge-transfer character.

Like electron transfer from P* to BPh_L, the movement of an electron from BPh_L^- to Q_A increases in speed with decreasing temperature. The time constant of the reaction drops from about 230 ps at 300 K to approximately 100 ps at 100 K, and then becomes essentially independent of temperature down to 4 K [76]. The back reaction between P^+ and Q_A^- also speeds up with decreasing temperature [156–158]. Its time constant is about 100 ms at 300 K, and about 30 ms at temperatures below 100 K. In *Ch. vinosum*, electron transfer from the low-potential cytochrome to P^+ decreases in speed with decreasing temperature down to about 100 K, but then becomes independent of temperature [159]. The curious temperature-dependences of these reactions have been rationalized in terms of the effects of nuclear motions on electron tunneling [76,160–167]. Because nuclei move more slowly than electrons, the overlap of nuclear vibrational wavefunctions of the reactants and products is a critical factor in determining the rate of electron transfer. Reactions that speed up with decreasing temperature generally are assumed to proceed most favorably from the lowest vibrational states of the reactants. In the case of the reaction between the cytochrome and P^+, motions of the tyrosine that bridges the gap between the hemes and P could be particularly important [102].

Extracting the nonheme Fe from the reaction center slows electron transfer from Q_A^- to Q_B by about a factor of 2 [168], a remarkably modest effect in view of the Fe's location between the two quinones (Fig. 4). Electron transfer from BPh_L^- to

Q_A is not affected if the Fe is replaced by Zn, but extracting the metal atom decreases the rate of this step 50-fold, suggesting that the electric field provided by the metal is important for electron transfer to the quinone [169]. Vibrations of BPh_L^- or Q_A^- in the electrical field of the metal atom could be critical for the conversion of electronic potential energy into vibrational energy [170]. It is the large free energy change which occurs in electron transfer from BPh_L^- to Q_A that renders charge separation in the reaction center essentially irreversible (Fig. 1).

Acknowledgements

I thank Drs. J. Deisenhofer and H. Michel for providing information on the *Rp. viridis* crystal structure and for helpful comments on the manuscript, D. Middendorf, C. Kirmaier, D. Holten, A. Warshel and N. Woodbury for additional helpful discussion, and the U.S. National Science Foundation and Department of Agriculture for financial support.

References

1 Duysens, L.N.M. (1954) Nature 173, 692–693.
2 Duysens, L.N.M. (1958) Brookhaven Symp. Biol. 11, 10–25.
3 Clayton, R.K. (1962) Photochem. Photobiol. 1, 201–210.
4 Kuntz, I.D., Loach, P.A. and Calvin, M. (1964) Biophys. J. 4, 227–249.
5 Parson, W.W. (1968) Biochim. Biophys. Acta 153, 248–259.
6 Norris, J.R., Uphaus, R.A., Crespi, H.L. and Katz, J.J. (1971) Proc. Natl. Acad. Sci. USA 68, 625–628.
7 McElroy, J.D., Feher, G. and Mauzerall, D.C. (1972) Biochim. Biophys. Acta 267, 363–374.
8 Feher, G., Hoff, A.J., Isaacson, R.A. and Ackerson, L.C. (1975) Ann. N.Y. Acad. Sci. 244, 239–259.
9 Fajer, J., Davis, M.S., Brune, D.C., Spaulding, L.D., Borg, D.C. and Forman, A. (1977) Brookhaven Symp. Biol. 28, 74–103.
10 Lendzian, F., Lubitz, W., Scheer, H., Bubenzer, C. and Mobius, K. (1981) J. Am. Chem. Soc. 103, 4635–4637.
11 Wasielewski, M.R., Norris, J.R., Crespi, H.L. and Harper, J. (1981) J. Am. Chem. Soc. 103, 7664–7665.
12 Lubitz, W., Isaacson, R.A., Abresch, E.C. and Feher, G. (1984) Proc. Natl. Acad. Sci. USA 81, 7792–7796.
13 Cusanovich, M.A. and Kamen, M.D. (1968) Biochim. Biophys. Acta 153, 376–396.
14 Case, G.D. and Parson, W.W. (1971) Biochim. Biophys. Acta 253, 187–202.
15 Dutton, P.L. (1971) Biochim. Biophys. Acta 226, 63–80.
16 Prince, R.C., Leigh, J.S. and Dutton, P.L. (1976) Biochim. Biophys. Acta 440, 622–636.
17 Bruce, B.D., Fuller, R.C. and Blankenship, R.E. (1982) Proc. Natl. Acad. Sci. USA 79, 6532–6536.
18 McElroy, J.D., Feher, G. and Mauzerall, D.C. (1970) Biophys. J. 10, 204a.
19 Leigh, J.S. and Dutton, P.L. (1972) Biochem. Biophys. Res. Commun. 46, 414–421.
20 Dutton, P.L. and Leigh, J.S. (1973) Biochim. Biophys. Acta 314, 178–190.
21 Feher, G. and Okamura, M.Y. (1978) in the Photosynthetic Bacteria (Clayton R.K. and Sistrom, W.R., eds.), pp. 349–386, Plenum Press, New York.
22 Loach, P.A. and Hall, R.L. (1972) Proc. Natl. Acad. Sci, USA 69, 786–790.
23 Imhoff, J.F., Truper, H.G. and Pfennig, N. (1984) Int. J. Syst. Bact. 34, 340–343.

24 Cogdell, R.J., Brune, D.C. and Clayton, R.K. (1974) FEBS Lett. 45, 344–347.
25 Okamura, M.Y., Isaacson, R.A. and Feher, G. (1975) Proc. Natl. Acad. Sci. USA 72, 3491–3495.
26 Halsey, Y. and Parson, W.W. (1974) Biochim. Biophys. Acta 347, 404–416.
27 Vermeglio, A. (1977) Biochim. Biophys. Acta 459, 516–524.
28 Wraight, C.A. (1977) Biochim. Biophys. Acta 459, 525–531.
29 Wraight, C.A. (1979) Biochim. Biophys. Acta 548, 309–327.
30 Wraight, C.A. (1981) Israel J. Chem. 21, 348–354.
31 Vermeglio, A. and Clayton, R.K. (1977) Biochim. Biophys. Acta 461, 159–165.
32 Shopes, R.J. and Wraight, C.A. (1985) Biochim. Biophys. Acta 806, 348–356.
33 Dutton, P.L., Leigh, J.S. and Wraight, C.A. (1973) FEBS Lett. 36, 169–173.
34 Arata, H. and Parson, W.W. (1981) Biochim. Biophys. Acta 638, 201–209.
35 Prince, R.C. and Dutton, P.L. (1976) Arch. Biochem. Biophys. 172, 329–334.
36 Rutherford, A.W. and Evans, M.C.W. (1980) FEBS Lett. 110, 257–261.
37 Clayton, R.K. and Straley, S.C. (1972) Biophys. J. 12, 1221–1234.
38 Slooten, L. (1972) Biochim. Biophys. Acta 275, 208–218.
39 Feher, G., Okamura, M.Y. and McElroy, J.D. (1972) Biochim. Biophys. Acta 267, 222–226.
40 Kleinfeld, D., Okamura, M.Y. and Feher, G. (1985) Biophys. J. 48, 849–852.
41 Lubitz, W., Abresch, E.C., Debus, R.J., Isaacson, R.A., Okamura, M.Y. and Feher, G. (1985) Biochim. Biophys. Acta 808, 464–469.
42 Wraight, C.A. and Stein, R. (1980) FEBS Lett. 113, 73–77.
43 Kleinfeld, D., Okamura, M.Y. and Feher, G. (1984) Biochim. Biophys. Acta 766, 126–140.
44 Mancino, L.J., Dean, D.P. and Blankenship, R.E. (1984) Biochim. Biophys. Acta 764, 46–54.
45 Arata, H. (1985) Biochim. Biophys. Acta 809, 284–287.
46 Blankenship, R.E. (1984) Photochem. Photobiol. 40, 801–806.
47 Vasmel, H. and Amesz, J. (1983) Biochim. Biophys. Acta 724, 118–122.
48 Wraight, C.A. and Clayton, R.K. (1974) Biochim. Biophys. Acta 333, 246–260.
49 Zankel, K.L., Reed, D.L. and Clayton, R.K. (1968) Proc. Natl. Acad. Sci. USA 61, 1243–1249.
50 Slooten, L. (1972) Biochim. Biophys. Acta 256, 452–466.
51 Parson, W.W. (1969) Biochim. Biophys. Acta 189, 384–396.
52 Parson, W.W., Clayton, R.K. and Cogdell, R.J. (1975) Biochim. Biophys. Acta 387, 265–278.
53 Parson, W.W. and Monger, T.G. (1977) Brookhaven Symp. Biol. 28, 195–212.
54 Holten, D., Windsor, M.W., Parson, W.W. and Thornber, J.P. (1978) Biochim. Biophys. Acta 501, 112–126.
55 Fajer, J., Brune, D.C., Davis, M.S., Forman, A. and Spaulding, L.D. (1975) Proc. Natl. Acad. Sci. USA 72, 4956–4960.
56 Fajer, J., Davis, M.S., Brune, D.C., Forman, A. and Thornber, J.P. (1978) J. Am. Chem. Soc. 100, 1918–1920.
57 Shuvalov, V.A. and Parson, W.W. (1981) Proc. Natl. Acad. Sci. USA 78, 957–961.
58 Blankenship, R.E., Schaafsma, T.J. and Parson, W.W. (1977) Biochim. Biophys. Acta 461, 297–305.
59 Hoff, A.J., Rademaker, H., van Grondelle, R. and Duysens, L.N.M. (1977) Biochim. Biophys. Acta 460, 547–554.
60 Werner, H.J., Schulten, K. and Weller, A. (1978) Biochim. Biophys. Acta 502, 255–268.
61 Levanon, H. and Norris, J.R. (1978) Chem. Rev. 78, 185–198.
62 Haberkorn, R. and Michel-Beyerle, M.E. (1979) Biophys. J. 26, 489–498.
63 Hoff, A.J. (1981) Qt. Rev. Biophys. 14, 599–665.
64 Rademaker, H. and Hoff, A.J. (1981) Biophys. J. 34, 325–344.
65 Schenck, C.C., Blankenship, R.E. and Parson, W.W. (1982) Biochim. Biophys. Acta 680, 44–59.
66 Boxer, S.G., Chidsey, C.E.D. and Roelofs, M.G. (1983) Annu. Rev. Phys. Chem. 34, 389–417.
6l Shuvalov, V.A. and Klimov, V.V. (1976) Biochim. Biophys. Acta 440, 587–599.
68 Tiede, D.M., Prince, R.C. and Dutton, P.L. (1976) Biochim. Biophys. Acta 449, 447–467.
69 Okamura, M.Y., Isaacson, R.A. and Feher, G. (1979) Biochim. Biophys. Acta 546, 394–417.
70 Rockley, M., Windsor, M.W., Cogdell, R.J. and Parson, W.W. (1975) Proc. Natl. Acad. Sci. USA 72, 2251–2255.
71 Kaufmann, K.J., Dutton, P.L., Netzel, T.L., Leigh, J.S. and Rentzepis, P.M. (1975) Science 188, 1301–1304.

72 Holten, D., Hoganson, C., Windsor, M.W., Schenck, C.C., Parson, W.W., Migus, A., Fork, R.L. and Shank, C.V. (1980) Biochim. Biophys. Acta 592, 461–477.
73 Peters, K., Avouris, P. and Rentzepis, P.M. (1978) Biophys. J. 23, 207–217.
74 Shuvalov, V.A., Klevanik, A.V., Sharkov, A.V., Matveetz, J.A. and Krukov, P.G. (1978) FEBS Lett. 91, 135–139.
75 Kirmaier, C., Holten, D. and Parson, W.W. (1985) Biochim. Biophys. Acta 810, 33–48.
76 Kirmaier, C., Holten, D. and Parson, W.W. (1985) Biochim. Biophys. Acta 810, 49–61.
77 Kirmaier, C., Holten, D. and Parson, W.W. (1985) FEBS Lett. 185, 76–82.
78 Godik, V.I. and Borisov, A. Yu. (1979) Biochim. Biophys. Acta 548, 296–308.
79 Godik, V.I, Kotova, E.A. and Borisov, A. Yu. (1982) Photobiochem. Photobiophys. 4, 219–226.
80 Woodbury, N.W. and Parson, W.W. (1984) Biochim. Biophys. Acta 767, 345–361.
81 Woodbury, N.W. and Parson, W.W. (1986) Biochim. Biophys. Acta 850, 197–210.
82 Blankenship, R.E. (1985) Photosynth. Res. 6, 317–333.
83 Nuijs, A.M., Vasmel, H., Joppe, H.L.P., Duysens, L.N.M. and Amesz, J. (1985) Biochim. Biophys. Acta 807, 24–34.
84 Braumann, T., Vasmel, H., Grimme, L.H and Amesz, J. (1986) Biochim. Biophys. Acta 848, 83–91.
85 Shuvalov, V.A., Amesz, J. and Duysens, L.N.M. (1986) Biochim. Biophys. Acta 851, 1–5.
86 Clayton, R.K. and Wang, R.T. (1971) Methods Enzymol. 23, 696–704.
87 Noel, H., Van der Rest, M. and Gingras, G. (1972) Biochim. Biophys. Acta 275, 219–230.
88 Jolchine, G. and Reiss-Husson, F. (1974) FEBS Lett. 40, 5–8.
89 Okamura, M.Y., Steiner, L.A. and Feher, G. (1974) Biochemistry 13, 1394–1403.
90 Prince, R.C., Tiede, D.M., Thornber, J.P. and Dutton, P.L. (1977) Biochim. Biophys. Acta 462, 467–490.
91 Clayton, B.J. and Clayton, R.K. (1978) Biochim. Biophys. Acta 501, 470–477.
92 Michel, H. (1982) J. Mol. Biol. 158, 567–572.
93 Pierson, B.K. and Thornber, J.P. (1983) Proc. Natl. Acad. Sci. USA 80, 80–84.
94 Kirmaier, C., Holten, D. and Blankenship, R.E. (1983) FEBS Lett. 158, 73–78.
95 Vasmel, H., Meiburg, R.F., Kramer, H.J.M., de Vos, L.J. and Amesz, J. (1983) Biochim. Biophys. Acta 724, 333–339.
96 Linn, L. and Thornber, J.P. (1975) Photochem. Photobiol. 22, 37–40.
97 Vasmel, H., Swarthoff, T., Kramer, H.J.M. and Amesz, J. (1983) Biochim. Biophys. Acta 725, 361–367.
98 Allen, J.P. and Feher, G. (1984) Proc. Natl. Acad. Sci. USA 81, 4795–4799.
99 Gast, P. and Norris, J.R. (1984) FEBS Lett. 177, 277–280.
100 Chang, C.-H., Schiffer, M., Tiede, D., Smith, U. and Norris, J. (1985) J. Mol. Biol. 186, 201–203.
101 Deisenhofer, J., Epp. O., Miki, K., Huber, R. and Michel, H. (1984) J. Mol. Biol. 180, 385–398.
102 Deisenhofer, J., Epp, O., Miki, K., Huber, R. and Michel, H. (1985) Nature 318, 618–624.
103 Youvan, D.C., Bylina, E.J., Alberti, M., Megush, H. and Hearst, J.E. (1984) Cell 37, 949–957.
104 Wiemken, V. and Bachofen, R. (1984) FEBS Lett. 166, 155–159.
105 Case, G.D., Parson, W.W. and Thornber, J.P. (1970) Biochim. Biophys. Acta 223, 122–128.
106 Parson, W.W. and Case, G.D. (1970) Biochim. Biophys. Acta 205, 232–245.
107 Prince, R.C. and Dutton, P.L. (1977) Biochim. Biophys. Acta 459, 573–577.
108 Williams, J.C., Steiner, L.A., Ogden, R.C., Simon, M.I. and Feher, G. (1983) Proc. Natl. Acad. Sci, USA 80, 6505–6509.
109 Williams, J.C., Steiner, L.A., Feher, G. and Simon, M.I. (1984) Proc. Natl. Acad. Sci. USA 81, 7303–7307.
110 Michel, H., Weyer, K.A., Gruenberg, H., Dunger, I., Oesterhelt, D. and Lottspeich, F. (1986) EMBO J. 5, 1149–1158.
111 Michel, H., Weyer, K.A., Gruenberg, H. and Lottspeich, F. (1985) EMBO J. 4, 1667–1672.
112 Bachmann, R.C., Gillies, K. and Takemoto, J.Y. (1981) Biochemistry 20, 4590–4596.
113 Valkirs, G.E. and Feher, G. (1982) J. Cell Biol. 95, 179–188.
114 Nabedryk, E., Tiede, D.M., Dutton, P.L. and Breton, J. (1982) Biochim. Biophys. Acta 682, 273–280.
115 Brunisholz, R.A., Wiemken, V., Suter, F. and Bachofen, R. (1984) Hoppe-Seyler's Z. Physiol. Chem. 365, 689–701.

116 Straley, S.C., Parson, W.W., Mauzerall, D.C. and Clayton, R.K. (1973) Biochim. Biophys. Acta 305, 597–609.
117 Feher, G., Isaacson, R.A., McElroy, J.D., Ackerson, L.C. and Okamura, M.Y. (1974) Biochim. Biophys. Acta 368, 135–139.
118 Nam, H.K., Austin, R.H. and Dismukes, G.C. (1984) Biochim. Biophys. Acta 765, 301–308.
119 Prince, R.C. (1978) Biochim. Biophys. Acta 501, 195–207.
120 Robert, B. and Lutz, M. (1986) Biochemistry 25, 2303–2309.
121 Bunker, G., Stern, E.A., Blankenship, R.E. and Parson, W.W. (1982) Biophys. J. 37, 539–551.
122 Eisenberger, D., Okamura, M.Y. and Feher, G. (1982) Biophys. J. 37, 523–538.
123 Cogdell, R.J., Parson, W.W. and Kerr, M. (1976) Biochim. Biophys. Acta 430, 83–93.
124 Vermeglio, A., Breton, J. Paillotin, G. and Cogdell, R.J. (1978) Biochim. Biophys. Acta 501, 514–530.
125 Thornber, J.P., Cogdell, R.J., Seftor, R.E.B. and Webster, G.D. (1980) Biochim. Biophys. Acta 593, 60–75.
126 McGann, W.J. and Frank, H.A. (1985) Biochim. Biophys. Acta 807, 101–109.
127 Cogdell, R.J., Monger, T.G. and Parson, W.W. (1978) Biochim. Biophys. Acta 408, 189–199.
128 Knapp, E.W. and Fischer, S.F. (1985) in Antennas and Reaction Centers of Photosynthetic Bacteria Michel-Beyerle, M.E. ed.), pp. 103–108, Springer-Verlag, Berlin.
129 Knapp, E.W., Fischer, S.F., Zinth, W., Sander, M., Kaiser, W., Deisenhofer, J. and Michel, H. (1985) Proc. Natl. Acad. Sci. USA 82, 8463–8467.
130 Parson, W.W., Scherz, A. and Warshel, A. (1985) in Antennas and Reaction Centers of Photosynthetic Bacteria (Michel-Beyerle, M.E., ed.), pp. 122–130, Springer-Verlag, Berlin.
131 Woodbury, N.W., Becker, M., Middendorf, D. and Parson, W.W. (1985) Biochemistry 24, 7516–7521.
132 Martin, J.-L., Breton, J., Hoff, A., Migus, A. and Antonetti, A. (1986) Proc. Natl. Acad. Sci. USA 83, 957–961.
133 Shuvalov, V.A. and Parson, W.W. (1981) Biochim. Biophys. Acta 638, 50–59.
134 den Blanken, H.J. and Hoff, A.J. (1982) Biochim. Biophys. Acta 681, 365–374.
135 den Blanken, H.J. and Hoff, A.J. (1983) Chem. Phys. Lett. 98, 255–262.
136 Zinth, W., Sander, M., Dobler, J, Kaiser, W. and Michel, H. (1985) in Antennas and Reaction Centers of Photosynthetic Bacteria (Michel-Beyerle, M.E., ed.), pp. 97–102, Springer-Verlag, Berlin.
137 Breton, J. (1985) Biochim. Biophys. Acta 810, 235–245.
138 Parson, W.W. (1982) Annu. Rev. Biophys. Bioeng. 11, 57–80.
139 Eccles, J. and Honig, B. (1983) Proc. Natl. Acad. Sci. USA 80, 4959–4962.
140 Feher, G. (1971) Photochem. Photobiol. 14, 373–387.
141 Vermeglio, A. and Clayton, R.K. (1978) Biochim. Biophys. Acta 449, 500–515.
142 Shuvalov, V.A. and Asadov, A.A. (1979) Biochim. Biophys. Acta 545, 296–308.
143 Hoff, A.J., Lous, E.J., Moehl, K.W. and Dijkman, J.A. (1985) Chem. Phys. Letts. 114, 39–43.
144 Robert, B., Lutz, M. and Tiede, D.M. (1985) FEBS Lett. 83, 326–330.
145 Netzel, T.L., Rentzepis, P.M., Tiede, D.M., Prince, R.C. and Dutton, P.L. (1977) Biochim. Biophys. Acta 460, 467–478.
146 Schenk, C.C., Parson, W.W., Holten, D. and Windsor, M.W. (1981) Biochim. Biophys. Acta 635, 383–392.
147 Paschenko, V.Z., Korvatovskii, B.N., Kononenko, A.A., Chamorovsky, S.K. and Rubin, A.B. (1985) FEBS Lett. 191, 245–248.
148 Breton, J., Martin, J.-L., Migus, A., Antonetti, A. and Orszag, A. (1986) Proc. Natl. Acad. Sci. USA 83, 5121–5125.
149 Arnold, W. and Clayton, R.K. (1960) Proc. Natl. Acad. Sci. USA 46, 769–776.
150 Shuvalov, V.A. and Klevanik, A.V. (1983) FEBS Lett. 160, 51–55.
151 Shuvalov, V.A. and Duysens, L.N.M. (1986) Proc. Natl. Acad. Sci. USA 83, 1690–1694.
152 Borisov, A.Y., Danielius, R.V., Kudzmauskas, S.P., Piskarskas, A.S., Razjivin, A.P., Sirutkaitis, V.A. and Valkunes, L.L. (1983) Photobiochem. Photobiophys. 6, 33–38.
153 Heitele, H. and Michel-Beyerle, M.E. (1985) in Antennas and Reaction Centers of Photosynthetic Bacteria (Michel-Beyerle, M.E., ed.), pp. 250–255, Springer-Verlag, Berlin.

154 Meech, S.R., Hoff, A.J. and Wiersma, D.A. (1985) Chem. Phys. Lett. 121, 287–292.
155 Boxer, S.G., Lockhart, D.J. and Middendorf, T.R. (1986) Chem. Phys. Lett. 123, 476–482.
156 Parson, W.W. (1966) Biochim. Biophys. Acta 131, 154–172.
157 McElroy, J.D., Mauzerall, D.C. and Feher, G. (1974) Biochim. Biophys. Acta 333, 261–277.
158 Hsi, E.S.P. and Bolton, J.B. (1974) Biochim. Biophys. Acta 347, 126–153.
159 DeVault, D. and Chance, B. (1966) Biophys. J. 6, 825–847.
160 Hopfield, J.J. (1974) Proc. Natl. Acad. Sci. USA 71, 3640–3644.
161 Jortner, J. (1976) J. Chem. Phys. 64, 4860–4867.
162 Blankenship, R.E. and Parson, W.W. (1979) in Photosynthesis in Relation to Model Systems (Barber, J., ed.), pp. 71–114, Elsevier, Amsterdam.
163 Warshel, A. (1980) Proc. Natl. Acad. Sci. USA 77, 3105–3109.
164 Sarai, A. (1980) Biochim. Biophys. Acta 589, 71–83.
165 Kakitani, T. and Kakitani, H. (1981) Biochim. Biophys. Acta 635, 498–514.
166 DeVault, D. (1984) Quantum Mechanical Tunneling in Biological Systems, Cambridge University Press, Cambridge, U.K.
167 Marcus, R. and Sutin, N. (1985) Biochim. Biophys. Acta 811, 265–322.
168 Debus, R., Feher, G. and Okamura, M.Y. (1986) Biochemistry 25, 2276–2287.
169 Kirmaier, C., Holten, D., Debus, R., Feher, G. and Okamura, M.Y. (1986) Proc. Natl. Acad. Sci. USA 83, 6407–6411.
170 DeVault, D. (1986) Photosynth. Res. 10, 125–137.
171 Seftor, R.E.B. and Thornber, J.P. (1984) Biochim. Biophys. Acta 764, 148–159.
172 Michel, H., Epp, O. and Diesenhofer, J. (1986) EMBO J. 5, 2445–2451.

J. Amesz (ed.) *Photosynthesis*

CHAPTER 4

The primary reactions of photosystems I and II of algae and higher plants

P. MATHIS and A.W. RUTHERFORD

Département de Biologie, Service de Biophysique, CEN Saclay 91191 Gif-sur-Yvette Cedex, France

1. Introduction

In photosynthetic organisms, the 'primary reactions' fulfill the objective of converting the energy of light into a primary form of chemical energy which lasts for a time compatible with ordinary biochemical processes, i.e. milliseconds. In these reactions a rather large fraction, approximately 40%, of the photon energy is stored as chemical free energy. The primary reactions can be viewed from two major perspectives. Firstly, from a photochemical point of view: pigment molecules are excited to their lowest excited singlet state which reacts in an electron transfer reaction, the first step of a process of charge separation. Secondly, from a biochemical point of view: the reactions take place in highly organized complexes, the reaction centres, which are made up of several classes of molecules which cooperate in fulfilling complementary roles: architectural support, light absorption, energy transfer and electron transfer [1–3]. Reaction centres are membrane-bound complexes, made of a few hydrophobic polypeptides which hold together, in a well-defined conformation, various pigments (chlorophylls and carotenoids) and redox centres (tetrapyrroles, quinones, iron-sulfur centers, etc). The reaction centres have a well-defined positioning with respect to the membrane plane.

Photosynthetic organisms have adopted a large variety of shapes, colors and living conditions. The primary reactions in all organisms, however, share a large number of basic properties, and the purple bacteria, which have been studied in great detail, can be used as a good general model system. In oxygenic photosynthetic organisms, for which water is the ultimate source of reducing power, there are two types of reaction centres, photosystem I and photosystem II (PS I and PS II), which operate in series for electron transfer (Fig. 1). This cooperation of two photoreactions is made necessary by the large energy gap for the electron to be transferred from water ($E_m \simeq +0.8$ V) to the terminal electron acceptor $NADP^+$ ($E_m = -0.3$ V).

All oxygenic organisms, ranging from cyanobacteria to algae and higher plants, contain PS I and PS II reaction centres, with only minor variations in spite of their large taxonomic and ecological diversity. Small variations will not be emphasized

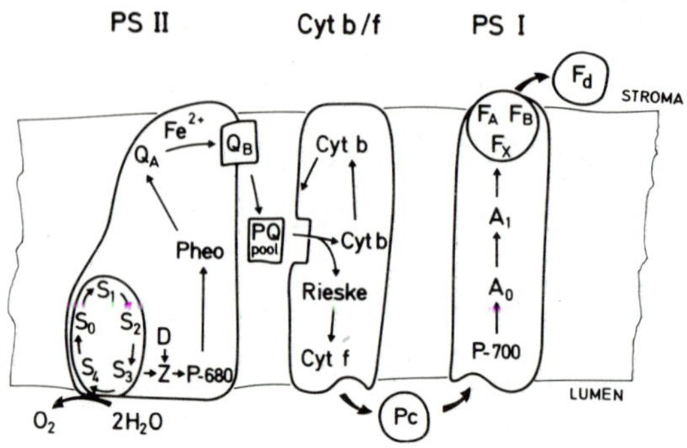

Fig. 1. A simplified scheme of the photosynthetic membrane, illustrating electron transfer from water to ferredoxin, which involves three protein complexes (the PS II reaction centre, the Cyt b_6/f complex, the PS I reaction centre) and two diffusible components, plastoquinone (PQ pool) and plastocyanin (Pc).

here and we will mainly focus on the general, functionally essential, properties. When appropriate we will also underline the analogies with non-oxygenic photosynthetic bacteria. Due to the extensive literature on the subject, citations will be generally restricted to articles published in the last few years and to recent reviews [4–7].

2. *Photosystem I reactions*

A number of experimental properties of oxygen-evolving photosynthetic organisms have been historically integrated into the concept of photosystem I reactions. We shall cite only four of them: (1) the stimulation of the rate of O_2 evolution under red or green light by far-red light, above 700 nm, which is unable to induce O_2 evolution by itself; (2) a small photoinduced bleaching of the absorption

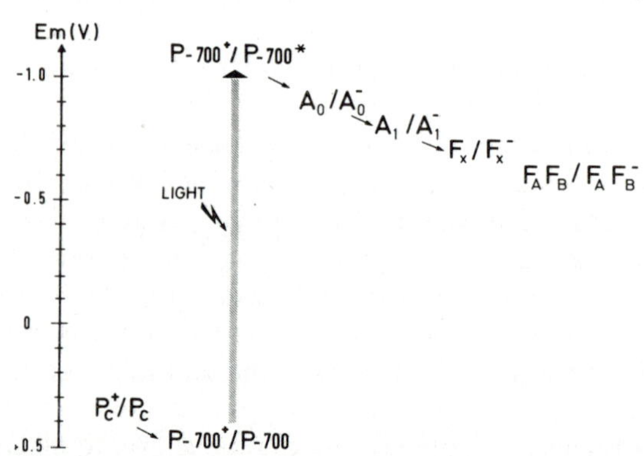

Fig. 2. A scheme of electron transfer in PS I. Redox centres are situated at their approximate or estimated midpoint potential E_m.

at 700 nm, which was interpreted as being due to the oxidation of a primary electron donor, P-700; a free radical EPR signal, Signal I, was attributed to P-700^+; (3) the photoinduced appearance of EPR signals characteristic of iron-sulfur proteins, at cryogenic temperatures; (4) the ability to reduce low-potential electron acceptors such as ferredoxin or viologens; this can be done even in the presence of inhibitors of the PS II reactions, such as DCMU, provided an artificial electron donor is added.

A coherent interpretation for many experimental results was provided by the concept of a PS I reaction centre. This centre has now been isolated, albeit perhaps not in a definitely pure state. It is made up of a few hydrophobic polypeptides, the primary donor (P-700), several electron acceptors (Fig. 2), and about 50 molecules of pigment (chlorophyll *a* and β-carotene). This composition is analogous to that of other types of reaction centres.

2.1. The primary donor P-700

2.1.1. Basic properties of P-700

In one of the first applications of differential absorption spectroscopy to photosynthetic membranes, Kok [8] observed a small light-induced bleaching at 700 nm. The bleaching can also be induced by addition of an oxidizing agent. It is now clearly associated with the photooxidation of P-700, the primary electron donor of PS I. This species cannot be isolated as a pure molecular entity. Its absorption spectrum is not known, but we know the difference spectrum due to its oxidation (i.e. P-700^+ minus P-700), which includes large bleachings at 700 ($\pm$ 3) nm and 430 nm, and small positive bands at 810 and 450 nm [9]. At low temperature, large narrow bands develop at 690 nm (positive) and 680 nm (negative) [10,11]. In chloroplasts, P-700 is present at a ratio of one per about 400 chlorophylls. In the purified PS I particles, which presumably are closest to the structure of a PS I reaction centre, there is one P-700 per about 50 chlorophylls. The E_m of P-700 is about +490 mV [12].

2.1.2. P-700: a chlorophyll species

The chemical nature of P-700 is difficult to establish. The absorption bleachings correspond approximately to the peaks of Chl *a*, which appears to be virtually the only tetrapyrrolic pigment in purified PS I particles. It has thus been assumed that P-700 is Chl *a* bound to a protein. A few recent results, however, may require this hypothesis to be refined. An examination of the spectroscopic and redox properties of P-700 led Wasielewski et al. [13] to propose that P-700 could be the enol form of Chl *a* where enolization was of the keto ester on ring V. This has not been confirmed by chemical extraction. Extraction experiments, however, have evidenced two other chlorophyll derivatives. A species named Chl-RC I has been isolated from PS I, at a nearly 1/1 molar ratio with P-700 and its structure shown to be a chlorinated derivative. It is not yet clear whether Chl-RC I is a native constituent of PS I or an extraction artefact. Chl-RC I has not been obtained in a recent chemical analysis by HPLC, which instead revealed two Chl *a'* per P-700 [14].

Chl *a*′ is the C 10 epimer of Chl *a*: at C 10 the two substituents (R_1 = H and R_2 = $COOCH_3$) are inverted with respect to the molecular plane. The same stoichiometry has been found in various PS I preparations and the spectroscopic properties of a Chl *a*′ dimer seem to resemble those of P-700.

2.1.3. *P-700: probably a dimer of chlorophyll*

A dimeric nature of P-700 was first proposed to explain its long wavelength of absorption and its circular dichroism spectrum, which can be attributed to chlorophyll–chlorophyll exciton interaction [15]. When two chlorophylls are close together, excitation by light tends to promote electronic transitions which belong to the pair and not to the individual molecules. In particular for the Q_Y transition, instead of one transition at the frequency ν, an excitonic interaction will create two transitions at $\nu + d\nu$ and $\nu - d\nu$. The latter is considered to be the transition at 700 nm, whereas the former could be of low intensity.

The dimer hypothesis received strong support from EPR studies. The EPR signal of P-700^+ is clearly narrowed compared to that of Chl a^+, with a halfwidth of 7.1 instead of 9 G. The linewidth is mainly due to hyperfine interactions of the unpaired electron with protons. Norris et al. [16] interpreted the narrowing in P-700^+ as being due to a delocalization of the unpaired electron over two chlorophyll molecules. It can be predicted that a complete delocalization over *n* molecules will decrease the linewidth by $\sqrt{n}$ compared to a monomer. However, a simple measurement of the EPR linewidth cannot give an unambiguous answer. As shown later by replacement of ^1H by ^2H, which has no nuclear spin, and of ^{12}C by ^{13}C, which provides a nuclear spin distributed over the whole molecule, the EPR linewidths of monomeric Chl a^+ in vitro and of P-700^+ in algae become identical [17]. It was thus concluded that in P-700^+ the unpaired electron was localized on only one chlorophyll molecule on the EPR time scale. Recent ENDOR data have been interpreted similarly [18].

Can this be reconciled with the proposal, based on optical studies, of a dimeric nature for P-700? A satisfactory hypothesis is to consider that two chlorophyll molecules, bound to the reaction centre polypeptides, make up P-700. The molecules are close enough to provide an electronic interaction in their ground neutral state. In the oxidized state the interaction is lost and the unpaired electron is localized preferentially on one of the two chlorophylls (this localization is corroborated by electron spin echo measurements on P-700^+ and Chl a^+ [19]). This is quite plausible if the interactions with the surroundings are dissymmetric, giving a different electronic affinity to the two chlorophylls. The recent use of absorption detected magnetic resonance (ADMR) allowed the triplet minus ground state absorption difference spectrum to be obtained. This spectrum led to the conclusion that the optical properties of P-700 resemble more those of a chlorophyll *a* dimer than of a monomer [20]. The triplet EPR spectrum by itself, however, has the properties of a monomer [21,22]. It thus appears that the ground state electronic interaction largely disappears in the triplet state, which can be viewed as one ground state Chl *a* and one triplet Chl *a*.

A more precise idea of the structure of P-700 can be based on that of synthetic models and of the primary donor P-960 in the purple bacterium *Rhodopseudomonas viridis* (see Chapter 3). For the sake of comparison, let us remember that in P-960 the two BChl *b* molecules are in strong electronic interaction and that in P-960$^+$ the unpaired electron is shared, although perhaps not equally (at variance with the situation in *Rhodospirillum rubrum*), between the two BChl *b* molecules. P-700 could be even more dissymmetric. For the moment we do not know the ligands to the chlorophylls in P-700. The arrangement of the two chlorophyll molecules with respect to each other could, in principle, be deduced from the optical data and the exciton theory. This deduction is hampered by two unknowns: (i) what is the reason for the large shift of the Q_Y absorption band from 663 nm (in some solvents) to 700 nm in P-700: is it due to chlorophyll–chlorophyll or to chlorophyll–protein interaction? (ii) in the exciton theory we need to know the intensity and position of both the high- and the low-frequency transitions: in P-700 the high-frequency transition has not yet been safely attributed. Based on the proposed dimeric structure of P-700, several model compounds have been prepared, either by spontaneous aggregation of chlorophyll or by means of chemical links holding two molecules in a rather well-defined configuration. Some of them have spectroscopic properties which are rather similar to those of P-700, but detailed information on P-700 cannot be inferred from these studies (reviewed in Refs. 23 and 24).

2.2. *Sequence of electron acceptors*

In the photosynthetic reactions, the primary electron donor P-700 becomes excited to its lowest excited singlet state and reacts by transferring an electron to the primary electron acceptor. The electron is then further transferred among a set of electron carriers arranged in order of increasing redox potentials (Fig. 2). This set of molecules is often viewed as a linear chain, a view which may not be the case in PS I. A photochemical description of these events would follow the electron path from the first (more primary) acceptor to more remote (secondary) acceptors. This is not possible because of the uncertainties concerning the early acceptors. We shall thus describe the more remote acceptors first and then move closer to the primary photoreaction.

2.2.1. *Terminal acceptors*

In all oxygen-evolving organisms, the PS I reaction centres finally reduce a water-soluble ferredoxin. This small protein of around 10 kDa has a (2Fe-2S) cluster and a rather low midpoint reduction potential of −400 mV. Ferredoxin binds to the PS I centre; after reduction it participates both in linear electron flow to NADP$^+$, via ferredoxin-NADP reductase, and in cyclic electron flow around the PS I centre. Two membrane-bound iron-sulfur centres, designated Centre A (or F_A) and Centre B (or F_B), appear to be the terminal acceptors in the reaction centre. Their mode of functioning is not clearly established and their structure is not well known, mainly because they cannot be extracted without their complete denaturation. F_A and F_B can be photoreduced at low temperature in cells or in purified PS I centres. Characteristic EPR spectra are thus obtained with *g* values of 1.86, 1.94, 2.05 for F_A^-, and 1.89, 1.92, 2.05 for F_B^-.

Detailed studies on F_A and F_B have been hampered by the property that their EPR spectra are not additive. This property has been attributed to a magnetic interaction between reduced F_A and F_B, indicating that they are very close to each other. The E_m values of F_A and F_B are −540 and −590 mV respectively in spinach PS I particles. Their E_m values are always in that range, but their relative values vary in different plant species; for example, F_A has a more negative E_m than F_B in barley and in a halophilic alga. The shape and temperature dependence of the EPR spectra of F_A^- and F_B^- are typical of iron-sulfur proteins. They are considered to be 4Fe-4S centers, since after modification by dimethyl sulfoxide their spectrum is characteristic of 4Fe-4S centres and because their Mössbauer spectra are also in agreement with that attribution. The presence of 10–12 Fe-S pairs in each PS I centre is compatible with this assignment (for reviews, see Refs. 25 and 26).

The precise biological role of F_A and F_B is not firmly established. Both of them can be photoreduced at room temperature. At low temperature (below 77 K) illumination induces the irreversible transfer of one electron from P-700 to one of the iron-sulfur centres. In spinach PS I, F_A is photoreduced; but if F_A is reduced chemically, F_B is photoreduced. This behaviour fits the respective E_m of the centres and the hypothesis of a linear arrangement in electron transfer: P-700...F_B...F_A. In other species, however, F_B can be photoreduced first, while in some cases illumination produces reaction centres with either F_A or F_B reduced. In the temperature range from 215 to 25 K the temperature of illumination influences the nature of the terminal acceptor, as also does the addition of glycerol to the medium [4,5,25,26]. Attempts to observe electron transfer between F_A and F_B have failed [27]. To elucidate the respective roles of F_A and F_B, attempts have been made to denature or inactivate one of them specifically by chemical modifiers [28,29]. These attempts were partly successful but they gave contradictory answers on whether F_A could be photoreduced when F_B was inactivated. Altogether the present data do not allow a decision as to whether F_A and F_B are arranged sequentially for electron transfer or if they operate in parallel. It has been speculated that, in a 'parallel' model, they might be involved in cyclic or non-cyclic electron transfer, but experimental data are lacking for that proposal.

Flash absorption studies at room temperature have revealed a species named P-430 which behaves as the terminal acceptor of the PS I centre. Ke and coworkers have performed a large number of experiments which fit well with that view: the matching kinetic behaviour of P-700 and P-430, the effect of exogenous electron carriers, and redox titrations [9]. The reduction of P-430 induces weak absorption changes with a negative peak at 430 nm ($\Delta\epsilon$ = 13000 $M^{-1} \cdot cm^{-1}$) compatible with the reduction of an iron-sulfur protein. Both P-430 (observed at physiological temperature) and F_A or F_B (observed at low temperature by EPR) behave as terminal acceptors of the PS I reaction centre. Their identification is highly plausible but not completely substantiated. P-430 could be either F_A or F_B, or both, since kinetic absorption spectroscopy is probably not able to distinguish two closely related iron-sulfur proteins. In steady-state measurements, some spectral data attributed to P-430 might have included a contribution of the low-potential species F_X.

The kinetic behaviour of F_A and F_B is poorly understood. Their time of reduction is difficult to measure. At low temperature, a value of less than 1 ms has been obtained [30,31], but the kinetics are probably highly heterogeneous. At room temperature, the absorption bleaching at 430 nm rises in less than 0.1 μs [9], but its attribution to P-430 is uncertain. The kinetics of the back-reaction between P-700^+ and the reduced terminal acceptors have been studied in great detail. At room temperature, the back-reaction has a $t_{1/2} \simeq 30$ ms [9]. This $t_{1/2}$ increases upon cooling, indicating a very large activation energy of 67 $kJ \cdot mol^{-1}$ for the reaction [32]. At low temperature the kinetics are very complex and are best interpreted, in electron transfer theory, by assuming a distribution of distances between P-700^+ and F_A^- [32], in agreement with the results on forward electron transfer. A study of the rate of the back-reaction at a given temperature after illumination at different temperatures indicated that the photoreduction of F_A or F_B is accompanied by a change in conformation, which takes place above 150 K and slows down the back-reaction [27].

Direct electron transfer from P-430 to exogenous acceptors (methyl or benzyl viologen, safranine T, etc.) has been demonstrated; the reaction also occurs with ferredoxin [9]. Many other acceptors can accept electrons from PS I but their site of reaction is not known. Recently, methyl purple has been introduced as a specific PS I acceptor with useful spectroscopic properties [33].

2.2.2. Centre X, an intermediate acceptor?

Below 77 K the photoinduced electron transfer between P-700 and F_A (or F_B) is essentially irreversible. It became apparent, however, that a fraction of the photooxidized P-700 is re-reduced in hundreds of milliseconds. This proportion of centres undergoing the reversible reaction greatly increases at low redox potential, when F_A and F_B are chemically reduced. This result was taken as evidence for the existence of an electron acceptor (originally named X, but designated F_X in this review), more primary than F_A and F_B [34,35]. The EPR spectrum of F_X^- has been obtained; it is characterized by three broad lines at g = 1.78, 1.88 and 2.08. F_X has been considered as an iron-sulfur centre of unusual structure (for review, see Ref. 26). This attribution is consistent with data from Mössbauer spectroscopy. In redox titrations, the appearance of the reversible signals associated with the photoreduction of F_X occurs as F_B (or F_A in some cases) goes reduced. A titration of the EPR signal of F_X^- gave an approximate value of -705 mV for the F_X/F_X^- couple [36]. On the basis of the previously described experiments, it has usually been assumed that F_X is a more primary electron carrier, located before F_A or F_B.

At room temperature, flash absorption studies revealed that an electron acceptor designated A_2 was functioning under conditions where F_A and F_B were presumably reduced [37]. The state (P-700^+, A_2^-) is formed upon flash excitation and recombines with $t_{1/2} \simeq 250$ μs. The difference spectrum due to its formation was analysed into contributions of P-700^+ and A_2^-. The latter includes mainly a small and broad bleaching around 430 nm, and perhaps some absorption shifts in the red. These absorption properties, together with the disappearance of the A_2 absorption signal when iron-sulfur proteins are denatured [38,39], indicate that A_2 may be an iron-sulfur centre.

In early work, it was proposed that A_2 and F_X were the same chemical species detected under different experimental conditions [37]. This proposal fits most of the present data, although F_X has also been proposed to be identified with P-430. In recent work [40,41] it was shown that a mild treatment with lithium dodecyl sulfate denatures F_A and F_B, but not F_X. In such preparations, a flash-induced charge separation at room temperature decays with $t_{1/2}$ = 1.2 ms. This could be equivalent to the 250 μs decay observed earlier which occurs when F_A and F_B are chemically reduced, the kinetic difference originating possibly in an electrostatic effect of F_A^- and F_B^- on the lifetime of (P-700$^+$, F_X^-). An important difference between the behaviour of A_2 and F_X resides in the efficiency of their light-induced reduction: a saturating flash reduces A_2 in all the reaction centers at room temperature [37], whereas F_X is reduced in only 10–15% of the centres at low temperature [31]. A good correlation has been established between F_X and A_2 by studying the effect of time in darkness after illumination at 3°C, in PS I particles [42].

2.2.3. *Primary acceptors: A_0 , A_1*

In the present state of our knowledge two electron acceptors are implicated between P-700 and the bound iron-sulfur centres. They are named A_0 and A_1 because they have not been chemically identified. The first evidence for the existence of these acceptors came from kinetic absorption studies, under conditions where F_X (or A_2) was reduced or inactive [37,39]. An important development resulted from the discovery of a spin-polarized triplet state [21,22] and the hypothesis of its formation as a result of a back-reaction between P-700$^+$ and a reduced primary acceptor. The analysis of the triplet signal versus the extent of reduction of the acceptors [43,44] led to the present hypothesis of two acceptors, A_0 and A_1, arranged sequentially.

The species A_0 is assumed to be the direct partner of P-700 in the primary photochemistry: P-700*...$A_0 \rightarrow$ (P-700$^+$, A_0^-). The study of A_0 has been performed using three different methods.

(i) A recent study using picosecond flash absorption (Ref. 45; see reference to previous picosecond work therein) has shown that the state (P-700$^+$, A_0^-) lasts for nanoseconds under reducing conditions. The difference spectrum indicated that A_0 is a chlorophyll species. Under non-reducing conditions, however, the absorption of A_0^- was not observed and it was supposed that A_0^- reduces A_1 in less than 50 ps. That a chlorophyll molecule can function as the very low-potential A_0 ($E_m \leqslant$ $-$ 1.0 V) is in agreement with the known redox properties of chlorophyll in vitro [46].

(ii) Illumination of PS I particles under reducing seems to allow A_0^- to be accumulated. A difference absorption spectrum has been obtained and associated with the reduction of A_0 [47,48]; it includes a bleaching at 670 nm, suggestive of a Chl *a*. A radical formed under similar conditions has EPR and ENDOR spectra compatible with a chlorophyll *a* anion [49,50]. However, the absorption bleaching at 670 nm does not agree with the 693 nm bleaching obtained in picosecond studies, leaving doubts about the nature of the accumulated species.

(iii) When forward electron transfer is blocked after A_0, the primary biradical follows behaviour previously studied in bacterial reaction centres (see Chapter 3, this volume): initially formed as a singlet state it lasts for a rather long time, which permits an oscillation of spin dephasing (giving a triplet biradical) and rephasing (giving a singlet biradical), and eventually recombines into a neutral singlet or triplet state. This situation occurs when A_1 is (photo-)chemically reduced, or rendered inactive by a treatment with a detergent [11,21,22,43,44]. The triplet state thus populated by the so-called radical-pair mechanism is characterized by a largely unequal population of the spin sublevels which renders it relatively easy to detect by EPR at low temperature. Absorption studies have shown that the triplet state is localized on P-700 [11,20] and that its lifetime is highly temperature-dependent: 10 μs at 294 K and 800 μs below 80 K [39]. At low temperature the triplet sublevels decay with different kinetics [20,51]. The lifetime of the biradical (P-700^+, A_0^-) is about 50 ns, as measured by flash absorption [52]. Delayed emission arising from PS I presumably originates in the recombination of (P-700^+, A_0^-) forming a singlet excited state or the triplet state which can decay by phosphorescence. The influence of external magnetic fields on the light emission [53,54] is consistent with the idea that the recombination produces a chlorophyll singlet excited state.

The species A_1 is as poorly known as A_0. Flash absorption studies revealed that, at low temperature, P-700 is photooxidized by a flash and re-reduced mostly with $t_{1/2}$ = 120 μs [31]. This reaction is much faster than the back-reactions involving iron-sulfur centres, and EPR measurements indicated that F_X was not the partner of P-700. This reaction is much slower than the back-reaction with A_0^- and the difference spectrum is clearly different from that of ^{3}P-700. The 120-μs phase of P-700^+ reduction is in the range expected for the back-reaction between P-700^+ and A_1^-. When F_X is reduced the 120-μs decay is replaced by a 20-μs decay, with a similar difference spectrum: this was attributed to the decay of (P-700^+, A_1^-) accelerated by a coulombic repulsion due to F_X^- [55]. Difference spectra due to the formation of (P-700^+, A_1^-) [31] or to the accumulation of A_1^- [48] show that the reduction of A_1 induces only very weak absorption changes in the range 350–900 nm, proving that A_1 is not a tetrapyrrolic molecule. After accumulation of A_1^- its EPR spectrum has been measured: it has a g value of $\simeq$ 2.005, a 10–11 G width, and a clear anisotropic character [43,44,56]. The organization of electron carriers has been probed by means of CIDEP (chemically induced dynamic electron polarization). The details of this process are outside the scope of this review. In brief, the transient perturbation of the EPR signal of P-700^+ and A_1^- is measured and analysed in terms of interaction between the photoreactants and of dynamic properties: chemical decay and spin-lattice relaxation. The interpretation of the early CIDEP measurements suffered from the lack of information at that time concerning the nature and number of electron acceptors. More recent work [56–59] is interpreted in a more realistic manner: two acceptors A_0 and A_1 are involved, preceding F_X, and the properties calculated for their radical anions agree with those obtained in steady-state EPR. The chemical nature of A_1 is not yet known. The most commonly envisioned hypothesis is that it is a quinone. This is consistent with absorption and EPR data, and with the occurrence of quinones (mainly vitamin

K_1) bound to purified PS I particles [60]. Recent work by flash absorption [61] strongly supports the hypothesis that A_1 is a vitamin K_1. A_1 must reduce the bound iron-sulfur centres and thus have a very low reduction potential. The E_m for the one-electron reduction of quinones can be rather low in purely apolar media [62]. These values are perhaps compatible with the functional properties of A_1, which is probably located in a hydrophobic protein and must have an E_m of about −0.9 V.

2.2.4. Overview of primary reactions and of electron acceptors

The quantum yield of charge separation in PS I is close to 100%. It apparently decreases under some conditions, but this is probably due to impaired energy transfer from the antenna to P-700 [52,63]. Fluorescence is a source of energy loss which could be used to gain valuable insights into primary reactions. Unfortunately, fluorescence of the PS I reaction centre is weak and its analysis is complicated by the emission from the Chl *a*/*b* PS I antenna and from a species emitting strongly around 730 nm at low temperature. The fluorescence lifetime in PS I mainly reflects properties of the PS I antenna (reviewed in Ref. 64). Hole-burning experiments in the P-700 absorption band have led to the conclusion that excited P-700 has a lifetime of at least 50 ps [65]. The fluorescence yield in PS I is dramatically influenced by temperature, and less strongly by the redox state of P-700 and of the acceptors (see Ref. 66 for a review on energy trapping).

Kinetics of electron transfer have been measured for the electron return from all the reduced acceptors to oxidized P-700. The rates of the forward steps, however, are poorly known in the absence of convincing kinetic absorption data. Electron spin echo provides a submicrosecond time resolution. A decay phase of 170 ns has been attributed to the electron transfer from F_X^- to F_A or F_B [67], but it could also be attributed to the reoxidation of A_1^-.

There are indications that the rates of electron transfer are very heterogeneous in PS I, especially at low temperature. This seems to be the best way to interpret several observations: (i) a large range of yields for the photoreduction of F_A at 10 K [31]; (ii) if F_X is on the main path of electron transfer, a small fraction of it can be photoreduced with a high yield and a large fraction is apparently not reducible at low temperature [30,31]; (iii) the triplet state of P-700 is populated under conditions where A_1 and F_X are oxidized, specially in particles prepared with Triton [43,61]. These observations can be explained by a competition between forward and backward reactions, and by a large distribution of their relative rates. At low temperature a number of conformations appear to be frozen in, as in several other biological systems such as hemoglobin and bacterial reaction centres.

2.3 Electron donation to P-700

After its photooxidation, P-700 stays oxidized for more than a few microseconds. It is re-reduced by the soluble copper protein plastocyanin or, in cyanobacteria and some algae, by the soluble cytochrome *c*-553. The relationship between plastocyanin and P-700 has been mainly studied through kinetic analysis of the P-700 absorption changes after a flash (Refs. 68,69, and references therein). A major phase of reduction, with $t_{1/2}$ = 12 μs, was attributed to electron donation by plastocyanin bound to the reaction centre. A slower phase, with $t_{1/2} \simeq 250$ μs in chloroplasts, appears to be of second order and to result from a diffusion-limited reaction; its rate is influenced by the concentration of reduced plastocyanin, by the viscosity and temperature. A very slow phase of small amplitude is not understood. Electron transfer from plastocyanin to P-700 is inhibited at low temperature as a result of two phenomena: a decrease in the amount of bound plastocyanin and a slowing down of the diffusion-controlled reaction [69]. The kinetic properties of electron transfer from plastocyanin to P-700 are very similar to those from soluble Cyt c_2 to the reaction centre of *Rhodopseudomonas* (*Rhodobacter*) *sphaeroides* [70]: in particular the kinetic analysis indicates two states of binding: a 'close' state with a fast transfer, and a 'distant' state with a slower transfer [69]. In some green algae, the thylakoids contain a soluble Cyt *c*-553, in addition to plastocyanin; their respective concentrations are greatly influenced by the concentration of Cu or Fe in the medium [71]. This cytochrome appears to function like plastocyanin [72].

The structure of plastocyanin is known at a highly refined level, which allows interesting hypotheses on which part of the molecule is involved in interactions permitting electron transfer [73]. Several areas on the surface of the molecule have been modified with chemical reagents, which can change the binding and reactivity [74], which are highly sensitive to electrical interactions, as shown by the influence of cations on the rate of electron transfer (see e.g. Refs. 68 and 75).

2.4. *Structure of the PS I reaction centre*

2.4.1. *Polypeptides and redox centres*

The redox centres, which we have described above, are held together in a specific and rather stable conformation by a few polypeptides (Fig. 3). In PS I it has not been possible to separate the polypeptides while keeping some of their functional

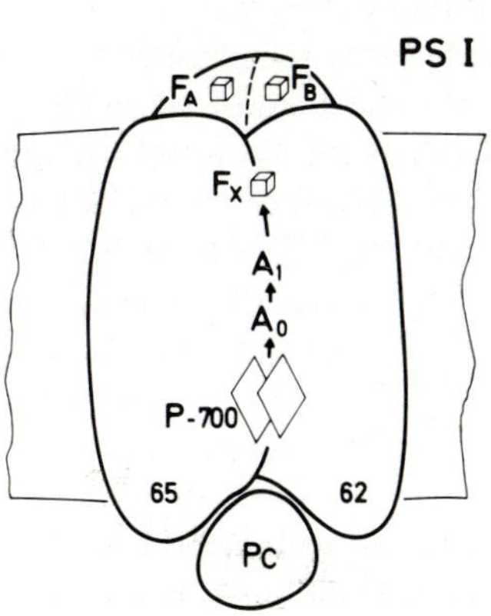

Fig. 3. Tentative representation of the structure of the PS I reaction centre. The two large polypeptides are supposed to hold the primary donor P-700 and the first acceptors (see text).

character. In order to understand the association between redox centres and polypeptides the most widely used approach has thus been to prepare rather large PS I particles which perform many steps of electron transfer and which comprise many polypeptides, and to peel them off progressively [76]. The most intact particles reveal a large number of polypeptides upon electrophoresis under denaturing conditions. Primary photochemistry takes place in the 'core' of PS I, also named CP1, which consists of polypeptides of 60–70 kDa [77–79]. In many cases, two polypeptides appear on the gels with slightly different molecular weight. It has been thought that one of them might be a degradation product of the other, but an analysis of the chloroplast DNA showed the existence of two genes, coding for two homologous polypeptides of 83.2 and 82.5 kDa [80] which are probably the precursors of the two polypeptides of CP1 particles. The stoichiometry of these large polypeptides in CP1 is not fully established, although most studies indicate two polypeptides per P-700. The location of P-700 is a matter of conjecture: it could be within one subunit or at the interface between the two. The likely dimeric nature of P-700 might reflect the presence of one Chl molecule in each of the two polypeptides, as is the case in reaction centres of purple bacteria. CP1 also contains A_0 [21,39] and perhaps also A_1 [60], although not a full complement of it and not in an active form. The iron-sulfur centres F_A and F_B are certainly associated with small polypeptides but some authors place them with polypeptides of about 15–19 kDa [81,82], whereas there are good arguments for associating them with smaller polypeptides around 8 kDa [83]. A polypeptide of 20 kDa seems to be involved in the electron transfer from plastocyanin to P-700 in higher plants [76], but a similar subunit is absent in a cyanobacterial PS I particle which efficiently oxidizes the soluble Cyt *c*-553 [84]. Centre F_X has been suggested to be associated with the core of two large subunits [83], and recent experiments of selective denaturation strongly favor that hypothesis [40]. As a general comment, it appears that much progress still has to be made to reach a precise view of the anchoring of redox centres in PS I.

2.4.2. Photosystem I light-harvesting antenna

PS I particles can be grossly divided into two classes with antenna sizes of 100–150 and 50–60 Chl molecules [85,86]. The biggest particles contain a peripheral antenna, often named LHCI or CPO, with about 40 Chls *a* and *b* carried by polypeptides of MW 20–25 kDa [87–89]. This peripheral antenna can be disconnected, leaving a PS I core with 50–60 Chl *a* [77–79,86,90,91]. It seems that in this core, all of the Chls (and also about 8 molecules of β-carotene) are associated with the two large polypeptides. The Chl content can be further decreased by treatments with detergents or organic solvents, but this removal is apparently progressive and non-specific. It thus seems that a reaction centre such as in purple bacteria, in which there is practically no pigment having a pure antenna function, is not present in PS I and that the PS I reaction centre has an intrinsic antenna of $\simeq$ 50 Chls. How these Chls are bound by the two large subunits is not yet known. The recent elucidation of the primary structure of these apoproteins [80] should allow better-founded speculations.

2.4.3. Organization of the reaction center in the membrane

The PS I centre performs electron transfer from the inside to the outside of the thylakoid, as shown by various functional studies and by the formation of an electrical membrane potential [92]. More details on these structural properties have been obtained by two types of method.

(i) The orientation of various redox centres has been probed by absorption spectroscopy and by EPR, with oriented samples. EPR was used for the iron-sulfur centres, which all have anisotropic *g* values and a well-defined orientation in the membrane [93–95]. Absorption studies showed that the Q_Y transition of P-700 lies flat in the membrane [96]. Photoselection studies additionally revealed that a β-carotene molecule has a well-defined position with respect to P-700 [97]. The proteins of the reaction centre are also oriented, with the axis of their α-helices tilted at $\simeq 35°$ from the normal to the membrane plane [98].

(ii) The transverse organization has been probed with antibodies, with impermeant chemical modifiers and with proteases [76,92,99,100]. It appears that the reaction centre spans the membrane. Some of the subunits traverse the membrane, but a majority are apparently accessible only from the lumenal or from the stroma side.

3. Photosystem II reactions

3.1. Introduction

Photosystem II (PS II) is a pigment-protein complex which spans the thylakoid membrane. When excited by light it extracts electrons from water, resulting in the release of molecular oxygen and of protons on the inside of the membrane. The electrons are delivered to the other side of the membrane where plastoquinone (PQ) is reduced with the uptake of two protons. The reduced plastoquinone (PQH_2) acts as a source of electrons for other electron transfer reactions and the proton gradient established directly by PS II photochemistry represents a significant fraction of the stored energy obtained from the light.

PS II has probably been subjected to more investigation than any other photosystem. The unique photodriven water-splitting enzyme, the source of most atmospheric oxygen, has been an enticing yet elusive subject for research. In addition, PS II is the site of action of a large number of commercial herbicides and it is probably the site of damage when plants are exposed to high light intensities (photoinhibition). These reasons, together with the relative ease in measuring PS II photochemistry by luminescence and fluorescence, phenomena almost wholly associated with PS II, have provided a formidable literature on the subject (see Refs. 6 and 7 for recent reviews).

Despite this considerable attention, PS II has remained rather mysterious. Recently, advances in biochemical techniques, which have led to the routine preparation of PS II, free from PS I contamination, have allowed the use of more direct measurements of PS II photochemistry (particularly absorption spectrophotome-

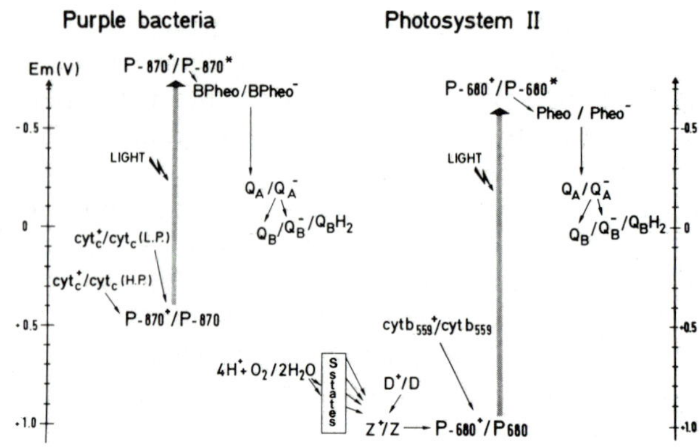

Fig. 4. A comparison of electron transfer in reaction centres of purple bacteria and PS II. Components are situated at their approximate or estimated midpoint potential. For PS II, these are discussed in the text; for purple bacteria, see Chapter 3.

try and EPR). Thus a number of the more enigmatic aspects of PS II have been demystified in recent years. There are, however, plenty of areas which remain obscure. In this section we will present a simplified picture of PS II but some of the major unanswered questions will be pointed out.

3.2. PS II photochemistry – a comparative view

When the components of the PS II reaction centre are drawn on a redox scale and compared in this way to those of the purple bacterial reaction centre, a remarkable similarity can be seen between the electron acceptors in each system (Fig. 4). The chemical natures of these components are extremely similar, being made up of a complex of two quinones, an iron atom and a pheophytin (a bacteriopheophytin in bacteria). The donor side of PS II in the redox scheme is, however, not comparable to that in bacteria. P-680 may appear to be structurally similar to P-870 in bacteria in that it is made up of chlorophyll (bacteriochlorophyll in bacteria) and that is acts as the primary electron donor; however, the P-680/P-680$^+$ redox couple is approximately 600–800 mV more oxidizing than the equivalent bacterial redox couple P-870/P-870$^+$, E_m = +450 mV). In addition, PS II has an array of high-potential components which make up the O_2-evolving enzyme and which are clearly unique to that system.

3.3 The electron acceptor side

3.3.1. The quinone-iron complex

(a) Q_A, the first quinone acceptor. A great deal of work on this component has been done using fluorescence as a probe of its redox state. When it is oxidized the fluorescence yield is found to be low (F_0), but when it is reduced the fluorescence yield is high (F_{max}). Thus the component was designated Q for quencher of fluo-

rescence. By a happy coincidence Q turned out to be a quinone. Being the first of two quinones it is designated Q_A.

When Q_A undergoes reduction it gives rise to a number of changes detectable by absorption spectrophotometry. The first of these to be identified was a change at 320 nm [101], another was a bandshift at 550 nm [102]. The full absorption spectrum was obtained by Van Gorkom [103] and by its similarities to the in vitro spectrum of plastosemiquinone anion its identity was established. The change around 550 nm (C-550) was attributed to a bandshift of a pheophytin molecule close to Q_A^- [103]. Extraction and reconstitution experiments have supported the identity of Q_A as a plastoquinone molecule [103]. Strangely, the presence of carotenoid seems to be required for the bandshift at 550 nm to take place even though carotenoid itself plays no direct role in electron transfer on the acceptor side [104].

Q_A is a plastoquinone molecule which is rather firmly bound to the PS II reaction centre protein. As a result of this binding and the influence of the protein, the chemistry of Q_A is very different from that of a free plastoquinone at physiological pH values. The free plastoquinone in the thylakoid membrane, for example, undergoes reduction as a two-electron, two-proton event, since the semiquinone is highly unstable. In contrast, the plastoquinone that makes up Q_A undergoes a single reduction, forming a stable unprotonated semiquinone. Under normal circumstances Q_A^- probably never undergoes a further reduction step; however, in reducing conditions, continuous illumination can force Q_A to become fully reduced [105].

The normal one-electron reduction of Q_A occurs with a midpoint potential lower than 0 mV but the actual value is still a subject of some controversy (see Section 3.3.3 below). The value often cited by those not wishing to get bogged down in that controversy is that obtained by measuring the redox potential dependence of Cyt *b*-559 photooxidation at 77 K in chloroplasts [106]. A value was obtained which was pH-dependent at pH values below pH 8.6. The E_m value at and above pH 8.6 (the pK of Q_A^-) was −130 mV. This value is usually considered the operative E_m, since Q_A^- is not protonated on a functional time scale. This assumption was also made earlier for the E_m of Q_A/Q_A^- in purple bacteria [107]. Arguments for and against the use of the E_m-pK are discussed in detail in a recent review [108].

The first EPR signal which was attributable to Q_A^- was reported by Klimov et al. [109] in PS II preparations from which iron had been removed. The signal was a 9-G-wide free radical centred at $g = 2.0044$, and is typical of a semiquinone. In centres where the iron was still present, a broader EPR signal was present [110,105]. This signal at $g = 1.82$ is very similar to a signal attributed to the semiquinone anion interacting with the iron, $Q_A^-\ Fe^{2+}$, in purple bacteria which had been discovered several years earlier (see Chapter 3). In PS II the semiquinone-iron signal at $g = 1.82$ was found to be only one of two interconvertible forms [105]. The second form, having a broad resonance centred at $g = 1.9$, dominates at high pH and is changed into the $g = 1.82$ form by lowering the pH or by the binding of some herbicides. The two different EPR signals presumably represent a slight structural modification which affects the interaction between the iron and the semiquinone. The $g = 1.9$ form of the $Q_A^-\ Fe^{2+}$ is also found in some species of purple bacteria, in particular in *R. rubrum* [111].

In forward electron transfer, Q_A is reduced very rapidly after a flash. Direct measurements of the kinetics of Q_A^- formation have not been reported, although estimations of hundreds of picoseconds have come from indirect measurements [112].

Q_A^- donates an electron to the next acceptor, Q_B, with kinetics that depend on the redox state of Q_B (see part b of this section). The electron transfer step between Q_A^- and Q_B is inhibited by a number of commercial herbicides, of which DCMU is the most commonly used in studies of PS II. If forward electron transfer is blocked by DCMU, the lifetime of Q_A^- formed by illumination depends on the availability and stability of positive charges on the donor side of PS II with which the electron on Q_A^- can recombine. If the positive charge is removed from the reaction centre by an electron being provided from an exogenous electron donor (e.g. high concentrations of NH_2OH or ANT2p) then Q_A^- (in the presence of DCMU) is stably trapped for many minutes. This reflects the fact that although Q_A^- is a rather low-potential species it is hidden within the protein from contact with the PQ pool other than through the native Q_B reactions.

(b) Q_B, the second quinone acceptor. In forward electron transfer Q_A^- donates to an electron acceptor designated Q_B. The reduction of Q_B to Q_B^- gives rise to characteristic absorption changes in the UV which are similar to those which occur when Q_A is reduced and which are characteristic of plastosemiquinone formation [113]. The full optical Q_B^- minus Q_B spectrum is similar in many respects to that of Q_A^- minus Q_A but the bandshifts in the blue and green parts of the spectrum are different [114], indicating that Q_B^- is different from Q_A^- in terms of its proximity to the pheophytin, the spectrum of which is electrochromically shifted.

The redox properties of Q_B are also unlike those of plastoquinone in the pool. The semiquinone form, Q_B^-, is tightly bound to a protein of the reaction centre and is thus stabilized. Q_B^- is much more stable than Q_A^-, since forward electron transfer does not take place from Q_B^-. The lifetime of Q_B^-, like that of Q_A^- in the presence of DCMU, is determined by the stability and availability of positive charges on the donor side. For example, Q_B^- recombination occurs with S_2 or S_3 (Ref. 115, and see section 3.5) with a $t_{1/2}$ of approximately 30 s [116] but when Q_B^- is present in centres where the stable S states, S_0 and S_1, are present, Q_B^- is stable for hours. This probably explains why a certain proportion of Q_B^- is present even in PS II which has been dark-adapted for long periods. A number of measurements have indicated the involvement of proton uptake when Q_B is reduced to semiquinone form [117], although the optical spectrum is more compatible with Q_B^- being the unprotonated anion. This can be explained by the protonation of a group on the protein close to Q_B^-, as first proposed in purple bacterial reaction centres to explain similar phenomena [118].

Unlike Q_A^-, Q_B^- can accept a second electron in a physiological reaction. The kinetics of electron transfer from Q_A^- to Q_B are faster than those of Q_A^- to Q_B^- [116,119]. Half-times of $\simeq 100$ and $\simeq 200$ μs have recently been reported for Q_A^- to Q_B and Q_A^- to Q_B^- respectively [116]; however, values significantly different from these have also been reported [119].

The second reduction of Q_B is accompanied by a true protonation forming the hydroquinone, Q_BH_2. This is then thought to leave the binding site on the reaction centre and to become part of the membrane pool. The vacant site on the reaction centre is then available for binding of an unreduced plastoquinone molecule from the pool becoming Q_B, ready to accept further electrons from Q_A^- [120].

The affinity of the Q_B binding site is low for PQ, lower for PQH_2, but high for Q_B^- [120,121]. The mechanism of action of DCMU-like electron transfer inhibitors is thought to be the binding of the herbicide in or close to the Q_B site on the protein in competition with PQ binding [120,121]. Since Q_B^- is tightly bound to the protein, electron transfer from Q_A^- to Q_B^- is not blocked by these herbicides [122]. However, the addition of herbicides to centres in which Q_B^- is present results in formation of Q_A^- due to herbicide binding in competition with Q_B. Normally in the equilibrium reaction, $Q_A\ Q_B^- \rightleftharpoons Q_A^-\ Q_B$, very little Q_A^- is present because of the difference in the functional redox potential between the Q_A/Q_A^- couple (−130 mV) and the Q/Q_B^- couple (probably around 100 mV more oxidizing than Q_A/Q_A^-). However, when an inhibitor is present it competes with PQ binding for the Q_B site and thus the equilibrium is displaced to the right i.e. Q_A^- is formed ($Q_A\ Q_B^-$ + inhibitor $\rightleftharpoons Q_A^-\ Q_B$ + inhibitor $\rightleftharpoons Q_A^-$ inhibitor + PQ).

Plants which have developed a resistance to these kinds of herbicides have a single amino acid change in a reaction centre protein which is presumably in the region of the Q_B binding site. In addition to imparting resistance to specific herbicides by lowering their binding affinity, the amino acid change results in the kinetics of electron transfer from Q_A^- to Q_B becoming slower. In addition, the affinity of the Q_B site for PQ is also affected by the herbicide resistance [123,124].

When a series of flashes is given to PS II in the absence of inhibitors, Q_B^- is formed stably on the first flash. On the second flash, Q_BH_2 is formed and is replaced by a PQ from the pool. These reactions give rise to the characteristic period of 2 oscillations of electron transfer out of the PS II reaction centre when excited by a series of flashes. Many different kinds of electron transfer phenomena reflect this 2-electron gate. These include absorption changes from the semiquinone, Q_B^- [113], differential kinetics of electron transfer from Q_A^- to Q_B and from Q_A^- to Q_B^- measured by fluorescence [116,119], the extent of Q_A^- formed by addition of DCMU after the flash, measured by fluorescence and luminescence [125] and differential DCMU binding to centres where either Q_B^- or Q_B is present [122]. It is also of note that since electrons arrive in the PQ pool two at a time under some conditions they also arrive at PS I with an oscillating 0,2,0,2 pattern. Thus electron transport through PS I measured after a series of flashes can be used as a measure of the redox state of Q_B [126,127]. Paradoxically it was by measuring PS I that the existence of Q_B was first indicated [127]. An identical 2-electron gating mechanism of electron transfer through an analogous Q_B component was found some years later in the purple bacterial reaction centre (see Chapter 3).

In PS II the E_m values for $Q_B/Q_B^-(H^+)$ and $Q_B^-(H^+)/Q_BH_2$ redox couples are not known. Estimations of the E_m values of these couples from kinetic parameters vary depending on the value taken for the Q_A/Q_A^- couple. Taking −130 mV for Q_A/Q_A^-, a pH-dependent value for $Q_B/Q_B^-(H^+)$ of close to 0 mV has been estimated, with the E_m of $Q_B^-(H^+)/Q_BH_2$ being somewhat lower than this [116,117].

There is only one rather preliminary report of an EPR signal arising from Q_B^- [128]. The signal is broad, centred on $g = 1.92$ and is attributed to a semiquinone interacting with Fe^{2+}. The equivalent signals reported in purple bacteria are centred at $g = 1.82$ (e.g. Ref. 129); however, a signal from Q_B^-Fe in *R. rubrum* has been found to be similar to that observed in PS II (Rutherford, A.W. and Beijer, C., unpublished).

(c) The iron-Q_{400} As described in the previous sections, Q_A^- and probably Q_B^- are close to a ferrous iron atom. The first indication of this association came from the observation of an EPR signal arising from the pheophytin acceptor in its reduced form [130]. The signal showed a splitting which is remarkably similar to that seen in purple bacteria from reduced bacteriopheophytin in centres where $Q_A^-Fe^{2+}$ was present [131]. The presence of both semiquinone and iron was required for the splitting of the bacteriopheophytin signal to occur. By analogy, the existence of $Q_A^-Fe^{2+}$ was postulated in PS II [130]. Extraction and reconstitution experiments supported this hypothesis [109]. Direct observation of the Q_A^-Fe signals came somewhat later [110,132,105].

In the bacterial reaction centre the iron is situated between the two quinones [133], hence the almost identical interactions between Q_A^- and Fe^{2+} and between Q_B^- and Fe^{2+}. The distance between the quinone and the iron estimated from considerations of the magnetic interaction [134,135] was verified as being 7 Å by X-ray crystallography [133]. This value can probably be directly applied to the distance between Q_A^- and Fe^{2+} in PS II, since the EPR signal is so similar to that in bacteria.

The function of the iron remains unknown in both the bacterial reaction centre and PS II. In bacteria the iron can be replaced by other divalent transition metals with no apparent effect on the electron transfer reactions [136]. Removal of the metal slows down (by 2-fold) the electron transfer rate from Q_A^- to Q_B but does not block electron transfer [136]. Despite these observations the conservation of this metal within the quinone complex throughout the evolutionary processes that separate the purple bacteria from higher plants indicates an important role for this component. As yet we are still ignorant of that role.

Recently, Petrouleas and Diner [137] showed that the Fe^{2+} could be oxidized by the addition of oxidants (e.g. ferricyanide) to PS II preparations. Illumination of the oxidized samples led to reduction of Fe^{3+} to Fe^{2+}. The Fe^{3+} gives rise to EPR signals at $g = 8$ and $g = 6$ which disappear upon illumination at 200 K [137] or with a single flash at room temperature [138]. Whether the change in oxidation state of the iron has any physiological significance is not known. However, it does supply the explanation to one of the long-standing PS II photochemical mysteries. It has been observed in fluorescence studies that an extra acceptor is present in samples incubated with ferricyanide and that this component accepts electrons from Q_A^- rather rapidly and is insensitive to DCMU. This effect titrates with a midpoint of $\simeq$ 400 mV (pH 7.0) [139]. This high-potential acceptor has been designated Q_{400}. Q_{400} has now been identified as Fe^{3+} and this provides a very tidy explanation for the phenomena [137].

More recently it has been shown that the Fe^{2+} in PS II can be oxidized by the unstable semiquinone form of some exogenous quinone acceptors [139]. The unstable semiquinone is formed on the first flash of a series, and it then oxidizes the Fe^{2+} to Fe^{3+}, forming the hydroquinone. On the second flash the iron^{3+} is rapidly rereduced. This period of two oscillation of iron oxidation and reduction can be observed by EPR [138].

3.3.2. Pheophytin – the intermediate electron acceptor

In purple bacteria a number of different lines of evidence led to the conclusion that bacteriopheophytin (BPh) acts as an electron carrier between the primary donor and Q_A (Chapter 3). When Q_A is reduced illumination results in the photoaccumulation of reduced bacteriopheophytin, detected by its characteristic absorption changes and by an EPR signal split due to its interaction with $Q_A^-Fe^{2+}$. At temperatures too low for rapid photoaccumulation of BPh^- to take place, illumination results in formation of a triplet state of the primary donor P-870 which has a polarization pattern characteristic of its formation by recombination of a radical pair. When BPh is reduced this triplet state cannot be formed. The most direct proof that BPh acts as a primary acceptor comes from the direct observation by absorption spectroscopy of BPh reduction within a few picoseconds after the flash. The BPh is reoxidized in 200 ps by electron transfer to Q_A or, if Q_A is already reduced, by recombination in 14 ns (see Chapter 3).

Photoaccumulation of reduced pheophytin ($Pheo^-$) in PS II under reducing conditions was first reported by Klimov et al. [140] and it was proposed that Pheo might play a role as an intermediate acceptor analogous to that of BPh in purple bacteria. Support for the analogy came from the observation of a split $Pheo^-$ EPR signal [109] and from the observation of a characteristically spin-polarized triplet signal from P-680 [141]. The ability to form the triplet is lost when pheophytin is photoaccumulated. Since the splitting of the $Pheo^-$ EPR signal induced by Q_A^-Fe is very similar to that seen in the purple bacterium *Chromatium vinosum* it is probable that the spatial relationship between these components is similar to that in purple bacteria.

Flash kinetic absorption measurements have provided further support for its role as an intermediate acceptor. In PS II in which Q_A^- was reduced, the formation of some $Pheo^-$ which decayed in only a few nanoseconds was observed [142] and recently a change attributed to $Pheo^-$ decaying in 200 ps was reported in particles in which Q_A was oxidized before flash illumination [112].

The reduced minus oxidized difference spectrum of $Pheo^-$ photoaccumulated in PS II, showing large changes in the blue and red and in particular a small bleaching at 545 nm, is typical of Pheo reduction compared to spectra obtained in vitro [109,140]. The EPR spectrum from $Pheo^-$ is split by $Q_A^-Fe^{2+}$ when it is present but is a featureless free radical signal centred at $g = 2.0030$ when $Q_A^-Fe^{2+}$ is absent [130]. Redox titrations of the $Pheo/Pheo^-$ couple have given values of approximately -610 mV [143,144], which are similar to those seen for this couple in vitro.

The photoaccumulation of $Pheo^-$ in PS II is accompanied by a large decrease in the level of fluorescence [140,109]. This observation led to the hypothesis that variable fluorescence (i.e. the fluorescence increase associated with Q_A reduction) is in fact luminescence from $P680^+$ - $Pheo^-$ recombination [140]. However, questions have been raised concerning this hypothesis, based largely on fluorescence lifetime measurements [145]. It is quite possible that, when Q_A is reduced, the excitation resides preferentially on the antenna chlorophylls, because of a lower extent of charge separation between P-680 and Pheo. Variable fluorescence would then originate in the antenna and not as a reaction center luminescence.

3.3.3. Other possible acceptors and heterogeneity

(a) Introduction. There are numerous examples of electron acceptors, other than those described above, which have been proposed to exist in PS II. Since the analogy to bacteria is so compelling on the electron acceptor side of PS II, there is a temptation to disregard these reports and to hope that increased understanding will allow them to be painlessly integrated into the analogy model. Exactly this happened recently for Q_{400} (see above). However, the explanations for some of these effects may not be so easy to find. The complexity of plants compared to bacteria and the tendency to work with PS II in unfractionated membrane systems rather than the isolated (and, perhaps, homogenized) reaction centres of bacteria may give rise to observations in PS II which are not analogous to those in bacterial reaction centres.

(b) Q_α and Q_β heterogeneity. A minority population (30%) of PS II centres, designated β centres, has been proposed to exist in the stromal lamellae (see Refs. 6, 146, 147 for recent reviews). These centres have an antenna system different from that of the majority of centres (designated PS II α centres), which are proposed to be located largely in the membranes of the grana stacks. The different antenna in these centres results in slower kinetics for Q reduction when measured by fluorescence induction. Strangely, the E_m of Q_A/Q_A^- in β centres measured by fluorescence is much higher (E_m=120 mV [148,149]) than in PS II α. This value is higher than the E_m of the PQ pool and it is therefore no surprise to find that the characteristic period of 2 oscillations associated with Q_B function does not seem to take place in β centres [149]. A different acceptor system has been suggested to be functioning instead of the quinone-iron complex [6,147]. Little evidence for this exists. Nearly all the evidence for the existence of the PS II β centres comes from fluorescence measurements in the presence of DCMU. Reinvestigation of some of the PS II phenomena has recently led to the proposal that some effects are due to an interconvertible heterogeneity in the affinity for DCMU – with 'PS II β' phenomena arising from centres which are less sensitive to DCMU [150]. Normal PS II centres can show a low affinity for DCMU if Q_B^- is present (see above) or if Q_{400} (Fe^{3+} see above) is present, and it has been suggested that some of the PS II β phenomena could arise from a small proportion of centres where Q_{400} (Fe^{3+}) is present even in the absence of oxidants [7].

Verification of the existence (or non-existence) of PS II β centres requires biochemical advances to produce preparations free from PS II α which could be used for more unequivocal analysis. Until then speculation on the functional role of β centres may be premature.

(c) The Q_H and Q_L phenomena. Many redox titrations of Q_A, monitored in a variety of different ways, have given titration curves with 2 waves of reduction (at 0 mV and −275 mV). Two main types of explanation have been put forward: firstly, that all the centres have two different types of acceptor (Q_{High} and Q_{Low}); secondly, that there are two populations of PS II with different redox properties for their respective Q_A acceptors (see Refs. 147, 148 and references therein). From the analogy to bacteria it is difficult to accept the existence of an extra acceptor. Indeed, in *Rps. viridis* chromatophores, 2-step titrations of Q_A reduction have been obtained [151] and no extra acceptors are present in the model of its reaction centre from X-ray crystallography [133]. The 2 waves might be explained if in some centres Q_A is less accessible to reduction than in others (i.e. an artifact of the titration). Alternatively, Q_A may really have a lower potential in a fraction of the centres. In fact it is clear that an interconvertible structural heterogeneity of $Q_A^-Fe^{2+}$ does exist [105] and this could also be related to functional and redox heterogeneity.

Although our current understanding of the Q_H/Q_L phenomenon is not yet clear (in fact the low-potential wave is absent in some titrations; e.g. Ref. 152), it seems that the more plausible explanations require no modification of the structural model of PS II based on the bacterial reaction centre.

(d) Are there electron acceptors other than Pheo functioning prior to Q_A? A number of phenomena have led to the suggestion that extra electron acceptors exist other than Q_A and Pheo.

(1) Eckert and Renger [153] measured P-680$^+$ formation and rereduction at 690 nm with microsecond time resolution and found a phase upon the second of two closely spaced flashes, which was attributed to reduction of an acceptor other than Q_A. The flash spacing was such that Q_A^- was expected to be still reduced when the second flash was given. The electron on the putative acceptor, designated X_a^-, recombined with P-680$^+$ with a $t_{1/2}$ of 35 μs and its reduction was not associated with a transmembrane potential.

(2) Joliot and Joliot [154] observed a slow phase of fluorescence yield in addition to that normally associated with Q_A reduction when chloroplasts were illuminated in the presence of DCMU. This effect was more marked when NH_2OH was used as an electron donor in place of the native system. These effects were attributed to an acceptor designated Q_2. Reduction of Q_2 corresponded to significant oxidation of NH_2OH and, as with X_a, was not associated with the generation of a membrane potential. In addition it was shown not to be a quinone, since it lacked a 320 nm absorption change and did not induce a bandshift on the pheophytin (no 550 nm change) [155]. The photoreduction of Q_2 in the presence of Q_A^- and DCMU required many flashes and its reoxidation was much more rapid than was that of Q_A^- [154]. A fluorescence increase which was not associated with Q_A reduction by Q_B^- upon DCMU addition (originally designated the ‘non B’ effect) probably also reflects Q_2 reduction [155]. Evidence was provided indicating that Q_2 (or the non B-quencher) was present in centres which contained both Q_A and Q_B [155a].

(3) Meiburg et al. [156] observed luminescence decaying with phases of 10 μs and 60 μs which was not affected by a strong external electric field across the

membrane, indicating that the charge separation responsible for the luminescence occurred parallel to the membrane. Such an arrangement would be predicted for X_a and Q_2 and thus the luminescence was attributed to P-680$^+$ recombination with the X_a^-/Q_2^- acceptor. Unlike the fluorescence effects attributed to Q_2, it was suggested that the luminescence arose in only a fraction of the centres but that Q_2 function and that of Q_A may be interconvertible.

(4) Evans et al [157] did redox titrations of the P-680 triplet state and found that it was formed at increased yield upon reduction of a component with an E_m (pH 10) of −450 mV. This reduction step did not seem to be associated with the 'Q_L' wave of Q_A reduction in contrast to an earlier suggestion [141]. To account for this effect an acceptor, U, was proposed to exist between Pheo and Q_A.

Recently, Brettel et al. [158] have investigated the existence of electron acceptors functioning earlier than Q_A by performing well-resolved flash absorption measurements of P-680$^+$ at 820 nm using pairs of closely spaced flashes. In chloroplasts in which the donor complex was intact, it was found that the second flash, given 60 ns after the first flash (a condition where P-680$^+$ is largely reduced to P-680 and Q_A^- remains reduced) produced no changes attributable to charge separation stable for longer than a few nanoseconds. This work clearly rules out the existence of an electron acceptor (apart from Pheo) functioning with a quantum yield of more than 15% when Q_A is reduced. It is possible that the $X_a/Q_2/U$ phenomena reflect a low quantum yield, side-path acceptor. The existence of such an acceptor might even be reconciled with the bacterial reaction centre analogy, where the second Bph can be reduced with a low yield (Chapter 3). By a small stretch of the analogy it is possible to explain some of the $X_a/Q_2/U$ phenomena as a low quantum yield reduction of a second Pheo in the PS II reaction centre. This kind of effect also probably explains the 'A_0^-' EPR signal photoinduced when Pheo was reduced [159]. To improve the current hazy picture of this area of PS II, absorption and/or EPR spectra of the putative acceptors are required.

3.4. The electron donor side of PS II

3.4.1. P-680, the primary donor

The primary electron donor of PS II was detected as a flash-induced absorption change attributable to chlorophyll *a* oxidation [160]. Its bleaching maximum is close to 680 nm and it is thus designated P-680 (it is also called Chl a_{II}). The fluorescence of the light-harvesting chlorophylls interferes with measurements at this wavelength, thus many kinetic studies of P-680$^+$ have been done by measuring the smaller broad absorption increase at around 820 nm [161,162]. This broad absorption in the near infrared is probably responsible for the fact that P-680$^+$ is a quencher of chlorophyll fluorescence.

Since P-680$^+$ is rapidly reduced by the native electron donor system on a time scale which ranges from 50 to 250 ns (see the next section), most studies of P-680$^+$ have been done under conditions in which the secondary donors are inhibited (e.g. Tris-washing, NH_2OH treatment, detergent treatment, extremes of pH). Even under these conditions P-680$^+$ is short-lived, being reduced either more slowly by the damaged donor system or by recombination with electrons from the acceptor side (P-680$^+$ Q_A^- recombination takes place with a $t_{1/2}$ of 100 μs at 20°C [163]).

The very high oxidizing power of P-680$^+$ leads to it oxidizing a number of different donors. Some of the components oxidized by P-680$^+$ are not functional as electron donors under physiological conditions but are oxidized when the native rapid donor system is not operational. Such donors include Cyt *b*-559, Chl and carotenoid and their properties are dealt with in a subsequent section.

The non-physiological oxidation of close-by components by P-680$^+$ has led to some confusion. In particular photoinduced free radical EPR signals were at first misassigned to P-680$^+$. However, time-resolved EPR spectra of P-680$^+$ have been reported and the *g* value of $g = 2.0027$ and linewidth of 8 G [164,165] are compatible with oxidation of a single Chl *a* molecule.

An EPR signal attributable to the spin-polarized triplet state of P-680 (^{3}P-680) formed by recombination of P-680$^+$ Pheo$^-$ was detected by illuminating PS II preparations at liquid helium temperature [141]. The zero-field splitting parameters are identical to those of triplets of monomeric Chl in vitro. The triplet minus singlet spectrum of P-680 measured by absorption-detected magnetic resonance indicated a monomeric triplet but a dimeric ground state [166].

The E_m of P-680/P-680$^+$ couple has not been measured due to its high potential. However, estimates have been made based on the activation energy of the back reactions and the measured E_m of the electron acceptors (P-680/P-680$^+$, $E_m \simeq +$ 1.1 V [143]). Such a high potential is required to oxidize water ($2H_2O/O_2 + 4H^+$, E_m pH 7.0=820 mV). From in vitro redox studies it was suggested that P-680$^+$ could be a ligated monomeric chlorophyll [167].

3.4.2. Z, the electron donor to P-680$^+$

In native PS II, P-680$^+$ reduction at room temperature takes place due to electron donation which occurs with kinetic phases in the tens and hundreds of nanoseconds range and, to a much lesser extent, with phases in the microsecond range. This reaction has been measured by absorption changes at 820 nm and 680 nm [161,162] and by the decay of the fluorescence quenching associated with P-680$^+$ [168,169].

The variation in the kinetics observed for this reaction are related to charge accumulation on the oxygen-evolving enzyme, as recently clearly demonstrated by Brettel et al [162], who measured 820 nm absorption changes as a function of flash number. When S_0 or S_1 was present, P-680$^+$ was reduced largely with a $t_{1/2}$ of 50 ns, while, when S_2 or S_3 was present, P-680$^+$ reduction slowed down ($t_{1/2}$ 240 ns). This was attributed to a coulombic effect of the positive charge present in the S_2 and S_3 states (Ref 162, and see Section 3.5). When the O_2-evolving enzyme is destroyed by, for example, treatment with Tris, the donation of electrons to P-680$^+$ is slowed down dramatically. This reaction is sensitive to the pH ($t_{1/2}$ = 4–6 μs pH 7.0, 14 μs pH 5.0) [170].

The simplest picture of donation to P-680$^+$ is one in which a single component, Z, acts as a carrier for electrons between P-680 and the oxygen-evolving complex. Time-resolved absorption spectra of Z$^+$ minus Z in Tris-inhibited [171,172] and in native [173] preparations are similar and they are compatible with the oxidation of a hydroquinone to a semiquinone cation. EPR spectra of Z^+ in the native (Signal II vf) and Tris-inhibited form (Signal II f) are also similar [174]. The unusual lineshape and the g value ($g = 2.0045$) led to the proposal that Z^+ could be a plastosemiquinone cation [175]. This was also supported by redox arguments, since QH_2^+/QH_2 couples are expected to be very oxidizing. A problem with the assignment of Z to a hydroquinone is that extraction experiments do not indicate the presence of suffcient quinone in PS II to account for quinones other than Q_A and Q_B (e.g. Ref. 176).

In solution, semiquinone cations are only stable at very low pH. Thus it is assumed that the protein provides a binding site which has no basic amino acids but which involves amino acids which hydrogen bond to the oxygen of the -OH group of the semiquinone cation [175]. Such a site must be highly inaccessible to the ambient medium and this may partially explain the difficulties encountered in attempts to extract the semiquinone.

By comparisons of the EPR and partial ENDOR spectra of Signal II with immobilized semiquinone cations in vitro, the characteristic lineshape of Signal II was explained as arising largely from hyperfine interaction due to a single methyl group at position 2 on the quinone ring [175]. Orientation data supported this assignment [177]. However, Brok et al. [178] reinterpreted the hyperfine interaction as arising from the methylene group at position 5 on the quinone ring and both hydroxyl groups. This discrepancy may be resolved by more detailed ENDOR studies.

Z^+ is reduced in the native system by electrons coming from the water-splitting enzyme and the kinetics of Z^+ reduction are affected by the charge storage state (the so-called S states) of the enzyme. The differential kinetics of Z^+ reduction for each S state were first observed by EPR [179] and correspond to the values obtained for the kinetics of S state turnover measured by absorption changes in the UV [180].

The E_m of Z/Z^+ has not been measured because of its high potential but it has been estimated (+1.12 V) to be about 25 mV more negative than that of P-680/P-680^+ from equilibrium considerations [181]. The E_m of Z/Z^+ in Tris-inhibited PS II is estimated to be 200 mV lower compared to the native Z/Z^+ couple [181a].

From kinetic arguments it has been suggested that an electron carrier may function between P-680 and Z [162,172]. Although there is little evidence for this and no direct measurements of such a component have been made, its existence cannot be ruled out at present due to our limited knowledge of this area of the reaction centre.

3.4.3. D, the component associated with Signal II slow

The EPR spectrum arising from Z^+, the highly reactive and short-lived electron donor to P-680, is almost identical to that of another component, D^+ [182]. Unlike Z^+, however, D^+ is extremely stable and is present in the dark in all untreated plant material. Various treatments remove the signal, presumably either by direct electron donation (incubation with DCIP + ascorbate) or by exposing this highly oxidizing component to the environment. When reduced, D can act as an electron donor, reducing S_2 and S_3 ($t_{1/2}$=2 s [182,183]) in intact PS II or reducing Z^+ ($t_{1/2}$ a few ms at pH 8.5) in Tris-washed material [184]. The donation of electrons to S_2 and S_3 is responsible for rapid deactivation of these states in a fraction of centres on the first (and second) flash given to dark-adapted chloroplasts, giving rise to an apparent population of S_0 in the dark [185]. It has also been proposed that D^+ may act as an electron acceptor from S_0 to form S_1 in the dark [186].

Although its EPR spectrum is the same as that of Z^+ it seems to be in an inequivalent position relative to the manganese of the O_2-evolving enzyme since its EPR microwave power saturation characteristics are different but become similar upon removal of the Mn [187,200].

A signal lacking the characteristic line shape of Signal II slow was obtained by oxidation and attributed to D^+ in an environment modified by the strong oxidant potassium iridate [188]. It had a redox potential at pH 8.5 (oxidizing direction only) of 760 mV. The physiological role of D is not understood.

3.4.4. Other electron donors in PS II

(a) Cytochrome b-559. Cyt *b*-559 co-isolates with PS II and it donates electrons to P-680$^+$ at temperatures below 120 K down to 4 K. At higher temperatures the native donor system of the O_2-evolving enzyme functions instead. When the S_2 or S_3 states are present, the cytochrome acts as a donor to P-680$^+$ at 220 K [189]. There are reports that Cyt *b*-559 can be photooxidized, or even photoreduced at room temperature (reviewed in Ref. 7). In most cases these reactions are artificially induced, in other cases the reactions observed involve the cytochromes of the *b*/*f* complex.

The cytochrome is normally in its reduced state in intact dark-adapted material but is very sensitive to changes in its environment, changing from its high-potential form ($E_m \simeq 380$ mV) to its low-potential form ($E_m \simeq 80$ mV). The redox state of the cytochrome seems to have no direct relationship to the function of the O_2-evolving enzyme (e.g. Ref. 190). The oxidized Cyt *b*-559 gives rise to characteristic low-spin haem EPR signals [191].

Since the cytochrome is in its reduced form in vivo and since it is in close contact with the reaction centre, it might have been expected that it would donate electrons to the highly oxidizing intermediates formed in O_2 evolution. However, no role in deactivation of the S states has been demonstrated.

(b) Carotenoid. Carotenoid can be photooxidized by PS II under certain conditions. The carotenoid oxidation is detected by characteristic bleachings in the range around 500 nm and by an absorption increase at 990 nm. Carotenoid is photooxidized in a small proportion of PS II centers at low pH or at temperatures below 77 K [192]. The quantum yield increases to about 80% even at room temperature if lipophilic anions are present [192,193]. The phenolic herbicides, which block electron transfer between Q_A^- and Q_B, also induce this effect, presumably due to their lipophilic anion character [194]. The kinetics of carotenoid oxidation are rapid and may indicate a close association with P-680$^+$ or Z^+ reduction. It has been suggested that carotenoid shares a common electron transfer pathway with Cyt *b*-559 [7,194].

(c) Chlorophyll. The stable photooxidation of a chlorophyll molecule has been observed at 77 K when Cyt *b*-559 was already oxidized [195]. The EPR signal from oxidized chlorophyll accounts for one spin per centre in PS II particles [196]. This amount is limited by the capacity of the electron acceptor at low temperature and it is possible that the chlorophyll oxidation is non-specific.

3.5. Photochemical electron transfer in PS II – an overview

Upon light excitation of dark-adapted PS II, the primary charge separation takes place, forming P-680$^+$ and Pheo$^-$. This probably happens in a small number of picoseconds. Electron transfer from Pheo$^-$ to Q_A occurs in a few hundred picoseconds, stabilizing the separated charges [112]. If Q_A is already reduced the (P-680$^+$ Pheo$^-$) radical pair can still be formed, although perhaps with a low quantum yield (see Ref. 145), but now it lasts for a few nanoseconds [142] and gives rise to some recombination luminescence or, at low temperature, populates the triplet state of P-680 [141], which itself decays with a $t_{1/2}$ of around 1 ms [166].

After formation of (P-680$^+$ Pheo $Q_A{}^-$), P-680$^+$ is reduced rapidly by an electron from Z with a $t_{1/2}$ of 50 ns [161,162]. The state (P-680$^+$ $Q_A{}^-$) recombines with a $t_{1/2}$ of 2 ms below 77 K. If Z is already oxidized, it recombines with a $t_{1/2}$ of 100–200 μs at 20°C.

When (Z$^+$ P-680 Pheo $Q_A{}^-$) is formed the forward electron transfer from $Q_A{}^-$ to Q_B takes place with kinetics of $\simeq$ 100 μs at room temperature [116]. When $Q_B{}^-$ is present in the dark (in 30% of the centres in dark-adapted chloroplasts) the Q^- to $Q_B{}^-$ transfer kinetics have a $t_{1/2}\simeq$200 μs [116].

The manganese of the O_2-evolving enzyme donates to Z$^+$ with a $t_{1/2}$ of 100 μs ($S_1 \rightarrow S_2$). On subsequent flashes the reduction of Z$^+$ is slower ($S_2 \rightarrow S_3$ 350 μs, $S_3 \rightarrow S_0$ 1 ms) due to the charge on the S_2 and S_3 states, while the S_0 to S_1 reduction of Z$^+$ is faster ($t_{1/2}$ 30 μs) [180]. This is also reflected by changes in the kinetics of Z to P-680$^+$ electron transfer ($S_0 \rightarrow S_1$ and $S_1 \rightarrow S_2$ $t_{1/2}$ 50 ns, $S_2 \rightarrow S_3$, $S_3 \rightarrow S_0$ $t_{1/2}$ 250 ns) [162].

The (S_2Z P-680 Pheo $Q_A{}^-$) state back-reacts in approximately 1 s (e.g. Ref. 197), while the (S_2Z P-680 Pheo $Q_AQ_B{}^-$) state back-reacts in $\simeq$ 30 s (e.g. Ref. 116).

At temperatures below approximately −30°C (240 K) electron transfer from $Q_A{}^-$ to Q_B becomes gradually blocked [198]. The transitions from S_1 to S_2, S_2 to S_3 and S_3 to S_0 are blocked by low temperature at 140 K, 240 K and 250 K respectively [199]. Donation from Z to P-680$^+$ is reported to be blocked at 240 K [200] but since this measurement was done with repetitive flashes this value could reflect the electron transfer block on the acceptor side between $Q_A{}^-$ and Q_B. It is more reasonable to suppose that donation from Z can occur at much lower temperature, since S_1 can donate (presumably via Z) down to 140 K. Below 140 K, if Cyt *b*-559 is oxidized prior to illumination, chlorophyll is oxidized [195,196]. Under some conditions Signal II can be stably photoinduced at low temperature; this has been attributed to Z$^+$ [201]. Most of these observations can, however, probably be attributed to D$^+$.

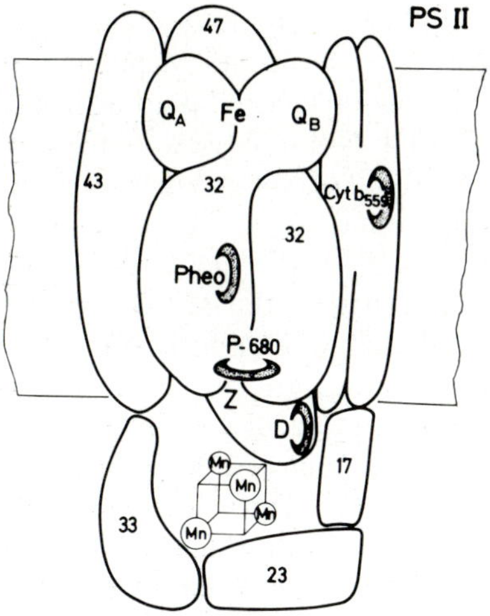

Fig. 5. A possible structure of the PS II reaction centre. The model leans heavily on the analogy with the bacterial reaction centre. Discussion of the location of the chromophores within the polypeptides is given in the text. The orientation of some of the components is shown. The role of the extrinsic polypeptides and the possible structure of the manganese cluster are discussed in Chapter 6.

3.6. Structural aspects

PS II is made up of a cluster of polypeptides, several of which span the membrane. The location of the various components dealt with in previous sections is in many cases not clearly demonstrated. At present, then, models of the structure of the PS II reaction centre are rather speculative. The model shown in Fig. 5 is no exception.

For some time, evidence accumulated that two chlorophyll-containing polypeptides with apparent molecular masses of 47 and 43 kDa were the major subunits which made up the core of the PS II reaction centre. The core was even further divided and the part containing the 47 kDa peptide seemed to retain the reaction centre activity while the 43 kDa one, having no activity, was attributed to a core antenna subunit (see Chapter 11). However, at the same time spectroscopic studies showed that the electron acceptor complex of PS II was in many respects almost identical to that in purple bacteria. When homologies in the primary structure were looked for between the reaction centre polypeptides of purple bacteria and the polypeptides associated with PS II, somewhat surprisingly (at that time) similar sequences were observed in the two PS II polypeptides of molecular mass 32 kDa (known as D_1 and D_2) [202]. One of the 32 kDa polypeptides (D_1) had been previously well characterized as the rapidly turning-over, herbicide-binding protein and was considered to be the site for Q_B binding. The possibility thus arose that the two 32 kDa proteins (D_1 and D_2) were the PS II equivalent to the L and M polypeptides of the purple bacterial reaction centre.

When the crystal structure of the *Rps. viridis* reaction centre was published [133], along with the primary structure of the L and M polypeptides, a basis was provided for a model of the PS II reaction centre based on sequence homologies with the 32 kDa polypeptides [203–204]. Not only were remarkable homologies of secondary structure obtained but also all of the specific amino acid changes associated with herbicide resistance were found to be clustered around the predicted Q_B binding site [204].

Experimentally some support exists for the 32 kDa polypeptides being reaction centre core proteins. Firstly, although often poorly stained on polyacrylamide gels it seems likely that these polypeptides are present in all functioning core PS II preparations. Secondly, the polypeptide discovered by Metz et al. [205] to be modified (34 kDa $\rightarrow$ 36 kDa) in a mutant of a green alga which had a PS II donor side lesion was recently shown to be the herbicide-binding protein (i.e. 32 kDa or D_1 polypeptide) [206]. This indicates both a donor and an acceptor side role for this polypeptide, as might be predicted for a reaction centre core subunit.

The model in Fig. 5 is based on the X-ray structure of the purple bacterial reaction centre. Since no analogies to Z and D are present in purple bacteria it is reasonable to suggest that these components originate in polypeptides other than the 32 kDa (D_1 and D_2) polypeptides. An obvious candidate is the 47 kDa polypeptide which forms part of the PS II core.

A major problem in this model is the location of P-680 itself. The conservation of the histidines associated with the bacteriochlorophyll dimer of the bacterial reaction centre in the 32 kDa (D_1 and D_2) polypeptides of PS II indicates a close structural analogy for this part of the reaction centre [203–204]. However, some experimental evidence exists which weighs against such a close analogy for P-680. Firstly, there are some indirect estimations of P-680 position in the membrane relative to the pheophytin. Time-resolved photovoltage measurements [207], electric field effects on the charge separation [208] and EPR interaction data [209] have all been interpreted as indicating that the P-680 to pheophytin distance is much smaller in PS II than in the bacterial reaction centre. Secondly, orientation measurements of P-680 using the anisotropy of the triplet state indicate that at least one of the two chlorophylls thought to make up P-680 is oriented flat in the membrane [210]. The bacteriochlorophylls that make up the P-870 dimer in purple bacteria are known to be perpendicular to the membrane [211]. While all of these observations may eventually be explained away within the framework of the analogy with the bacterial reaction centre, at present they weigh against an exact analogy at the level of P-680. Nevertheless, in Fig. 5, P-680 is placed at the interface between the two 32 kDa polypeptides as in the bacterial system.

Cyt *b*-559 is closely associated with the PS II reaction centre. A structural model has recently appeared in which the haem, which is oriented perpendicular to the membrane, is liganded to two histidines each on different membrane-spanning polypeptides (9 kDa). Changes of the relative orientation of the imidazole rings from parallel to perpendicular have been proposed to be responsible for the high-potential to low-potential redox form transition [212]. It is still not clear whether 1 or 2 cytochromes are present per reaction centre. The conflicting reports may be due to the loss of the cytochrome during purification of PS II reaction centres. Interestingly, during development of the photosynthetic apparatus the Cyt *b*-559 is the first membrane protein of the PS II reaction centre to be put in place and it may act as an anchor for subsequent reaction centre assembly (see Chapter 6).

The orientation of some of the chromophores has been determined and is included in Fig. 5 [210,213,214]. It is of interest that the ultra-rapid electron transfer reaction that takes place between P-680 and Pheo occurs between chromophores that are perpendicular to each other. This is also the case in purple bacteria [211].

Also shown in Fig. 5 are the 3 extrinsic polypeptides which bind to the inner surface of the PS II reaction centre and which are involved in the chloride binding associated with the water-splitting reaction. These polypeptides may make up a pocket surrounding the manganese atoms of the O_2-evolving enzyme. The four manganese atoms are probably bound to the PS II reaction centre and may be arranged as a distorted cube (see Chapter 6).

References

1 Okamura, M.Y., Feher, G. and Nelson, N. (1982) in Photosynthesis: Energy Conversion by Plants and Bacteria (Govindjee, ed.) Vol. 1, pp. 195–272, Academic Press, New York.
2 Parson, W.W. and Ke, B. (1982) in Photosynthesis: Energy Conversion by Plants and Bacteria (Govindjee, ed.) Vol. 1, pp. 331–355, Academic Press, New York.
3 Mathis, P. and Paillotin, G. (1981) in The Biochemistry of Plants, (Hatch, M.D. and Boardman, N.K., eds.) Vol. 8, pp. 97–161, Academic Press, New York.
4 Rutherford, A.W. and Heathcote, P. (1985) Photosynth. Res. 6, 295–316.
5 Sétif, P. and Mathis, P. (1986) in Encyclopedia for Plant Physiology, New Series, (Staehelin, A. and Arntzen, C.J., eds.) Vol. 19, pp. 476–486 Springer Verlag, Berlin.
6 Van Gorkom, H.J. (1985) Photosynth. Res. 6, 97–112.
7 Diner, B.A. (1986) in Encyclopedia for Plant Physiology, New Series, (Staehelin, A. and Arntzen, C.J., eds.) Vol. 19, pp. 422–436, Springer Verlag, Berlin.
8 Kok, B. (1956) Biochim. Biophys. Acta 22, 399–401.
9 Ke, B. (1973) Biochim. Biophys. Acta 301, 1–33.
10 Lozier, R.H. and Butler, W.L. (1974) Biochim. Biophys. Acta 333, 465–480.
11 Sétif, P., Hervo, G. and Mathis, P. (1981) Biochim. Biophys. Acta 638, 257–267.
12 Sétif, P. and Mathis, P. (1980) Arch Biochem. Biophys. 204, 477–485.
13 Wasielewski, M.R., Norris, J.R., Shipman, L.L., Lin, C.-P. and Svec W.A. (1981) Proc. Natl. Acad. Sci. USA 78, 2957–2961.
14 Watanabe, T., Kobayashi, M., Hongu, A., Nakazato, M., Hiyama, T. and Murata, N. (1985) FEBS Lett. 191, 252–256.
15 Philipson, K.D., Sato, V.L. and Sauer, K. (1972) Biochemistry 11, 4591–4595.
16 Norris, J.R., Uphaus, R.A., Crespi, H.L. and Katz, J.J. (1971) Proc. Natl. Acad. Sci. USA 68, 625–628.
17 Wasielewski, M.R., Norris, J.R., Crespi, H.L. and Harper, J. (1981) J. Am. Chem. Soc. 103, 7664–7665.
18 O'Malley, P.J. and Babcock, G.T. (1984) Proc. Natl. Acad. Sci. USA 81, 1098–1101.
19 Dikanov, S.A., Astashkin, A.V. and Tsvetkov, Yu.D. (1983) Chem. Phys. Lett. 101, 206–210.
20 Den Blanken, H.J. and Hoff, A.J. (1983) Biochim. Biophys. Acta 724, 52–61.
21 Rutherford, A.W. and Mullet, J.E. (1981) Biochim. Biophys. Acta 635, 225–235.
22 Frank, H.A., McLean, M.B. and Sauer, K. (1979) Proc. Natl. Acad. Sci. USA 76, 5124–5128.

23 Wasielewski, M.R. (1982) in Light-reaction path of Photosynthesis (Fong, F.K., ed.) pp. 234–276, Springer Verlag, Berlin.
24 Boxer, S.G. (1983) Biochim. Biophys. Acta 726, 265–292.
25 Malkin, R. and Bearden, A.J. (1978) Biochim. Biophys. Acta 505, 147–181.
26 Evans, M.C.W. (1982) in Iron-Sulfur Proteins (Spiro, T., ed.) Vol. 4, pp. 250–284, John Wiley, New York.
27 Chamorovsky, S.K. and Cammack, R. (1982) Biochim. Biophys. Acta 679, 146–155.
28 Golbeck, J.H. and Warden, J.T. (1982) Biochim. Biophys. Acta 681, 77–84.
29 Malkin, R. (1984) Biochim. Biophys. Acta 764, 63–69.
30 Crowder, M.S. and Bearden, A. (1983) Biochim. Biophys. Acta 722, 23–35.
31 Sétif, P., Mathis, P. and Vänngard, T. (1984) Biochim. Biophys. Acta 767, 404–414.
32 Ke, B., Demeter, S., Zamaraev, K.I. and Khairutdinov, R.F. (1979) Biochim. Biophys. Acta 545, 265–284.
33 Graan, T., Ort, D.R. and Prince, R.C. (1985) Anal. Biochem. 144, 193–198.
34 McIntosh, A.R., Chu, M. and Bolton, J.R. (1975) Biochim. Biophys. Acta 376, 308–314.
35 Evans, M.C.W., Sihra, C.K., Bolton, J.R. and Cammack, R. (1975) Nature 256, 668–670.
36 Chamarovsky, S.K. and Cammack, R. (1982) Photobiochem. Photobiophys. 4, 195–200.
37 Sauer, K., Mathis, P., Acker, S. and Van Best, J.A. (1978) Biochim. Biophys. Acta 503, 120–134.
38 Golbeck, J.H., Velthuys, B.R. and Kok, B. (1978) Biochim. Biophys. Acta 504, 226–230.
39 Mathis, P., Sauer, K. and Remy, R. (1978) FEBS Lett. 88, 275–278.
40 Golbeck, J.H. and Cornelius, J.M. (1986) Biochim. Biophys. Acta 849, 16–24.
41 Warden, J.T. and Golbeck, J.H. (1986) Biochim. Biophys. Acta 849, 25–31.
42 Sétif, P., Mathis, P., Lagoutte, B. and Duranton, J. (1984) in Advances in Photosynthesis Research, Vol. 1, (Sybesma, C., ed.) pp. 589–592, Martinus Nijhoff/Dr. W. Junk, The Hague.
43 Bonnerjea, J. and Evans, M.C.W. (1982) FEBS Lett. 148, 313–316.
44 Gast, P., Swarthoff, T., Ebskamp, F.C.R. and Hoff, A.J. (1983) Biochim. Biophys. Acta 722, 163–175.
45 Nuijs, A.M., Shuvalov, V.A., Van Gorkom, H.J., Plijter, J.J. and Duysens, L.N.M. (1986) Biochim. Biophys. Acta 850, 319–323.
46 Fujita, I., Davis, M.S. and Fajer, J. (1978) J. Am. Chem. Soc. 100, 6280–6282.
47 Swarthoff, T., Gast, P., Amesz, J. and Buisman, H.P. (1982) FEBS Lett. 146, 129–132.
48 Mansfield, R.W. and Evans, M.C.W. (1985) FEBS Lett. 190, 237–241.
49 Heathcote, P., Timofeev, K.N. and Evans, M.C.W. (1979) FEBS Lett. 101, 105–109.
50 Forman, A., Davis, M.S., Fujita, I., Hanson, L.K., Smith, K.M. and Fajer, J. (1981) Isr. J. Chem. 21, 265–269.
51 Sétif, P., Quaegebeur, J.-P. and Mathis, P. (1982) Biochim. Biophys. Acta 681, 345–353.
52 Sétif, P., Bottin, H. and Mathis, P. (1985) Biochim. Biophys. Acta 808, 112–122.
53 Voznyak, V.M., Ganago, I.B., Moskalenko, A.A. and Elfimov, E.I. (1980) Biochim. Biophys. Acta 592, 364–368.
54 Sonneveld, A., Duysens, L.N.M. and Moerdijk, A. (1981) Biochim. Biophys. Acta 636, 39–49.
55 Sétif, P. and Mathis, P. (1986) Photosynth. Res. 9, 47–54.
56 Thurnauer, M.C. and Gast, P. (1985) Photobiochem. Photobiophys. 9, 29–38.
57 McCracken, J.L. and Sauer, K. (1983) Biochim. Biophys. Acta 724, 83–93.
58 Furrer, R. and Thurnauer, M.C. (1983) FEBS Lett. 153, 399–403.
59 Manikowski, H., McIntosh, A.R. and Bolton, J.R. (1984) Biochim. Biophys. Acta 765, 68–73.
60 Interschick-Niebler, E. and Lichtenthaler, H.K. (1981) Z. Naturforsch. 36 c, 276–283.
61 Brettel, K., Sétif, P. and Mathis, P. (1986) FEBS Lett. 203, 220–224.
62 Prince, R.C., Dutton, P.L. and Bruce, J.M. (1983) FEBS Lett. 160, 273–276.
63 Il'Ina, M.D. and Borisov, A.Y. (1982) Biokhimiya 47, 1954–1962.
64 Karukstis, K.K. and Sauer, K. (1983) J. Cell. Biochem. 23, 131–158.
65 Maslov, V.G., Chunaev, A.S. and Tugarinov, V.V. (1982) Molekul. Biol. 15, 1016–1027.
66 Van Grondelle, R. (1985) Biochim. Biophys. Acta 811, 147–195.
67 Thurnauer, M.C., Rutherford, A.W. and Norris, J.R. (1982) Biochim. Biophys. Acta 682, 332–338.
68 Haehnel, W., Pröpper, A. and Krause, H. (1980) Biochim. Biophys. Acta 593, 384–399.

69 Bottin, H. and Mathis, P. (1985) Biochemistry 24, 6453–6460.
70 Overfield, R.E., Wraight, C.A. and Devault, D. (1979) FEBS Lett. 105, 137–142.
71 Wood, P.M. (1978) Eur. J. Biochem. 87, 9–19.
72 Nanba, M. and Katoh, S. (1983) Biochim. Biophys. Acta 725, 272–279.
73 Guss, J.M. and Freeman, H.C. (1983) J. Mol. Biol. 169, 521–563.
74 Burkey, K.O. and Gross, E.L. (1981) Biochemistry, 20, 5495–5499.
75 Tamura, N., Itoh, S., Yamamoto, Y. and Nishimura, M. (1981) Plant Cell Physiol. 22, 603–612.
76 Bengis, C. and Nelson, N. (1977) J. Biol. Chem. 252, 4564–4569.
77 Thornber, J.P., Alberte, R.S., Hunter, F.A., Shiozawa, J.A. and Kan, K.S. (1976) Brookhaven Symp. Biol. 28, 133–148.
78 Sétif, P., Acker, S., Lagoutte, B. and Duranton, J. (1980) Photosynth. Res. 1, 17–27.
79 Vierling, E. and Alberte, R.S. (1983) Plant Physiol. 72, 625–633.
80 Fish, L.E., Kück, U. and Bogorad, L. (1985) J. Biol. Chem. 260, 1413–1421.
81 Moller, B.L., Hoyer-Hansen, G. and Hiller, R.G. (1981) in Photosynthesis III. Structure and Molecular Organisation of the Photosynthetic Apparatus (Akoyunoglou, G., ed.) pp. 245–256, Balaban International Science Services, Philadelphia.
82 Bonnerjea, J., Ortiz, W. and Malkin, R. (1985) Arch. Biochem. Biophys. 240, 15–20.
83 Lagoutte, B., Sétif, P. and Duranton, J. (1984) FEBS Lett. 174, 24–29.
84 Takahashi, Y. and Katoh, S. (1982) Arch. Biochem. Biophys. 219, 219–227.
85 Wollman, F.A. (1986) in Encyclopedia for Plant Physiology, Vol. 19 New Series, (Staehelin, A. and Arntzen, C.J., eds.) pp. 487–495, Springer Verlag, Berlin.
86 Mullet, J.E., Burke, J.J. and Arntzen, C.J. (1980) Plant Physiol. 65, 814–822.
87 Wollman, F.-A. and Bennoun, P. (1982) Biochim. Biophys. Acta 680, 352–360.
88 Haworth, P., Watson, J.L. and Arntzen, C.J. (1983) Biochim. Biophys. Acta 724, 151–158.
89 Lam, E., Ortiz, W. and Malkin, R. (1984) FEBS Lett. 168, 10–14.
90 Shiozawa, J.A., Alberte, R.S. and Thornber, J.P. (1974) Arch. Biochem. Biophys. 165, 388–397.
91 Nechushtai, R., Muster, P., Binder, A., Liveanu, V. and Nelson, N. (1983) Proc. Natl. Acad. Sci. USA 80, 1179–1183.
92 Trebst, A. (1974) Annu. Rev. Plant Physiol. 25, 423–458.
93 Aasa, R., Bergström, J. and Vänngard, T. (1981) Biochim. Biophys. Acta 637, 118–123.
94 Prince, R.C., Crowder, M.S. and Bearden, A.J. (1980) Biochim. Biophys. Acta 592, 323–337.
95 Hootkins, R. and Bearden, A. (1983) Biochim. Biophys. Acta 723, 16–29.
96 Breton, J. and Vermeglio, A. (1982) in Photosynthesis: Energy conservation by plants and bacteria (Govindjee, ed.) Vol. 1, pp. 153–194, Academic Press, New York.
97 Schaffernicht, H. and Junge, W. (1981) Photochem. Photobiol. 34, 223–232.
98 Nabedryk, E., Biaudet, P., Darr, S., Arntzen, C.J. and Breton, J. (1984) Biochim. Biophys. Acta 767, 640–647.
99 Ortiz, W., Lam, E., Chollar, S., Munt, D. and Malkin, R. (1985) Plant Physiol. 77, 389–397.
100 Ryrie, I.J., Anderson, J.M. and Glare, T. (1985) Photobiochem. Photobiophys. 5, 145–157.
101 Stiehl, H.H. and Witt, H.T. (1968) Z. Naturforsch. 23b, 220–224.
102 Knaff, D.B. and Arnon, D.I. (1969) Proc. Natl. Acad. Sci. USA 63, 963–969.
103 Van Gorkom, H.J. (1974) Biochim. Biophys. Acta 347, 439–442.
104 Cox, R.P. and Bendall, D.S. (1974) Biochim. Biophys. Acta 347, 49–59.
105 Rutherford, A.W. and Zimmermann, J.J. (1984) Biochim. Biophys. Acta 767, 168–175.
106 Knaff, D.B. (1975) FEBS Lett. 60, 331–335.
107 Prince, R.C. and Dutton, P.L. (1976) Arch. Biochem. Biophys. 172, 329–334.
108 Crofts, A.R. and Wraight, C.A. (1983) Biochim, Biophys. Acta 726, 149–185.
109 Klimov, V.V., Dolan, E., Shaw, E.R. and Ke, B. (1980) Proc. Natl. Acad. Sci. USA 77, 7227–7231.
110 Nugent, J.H.A., Diner, B.A. and Evans, M.C.W. (1981) FEBS Lett. 124, 241–244.
111 Prince, R.C. and Thornber, J.P. (1977) FEBS Lett. 81, 233–237.
112 Nuijs, A.M., Van Gorkom, H.J., Plijter, J.J. and Duysens, L.N.M. (1986) Biochim. Biophys. Acta 848, 167–175.
113 Mathis, P. and Haveman, J. (1977) Biochim. Biophys. Acta 461, 167–181.
114 Schatz, G.H. and Van Gorkom, H.J. (1985) Biochim. Biophys. Acta 810, 283–294.

115 Rutherford, A.W., Crofts, A.R. and Inoue, Y. (1982) Biochim. Biophys. Acta 682, 457–465.
116 Robinson, H.H. and Crofts, A.R. (1983) FEBS Lett. 153, 221–226.
117 Robinson, H.H. and Crofts, A.R. (1984) in Advances in Photosynthesis Research (Sybesma, C., ed.) Vol. I, pp. 477–480, Martinus Nijhoff/Dr. W. Junk, The Hague.
118 Wraight, C.A. (1979) Biochim. Biophys. Acta 548, 309–327.
119 Bowes, J.M. and Crofts, A.R. (1980) Biochim. Biophys. Acta 590, 373–384.
120 Velthuys, B.R. (1981) FEBS Lett. 126, 277–281.
121 Wraight, C.A. (1981) Isr. J. Chem. 21, 348–354.
122 Lavergne, J. (1982) Biochim. Biophys. Acta 682, 345–353.
123 Hirschberg, J. and McIntosh, L. (1983) Science 222, 1346–1348.
124 Erickson, J.M., Rahire, M., Rochaix, J.-D. and Mets, L. (1985) Science 228, 204–207.
125 Velthuys, B.R. and Amesz, J. (1974) Biochim. Biophys. Acta 333, 85–94.
126 Bouges-Bocquet, B. (1973) Biochim. Biophys. Acta 314, 250–256.
127 Hangulov, S.V., Goldfield, M.G. and Blumenfeld, L.A. (1975) Studia Biophys. 48, 85–92.
128 Rutherford, A.W., Zimmermann, J.L. and Mathis, P. (1984) in Advances in Photosynthesis Research (Sybesma, C., ed.) Vol. 1, pp. 445–448, Martinus Nijhoff/Dr. W. Junk, The Hague.
129 Wraight, C.A. (1978) FEBS Lett. 93, 283–288.
130 Klimov, V.V., Dolan, E. and Ke, B. (1980) FEBS Lett. 112, 97–100.
131 Tiede, D.M., Prince, R.C. and Dutton, P.L. (1976) Biochim. Biophys. Acta 449, 447–467.
132 Rutherford, A.W. and Mathis, P. (1983) FEBS Lett. 154, 328–334.
133 Deisenhofer, J., Epp, O., Miki, K., Huber, R. and Michel, H. (1984) J. Mol. Biol. 180, 385–398.
134 Butler, W.F., Calvo, R., Fredkin, D.R., Isaacson, R.A., Okamura, M.Y. and Feher, G. (1984) Biophys. J. 45, 947–973.
135 Dismukes, G.C., Frank, H.A., Friesner, R. and Sauer, K. (1984) Biochim. Biophys. Acta 764, 253–271.
136 Debus, R.J., Feher, G. and Okamura, M.Y. (1986) Biochemistry 25, 2276–2287.
137 Petrouleas, V. and Diner, B.A. (1986) Biochim. Biophys. Acta 849, 264–275.
138 Zimmermann, J.L. and Rutherford, A.W. (1986) Biochim. Biophys. Acta 851, 416–423.
139 Ikegami, I. and Katoh, S. (1973) Plant Cell Physiol. 14, 829–836.
140 Klimov, V.V., Klevanik, A.V., Shuvalov, V.A. and Krasnovsky, A.A. (1977) FEBS Lett. 82, 183–186.
141 Rutherford, A.W., Paterson, D.R. and Mullet, J.E. (1981) Biochim. Biophys. Acta 635, 205–214.
142 Shuvalov, V.A., Klimov, V.V., Dolan, E., Parson, W.W. and Ke, B. (1980) FEBS Lett. 118, 279–282.
143 Klimov, V.V., Allakhverdiev, S.I., Demeter, S. and Krasnovskii, A.A. (1979) Dokl. Akad. Nauk. SSSR 249, 227–230.
144 Rutherford, A.W., Mullet, J.E. and Crofts, A.R. (1981) FEBS Lett. 123, 235–237.
145 Holzwarth, A.R. (1986) in Encyclopedia of Plant Physiology, New Series Vol. 19 (Staehelin, A. and Arntzen, C.J., eds.) pp. 299–309, Springer Verlag, Berlin.
146 Melis, A. (1985) Biochim. Biophys. Acta 808, 334–342.
147 Black, M.T., Brearley, T.H. and Horton, P. (1986) Photosynth. Res. 8, 193–207.
148 Horton, P. (1981) Biochim. Biophys. Acta 635, 105–110.
149 Thielen, A.P.G.M. and Van Gorkom, H.J. (1981) FEBS Lett. 129, 205–209.
150 Hodges, M. and Barber, J. (1986) Biochim. Biophys. Acta 848, 239–246.
151 Rutherford, A.W., Heathcote, P. and Evans, M.C.W. (1979) Biochem. J. 182, 515–523.
152 Diner, B.A. and Delosme, R. (1983) Biochim. Biophys. Acta 722, 443–451.
153 Eckert, H.J. and Renger, G. (1980) Photochem. Photobiol. 31, 501–511.
154 Joliot, P. and Joliot, A. (1981) in Photosynthesis III (Akoyunoglou, G., ed.) pp. 885–899, Balaban International Science Services, Philadelphia.
155 Joliot, P. and Joliot, A. (1981) FEBS Lett. 134, 155–158.
155a Lavergne, J. (1982) Photobiochem. Photobiophys. 3, 257–271.
156 Meiburg, R.F., Van Gorkom, H.J. and Van Dorssen, R.J. (1984) Biochim. Biophys. Acta 765, 295–300.
157 Evans, M.C.W., Atkinson, Y.E. and Ford, R.C. (1985) Biochim. Biophys. Acta 806, 247–254.

158 Brettel, K., Schlodder, E. and Witt, H.T. (1985) Photobiochem. Photobiophys. 9, 205–213.
159 Rutherford, A.W. (1981) Biochem. Biophys. Res. Commun. 102, 1065–1070.
160 Döring, G., Stiehl, H.H. and Witt, H.T. (1967) Z. Naturforsch. 22b, 639–644.
161 Van Best, J.A. and Mathis, P. (1978) Biochim. Biophys. Acta 503, 178–188.
162 Brettel, K., Schlodder, E. and Witt, H.T. (1984) Biochim. Biophys. Acta 766, 403–415.
163 Conjeaud, H., Mathis, P. and Paillotin, G. (1979) Biochim. Biophys. Acta 546, 280–291.
164 Ke, B., Inoue, H., Babcock, G.T., Fang, Z.-X. and Dolan, E. (1982) Biochim. Biophys. Acta 682, 297–306.
165 Malkin, R. and Bearden, A.J. (1975) Biochim. Biophys. Acta 396, 250–259.
166 Den Blanken, H.J., Hoff, A.J., Jongenelis, A.P.J.M. and Diner, B.A. (1983) FEBS Lett. 157, 21–27.
167 Davis, M.S., Forman, A. and Fajer, J. (1979) Proc. Natl. Acad. Sci. USA 76, 4170–4174.
168 Mauzerall, D. (1972) Proc. Natl. Acad. Sci. USA 69, 1358–1362.
169 Sonneveld, A., Rademaker, H. and Duysens, L.N.M. (1979) Biochim. Biophys. Acta 548, 536–551.
170 Conjeaud, H. and Mathis, P. (1980) Biochim. Biophys. Acta 590, 353–359.
171 Dekker, J.P., Van Gorkom, H.J., Brok, M. and Ouwehand, L. (1984) Biochim. Biophys. Acta 764, 301–309.
172 Diner, B.A. and De Vitry, C. (1984) in Advances in Photosynthesis Research (Sybesma, C., ed). Vol. I, pp. 407–411, Martinus Nijhoff/Dr. W. Junk, The Hague.
173 Lavergne, J. (1984) FEBS Lett. 173, 9–14.
174 Blankenship, R.E., Babcock, G.T., Warden, J.T. and Sauer, K. (1975) FEBS Lett. 51, 287–293.
175 O'Malley, P.J. and Babcock, G.T. (1984) Biochim. Biophys. Acta 765, 370–379.
176 De Vitry, C., Carles, C. and Diner, B.A. (1986) FEBS Lett. 196, 203–206.
177 O'Malley, P.J., Babcock, G.T. and Prince, R.C. (1984) Biochim. Biophys. Acta 766, 283–288.
178 Brok, M., Ebskamp, F.C.R. and Hoff, A.J. (1985) Biochim. Biophys. Acta 809, 421–428.
179 Babcock, G.T., Blankenship, R.E. and Sauer, K. (1976) FEBS Lett. 61, 286–289.
180 Dekker, J.P., Plijter, J.J., Ouwehand, L. and Van Gorkom, H.J. (1984) Biochim. Biophys. Acta 767, 176–179.
181 Bouges-Bocquet, B. (1980) Biochim. Biophys. Acta 594, 85–103.
181a Yerkes, C.T., Babcock, G.T. and Crofts, A.R. (1983) FEBS Lett. 158, 359–363.
182 Babcock, G.T. and Sauer, K. (1973) Biochim. Biophys. Acta 325, 504–519.
183 Velthuys, B.R. and Visser, J.W.M. (1975) FEBS Lett. 55, 109–112.
184 Boussac, A. and Etienne, A.L. (1982) Biochem. Biophys. Res. Commun. 109, 1200–1205.
185 Vermaas, W.F.J., Renger, G. and Dohnt, G. (1984) Biochim. Biophys. Acta 764, 194–202.
186 Zimmermann, J.-L. and Rutherford, A.W. (1985) Physiol. Vég. 23, 425–434.
187 Yocum, C.F., Yerkes, C.T., Blankenship, R.E., Sharp, R.R. and Babcock, G.T. (1981) Proc. Natl. Acad. Sci. USA 78, 7507–7511.
188 Boussac, A. and Etienne, A.L. (1984) Biochim. Biophys. Acta 766. 576–581.
189 Vermeglio, A. and Mathis, P. (1973) Biochim. Biophys. Acta 314, 57–65.
190 Briantais, J.-M., Vernotte, C., Miyao, M., Murata, N. and Picaud, M. (1985) Biochim. Biophys. Acta 808, 348–351.
191 Malkin, R. and Vänngård, T. (1980) FEBS Lett. 111, 228–231.
192 Schenck, C.C., Diner, B., Mathis, P. and Satoh, K. (1982) Biochim. Biophys. Acta 680, 216–227.
193 Velthuys, B.R. (1981) FEBS Lett. 126, 272–276.
194 Mathis, P. and Rutherford, A.W. (1984) Biochim. Biophys. Acta 767, 217–222.
195 Visser, J.W.M., Rijgersberg, C.P. and Gast, P. (1977) Biochim. Biophys. Acta 460, 36–46.
196 De Paula, J.C., Innes, J.B. and Brudvig, G.W. (1985) Biochemistry 24, 8114–8120.
197 Lavergne, J. and Etienne, A.-L. (1980) Biochim. Biophys. Acta 593, 136–148.
198 Joliot, A. (1974) Biochim. Biophys. Acta 357, 439–448.
199 Inoue, Y. and Shibata, K. (1978) FEBS Lett. 85, 193–197.
200 Warden, J.T., Blankenship, R.E. and Sauer, K. (1976) Biochim. Biophys. Acta 423, 462–478.
201 Nugent, J.H.A. and Evans, M.C.W. (1979) FEBS Lett. 101, 101–104.
202 Hearst, J.E. and Sauer, K. (1984) in Advances in Photosynthesis Research (Sybesma, C., ed.) Vol. III, pp. 355–359, Martinus Nijhoff/Dr. W. Junk, The Hague.

203 Deisenhofer, J., Michel, H. and Huber, R. (1985) Trends Biochem. Sci. 10, 243–248.
204 Trebst, A. (1986) Z. Naturforsch. 41c, 240–245.
205 Metz, J.G., Wong, J. and Bishop, N.I. (1980) FEBS Lett. 114, 61–66.
206 Metz, J.G., Bricker, T.M. and Seibert, M. (1985) FEBS Lett. 185, 191–196.
207 Trissl, H.W., Kunze, U. and Junge, W. (1982) Biochim. Biophys. Acta 682, 364–377.
208 Meiburg, R.F., Van Gorkom, H.J. and Van Dorssen, R.J. (1983) Biochim. Biophys. Acta 724, 352–358.
209 Rutherford, A.W. and Thurnauer, M.C. (1982) Proc. Natl. Acad. Sci. USA 79, 7282–7287.
210 Rutherford, A.W. (1985) Biochim. Biophys. Acta 807, 189–201.
211 Tiede, D.M. (1986) in Photosynthesis III. Photosynthetic membranes and Light Harvesting Systems. Encyclopedia of Plant Physiology, New Series, Vol. 19 (Staehelin, A. and Arntzen, C.J., eds.) pp. 344–352, Springer Verlag, Berlin.
212 Babcock, G.T., Widger, W.R., Cramer, W.A., Oertling, W.A. and Metz, J.G. (1985) Biochemistry 24, 3638–3645.
213 Ganago, I.B., Klimov, V.V., Ganago, A.O., Shuvalov, V.A. and Erokhin, Y.E. (1982) FEBS Lett. 140, 127–130.
214 Bergström, J., Vänngard, T. (1982) Biochim. Biophys. Acta 682, 452–456.

J. Amesz (ed.) *Photosynthesis*

CHAPTER 5

Electron paramagnetic resonance in photosynthesis

A.J. HOFF

Department of Biophysics, Huygens Laboratory of the State University, P.O. Box 9504, 2300 RA Leiden, The Netherlands

1. Introduction

Photosynthesis is a garden of Eden for the electron paramagnetic resonance (EPR) spectroscopist. Practically all aspects of EPR spectroscopy come to the fore, individually or in combination, in the various photosynthetic systems of plants and bacteria, in intact cells or in isolated subcellular particles or purified reaction center proteins. In this chapter I will highlight a few of the more salient applications of EPR in photosynthesis, chosen for their potential and impact. I will assume that the reader is more or less familiar with photosynthesis and not at all familiar with EPR spectroscopy. Therefore in Sections 2 and 3 an EPR primer is presented, mostly to develop an intuitive feeling for the principles, a discriminating ear for the jargon and a panoramic eye for the mathematical formulation. To avoid reader's rigor mortis, the presentation is kept elemental and quite unrigorous, but hopefully of practical use. In addition I have tried to avoid the abstruse nomenclature common to the photosynthetic literature.

Sections 4–8 provide a bird's eye view of the applications of EPR in photosynthesis. It is expressly not the intention to review the results in detail. These, and comprehensive literature references, can be found in a number of recent reviews (see reference list). Here, only key references will be given, and references to literature not yet reviewed. Thus, by sacrificing mathematical and bibliographical rigor to readability, I hope the reader will share some of the enjoyment that spectroscopists experience in studying the physical processes of photosynthesis.

2. Magnetic resonance for the layman

Spectroscopy may be defined as the method of monitoring the absorption or emission of electromagnetic or particulate radiation as a function of wavelength or frequency. Of the multifarious forms of spectroscopy, optical spectroscopy is probably the most familiar to biologists and biochemists. In this spectroscopy the interaction of the electric vector of electromagnetic radiation – usually but not necessarily in the near-ultraviolet, visible or near-infrared region of the spectrum – with matter is monitored. The interaction is a result of the electric polarizability of the material. Electric transition dipole moments are induced by the oscillating field, which result in jumps of electrons to higher (absorption) or lower (stimulated emission) electronic orbitals whenever particular 'resonance' frequencies are hit. These frequencies, of course, correspond to the energy difference between the electronic orbitals. In absorption spectroscopy, the electrons of a system are normally in the ground state: the electrons jump to an empty energy level and the intensity of the absorptive resonance is proportional to the total number of molecules. In emission spectroscopy, the intensity is proportional to the number of molecules in which the electrons occupy an excited state, mostly because of earlier absorption of 'light'.

A close relative of optical spectroscopy is magnetic resonance spectroscopy. Here, the interaction of the magnetic vector of electromagnetic radiation with matter is monitored. Now, the absorption or emission of radiation results from interaction with an intrinsic magnetic (dipole) moment; the induced moment resulting from the magnetic polarizability is far too small to play a role (except in circular dichroism).

What are these magnetic moments? Primarily and ubiquitously, they are the nuclear magnetic moments (spin) of protons. Normally, there is no macroscopic magnetic moment associated with the ensemble of nuclear spin magnetic moments. However, when a laboratory magnetic field is applied, these nuclear magnetic dipoles orient themselves parallel or antiparallel to the field. Parallel corresponds to a higher energy, because, just as for bar magnets in a magnetic field, work is needed to turn a spin from antiparallel to parallel. (The difference in energy between parallel and antiparallel positions for the spins is often called Zeeman energy.) Therefore, in equilibrium more protons are antiparallel than parallel, and the difference generates a macroscopic nuclear paramagnetic moment. An oscillating (electro)magnetic field may now, just as in optical spectroscopy, cause 'spins' to jump from the antiparallel (lower energy) to the parallel (higher energy) position: energy is absorbed from the oscillating field. The energy difference, however, is slight: at room temperature and in a laboratory magnetic field of 10 kG (1 tesla) the fractional population difference (to which the net absorption, i.e. the sensitivity, is proportional) of the two energy states is about 10^{-5}. Moreover, the small energy difference between the two states makes the resonance frequency (hence the sensitivity of detectors) low; e.g. in 10 kG only 42 MHz for protons and even lower for other nuclei. Together, the two effects make nuclear magnetic resonance (NMR) a very insensitive spectroscopic technique.

A much more powerful intrinsic magnetic moment is possessed by a single 'free' electron: its 'spin' is more than 1000 times larger than that of protons. Thus, electron (para)magnetic resonance (EPR) is much more sensitive than NMR. However, in biological material 'free' electrons are rare. They are found only in certain transition metals present in metalloproteins, in so-called radicals, and in triplet states. This is because electrons normally are 'paired'. In every electronic orbital one finds one 'parallel' and one 'antiparallel' electron, so that the sum of their

TABLE 1

Three spectroscopic techniques

	Optical spectroscopy	EPR	NMR
Frequency	10^{14} Hz	10^{10} – 10^{11} Hz	10^{6} – $5{\cdot}10^{8}$ Hz
Applicability	All chromophores	All paramagnetic centers, especially transition metals, radicals, triplets	Most nuclei, in biology primarily ^{1}H, ^{13}C and ^{31}P
Relative sensitivity	1000	1	0.001
Density of spectral information	Low	Low	High
Selectivity	Low	High	Low
Experimental technique	Optical components	Microwave	Radiofrequency
Cost (k$)	$\simeq$ 20 and up	$\simeq$ 100 and up	$\simeq$ 100 and up

magnetic moments is zero: the material is diamagnetic.

The transition metals have a d-orbital that may contain up to 5 unpaired electrons; It is sometimes energetically more favorable to pile up a number of unpaired electrons that are then in different sub-orbitals. Such metal centers may have considerable magnetic moments (they are paramagnetic), and therefore strong EPR signals.

Radicals are molecules which are normally diamagnetic, but which for one reason or another (because of chemical reactions or photolysis) have lost or gained one electron. The pairing balance is therefore lost; one electron is unpaired and possesses a magnetic moment which in a magnetic field interacts with electromagnetic radiation as described above.

Triplet states are generally molecules in which one electron is promoted to a higher electronic orbital (e.g. by the absorption of light) under spin reversal, i.e. it is no longer paired with its partner in the orbital it has left and the two electrons have a combined magnetic moment what is twice that of a radical. Some molecules, e.g. dioxygen, have a triplet ground state.

Because the presence of an unpaired electronic magnetic moment is relatively rare, EPR spectroscopy is a highly selective technique: one may, for example, pick out the active center from its protein environment. This, and its higher sensitivity compared to NMR, makes it a useful technique that may provide detailed information on the structure of key biological molecules and on their interactions with the environment. Table 1 compares the ‘pros’ and ‘cons’ of optical, NMR and EPR spectroscopy.

3. *Physics of EPR*

3.1. *Basic principles*

As in all spectroscopic techniques, EPR has its own ‘jargon’. This is not the place to go into any detail with regard to theory or experimental techniques. A number of excellent textbooks are available, some of which (such as the monographs by Feher [1] and by Swartz et al. [2]) stress the biological applications. Yet to facilitate the discussion a certain minimum of physical notions and notations is needed. In Table 2 the basics of 'free' electron EPR are tabulated.

In actual practice the unpaired electron is not free. It is generally associated with one or more nuclei, which may have a nuclear spin magnetic moment. This moment generates a magnetic field at the location of the unpaired electron, due to the so-called contact or Fermi hyperfine interaction (the electron has a finite probability of penetrating to the atomic nucleus) and to the through-space dipolar interaction between nuclear and electronic magnetic spin moment, represented by

TABLE 2

Glossary of electron paramagnetic resonance

Electronic magnetic moment	$-\mu_e = \gamma \hbar S = g\beta S$
Electronic Bohr magneton	$\beta = \dfrac{e\hbar}{2m} = 9.274\times10^{-28}$ J G^{-1} = 9.274×10^{-24} C s^{-1}m^2
Planck's constant	$h = 2\pi\hbar = 6.62618 \times 10^{-34}$ J s
Gyromagnetic ratio = ratio between magnetic moment due to orbital motion and intrinsic magnetic moment (spin)	$\gamma = 1.7608 \times 10^7$ rad s^{-1} G^{-1}; $\gamma/2\pi = 2.8025$ MHz G^{-1}
g value = proportionality factor	$g = \gamma\beta^{-1}\hbar = 2.00232$ for free electron
Magnetic spin vector = quantum mechanic operator	$S = (S_x, S_y, S_z)$
Projection of S on magnetic field $B = B_o z$	m_S, magnetic quantum number
Energy of electron in field B	$E = -\mu_e B = +\gamma h B_0 m_S$ $= +\frac{1}{2} g \beta B_0$ ($m_S = +\frac{1}{2}$) 'parallel' or 'up' $= -\frac{1}{2} g \beta B_0$ ($m_S = -\frac{1}{2}$) 'antiparallel' or 'down'
Energy difference between 'up' and 'down'	$\Delta E = g\beta B_0 = h\nu = \hbar\omega_0$
Resonance frequency	$\nu = g\beta h^{-1} B_0 = (\gamma/2\pi) B_0$ $\omega_0 = 2\pi\nu = \gamma B_0$
Ratio of population of 'up' and 'down' states (T is temperature)	$N\uparrow/N\downarrow = e^{-\Delta E/kT} = e^{-g\beta B_0/kT} = e^{-h\nu/kT}$
Boltzmann's constant	$k = 1.38044 \times 10^{-23}$ J K^{-1}
Relative population difference (N = total number of 'spins', B_0 = 3.3 kG)	$\dfrac{\Delta N}{N} = \dfrac{N\downarrow - N\uparrow}{N\downarrow + N\uparrow} = \dfrac{1 - e^{-\Delta E/kT}}{1 + e^{-\Delta E/kT}} \simeq \dfrac{g\beta B_0}{2kT}$ $\simeq 10^{-3}$, T = 300 K; $\simeq 0.2$, T = 1 K

Units of magnetic field (in fact the magnetic flux density) are the older gauss (G) and the SI unit tesla (T). 1 G $\leftrightarrow$ 10 mT = 10^{-2} J m^{-2} A^{-1}.

TABLE 3

Interactions

I. Electronic Zeeman energy with the spin-orbit interaction incorporated in the anisotropic g value	$\hat{E} = \beta B \cdot \mathbf{g} \cdot S$
For $B//i$ ($i = x,y,z$, the principal axes of the g tensor)	$\boldsymbol{E} = \pm\beta g_{ii} B_0,\ i = x,y,z,\ S = \frac{1}{2}$
II. Isotropic hyperfine interaction (contact term)	$\hat{E} = aI \cdot S$
First-order approximation (valid for high B_0)	$\boldsymbol{E} = aI_z \cdot S_z$
I is nuclear spin operator	$= am_I \cdot m_S = \pm\frac{1}{4}a$ for $I = \frac{1}{2},\ S = \frac{1}{2}$
Anisotropic hyperfine interaction (dipolar term)	$\hat{E} = I \cdot \mathbf{A} \cdot S$
x',y',z': principal axes of hyperfine interaction tensor **A**	$\boldsymbol{E} = \pm\frac{1}{4}a_{jj}\ (j = x',y',z'),\ I = S = \frac{1}{2},\ B//j$
III. Electronic dipole–dipole interaction tensor **D**	$\hat{E} = S \cdot \mathbf{D} \cdot S,\ S = S_1 + S_2$
Fine structure or zero-field parameters D,E	$= D(S_z^2 - \frac{1}{3}S^2) + E(S_x^2 - S_y^2)$
	$S^2 = S(S + 1)$
	$= 0$ singlet state, antiparallel spins
	$= 2$ triplet state, parallel spins
Axial symmetry ($E = 0$)	$\boldsymbol{E} = -D(\cos^2\theta - \frac{1}{3}),\ m_S = 0$
θ, angle between B and dipolar z axis	$= \frac{1}{2}D(\cos^2\theta - \frac{1}{3}) \pm g\beta B_0,\ m_S = \pm 1$
IV. Isotropic exchange interaction, J = scalar	$\hat{E} = -J(\frac{1}{2} + 2S_1 \cdot S_2)$
	$\boldsymbol{E} = -J(S^2 - 1),\ S_1 = S_2 = \frac{1}{2}$

$\hat{E}$ is the energy operator. To calculate the actual energy $\boldsymbol{E}$ the magnetic quantum numbers are substituted for the spin vectors S_z, I_z. For $S = \frac{1}{2}$, $m_S = \pm\frac{1}{2}$; for $S = 1$, $m_S = 0, \pm 1$, etc.

the spin vectors I and S, respectively. This field adds or subtracts from the applied magnetic field, changing the resonance frequency.

Secondly, the electron is a moving electric charge that creates an 'orbital' magnetic moment. If the electronic orbital is filled with two electrons of opposite spin, the two orbital moments cancel. Often, however, orbital degeneracy is lifted and a net orbital magnetic moment exists, which interacts with the spin moment. This causes a change in the g value (Table 2) that generally will depend on the direction of the applied magnetic field with respect to the molecule. Thus, the g value and therefore the resonance frequency become anisotropic, and must be represented by a 3×3 matrix, often called the g tensor.

Thirdly, when two unpaired electrons are sufficiently close, as for example in the triplet state, the two magnetic dipoles interact magnetically: the magnetic dipole–dipole interaction. The interaction can be viewed as an additional local magnetic field and therefore it changes the resonance frequency. The local field depends on the angle between the dipolar axes of the two unpaired electrons and the applied field, and therefore the change in resonance frequency is anisotropic.

Fourthly, two unpaired electrons interact because of the overlap of their electronic orbitals. This gives rise to the so-called exchange energy, which again changes the resonance frequency of the individual electrons compared to that of the free electron. In Table 3 the four interactions are tabulated, together with their mathematical expressions. We have neglected the nuclear Zeeman interaction as this is more than six hundred times smaller than the electronic Zeeman interaction and, to first order, does not influence the EPR resonance.

3.2. The EPR spectrum

The first-order resonance frequency is given by adding the energies of the four interactions of Table 3 for the two $m_S = +\frac{1}{2}$ and $-\frac{1}{2}$ states (for $S = \frac{1}{2}$ 'doublet' systems) or the $m_S = 0, \pm 1$ states (for $S = S_1 + S_2 = 1$ 'triplet' systems). Thus, for isotropic g and a:

$$E = \pm\tfrac{1}{2}g\beta B_0 \pm \tfrac{1}{4}a \qquad \text{doublet with one proton}$$

$$\left.\begin{aligned} E_1 &= -\tfrac{2}{3}D - J \\ E_{2,3} &= \pm g\beta B_0 \pm \tfrac{1}{2}a + \tfrac{1}{3}D - J \end{aligned}\right\} \quad \begin{aligned}&\text{triplet with one proton}\\ &B//z \text{ dipolar axis, axial symmetry}\end{aligned}$$

The resulting energy levels are depicted in Fig. 1a,b, where the energy is plotted against magnetic field. The resonance frequencies depend, of course, on the field. For experimental reasons, in EPR spectroscopy one keeps the frequency of the microwave field constant (usually close to 9 GHz, X-band, or to 35 GHz, Q-band), and varies the field B_0. The resulting transitions (corresponding to $\Delta E = h\nu$ (microwaves)) are indicated with arrows, and displayed in a so-called stick spectrum (Fig. 1c).

The transition probabilities from the $m_S = +\frac{1}{2}$ to the $m_S = -\frac{1}{2}$ and from the $m_S = -\frac{1}{2}$ to the $m_S = +\frac{1}{2}$ level in Fig. 1 are equal, so that resonance will only take

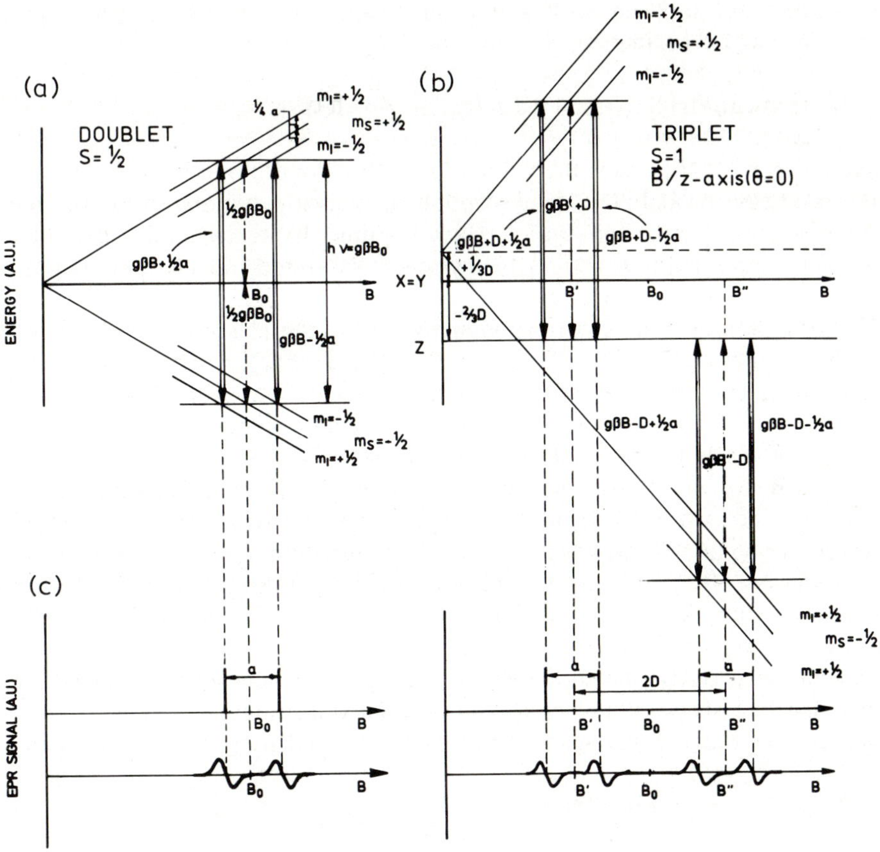

Fig. 1. (a) Energy levels of an $S = \frac{1}{2}$, $I = \frac{1}{2}$ spin system as a function of magnetic field. Double arrows indicate the transitions that are allowed according to the selection rules $\Delta m_S = 1$, $\Delta m_I = 0$ for a fixed microwave frequency ν of quantum energy $h\nu$. The hyperfine levels are drawn according to the high-field approximation (not accurate for $B_0 \lesssim a, D$). (b) Same for $S = 1$, $I = \frac{1}{2}$ spin system, with axial dipolar interaction tensor ($E = 0$), $B//z$-dipolar axis and $J >> g\beta B_0$. Note the energy splitting in zero magnetic field. (c) Upper trace, stick spectra of the EPR transitions; lower trace, derivative EPR spectrum as recorded with a field-modulation spectrometer.

place when the populations in the upper $+\frac{1}{2}$ level ($N\uparrow$) and in the lower $-\frac{1}{2}$ level ($N\downarrow$) are different. Normally the system is in Boltzmann equilibrium (Table 2):

$$N\uparrow/N\downarrow = \exp(-\Delta E/kT) = \exp(-h\nu_0/kT)$$

At $T = 300$ K and $\nu_0 = 9$ GHz, $N\uparrow/N\downarrow \sim 0.999$.

At resonance, energy absorption from the alternating electromagnetic field would come quickly to an end due to equalization of $N\uparrow$ and $N\downarrow$ if the upper spins did not have an independent means of falling back. This comes about by the same mechanism that establishes Boltzmann equilibrium, i.e. the contact with the lattice (the molecular environment). We thus have

$$\mathrm{d}n/\mathrm{d}t = (\mathrm{d}n/\mathrm{d}t)_{\text{microwaves}} + (\mathrm{d}n/\mathrm{d}t)_{\text{lattice}} = -2Pn + (n_0 - n)/T_1$$

where $n = N\uparrow - N\downarrow$, n_0 is the value of n in the absence of a microwave field, P is the transition probability for absorption or emission stimulated by the microwave field, and T_1 is the spin-lattice relaxation time. In steady state, $\mathrm{d}n/\mathrm{d}t = 0$ and $n = n_0/(1 + 2PT_1)$; for $P >> T_1$ the system is saturated, i.e. the population difference between the upper and the lower level approaches zero and the intensity of the ESR line decreases correspondingly. Using the classical treatment of the resonance phenomenon it can be shown that $P = \frac{1}{2}\gamma^2 B_1^2 T_2$, where B_1 is the amplitude of the microwave field, and T_2, the transverse or spin-spin relaxation time, is related to the width of the energy levels. For a homogeneous, Lorentzian EPR line the half-width at half-height is given by $\Delta\omega = T_2^{-1}$.

The information of an EPR spectrum is contained in (i) the resonance frequency (in practice the g value), (ii) the line shape or spectral structure, and (iii) the relaxation behaviour. Comparison of the characteristic parameters with those of known species often leads to identification of the paramagnetic entity under investigation. More fundamentally, they give insight into the magnetic interactions to which the unpaired electron is subjected, and thus into the structure of its environment. In actuality, the stick spectrum of Fig. 1c will consist of resonances with Lorentzian line shape in the case of individual transitions as in Fig. 1, or with Gaussian line shape when many different unresolved hyperfine interactions are

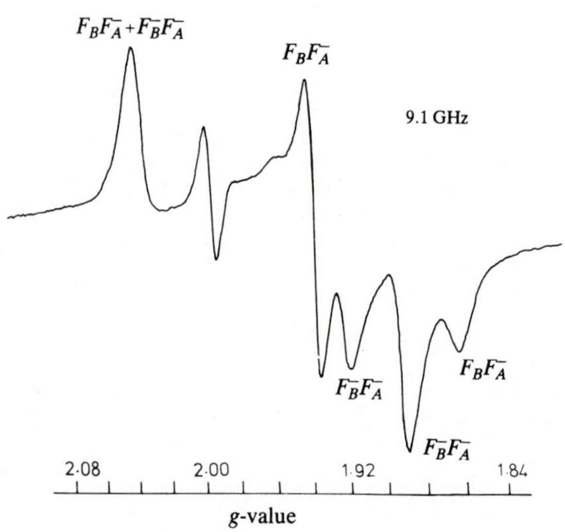

Fig. 2. Spectrum of the reduced PS I secondary F_A^- and F_B^- at 9.1 GHz (T = 5 K). PS I particles were reduced with 25 mM dithionite for 15 min in the dark and frozen under illumination. Incomplete reduction of F_B leaves visible part of the g = 1.86 line that is due to F_A^-. The signal at g = 2.00 is a residual P-700$^+$ line. From Ref. 16.

present. To enhance the signal-to-noise ratio, the magnetic field is usually modulated and the modulated absorption of microwaves is detected with a phase-sensitive (lock-in) detector, resulting in a derivative Lorentzian or Gaussian line shape. When anisotropic interactions are present in a randomly oriented system, the spectrum will consist of the envelope of the resonances for each particular orientation weighted according to the appropriate orientational distribution function. This is illustrated in Fig. 2, where spectra are displayed of the iron-sulfur proteins (ferredoxins) that function as electron acceptors in photosytem I and which have anisotropic g values.

3.3. *Electron nuclear double resonance, ENDOR*

ENDOR plays an important role in the identification of photo-induced radicals in photosynthesis. Therefore, a short discussion of this technique is given below.

Consider the expression for the energy of a one-electron ($S = \frac{1}{2}$, $m_S = \pm\frac{1}{2}$), one-proton ($I = \frac{1}{2}$, $m_I = \pm\frac{1}{2}$) system, now including the nuclear Zeeman interaction (whose sign is opposite that of the electron Zeeman interaction):

$$E = \pm\tfrac{1}{2}g\beta B_0 \pm \tfrac{1}{4}a \pm \tfrac{1}{2}g_n\beta_n B_0$$

The energy level diagram is displayed in Fig. 3a. Suppose the EPR transition $(+\frac{1}{2},+\frac{1}{2}) \rightarrow (-\frac{1}{2},+\frac{1}{2})$, where the bracketed numbers refer to m_S and m_I in this order, is semi-saturated. If we now apply radiofrequency (RF) power of frequency corresponding to the transition $(+\frac{1}{2},+\frac{1}{2}) \rightarrow (+\frac{1}{2},-\frac{1}{2})$, the uppermost level becomes somewhat less populated, because spins are transferred to the $(+\frac{1}{2},-\frac{1}{2})$ level. This means that the EPR transition becomes somewhat less saturated (we have opened another relaxation channel) and the intensity of the EPR line increases somewhat. Thus, if we monitor the intensity of the EPR line while scanning through the NMR transition with an RF source, we see an enhancement at frequencies $\nu_n = |\frac{1}{2}a \mp g_n\beta_n B_0|$, where g_n and β_n are the nuclear g factor and magneton, respectively. The plus sign refers to the transition $(-\frac{1}{2},+\frac{1}{2}) \rightarrow (-\frac{1}{2},-\frac{1}{2})$. The absolute value is taken because it may easily happen that the Zeeman term exceeds the hyperfine interaction. For one nuclear spin having $\frac{1}{2}a < g_n\beta_n B_0$ we observe ENDOR transitions spaced by the hyperfine interaction a, symmetrically located with respect to the free precession frequency of the nucleus (Fig. 3b). If $g_n\beta_n B_0 < \frac{1}{2}a$ we see two lines spaced by $2g_n\beta_n B_0$ at a frequency corresponding to $a/2$. In Fig. 3 a relaxation pathway T_x is assumed to exist between the $(+\frac{1}{2},-\frac{1}{2})$ and the $(-\frac{1}{2},+\frac{1}{2})$ level. Without this cross-relaxation, ENDOR would only be a transient phenomenon. An analogous pathway T_{xx} between the $(+\frac{1}{2},+\frac{1}{2})$ and $(-\frac{1}{2},-\frac{1}{2})$ levels is less probable because then a double spin flip has to occur. Nevertheless, the two ENDOR transitions are frequently of comparable magnitude. It should be kept in mind, though, that ENDOR intensities depend on a great number of experimental variables, and are not very suitable for quantitative measurements.

The unique advantage of ENDOR lies in its simplification of the resonance spectra and in its resolving power. For k sets of n_k equivalent nuclei with spin I_k

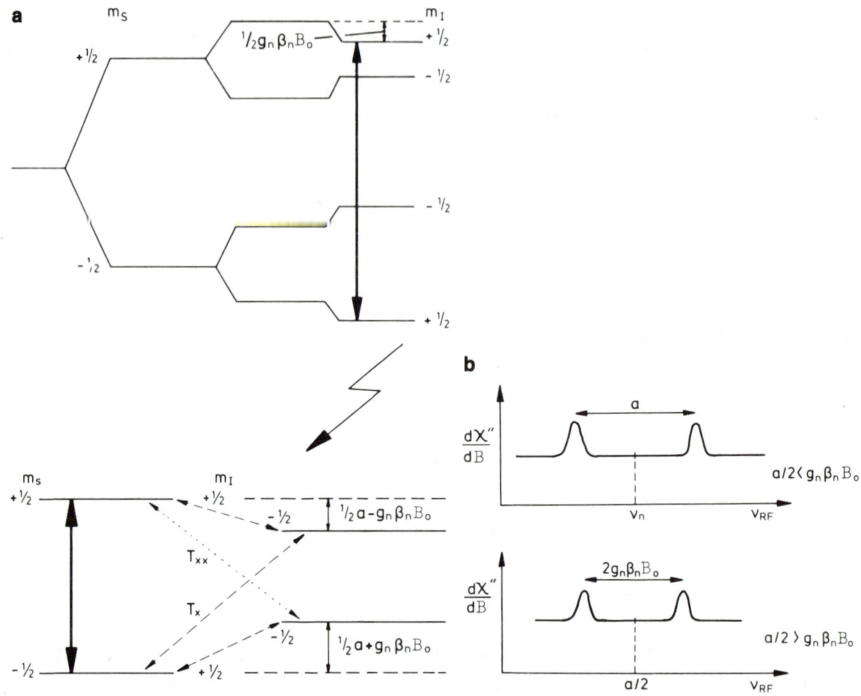

Fig. 3. (a) Term schemes of an $S = \frac{1}{2}$, $I = \frac{1}{2}$ spin system. Heavy arrows indicate EPR transitions with the selection rules $\Delta m_S = 1$, $\Delta m_I = 0$. In the lower scheme, light arrows indicate ENDOR transitions ($\Delta m_S = 0$, $\Delta m_I = 1$), the heavy arrow represents the half-saturated EPR transition, T_x is an allowed, T_{xx} a spin-forbidden relaxation pathway. (b) Schematic ENDOR spectrum for the spin system of Fig. 3a for $a/2 < g_n\beta_nB_0$ (upper trace) and $a/2 > g_n\beta_nB_0$ (lower trace).

there are $\Pi_k\ (2n_kI_k + 1)$ EPR lines, but only $2k$ ENDOR lines, which are usually very narrow (of the order of 100 kHz).

The EPR lines encountered in photosynthetic material often consist of the envelope of many hyperfine lines (they are inhomogeneously broadened Gaussians) and contain little structural information. From their ENDOR spectrum, however, several hyperfine coupling constants could be determined.

4. *EPR of primary reactants in photosynthesis*

From Sections 2 and 3 it will be clear that the shape of an EPR spectrum contains information on structure (g tensor, hyperfine splittings) and on interactions (fine structure parameters, exchange coupling). This has been exploited with much success in the study of the primary reactants in photosynthesis. In this section this will be highlighted for oxidized and reduced primary reactants in the bacterial reaction center and for the two plant photosystems. More extensive reviews on applications of EPR in photosynthesis are found in Refs. R1 and R2.

4.1. The primary electron donor

4.1.1. Bacterial photosynthesis

EPR has been instrumental in the demonstration that the primary electron donor, P, in bacterial reaction centers (RC) is a bacteriochlorophyll dimer. In Fig. 4 the EPR spectrum of P^+ of *Rhodobacter* (formerly *Rhodopseudomonas) sphaeroides* R-26 is compared to that of BChl a^+. The g values are identical, viz. 20026 ± 0.0001, and both lines are Gaussians, but the line width (ΔB) for BChl a^+ (ΔB = 13.0 ± 0.2 G) is $1.4 = \sqrt{2}$ times larger than that of P^+ (ΔB = 9.4 ± 0.2 G) [3,4]. This and similar observations on plant material led to the hypothesis [3] that the unpaired electron is shared by two identical BChl a molecules. This would lead to a halving of the hyperfine interactions (the unpaired electron would spend only half as much time on each nucleus), and to a reduction of $\sqrt{2}$ in the EPR line width (in general, for a sharing over n molecules a factor $\sqrt{n}$). The above suggestion was tested by ENDOR experiments, in which the hyperfine coupling is measured directly. For the strongest ENDOR line at 77 K the predicted halving was indeed observed [5–7]. In later proton ENDOR work [8], it was found that at higher temperatures, where more ENDOR lines are resolved, the ratio a(RC)/a(BChl a^+) for the individual couplings was generally not precisely 0.5; the ratio averaged over all observed couplings, however, was very close ot 0.5. This result was rationalized in terms of a description of P^+ as a supermolecule, consisting of two closely coupled BChls. The details of the structure then determine the actual 'density' of the unpaired electron on each nucleus. A nice complement to the proton ENDOR work was that in which the ENDOR spectrum of the pyrrole nitrogens was measured [9]. The ^{15}N hyperfine couplings agreed well with the supermolecule concept.

For the BChl b-containing bacterium *Rhodopseudomonas viridis* the observed reduction in line width of P^+ was not a factor of 1.4 but of 1.2. This is remarkable,

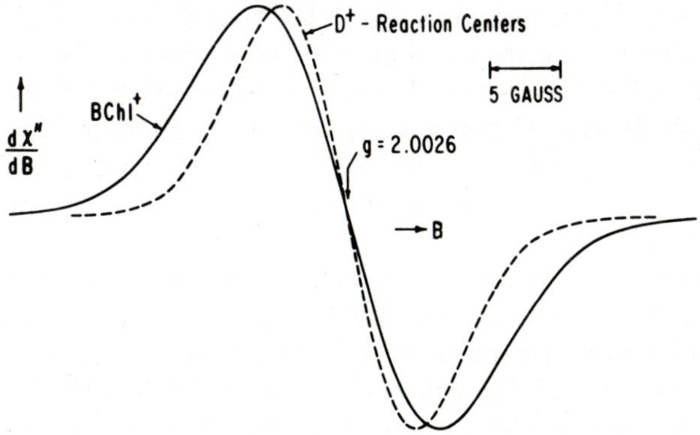

Fig. 4. EPR spectra of P-870$^+$ (D$^+$, dashed line) and BChl a^+ (drawn line) at 77 K. From Ref. 7.

because the crystal structure very clearly shows a BChl dimer [10]. The observation might be rationalized by assuming that certain protons have somewhat different bond angles in P^+ than in monomeric BChl b^+ [11,12], or by assuming that the dimer in *Rps. viridis* is less symmetric on an atomic scale than that in *Rb. sphaeroides* R-26.

Other photosynthetic bacteria show grosso modo the same EPR signal of P^+ as *Rb. sphaeroides,* and presumably their primary donor also consists of a dimeric BChl complex.

4.1.2. Photosystem I

The primary donor of plant photosytem I (P-700 in the oxidized state) gives rise to a Gaussian EPR line, with $g = 2.0025 \pm 0.0001$ and $\Delta H = 7.2 \pm 0.1$ G [13], i.e. the line width is about $\sqrt{2}$ smaller than that of the Chl a^+ radical in vitro. (Chl *a* in CH_3OH/50% glycerol, Fe^{3+} oxidized: $\Delta H = 9.7 \pm 0.1$ G; I_2 oxidized: 9.5 ± 0.1 G; A.J. Hoff, unpublished results.) The suggestion that the primary donor is a dimer (see Section 4.1) is corroborated by some [14,15] but not all [16,17] proton ENDOR experiments.

The line width in protonated material and the proton ENDOR spectra measure primarily the unpaired electron density on those carbon atoms that are close to a proton – one (α protons) or two (β protons) bonds away. A non-symmetric sharing could adventitiously lead to a halving of spin-density on a few carbons, whereas other carbons, not directly connected to an α or β proton, would have more or less 'spin' density. This was elegantly checked by Wasielewski et al. [18], who measured EPR signals of material 91% enriched in ^{13}C ($I = \frac{1}{2}$), substituting for ^{12}C ($I = 0$) and 99% enriched in ^2H ($I = 1$, low hyperfine coupling). Now the carbon nuclei themselves have a hyperfine interaction with the unpaired electron, and all would contribute to ΔB, whereas the ^2H contribution is small (about 20%). Interestingly, for PS I the ratio of the second moment (which is a measure of the line width for non-Gaussian lines) of ^{13}C-enriched PS I to ^{13}C-enriched monomeric Chl *a* was close to one, whereas the same experiment for the bacterium *Rhodospirillum rubrum* showed the ratio of 2 expected for a symmetric dimer. Thus, in PS I either the unpaired electron sits mainly on one Chl *a* of a dimer that has enough spin density on a few selected carbons on both Chl *a*s to lower the line width of ^{12}C P-700$^+$ due to interaction with α,β protons with a factor 1.3, or the primary donor is not a chlorophyll *a* molecule. It was suggested that P-700 is a monomeric Chl *a* enol, in which the ring V β-keto ester of Chl *a* is enolized [19]. Optical absorption difference spectra, however, do not support this contention [20].

4.1.3. Photosystem II

The primary donor of photosystem II (P-680) is much more difficult to observe with EPR than that of PS I, because in normally functioning PS II the photooxidized donor is very rapidly (within at most a few hundred ns) rereduced by an electron donor called Z [21,22] (see for a review Refs. R3 and R4). This (re)reduction can be slowed down by various treatments and by cooling to low temperature. In intact chloroplasts in which P-700$^+$ is fully oxidized chemically or by preillumination, at low temperature (< 100 K) an additional signal is reversibly photoinduced, with $g = 2.0026$ and ΔB between 6 and 8 G [23]. PS II particles showed a 9 G wide signal [24]. At room temperature, illumination induced in purified PS II particles an EPR signal with g value and ΔB close to that of P-700$^+$ [25]. When reduction of P-680$^+$ was slowed by hydroxylamine treatment, flash EPR revealed a 7 – 8 G wide line attributed to P-680$^+$ [26].

From the above experiments one might conclude that P-680 is similarly to P-700 a Chl *a* dimer. However, the optical absorbance difference spectrum (oxidized-minus-reduced) gives a bleaching at 680 nm, much closer to monomeric Chl *a* than the corresponding bleaching for PS I, which is at 700 nm, i.e. red-shifted as expected for a (parallel) dimer. In addition, the redox midpoint potential (E_m) of P-680 should be about +1.0 V or even higher, since the donor side of PS II oxidizes water ($E_m(H_2O) = +0.84$ V). This high value is not far from the E_m of the Chl *a*/Chl a^+ couple in vitro (0.8 V [27]), and much more positive than that of P-700, which is about 0.45 V. These considerations led Davis et al. [27] to suggest that P-680$^+$ might be a Chl *a* monomer, whose EPR line width is narrowed by ligand effects. Thus, the question whether P-680 is a Chl *a* monomer or dimer seems to be still open (see, however, Section 5).

4.2. *The primary acceptor*

4.2.1. *Purple bacteria*

There is some confusion as to which electron acceptor should be called primary. Historically, in purple bacteria the quinone acceptor, Q_A, was so named. Later it was found that a BPh molecule accepts an electron before Q_A, and possibly even earlier acceptors, or charge transfer states, exist. Since the latter matter is still under debate (see Chapter 3), one might prudently keep the label 'primary' for the quinone acceptor with the understanding that it is the first 'stable' (on a time scale of ms) acceptor.

In native RCs the EPR signal of Q_A^- deviates considerably from that of semiquinone in vitro [28,29]. It was shown that detergent treatment of isolated RCs or cells, which removed or 'uncoupled' an iron atom that is present in stoichiometric amounts to P, led to a narrow EPR signal with all the characteristics of a semiquinone signal [30,31]. The very broad signal observed when iron was present was attributed to a magnetic coupling between the semiquinone radical and the high-spin Fe^{2+} ($S = 2$) iron ion. From magnetization measurements and from analysis of the line shape of the $Q_A^- \cdot Fe^{2+}$ signal the strength of the (antiferromagnetic) exchange coupling J was determined: $2J = -0.2 \pm 0.1$ cm$^{-1} \approx -10^3$ G [32,33]. From a computer simulation it was concluded that the coupling was anisotropic, probably because of an admixing of dipolar interaction. Also the g tensor of Fe^{2+} was determined [33].

The role of the iron in electron transport is still obscure. Electron transfer from the first to the second quinone acceptor in Fe-depleted RC is only a factor of two slower than in Fe-containing RC [34]. Fe can be replaced by other divalent metals without much affecting the lifetime of Q^- [34]. The iron may stabilize the semiquinone form of the secondary quinone acceptor through an effect on its redox potential and protonation.

(Photo)reduction of RCs, in which the iron was dissociated from the quinone, leads to a near-Gaussian X-band EPR line at $g = 2.0046 \pm 0.0002$ and $\Delta B = 8.1 \pm 0.5$ G [31]. At Q-band (35 GHz) the line shows a structure that is identical to that of immobilized ubisemiquinone. For perdeuterated material the Q_A^- signal at Q-band shows a resolved spectrum showing rhombic g anisotropy with principal g values 2.0024, 2.0056 and 2.0067 (± 0.0002) [35]. Recent ENDOR work on Q_A^- and Q_B^- has identified the ubisemiquinone proton hyperfine splittings and those of protons in hydrogen bonds between the keto groups of the quinone and protein residues [36,37]. The latter couplings give valuable information on the binding of the quinone to the protein matrix.

4.2.2. Green bacteria

In the green sulfur bacteria, typified by *Prosthecochloris aestuarii* and *Chlorobium limicola,* the primary acceptor is a ferredoxin-type molecule [38,39], similar to those found as acceptors in plant PS I (Section 4.2.3), with $E_m \approx -560$ mV. Its EPR signal is typical of an unpaired spin with strongly anisotropic g value, consisting of three bands at g = 2.005, 1.94 and 1.89 that correspond to the principal g values g_{xx}, g_{yy}, g_{zz}. In *P. aestuarii* the low-potential ferredoxin acceptor is followed by a second iron-sulfur protein acceptor with E_m more positive than −420 mV, having an EPR spectrum with slightly different g values [39].

In the recently discovered photosynthetic bacterium *Heliobacterium chlorum,* which has a BChl g complex as primary donor, the primary acceptor is also a ferredoxin-type molecule with g values 2.04, 1.94 and 1.88 [40]. An earlier acceptor could be photoaccumulated at very low (< −620 mV) redox potential. It had a near-Gaussian EPR line at $g = 2.0038$ with ΔB = 15 G at X-band and 18 G at Q-band [40].

The primary acceptor of the green gliding bacterium *Chloroflexus aurantiacus* is a menaquinone [41]. It appears that, in spite of its green color brought about by the presence of antenna BChl c pigments, the photosystem of this bacterium is very similar to that of the purple bacteria (Chapter 3).

4.2.3. Photosystem I

The primary acceptors of the two plant photosystems differ fundamentally from each other, no doubt because of their different redox midpoint potentials (about −100 to −200 mV for PS II, −705 to −730 mV for PS I [R3–R5]). In PS I two iron-sulfur (ferredoxin-type) proteins, F_A and F_B, with characteristic EPR spectrum in the reduced state (E_M between −450 and −550 mV), have been observed (Fig. 2) that function either parallel or in series (see Ref. R5 for a recent review). The shape of the spectra of the two ferredoxin-type acceptors and in particular their principal g values depend on whether one or both acceptors are reduced (Fig. 2). It is unlikely that this is due to a magnetic interaction, as the differences depend linearly on the microwave frequency, i.e. on the applied magnetic field (exchange and dipolar interactions are independent of field; Table 3) [16,42]. Possibly, Coulomb repulsion causes strain-induced g shifts.

A third species, F_X, the spectrum of which considerably deviates from that of a ferredoxin, is observed under highly reducing conditions [43,44]. From Mössbauer studies it was calculated that F_X is a 4Fe-4S iron-sulfur protein [45]. It is still not quite certain, however, whether under physiological conditions F_X really acts as an obligatory electron acceptor. In spite of the above-mentioned uncertainties, EPR is the only technique that is capable of furnishing detailed information on the various iron-sulfur protein acceptors; their optical absorbance difference spectra all show a rather uninformative weak band around 430 nm.

4.2.4. Photosystem II
The primary acceptor Q_A in PS II is a plastoquinone, PQ, as ascertained from optical absorbance difference spectroscopy [46]. Until recently, the EPR spectrum of the semiquinone escaped observation, and only the advent of preparation methods for PS II subchloroplast particles made its recording possible. As surmised earlier, the spectrum of the intact acceptor [47] very much resembled the very broad quinone-iron acceptor complex in purple bacteria, whereas in iron-depleted PS II particles the narrow spectrum typical of an immobilized semiquinone was found [48]. As in the bacterial photosystem, flash-induced reduction of Q_A, of the second quinone, Q_B, or of both resulted in somewhat different EPR spectra, indicative of structural changes that influence the magnetic interaction between the semiquinone and the iron, and/or between the two semiquinones [49].

4.3. The intermediary acceptor

4.3.1. Bacterial photosynthesis
The reduced intermediary acceptor I (BPh) is normally too short-lived to be observable by EPR. However, it can be photoaccumulated at cryogenic temperatures in isolated RCs of, for example, *Rb. sphaeroides* when reduced Cyt *c* is added, because of slow, irreversible electron donation to P^+ [50]. The resulting EPR signal is a Gaussian line at $g = 2.0036 \pm 0.0002$ of width $\Delta B = 12.9 \pm 0.3$ G [50], which is typical of the monomer BChl a^- and b^- anion radicals [51]. The ENDOR spectrum of the narrow signal of the intermediary acceptor was similar to both the ENDOR spectra of monomeric BChl a^- and BPh a^- [51], thus showing that the intermediate was a monomer but not allowing a choice between BChl a^- and BPh a^-. Optical absorbance difference spectroscopy of other BChl *a*-containing purple bacteria, however, quite clearly shows that it must be a BPh *a* molecule [52,53] (see below).

In some bacteria *(Chromatium minutissimum, C. vinosum, Rps. viridis)* I^- can be photoaccumulated at cryogenic temperatures because of the presence of a bound Cyt *c* that irreversibly donates an electron to P^+ [54–56]. In these bacteria, the EPR signal of BPh measured at 5 K shows two lines with splitting of 60 – 120 G. This splitting was attributed to exchange interaction of BPh^- with Q_A^- (which in all three bacterial species is a menaquinone); Fig. 5. At higher temperatures, 20 K and up, the two lines merge to a single line. Sometimes both types of signals are present at 20 K [52]. This can be attributed to the presence of a mixed population of singly

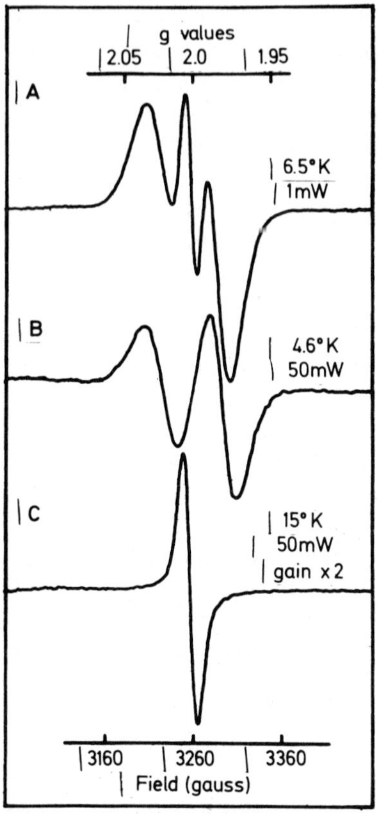

Fig. 5. EPR spectra from the reduced intermediary acceptor BPh a^- in RC of *C. vinosum*. Difference spectra of samples illuminated for 3 min at 200 K and non-illuminated samples. From Ref. 52.

and doubly reduced Q_A (the Q_A^{2-} species is diamagnetic, hence does not interact with BPh^-) [50]. The split line was not found in native RCs of *Rb. sphaeroides* R-26 but was observed for RCs in which the native ubiquinone was replaced by menaquinone [50]. The measured rates of double reduction of Q_A correspond well with the exchange interaction between I^- and Q_A^-, adopting a simple model for the relationship between the exchange interaction and the electron transport rate [57]. This is an illustration of how knowledge of the magnetic exchange interaction leads to insight into the electron transport properties.

Recent picosecond absorbance difference spectroscopy [58] and pigment extraction studies [59] have shown that the intermediary acceptor in green sulfur bacteria is a (probably monomeric) BChl *c* molecule. As yet no EPR data are available. In *H. chlorum* the intermediate is probably also a BChl *c*-like molecule [60].

4.3.2. Photosystem I

From kinetic optical absorbance difference spectroscopy it was concluded that in PS I before the iron-sulfur acceptors another earlier acceptor exists, labeled A_1 [61], which showed an EPR signal similar to that of monomeric Chl a^+ [62]. EPR spectroscopy of PS I particles that were depleted of the iron-sulfur acceptors by detergent treatment and in which the earliest acceptors were reduced by photoaccumulation, showed a complex spectrum that was explained as being composed of a Chl a^- monomer-type A_0^- spectrum and the spectrum of a later acceptor, A_1^-, which had some characteristics of a semiquinone spectrum [63–65]. Recent optical absorbance difference measurements confirmed that A_0 is a chlorophyll *a* molecule [66], whereas electron spin polarization experiments on PS I in deuterated algae supported the assignment of A_1 as a quinone-type molecule [67].

4.3.3. Photosystem II

Several years ago optical spectroscopy on PS II particles provided evidence that before PQ Pheo *a* functions as an earlier acceptor, with $E_m \approx -610$ mV [68]. By photoaccumulation it was established that the reduced intermediary acceptor has an EPR signal characteristic of monomeric Pheo a^- (g value 2.0033 ± 0.0003, ΔB = 12.6 ± 0.3 G) [48,69]. ENDOR work established a good agreement between methyl hyperfine splittings of $Pheo^-$ in vivo and monomeric $Pheo^-$ in vitro [70].

Recently, electron spin polarization and EPR data provided evidence that, at least at low temperatures under strongly reducing conditions, one or even two acceptors function between Pheo *a* and PQ [71,72]. The significance of these acceptors under physiological conditions, however, remains to be demonstrated.

5. The triplet state

In normal photosynthesis, in all photosystems the charge on the photoreduced intermediary acceptor is quickly transported to the next, primary, acceptor. When this acceptor is (photo)chemically prereduced or removed by extraction, however, this negative charge cannot be further transported, and recombines with the positive charge on the primary donor. The recombination product is either the singlet ground or excited state, or the triplet excited state of P. Although the triplet does not pay a role in normal photosynthesis, its properties, especially those measured by EPR, make it a versatile probe of pigment configuration in the RC and as such it deserves the considerable attention it has received over the years.

In the triplet state, the excited electron has the same spin orientation (parallel or antiparallel to the external field) as the electron in the original ground-state orbital, so that the state is paramagnetic with total spin $S = S_1 + S_2 = 1$. Its multiplicity (i.e. the number of quantum mechanically allowed projections of the spin vector S on the field B) is then $2S + 1 = 3$, with magnetic quantum numbers $m_S = 0, \pm 1$.

In general, the anisotropic dipole–dipole interaction between the two unpaired spins in the triplet state (Table 3) gives rise to a very broad EPR spectrum, with in the derivative representation characteristic peaks for directions of B parallel to the principal axes of the dipole–dipole interaction tensor **D**. For reaction centers the spectrum shows a peculiar distribution of lines that are in emission or in absorption (Fig. 6). This pattern is characteristic of the aforementioned mode of triplet formation by recombination [73]. It was first observed by Dutton et al. [74] in bacterial RCs, and later also in PS I and PS II particles (see Refs. R6 – R9 for reviews on the triplet state in photosynthesis).

The observation of the radical recombination triplet in all photosystems is direct evidence for the universality of structure in the reaction center. When forward transport is blocked the photoinduced radical pair lives long enough to generate a triplet configuration from the original singlet configuration of unpaired spins (for the known reactants this process takes of the order of 10 ns), and the decay rate via the triplet channel has (at least at low temperatures) a relatively high yield (for bacterial RCs close to 100%).

Formation of the donor triplet state, ^3P, gives rise to a bleaching of the long-wavelength absorbance of the primary donor and to certain intensity changes and shifts of the bands of the other pigments in the RC. This can be measured optically with flash difference spectroscopy, and much more accurately with the recently developed technique of absorbance-detected magnetic resonance (ADMR) of the triplet state in zero magnetic field [75] (reviewed in Ref. R10). The resulting spectra show that the triplet is probably localized on one Chl or BChl of the dimeric primary donor, in bacterial RCs as well as in those of PS I and PS II [76,77]. For RCs of *Rps. viridis* this has recently been corroborated by EPR of ^3P in single crystals (Refs. 78,79, and J.R. Norris, personal communication). From these data it was concluded that the triplet resides on the BChl *b* molecule closest to the BPh *b* acceptor. Localization of the triplet state would explain in a natural way why the values of the fine structure parameters *D* and *E* are practically equal (PS I and PS

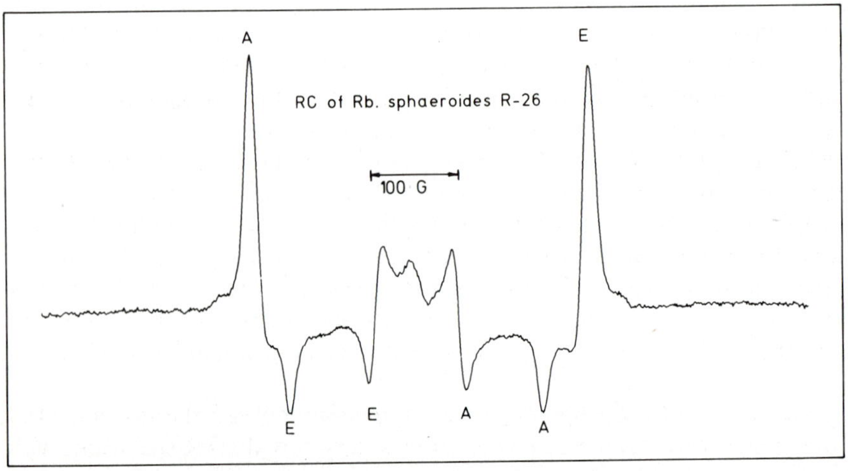

Fig. 6. Triplet EPR spectrum of ^3P of RCs of *Rb. sphaeroides* R-26 at 5 K. A, absorptive; E, emissive lines. Courtesy Mr. R. Evelo.

II) or close to (bacterial RC) the values found for monomeric triplet states of the chlorophylls in vivo, and why the low-temperature decay rates of the three triplet sublevels are virtually the same in vivo and in vitro.

The orientation of the transition moment of the long-wavelength absorbance of the primary donor with respect to the dipolar axes of the triplet state can be found from magnetophotoselection EPR or from EPR on oriented or crystalline material [80–85]. From EPR on single crystals of *Rps. viridis* it was found that the triplet x- and y-axes are very close to the pyrrole N–N axes of one of the BChls *b* making up the primary donor (Refs. 78,79 and J.R. Norris, personal communication).

6. *The oxygen-evolving complex*

6.1. *Manganese*

One of the most exciting new EPR signals in photosynthesis is that associated with the oxygen-evolving complex (OEC) of PS II. Manganese has long been implicated in the oxygen-evolving mechanism [R11]. In chloroplasts and in subchloroplast particles a spectrum was observed consisting of 16 to more than 19 hyperfine lines, extending over more than 1500 G, that was assigned to two or more magnetically coupled high-spin manganese atoms (see for reviews Refs. R12,R13 and Chapter 6]. Presumably, different oxidation states of an Mn cluster containing at least 3 Mn are responsible for three of the four so-called S states [86]. By flash EPR it was demonstrated that the multiline signal is associated with the S_2 state [87].

The multiline spectrum shows some resemblance to that of a mixed valence cluster Mn^{2+} – Mn^{3+} or Mn^{3+} – Mn^{4+} [88,89,97] and with a computer simulation of a tetramer spectrum such as $3Mn^{3+}$ – Mn^{4+} [90]. Precise agreement, however, is lacking so far. The form and number of the lines depend on the period of dark adaptation and the illumination temperature and are sensitive to the presence of inhibitors of O_2 evolution [91,92]. The signal is orientation-dependent, i.e. it has a fixed geometry with respect to the membrane [93].

Several interpretations of the two (perhaps even more) forms of the multiline signal have been advanced, making use of the temperature dependence of its intensity [91,94–98]. Further work may well resolve the still existing ambiguities and unexplained spectral forms, and allow a much deeper insight into the mechanism of oxygen evolution.

6.2. *Signal II*

One of the first EPR signals observed in photosynthetic material is the so-called Signal II. It has several kinetic forms, belonging to at least two different donor sites to P-680$^+$ but, judging from the identical spectral shape, to one chemical entity. Earlier work (reviewed in Ref. R1) established that one of the sites, Z, is part of the linear electron transport chain from the OEC to P-680, of which it is apparently the immediate donor [26]; the other donor is apparently connected to the OEC on a sidepath. The chemical identity of the Signal II radical has long been obscure. Recently, it was proposed that it is a plastosemiquinone cation [99,100]. This is an attractive suggestion in view of the high redox midpoint potentials of quinone cations (close to +1.0 V). The overall line shape, but not the line width, of model semiquinone cations supported the idea. The details of the assignment of the structure of Signal II to specific hyperfine couplings are still under debate [100–102], and the stoichiometry of plastoquinone content relative to P-680 is still uncertain [103,104]. (It is at present difficult to accomodate *two* plastoquinones on the donor side of PS II.)

7. *Electron spin polarization*

All the above EPR spectroscopy was carried out in the steady state. With the use of fast-response spectrometers, however, it was discovered a decade ago that when measured early (at room temperature a few μs) after a light flash, the EPR spectra of primary reactants in PS I [105,106] and in bacterial RC [107] showed EPR lines characteristic of systems out of Boltzmann equilibrium (Table 2). Part or all of these so-called spin-polarized lines may then either be in emission or show absorption that is enhanced or decreased compared to the equilibrium absorption (Fig. 7). Electron spin polarization occurs through magnetic interactions between two simultaneously induced donor-acceptor radicals (reviewed in Ref. R14). Thus, a study of spin-polarized EPR lines yields information on these magnetic interactions and therefore on the configuration (distance, relative orientation, etc.) of the radicals (see e.g. Ref. R14).

From EPR studies on oriented chloroplasts and PS I particles [108–110] and on perdeuterated bacterial RCs [35] it was concluded that anisotropic (dipolar) interactions played a major role, at least in oriented samples. Applying a theoretical

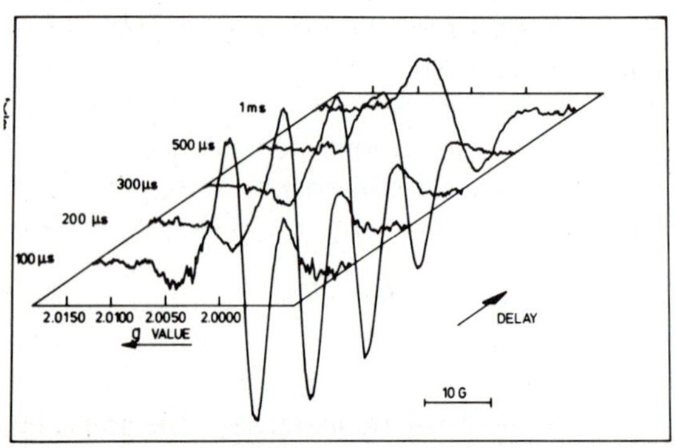

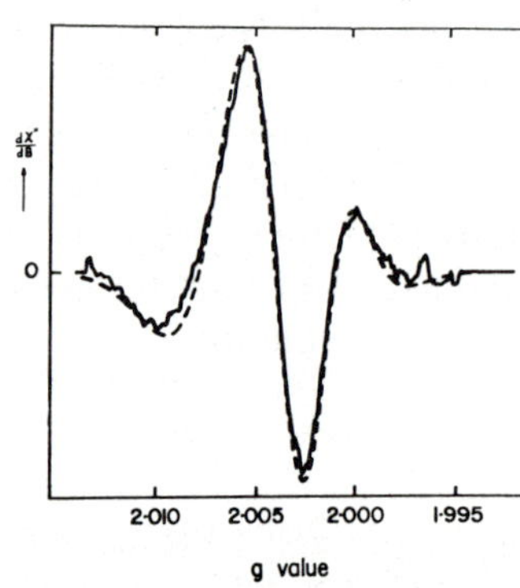

Fig. 7. Left, time evolution of the spin-polarized EPR signal at 20 K of RC of *Rhodospirillum rubrum* treated with sodium dodecyl sulfate to dissociate the $Q_A^- \cdot Fe^{2+}$ complex. Times are delays after a laser flash. In the 1 ms signal all polarization has decayed. Right, the 100 μs spin-polarized spectrum (—) and a theoretical simulation (---). From Refs. 128 and 111.

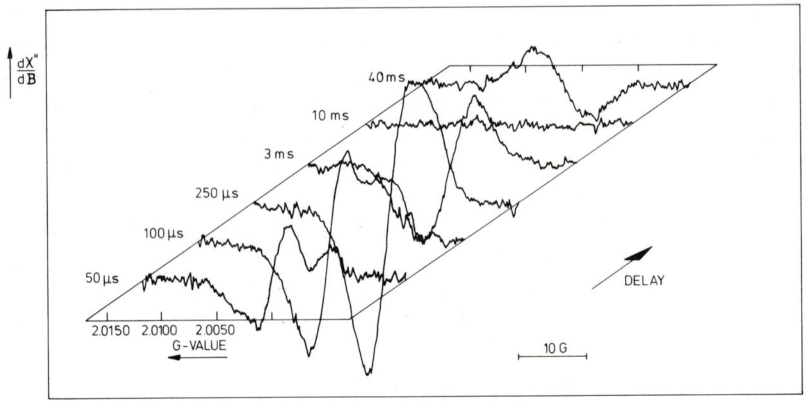

Fig. 8. Time evolution of the spin-polarized EPR signal of prereduced RCs (P I Q_A^-) treated as in Fig. 7. Note the inversion of the 3 ms spectrum with respect to the unpolarized 40 ms spectrum. The shoulder at low g value in the 50 μs spectrum is due to magnetic interaction with ^{3}P. From Ref. 128.

treatment of electron spin polarization that incorporates both the exchange and the dipolar magnetic interactions, Hore et al. [111] obtained from the experimental polarized EPR spectrum of bacterial RCs the fairly large ratio $D/J \approx 40$.

When the primary acceptor Q_A is prereduced, electron spin polarization can transfer by exchange interaction from BPh^- to Q_A^-, leading to an inversion of the EPR line of Q_A^- in RCs where Q_A^- was magnetically uncoupled from Fe^{2+} [112] (Fig. 8). From a phenomenological treatment [112–114] it was concluded that the exchange interaction $J(BPh^-Q_A^-)$ was 3 – 5 G, whereas $J(P^+BPh^-)$ was between 1 and 5 G. A more sophisticated treatment of the three-spin system $P^+BPh^-Q_A^-$ [115] led to $J(P^+BPh^-)$ between 0 and +8 G. (Note that for $J = 0$ polarization may develop if $D \neq 0$.) A positive value of J for a biradical state is unusual; it might be explained by some form of superexchange via an intermediate (possibly one of the accessory bacteriochlorophylls).

Regardless of the precise value of $J(P^+BPh^-)$ it is clear that the values obtained are much (some three orders of magnitude) smaller than expected from the rate of charge separation (2.8 ps [116]) when simple tunneling theory is applied [50,57,113,117,118]. This also seems to indicate the need for another intermediate (which need not function as a true electron acceptor but could act as a transmitting 'medium' via a superexchange mechanism).

8. *New techniques: ESE and RYDMAR*

In the last few years, pulsed EPR or electron spin echo (ESE) and reaction yield detected magnetic resonance (RYDMAR) techniques have been added to the arsenal of EPR techniques applied in photosynthesis. ESE combines high temporal resolution (currently 100 ns) with sensitivity to broad EPR signals, and it allows rapid and accurate determination of the spin-lattice and spin-spin relaxation times.

In addition, it is often possible to determine hyperfine couplings from modulations in the echo amplitude or by combining the microwave pulses with radiofrequency pulses, thus performing a pulsed ENDOR experiment. Recent applications of ESE to photosynthesis are discussed in Refs. 119–123 and in Ref. R14.

In RYDMAR (reviewed in Refs. R15 and R16) magnetic resonance is detected by monitoring the yield of a reaction product in the presence of microwaves resonant with the difference in energy between the levels of a coupled radical pair (which for two $S = \frac{1}{2}$ radicals has 1 singlet ($S = 0$) and 3 triplet ($S = 1$) energy levels). The yield of recombination and product formation is dependent on the relative population of the four levels, which is altered by microwave-induced transitions.

The first successful RYDMAR experiment on reaction centers was carried out by Bowman et al. [124] using laser flashes and pulsed X-band microwaves of high intensity. Recently, a sensitive RYDMAR technique was developed by Möhl et al. [125] using a combination of continuous illumination with weak magnetic fields (100 to 200 G) and low-intensity microwave radiation at about 300 MHz. Typical spectra are displayed in Fig. 9. From a simulation of these spectra and from their variation with microwave intensity it was concluded that $|D(P^+BPh^-| \leqslant 20$ G, $|2J(P^+BPh^-)| = 10.1 \pm 0.5$ G and the sum of the recombination rates to P*, P and

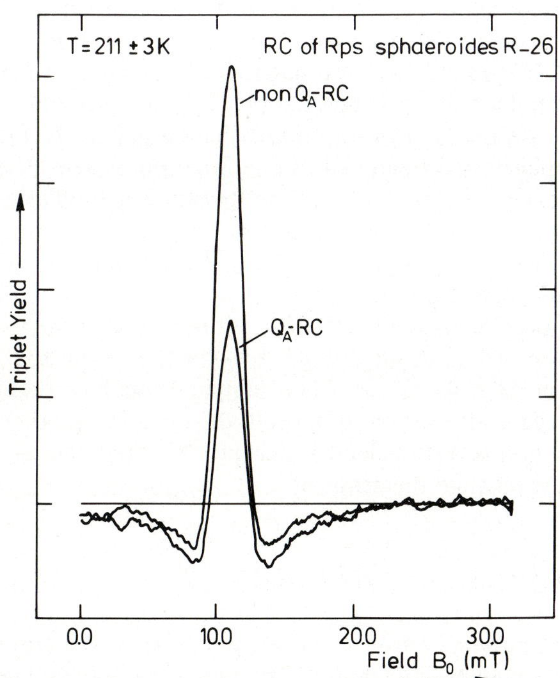

Fig. 9. RYDMAR signal of RCs of *Rb. sphaeroides* R-26 measured with low microwave power (40 W, frequency 307 MHz) at temperatures as indicated. Q_A-RC and non Q_A-RC: Q_A present and removed by extraction, respectively. From Ref. 125.

^{3}P, $k_S' + k_S + k_T = (0.26 \pm 0.01) \times 10^9\ s^{-1}$ at 211 K [125,129]. Note that the ratio of D and J differs significantly from that obtained from electron spin polarization [111]. The reason for this discrepancy is not yet clear; probably the RYMDAR values are the more accurate ones. In Q_A-depleted RC J is independent of temperature; this invalidates the two-step mechanism for charge separation [130].

9. *Conclusions and prospects*

The various applications of EPR spectroscopy in photosynthesis that were discussed in this chapter merely serve to illustrate its potential, and are far from an exhaustive literature survey. EPR (and ENDOR) spectroscopy has helped to identify the structure of primary and secondary reactants, and it has proved to be one of the few tools that can be used to measure the interactions between the primary reactants, which are of course crucial to electron transport. Much of the latter results are still uncertain, and also the relationship between magnetic interactions and electron transfer integrals is still only approximate. Future work will certainly focus on these aspects.

One of the most exciting developments in photosynthesis is the elucidation of the crystal structure of a bacterial reaction center by X-ray diffraction analysis [10]. Having single crystals is the dream of any EPR spectroscopist and the writer of this chapter is sure that we are just entering a new era of EPR spectroscopy on RC crystals. The first results are the determination of the g anisotropy of the primary donor of *Rb. sphaeroides* [126] and the beautiful work on the triplet state by Gast et al. [78,127], and many more will follow, in particular employing ENDOR spectroscopy. The results will allow a very precise determination of the position of protons (which are not seen by X-ray diffraction) and of the extent and the directionality of the magnetic interactions (which even on the basis of a refined crystal structure are very difficult to compute). These data will lead to a much better understanding of electron transfer, which may eventually result in the design of a solar energy cell based on the principles of photosynthesis.

Acknowledgements

The author could not have written this chapter without the enthusiasm and hard work of all those who worked with him in the past years on many of the problems discussed. He is indebted to his colleagues from the Biophysics Department and the Centre for the Study of Excited Molecules in Leiden for sharing their knowledge and for their support. The help of Mrs. Tineke Veldhuyzen in the fast and accurate preparation of the manuscript is greatly appreciated. Research in this laboratory was supported by the Netherlands Foundation for Chemical Research (SON), financed by the Netherlands Organization for the Advancement of Pure Research (ZWO).

Review articles

R1 Hoff, A.J. (1979) Applications of ESR in photosynthesis. Phys. Rep. 54, 75–200.

R2 Hoff, A.J. (1982) ESR and ENDOR of primary reactants in photosynthesis. Biophys. Struct. Mech. 8, 107–150.

R3 Zimmermann, J.-L. and Rutherford, A.W. (1985) The O_2-evolving enzyme of photosytem II. Recent advances. Physiol. Vég. 23, 425–434.

R4 van Gorkom, H.J. (1985) Electron transfer in photosystem II. Photosynth. Res. 6, 97–112.

R5 Rutherford, A.W. and Heathcote, P. (1985) Primary photochemistry in photosystem-I. Photosynth. Res. 6, 295–316.

R6 Levanon, H. and Norris, J.R. (1978) The photoexcited triplet state and photosynthesis. Chem. Rev. 78, 185–198.

R7 Hoff, A.J. (1982) ODMR spectroscopy in photosynthesis II. The reaction center triplet in bacterial photosynthesis. In Triplet State ODMR Spectroscopy (Clarke, R.H., ed.) pp. 367–425, John Wiley & Sons, Inc., New York.

R8 Schaafsma, T.J. (1982) ODMR spectroscopy in photosynthesis I. The chlorophyll triplet state in vivo and in vitro. In Triplet State ODMR Spectroscopy (Clarke, R.H., ed.) pp. 291–365, John Wiley & Sons, Inc., New York.

R9 Hoff, A.J. (1986) Triplets: phosphorescence and magnetic resonance. In Light Emission by Plants and Bacteria (Govindjee, Amesz, J. and Fork, D.C., eds.) pp. 225–265, Academic Press, New York.

R10 Hoff, A.J. (1986) Optically detected magnetic resonance (ODMR) of triplet states in vivo. In Photosynthesis III. Photosynthetic Membranes. Encyclopedia of Plant Physiology, New Series, Vol. 19 (Arntzen, C.J. and Staehelin, L.A., eds.) pp. 400–421, Springer Verlag, Berlin.

R11 Amesz, J. (1983) The role of manganese in photosynthetic oxygen evolution. Biochim. Biophys. Acta 726, 1–12.

R12 Dismukes, G.C. (1986) The metal centers of the photosynthetic oxygen-evolving complex. Photochem. Photobiol. 43, 99–115.

R13 Dismukes, G.C. (1986) The organization and function of manganese in the water-oxidizing complex of photosynthesis. In Manganese in Metabolism and Enzyme Function (Wedler, F.C. and Schram, V.L., eds.), Academic Press, New York, in the press.

R14 Hoff, A.J. (1984) Electron spin polarization of photosynthetic reactants. Q. Rev. Biophys. 17, 153–282.

R15 Hoff, A.J. (1986) Magnetic interactions between photosynthetic reactants. Photochem. Photobiol. 43, 727–746.

R16 Norris, J.R. and van Brakel, G. (1986) Energy trapping in photosynthesis as probed by the magnetic properties of reaction centers. In Photosynthesis III. Photosynthetic Membranes. Encyclopedia of Plant Physiology, New Series, Vol. 19 (Arntzen, C.J. and Staehelin, L.A., eds.) pp. 353–370, Springer Verlag, Berlin.

References

1 Feher, G. (1970) EPR with Applications to Selected Problems in Biology, Gordon and Breach, New York.

2 Swartz, H.M., Bolton, J.R. and Borg, D.C. (1972) Biological Applications of ESR, Wiley-Interscience, New York.

3 Norris, J.R., Uphaus, R.A., Crespi, H.L. and Katz, J.J. (1971) Proc. Natl. Acad. Sci. USA 68, 625–628.

4 McElroy, J.D., Feher, G. and Mauzerall, D.C. (1972) Biochim. Biophys. Acta 267, 363–374.

5 Feher, G., Hoff, A.J., McElroy, J.D. and Isaacson R.A. (1973) Biophys. J. 13, 67a.

6 Druyan, M.E., Norris, J.R. and Katz, J.J. (1973) J. Am. Chem. Soc. 95, 1682–1683.

7 Feher, G., Hoff, A.J., Isaacson, R.A. and Ackerson, L.C. (1975) Ann. N. Y. Acad. Sci. 244, 239–259.

8 Lubitz, W., Lendzian, F., Scheer, H., Gottstein, J., Plato, M. and Möbius, K. (1984) Proc. Natl. Acad. Sci. USA 81, 1401–1405.

9 Lubitz, W., Isaacson, R.A., Abresch, E.C. and Feher, G. (1984) Proc. Natl. Acad. Sci. USA 81, 7792–7796.

10 Deisenhofer, J., Epp, O., Miki, K., Huber, R. and Michel, H. (1984) J. Mol. Biol. 180, 385–398.

11 Fajer, J., Davis, M.S., Holten, J.A., Parson, W.W., Thornber, J.P. and Windsor, M.W. (1977) 4th Int. Congr. on Photosynth., Reading, UK, abstr. 108.

12 Fajer, J., Davis, M.S., Brune, D.C., Forman, A. and Thornber, J.P. (1978) J. Am. Chem. Soc. 100, 1918–1920.

13 Warden, J.T. and Bolton, J.R. (1974) Photochem. Photobiol. 20, 251–262.

14 Norris, J.R., Druyan, M.E. and Katz, J.J. (1973) J. Am. Chem. Soc. 95, 1680–1682.

15 Norris, J.R., Scheer, H. and Katz, J.J. (1975) Ann. N. Y. Acad. Sci. 244, 260–280.

16 Hoff, A.J. (1979) Phys. Rep. 54, 75–200.

17 Lendzian, F. (1982) Doctoral Thesis, Free University, Berlin.

18 Wasielewski, M.R., Norris, J.R., Crespi, H.L. and Harper, J. (1981) J. Am. Chem. Soc. 103 (1981) 7664–7665.

19 Wasielewski, M.R., Norris, J.R., Shipman, L.C., Lin, C.-P. and Svec, W.A. (1981) Proc. Natl. Acad. Sci. USA 78, 2957–2961.

20 den Blanken, H.J. and Hoff, A.J. (1983) Biochim. Biophys. Acta 724, 52–61.

21 van Best, J.A. and Mathis, P. (1976) Biochim. Biophys. Acta 503, 178–188.

22 Brettel, K., Schlodder, E. and Witt, H.T. (1984) Biochim. Biophys. Acta 766, 403–415.

23 Visser, J.W.M., Rijgersberg, C.P. and Gast, P. (1977) Biochim. Biophys. Acta 460, 36–46.

24 Ke, B., Sahu, S., Shaw, E. and Beinert, H. (1974) Biochim. Biophys. Acta 347, 36–48.

25 van Gorkom, H.J., Pulles, M.P.J. and Wessels, J.S.C. (1975) Biochim. Biophys. Acta 408, 331–339.

26 Ghanotakis, D.F. and Babcock, G.T. (1983) FEBS Lett. 153, 231–234.

27 Davis, M.S., Forman, A. and Fajer, J. (1979) Proc. Natl. Acad. Sci. USA 76, 4170–4174.

28 Feher, G. (1971) Photochem. Photobiol. 14, 373–387.

29 Leigh, J.S. and Dutton, P.L. (1972) Biochem. Biophys. Res. Commun. 46, 414–421.

30 Loach, P.A. and Hall, R.L. (1972) Proc. Natl. Acad. Sci. USA 69, 786–790.

31 Feher, G., Okamura, M.Y. and McElroy, J.D. (1972) Biochim. Biophys. Acta 267, 222–226.

32 Butler, W.F., Johnston, D.C., Shore, H.B., Fredkin, D.R., Okamura, M.Y. and Feher, G. (1980) Biophys. J. 32, 967–992.

33 Butler, W.F., Calvo, R., Fredkin, D.R., Isaacson, R.A., Okamura, M.Y. and Feher, G. (1984) Biophys. J. 45, 947–973.

34 Debus, R.J., Feher, G. and Okamura, M.Y. (1986) Biochemistry 25, 2276–2287.

35 Gast, P., de Groot, A. and Hoff, A.J. (1983) Biochim. Biophys. Acta 723, 52–58.

36 Lubitz, W., Abresch, A.C., Debus, R.J., Isaacson, R.A., Okamura, M.Y. and Feher, G. (1985) Biochim. Biophys. Acta 808, 464–469.

37 Feher, G., Isaacson, R.A., Okamura, M.Y. and Lubitz, W. (1985) In Antennas and Reaction Centers of Photosynthetic Bacteria (Michel-Beyerle, M.E., ed.), pp. 174–189, Springer-Verlag, Berlin.

38 Knaff, D.B. and Malkin, R. (1976) Biochim. Biophys. Acta 430, 244–252.

39 Swarthoff, T., Gast, P., Hoff, A.J. and Amesz, J. (1981) FEBS Lett. 130, 93–98.

40 Brok, M., Vasmel, H., Horikx, J.T.G. and Hoff, A.J. (1986) FEBS Lett. 194, 322–326.

41 Vasmel, H. and Amesz, J. (1983) Biochim. Biophys. Acta 724, 118–122.

42 Aasa, R., Bergström, J. and Vänngård, T. (1981) Biochim. Biophys. Acta 673, 118–123.

43 McIntosh, A.R., Chu, M. and Bolton, J.R. (1975) Biochim. Biophys. Acta 376, 308–314.

44 Heathcote, P., Williams-Smith, D.L. and Evans, M.C.W. (1978) Biochem. J. 170, 373–378.

45 Evans, E.H., Dickson, D.P.E., Johnson, C.E., Rush, J.D. and Evans, M.C.W. (1981) Eur. J. Biochem. 118, 81–84.

46 van Gorkom, H.J. (1974) Biochim. Biophys. Acta 347, 439–442.

47 Nugent, J.H.A., Diner, B.A. and Evans, M.C.W. (1981) FEBS Lett. 124, 241–244.

48 Klimov, V.V., Dolan, E., Shaw, E.R. and Ke, B. (1980) Proc. Natl. Acad. Sci. USA 77, 7227–7231.

49 Rutherford, A.W. and Zimmermann, J.-L. (1984) Biochim. Biophys. Acta 676, 168–175.

50 Okamura, M.Y., Isaacson, R.A. and Feher, G. (1979) Biochim. Biophys. Acta 547, 394–417.

51 Fajer, J., Davis, M.S., Brune, D.C., Forman, A. and Thornber, J.P. (1978) J. Am. Chem. Soc. 100, 1918–1920.
52 Tiede, D.M., Prince, R.C. and Dutton, P.L. (1976) Biochim. Biophys. Acta 449, 447–467.
53 van Grondelle, R., Romijn, J.C. and Holmes, N.G. (1976) FEBS Lett. 72, 187–192.
54 Shuvalov, V.A. and Klimov, V.V. (1976) Biochim. Biophys. Acta 440, 587–599.
55 Tiede, D.M., Prince, R.C., Reed, G.H. and Dutton, P.L. (1976) FEBS Lett. 65, 301–304.
56 Prince, R.C., Leigh, J.S. and Dutton, P.L. (1976) Biochim. Biophys. Acta 440, 622–636.
57 Hopfield, J.J. (1976) in Electrical Phenomena at the Biological Membrane Level (Roux, E., ed.) pp. 471–492, Elsevier, Amsterdam.
58 Nuijs, A.M., Vasmel, H., Joppe, H.L.P., Duysens, L.N.M. and Amesz, J. (1985) Biochim. Biophys. Acta 807, 24–34.
59 Braumann, T., Vasmel, H., Grimme, L.H. and Amesz, J. (1986) Biochim. Biophys. Acta 848, 83–91.
60 Nuijs, A.M., van Dorssen, R.J., Duysens, L.N.M. and Amesz, J. (1985) Proc. Natl. Acad. Sci. USA 82, 6865–6868.
61 Sauer, K., Mathis, P., Acker, S. and van Best, J.A. (1978) Biochim. Biophys. Acta 503, 120–134.
62 Baltimore, B.G. and Malkin, R. (1980) Photochem. Photobiol. 31, 485–490.
63 Gast, P. (1982) Doctoral Thesis, Leiden.
64 Gast, P., Swarthoff, T., Ebskamp, F.C.R. and Hoff, A.J. (1983) Biochim. Biophys. Acta 722, 163–175.
65 Bonnerjea, J. and Evans, M.C.W. (1982) FEBS Lett. 148, 313–316.
66 Mansfield, R.W. and Evans, M.C.W. (1985) FEBS Lett. 190, 237–241.
67 Thurnauer, A. and Gast, P. (1985) Photobiochem. Photobiophys. 9, 29–38.
68 Klimov, V.V., Klevanik, A.V., Shuvalov, V.A. and Krasnovskii, A.A. (1977) FEBS Lett. 82, 183–186.
69 Klimov, V.V., Dolan, E. and Ke, B. (1980) FEBS Lett. 112, 97–100.
70 Fajer, J., Davis, M.S., Forman, A., Klimov, V.V., Dolan, E. and Ke, B. (1980) J. Am. Chem. Soc. 102, 7143–7145.
71 Hoff, A.J. and Proskuryakov, I.I. (1985) Biochim. Biophys. Acta 808, 343–347.
72 Evans, M.C.W., Atkinson, Y.E. and Ford, R.C. (1985) Biochim. Biophys. Acta 806, 247–254.
73 Thurnauer, M.C., Katz, J.J. and Norris, J.R. (1975) Proc. Natl. Acad. Sci. USA 72, 3270–3274.
74 Dutton, P.L., Leigh, J.S. and Seibert, M. (1971) Biochem. Biophys. Res. Commun. 46, 406–413.
75 den Blanken, H.J. and Hoff, A.J. (1982) Biochim. Biophys. Acta 681, 365–374.
76 den Blanken, H.J. and Hoff, A.J. (1983) Biochim. Biophys. Acta 724, 52–61.
77 den Blanken, H.J., Hoff, A.J., Jongenelis, A.P.J.M. and Diner, B.A. (1983) FEBS Lett. 157, 21–27.
78 Gast, P., Wasielewski, M.R., Schiffer, M. and Norris, J.R. (1983) Nature 305, 451–452.
79 Norris, J.R., Budil, D.E., Crespi, H.L., Bowman, M.K., Gast, P., Lin, C.P., Chang, C.H. and Schiffer, M. (1985) in Antennas and Reaction Centers of Photosynthetic Bacteria (Michel-Beyerle, M., ed.) pp. 147–149, Springer Verlag, Berlin.
80 Frank, H.A., Bolt, J., Friesner, R. and Sauer, K. (1979) Biochim. Biophys. Acta 547, 502–511.
81 Trosper, T.L., Frank, H.A., Norris, J.R. and Thurnauer, M.C. (1982) Biochim. Biophys. Acta 679, 44–50.
82 Hales, B. and Das Gupta, A. (1979) Biochim. Biophys. Acta 548, 276–286.
83 Tiede, D.M. and Dutton, P.L. (1981) Biochim. Biophys. Acta 637, 278–290.
84 Frank, H.A., Machnicki, J. and Toppo, P. (1984) Photochem. Photobiol. 39, 429–432.
85 Rutherford, A.W. and Acker, S. (1986) Biophys. J. 49, 101–102.
86 Dekker, J.P., van Gorkom, H.J., Wensink, J. and Ouwehand, L. (1984) Biochim. Biophys. Acta 767, 1–9.
87 Dismukes, G.C. and Siderer, Y. (1980) FEBS Lett. 121, 78–80.
88 Dismukes, G.C. and Siderer, Y. (1981) Proc. Natl. Acad. Sci. USA 78, 274–278.
89 Brudvig, G.W., Casey, J.L. and Sauer, K. (1983) Biochim. Biophys. Acta 723, 366–371.
90 Dismukes, G.C., Ferris, K. and Watnick, P. (1982) Photobiochem. Photobiophys. 3, 243–256.
91 de Paula, J.C., Innes, J.B; and Brudvig, G.W. (1985) Biochemistry 24, 8114–8120.
92 Beck, W.F., de Paula, J.C. and Brudvig, G.W. (1985) Biochemistry 24, 3035–3043.
93 Rutherford, A.W. (1985) Biochim. Biophys. Acta 807, 189–201.

94 Hansson, Ö, Andréasson, L.-E. and Vänngård, T. (1984) in Advances in Photosynthesis Research (Sybesma, C., ed.) Vol. I, pp. 307–310, Nijhoff/Junk, The Hague.
95 de Paula, J.C. and Brudvig, G.W. (1985) J. Am. Chem. Soc. 107, 2643–2648.
96 Dismukes, G.C. and Damoder, R. (1985) Biophys. J. 47, 166.
97 de Paula, J.C., Beck, W.F. and Brudvig, G.W. (1986) J. Am. Chem. Soc. 108, 4002–4009.
98 Andréasson, L.E., Hansson, Ö. And Vänngård, T. (1983) Chem. Scripta 21, 71–74.
99 O'Malley, P.J. and Babcock, G.T. (1984) Biochim. Biophys. Acta 765, 370–379.
100 O'Malley, P.J., Babcock, G.T. and Prince, R. (1984) Biochim. Biophys. Acta 766, 283–288.
101 Brok, M., Ebskamp, F.C.R. and Hoff, A.J. (1985) Biochim. Biophys. Acta 809, 421–428.
102 Brok, M., Horikx, J.T.G. and Hoff, A.J. (1986) FEBS Lett. 203, 36–40.
103 Takahashi, T. and Katoh, S. (1986) Biochim. Biophys. Acta 848, 183–192.
104 de Vitry, C., Carles, C. and Diner, B.A. (1986) FEBS Lett. 196, 203–206.
105 Blankenship, R.E., McGuire, A. and Sauer, K., (1975) Proc. Natl. Acad. Sci. USA 72, 4943–4947.
106 McIntosh, A.R. and Bolton, J.R. (1976) Nature 263, 443–445.
107 Hoff, A.J., Gast, P. and Romijn, J.C. (1977) FEBS Lett. 73, 185–190.
108 Dismukes, G.C., McGuire, A., Blankenship, R. and Sauer, K. (1978) Biophys. J. 21, 239–256.
109 McCracken, J.L., Frank, H.A. and Sauer, K. (1982) Biochim. Biophys. Acta 679, 156–168.
110 McCracken, J.L. and Sauer, K. (1983) Biochim. Biophys. Acta 724, 83–93.
111 Hore, P.J., Watson, E.T., Pedersen, J.B. and Hoff, A.J. (1986) Biochim. Biophys. Acta 849, 70–76.
112 Gast, P. and Hoff, A.J. (1979) Biochim. Biophys. Acta 548, 520–535.
113 Hoff, A.J. and Gast, P. (1979) J. Phys. Chem. 83, 3355–3358.
114 Gast, P., Mushlin, R.A. and Hoff, A.J. (1982) J. Phys. Chem. 86, 2886–2891.
115 Hoff, A.J. and Hore, P.J. (1984) Chem. Phys. Lett. 108, 104–110.
116 Martin, J.-L., Breton, J., Hoff, A.J., Migus, A. and Antonetti, A. (1986) Proc. Natl. Acad. Sci. USA 83, 957–961.
117 Haberkorn, R., Michel-Beyerle, M.E. and Marcus, R.A. (1979) Proc. Natl. Acad. Sci. USA 76, 4185–4188.
118 Hoff, A.J. (1982) in Light Reaction Path of Photosynthesis (Fong, F.K., ed.) pp. 80–151, 322–326, Springer-Verlag, Berlin.
119 de Groot, A., Hoff, A.J., de Beer, R. and Scheer, H. (1985) Chem. Phys. Lett. 113, 286–290.
120 Hoff, A.J., de Groot, A., Dikanov, S.A., Astashkin, A.V. and Yu.D. Tsvetkov (1985) Chem. Phys. Lett. 118, 40–47.
121 de Groot, A., Evelo, R., Hoff, A.J., de Beer, R. and Scheer, H. (1985) 118, 48–54.
122 de Groot, A., Plijter, J.J., Evelo, R., Babcock, G.T. and Hoff, A.J. (1986) Biochim. Biophys. Acta 848, 8–15.
123 de Groot, A., Evelo, R. and Hoff, A.J. (1986) J. Magn. Resonance 66, 331–343.
124 Bowman, M.K., Budil, D.E., Closs, G.L., Kostka, A.G., Wraight, C.A. and Norris, J.R. (1981) Proc. Natl. Acad. Sci. USA 78, 3305–3307.
125 Möhl, K.W., Lous, E.J. and Hoff, A.J. (1985) Chem. Phys. Lett. 121, 22–27.
126 Allen, J.P. and Feher, G. (1984) Proc. Natl. Acad. Sci. USA 81, 4795–4799.
127 Gast, P. and Norris, J.R. (1985) FEBS Lett. 177, 297–280.
128 de Groot, A., Gast, P. and Hoff, A.J. (1984) in Advances in Photosynthesis Research (Sybesma, C., ed.) Vol. I, pp. 215–218, Nijhoff/Junk, The Hague.
129 Hunter, D.A., Hoff, A.J. and Hore, P.J. (1987) Chem. Phys. Lett. 134, 6–11.
130 Haberkorn, R., Michel-Beyerle, M.B. and Marcus, R.A. (1979) Proc. Natl. Acad. Sci. USA 76, 4185–4188.

J. Amesz (ed.) *Photosynthesis*

CHAPTER 6

The photosynthetic oxygen-evolving process

GERALD T. BABCOCK

Department of Chemistry, Michigan State University, East Lansing, MI 48824, U.S.A.

1. Introduction

The oxygen-evolving system in photosynthesis uses light energy to promote electrons from a plentiful substrate, water, to higher energy where they are ultimately used to reduce CO_2 to organic products. The protons liberated in water oxidation are released vectorially and contribute to the membrane free energy gradient that drives ATP synthesis. Fortunately, from a mammalian point of view, oxygen is released as a waste product. The chemistry occurring may be summarized by the following half-cell reaction:

$$\text{chemistry:}\quad 2H_2O \rightarrow \underset{\text{waste}}{O_2} + \underset{\text{ATP production}}{4H^+} + \underset{CO_2\ \text{fixation}}{4e^-}$$

At pH 5, which is a reasonable estimate of the thylakoid internal volume pH, the water/oxygen couple has an oxidation-reduction potential of +0.93 V. The energy required to drive this reaction is supplied by photon absorption in photosystem II which produces the oxidized reaction center chlorophyll, P-680$^+$

$$\text{Photochemistry:}\quad \text{P-680} \xrightarrow{h\nu} \text{P-680}^+ + e^-$$

The midpoint potential for the chlorophyll cation is estimated as +1.17 V [1].

The stoichiometries of the two reactions above raise an interesting mechanistic question: how is the *one*-electron photochemistry in PS II coupled to the *four*-electron water-splitting chemistry? The now-classical single turnover flash experiments of Joliot and co-workers [2], which showed period four oscillation in the O_2 yield with flash number, provided a clearcut answer: the PS II units function independently in accumulating the four oxidizing equivalents required to split water. This observation was quickly confirmed [3,4] and Kok provided the S-state model which is now widely used to summarize the situation:

$$S_0 \xrightarrow{h\nu} S_1 \xrightarrow{h\nu} S_2 \xrightarrow{h\nu} S_3 \xrightarrow{h\nu} S_4 \;\; (2H_2O \rightarrow O_2)$$
$$\uparrow\_\_\_\_\_\_\_\_\_\_\_\_\_\_\_\_\_\_\_\_\_\_\_\_\_\_\downarrow$$

Here S represents the water splitting center, the subscripts denote the number of stored oxidizing equivalents, and oxygen evolution occurs only after the S_4 state has been achieved. Although abstract, this model has proven to be extremely fruitful in suggesting further inquiries into the nature of the water-splitting process. The prediction of relatively stable, higher S-state intermediates implied a function for manganese, long known to be essential for water-splitting activity [5]. This is indeed the case and the S states are identified, although how closely is still somewhat uncertain, with a physical structure which includes protein-bound, redox-active manganese ions. From a biochemical viewpoint, Kok's model predicts that discrete structural units exist in the chloroplast membrane which carry out the charge-separating and water-oxidizing chemistry. This suggestion has also been borne out and the photosystem II/oxygen-evolving complex (PS II/OEC) is now known to be a multi-subunit catalytic complex plugged into the thylakoid membrane. Viewed from this perspective, the PS II/OEC appears as a biochemical structure of moderate complexity. Less than ten polypeptides are required for efficient photon absorption, quinone acceptor (Q_A, Q_B) reduction, and oxygen evolution. By comparison, mitochondrial cytochrome oxidase may contain up to 15 different subunits [6].

This review is intended to summarize the PS II/OEC unit in terms of its polypeptide and electron transfer cofactor composition, its electron transfer pathways and its mode of operation in producing oxygen. The photochemical aspects of its operation wil be dealt with only cursorily as these are treated in detail in Chapter 4 of this volume. There has been considerable review activity recently on specific aspects of PS II/OEC function, including articles on polypeptide composition [7–9], manganese function [10–12], electron transfer and O_2-evolving properties [13–18] and the chloride requirement [19].

2. *Oxygen evolution – the minimal unit*

Although oxygen-evolving PS II preparations have been available for some time [20,21], these early preparations were not widely used. The isolation by Stewart and Bendall [22] in 1979 of a purified oxygen-evolving complex from the cyanobacterium (blue-green alga) *Phormidium laminosum* marked the beginning of renewed interest in separating the PS II/OEC from the other multi-subunit protein complexes in the membrane.

The effort has proceeded in two stages. In the first, thylakoids were stripped of the PS I, Cyt b_6f and CF_0/CF_1 complexes, and membrane fragments containing only PS II/OEC and the Chl *a*/*b* light-harvesting complex (LHC) were isolated [23–25]. The essence of these preparation methods, which are generally quick (3–4 h from spinach leaves to the isolated fragments) and require only inexpensive reagents, involves stacking the thylakoids with divalent cations before and during detergent extraction. The stacking apparently leads to lateral concentration of PS II/OEC and LHC in the grana; selective solubilization of the stroma leads to an easily pelleted membrane fraction enriched to greater than 95% in PS II components.

The polypeptide and electron transfer cofactor compositions of typical preparations of these membrane fractions have been summarized (e.g. Refs. 26, 27) and Seibert and co-workers have carried out a comparative biochemical analysis of several different PS II preparations [28]. The Chl content is typically 225–250 per PS II, compared to 400 Chl/PS II in thylakoids [26,27]. Accompanying the reduction in the Chl/PS II ratio is a concomitant increase, in good preparations by a factor of two [29,30], in the oxygen rate on a per Chl basis. The pH optimum for oxygen evolution in the isolated fragments is shifted to acidic values, ≈pH 6 [24]. This reflects the fact that the membrane no longer provides a permeability barrier and that the water-splitting machinery, which is associated with the inner surface, equilibrates with the prevailing solution pH. A similar pH optimum is observed for oxygen evolution in sealed thylakoid membranes which have been everted during their isolation [31]. These observations have been incorporated into recent refinements of the procedures for preparing PS II/OEC fragments as the detergent solubilization produces more active particles when carried out at acidic pH [29,30].

Although the high rates of O_2 evolution in the membrane fragments suggest that the PS II/OEC units survive the detergent extraction, modification to the acceptor side of PS II has occurred. The exogenous acceptor requirements for maximal O_2 activity are more stringent, as a lipophilic quinone with moderately high redox potential (e.g., dichlorobenzoquinone [24]) is necessary at fairly high concentrations (≈300 μM) and van Gorkom and co-workers have provided evidence which suggests altered electron transfer in the Q_AQ_B region [32]. These observations suggest that the Q_B binding niche is accessible to detergent which may compete with endogenous PQ and exogenous acceptor for the site [33]. An additional complication is apparent in work by Petrouleas and Diner [34] and Zimmermann and Rutherford [35], which shows that certain exogenous acceptors are able to oxidize the Fe^{2+} associated with Q_A and Q_B.

The second stage in the resolution of the minimal PS II/OEC unit is in progress. Non-oxygen-evolving PS II cores, stripped of the LHC, had been isolated and characterized in a number of laboratories (e.g. Refs. 36–38). Recently several groups have described procedures by which a core preparation which retains high rates of O_2 evolution may be isolated either from higher plants [39–42] or from thermophilic cyanobacteria [43]. These procedures involve solubilizing PS II/OEC membrane fragments with non-ionic detergents and separating the core complex from LHC polypeptides by column chromatography [39,40], density gradient centrifugation [40,43], or salt fractionation and conventional centrifugation [41,42].

The core preparations have not yet been characterized in as great detail as the O_2-evolving membrane fragments. Nonetheless, several of their key properties are apparent (Table 1). The chlorophyll content is approximately the same as in non-O_2-evolving PS II cores. Reported oxygen rates (≈1000 μmoles O_2/mg Chl per h), however, are not as high as one might expect from the enhancement in their P-680 content. A likely cause for this behavior is modification to the PS II reducing side reactions as for the cruder PS II preparations. Inoue and co-workers [40] note that ferricyanide as an acceptor, in the presence of digitonin, provides maximal O_2 rates and that these rates are insensitive to DCMU. Ghanotakis and Yocum find

TABLE 1

Properties of PS II/OEC core complexes

Procedure (Ref.)	Organism	Polypeptides[a]	Maximal O_2 rates (μmoles O_2/mg Chl per h)	Chl/PS II	Mn/PS II
Tang, Satoh [39]	Spinach	47, 43, 33[b], 30[c], (23,17)[d], 22, 10[e]	150	–	–
Satoh et al. [43]	*Synechococcus*	47, 40, 35[b], 30[c], 18, 10[e]	300–400	50	3.2
Ikeuchi et al. [40]	Spinach	47, 43[f], 33[b], 34, 32, 30, (23,17)[d], 10[e]	550–850	60	4.0
Ghanotakis and Yocum [41]	Spinach	47, 43[f], 33[b], 34, 32, (23,17)[d], 20, 10[e]	900–950	60	4.0
Ghanotakis et al. [42]	(A) Spinach	47, 43, 33[b], 34, 32, (23,17)[d], 20, 10, 9[e]	1000–1100	60	4.0
	(B) Spinach[g]	47, 43, 33[b], 34, 32, (23,17)[d], 9[e]	1100	60	4.0

[a] MW from gel electrophoresis (kDa).
[b] Extrinsic polypeptide.
[c] Diffuse, may contain more than one polypeptide.
[d] Not isolated in these procedures but thought to be associated with PS II/OEC in higher plants (see text).
[e] May be lower molecular weight polypeptide(s) present [67].
[f] Runs as two bands owing to occurrence of proteolytic fragment.
[g] Obtained by subjecting preparation A to gel filtration chromatography.

reducing side properties more similar to those of the starting material [41]. A difference between these two preparations is that the former lacks a polypeptide in the 20 kDa range (Table 1), which suggests a role for this subunit in acceptor side reactions (Section 2.1.1).

2.1. Polypeptide composition and function in the PS II/OEC

Given the diversity of laboratories involved in their isolation, the PS II/OEC core preparations summarized in Table 1 show good consistency in terms of polypeptide content. Moreover, their composition is similar to that found in non-oxygen-evolving PS II cores [36–38] with the exception that they contain the ≈20 kDa polypeptide mentioned above and the water-soluble 33 kDa polypeptide implicated in maintaining the Mn content of the preparations.

2.1.1. Intrinsic polypeptides.

Although there is good agreement as to composition, the function of the various intrinsic polypeptides, particularly of the 47, 34 and 32 kDa subunits, is currently under hot debate. Both the 47 and 43 kDa peptides bind Chl. The 43 kDa may be removed without loss of photochemistry, however, and several groups have provided data which suggest that the 47 kDa contains the binding site for the reaction center Chl [44–50]. In this model of the PS II core, the intrinsic 32 kDa polypeptide, which is usually referred to as D-1 and which has been established as a locus of herbicide action in PS II [51], is postulated as providing the binding site for the secondary quinone acceptor Q_B. The intrinsic 34 kDa polypeptide, D-2, had been implicated by Metz, Bishop and co-workers in manganese binding [52,53], as they had observed decreased Mn levels in *Scenedesmus* mutants which appeared to contain an unprocessed form of the polypeptide (recent developments in this assignment are discussed below). Thus one view of the PS II/OEC core associates P-680 with the 47 kDa polypeptide, Q_B and herbicide binding with D-1, and manganese binding and the water-splitting site with D-2.

The principal alternative model for the roles of the intrinsic PS II polypeptides assigns to the 47 kDa protein only a role in light-harvesting and places the reaction center components in D-1 and D-2. Three lines of argument have been used to support this hypothesis. First, there is sequence homology between D-1 and D-2 [54] and between D-1 and D-2 and the L and M subunits of the bacterial reaction center [55]. Secondly, the crystal structure of the bacterial reaction center in *Rhodopseudomonas viridis* (Chapter 3) shows clearly that L and M form the core of the structure and are involved in binding the photochemically relevant BChls, the quinones and the bridging, acceptor side iron [56]. This led Michel, Deisenhofer and co-workers to suggest an analogous role for D-1 and D-2 in PS II. Trebst has considered this possibility in detail [57,58] and notes that if D-1 folds so that it crosses the membrane only five times (not seven as proposed [59]) in analogy with the known folding of L and M, then the location of amino acid replacements which confer herbicide resistance may be rationalized. He also points out that functionally important residues (e.g. the histidines involved in Fe binding) in L and M occur in similar positions in membrane-spanning helices in D-1 and D-2 in his folding scheme. Thirdly, there is remarkable similarity in the chromophore organization and photochemical routes in the bacterial reaction center compared to PS II [17,18,60]. Although it breaks down in certain aspects of the properties of the primary donors in the two systems, most strikingly in the apparent 90° rotation of P-680 in PS II relative to P-870 in bacteria (Ref. 61; see also 62), the analogy remains strong enough to suggest that the polypeptide organization in the two systems is also similar. Combining this idea with the Metz/Bishop data implicating D-2 (or D-1, see below) in manganese binding, one arrives at a model in which these two polypeptides, most likely in concert with the water-soluble 33 kDa subunit, not only bind the photochemical core of PS II but are also the locus of the water-splitting process. An interesting addition to this hypothesis is to identify the 20 kDa polypeptide mentioned above with the bacterial reaction center H subunit which also promotes electron transfer in the Q_AQ_B region [63].

The logic of the D-1/D-2 model is satisfying. Unfortunately, there are few data yet available to support it. Trebst has carried out trypsin digestion experiments and notes rapid disappearance of the 47 kDa polypeptide on SDS gels even though O_2 evolution remains active. This is not strong evidence for the D-1/D-2 model, however, as it is possible that the cleaved but still membrane-bound fragments retain activity [57]. In an interesting series of developments, it appears as if the 34 kDa polypeptide implicated in Mn binding and thought to be the D-2 polypeptide may actually be the D-1 subunit [64]. Affinity labeling work had shown that the Metz/Bishop peptide binds herbicide [65] and thus presumably Q_B. These data suggest, then, that D-1 is involved with cofactors which participate in electron transfer reactions on both the oxidizing and the reducing side of PS II. The data are weak, however, in showing that D-1 actually binds Mn; its lack of processing in the mutant may simply prevent Mn binding at its normal, but distant, site. Experiments are proceeding in a number of laboratories to test the folding patterns for D-1 suggested by Rao et al. [59] and by Trebst [58]. Data from Edelman's lab apparently conflict with Trebst's model [66], but more recent work provides support for the five-helix model, albeit with somewhat different surface-exposed domains (R. Sayre, B. Andersson and L. Bogorad, Cell, in the press). While the situation with respect to the functional roles of the intrinsic PS II polypeptides remains ambiguous, specific hypotheses are available and testable. (Recent data reported by Ki. Satoh at the 7th International Photosynthesis Congress (Providence, RI, August 1986) indicate that D-1 and D-2 constitute the reaction center polypeptide complex.)

In addition to the heavier intrinsic polypeptides of the PS II/OEC core, several lighter polypeptides are also isolated with the complex [67]. Two of these, with MW $\approx$ 6 and 10 kDa, are associated with cytochrome *b*-559. This heme species, although redox active, has not been shown to undergo light-induced electron transfer reactions at rates relevant to PS II/OEC function. Consequently its role in water splitting remains enigmatic. Herrmann, Cramer and co-workers have recently sequenced genes which code for both the 6 and the 10 kDa polypeptides [68,69]. Hydropathy plots show one membrane-spanning region for each and an analysis of the EPR indicated that the axial ligation for the isolated, low-potential form of the protein, and most likely for the native, membrane-bound high-potential form as well, involves two histidines. As both the 10 kDa and the 6 kDa polypeptides contain only a single histidine, these data suggest that the *b*-559 heme crosslinks separate 10 and/or 6 kDa polypeptides to form the holoprotein [70].

2.1.2. Extrinsic polypeptides

One peripheral polypeptide with MW = 33 kDa is isolated with the PS II/OEC core (Table 1). Two other extrinsic peptides with molecular masses of 17 and 23 kDa have been implicated in the O_2-evolving process, although they may be replaced in the core preparations by high concentrations of Ca^{2+} and Cl^- salts with preservation of O_2-evolution activity. The involvement of these three polypeptides in O_2 evolution was first suggested by the Tris extraction experiments on so-called inside-out chloroplast vesicles by Åkerlund et al. [71] and by cholate extraction of chloroplasts by Sayre and Cheniae [72]. A flurry of activity ensued and the situation with respect to these polypeptides is now reasonably clear. The biochemical properties of these polypeptides are discussed in detail in two recent reviews [7,8].

The 33 kDa polypeptide was originally purified and characterized by Murata and co-workers [8,73]. The N-terminal [74] and complete amino acid sequences [75] have been determined. An interesting aspect of this work, given the evidence supporting a role for this peptide in promoting Mn binding in the PS II/OEC core [8], is the similarity of part of its sequence to a region in Mn-superoxide dismutase that contains an aspartic acid ligand to the manganese [75]. In the 33 kDa polypeptide, however, this residue is replaced by cysteine, an unlikely candidate for an Mn ligand in the light of X-ray absorption fine structure (XAFS) data (Section 2.2.2) and the highly oxidizing potential maintained in intermediate oxidation states of the OEC. Although the 17 and 23 kDa peripheral proteins have been isolated and characterized in some detail [7,8], only the N-terminal sequence for the 17 kDa has appeared [76].

A major function of these extrinsic polypeptides is to promote interaction between anion and cation cofactors and the PS II/OEC core. (A second function – retardation of stored charge dissipation in the OEC – is discussed in Section 3.1.) Selective depletion of the lighter two polypeptides may be achieved by washing inside-out thylakoid vesicles [71] or PS II particles [77–80] with NaCl; depletion of the 33 kDa polypeptide occurs upon washing with Tris, urea or divalent cations [7,8,81]. If the counterion in the latter treatment is chloride, Mn is retained in the PS II/OEC core. On peptide depletion and removal of residual Ca^{2+} and Cl^-, oxygen evolution is inhibited [82]. Activity may be restored, provided Mn has not been perturbed, by readdition of Ca^{2+} and Cl^- [7,8,83]. The half-saturating concentrations of the ionic cofactors are lowered dramatically if the 17 and 23 kDa peptides are also rebound. The polypeptide rebinding process is complex but may be summarized as follows. The 33 and 23 kDa peptides rebind directly to the hydrophobic core [80]; Mn promotes 33 kDa rebinding [84] while the 33 kDa facilitates 23 kDa binding [85]. Binding of the 17 kDa polypeptide apparently occurs only when the 23 kDa is in place [80]. At least two attempts have been made to identify the intrinsic polypeptides to which the extrinsic 23 and 33 kDa peptides bind. Lundberg et al. used antibody techniques to implicate polypeptides of 10, 22 and 24 kDa [86], while Bowlby and Frasch used photoaffinity-labeled 33 kDa to identify 22, 24, 26, 28, 29 and 31 kDa polypeptides in the binding [87]. Some of the polypeptides in the latter study clearly arise from LHC components and their labeling may be fortuitous. Both sets of experiments were done on preparations more complex than those in Table 1, and it appears that repeating these experiments with the more resolved preparations will be useful. These have already proven useful in evaluating whether given polypeptides are fundamental to O_2 evolution; for example, Table 1 indicates that a 10 kDa polypeptide isolated by Tris extraction [88] is probably not essential. Many authors have used the data summarized above to draw models for the polypeptide/cofactor organization of PS II (e.g. Refs. 7, 8, 16, 19 and 89).

While the identity of the polypeptides present in the PS II/OEC core complex is now reasonably well-established, the stoichiometries of these species remain uncertain. Little is known regarding the concentration ratios per PS II/OEC unit of the intrinsic polypeptides and controversy exists regarding the stoichiometries of the extrinsic subunits [7]. In the latter case, however, the work is reaching agreement on between 1 and 2 copies of each of the water-soluble polypeptides per P-680. More quantitative work, most profitably with ^{14}C labeling, will be necessary.

2.2. *Electron transfer components*

Photon absorption in the PS II/OEC leads to charge separation in the PS II reaction center to generate the oxidized reaction center, P-680$^+$ (Ref. 18; Chapter 4, this volume.) The simplest scheme for subsequent electron transfer steps involves only the intermediate carrier, Z, and the Mn ensemble at the water-splitting site:

$$\text{P-680}^+ \longleftarrow \text{Z} \longleftarrow (\text{Mn})_4$$

The questions of branched pathways and of additional intermediates are often raised (Section 3.1), but only the components noted above have been detected directly as entities with distinct functional and spectroscopic properties (Table 2).

2.2.1. *P-680 and Z*

As opposed to P-700$^+$ in PS I and to the cation radicals of the bacterial reaction centers, P-680$^+$ is difficult to trap in its oxidized state – even at low temperatures its lifetime following photogeneration is only 3–4 ms [90] – and chemical oxidation so far has not been possible owing to the high P-680$^+$ midpoint potential [1]. Consequently the battery of techniques, particularly magnetic resonance, which has proven fruitful in unraveling the structures of the other reaction center chlorophylls has not been applied to P-680. Its spin-polarized triplet has been detected [61,91] and its unexpected parallel orientation with respect to the membrane plane postulated. The zero-field splitting parameters are almost identical to those of

TABLE 2

Selected properties of PS II/OEC components

Species	Detection	Identification	Function	Stoichiometry (/PS II)	Binding site
P-680	Optical/EPR	Exciton-coupled Chl *a* pair	Primary donor	1	47 kDa or D-1, D-2
D	EPR	Plastoquinone cation radical	?	1	47 kDa or D-1, D-2
Z	Optical/EPR	Plastoquinone cation radical	Intermediate electron carrier	1	47 kDa or D-1, D-2
Cyt *b*-559	Optical/EPR	$Fe^{2+/3+}$ low-spin protoheme	?	2	6, 10 kDa
Mn	Optical/EPR/XAS	Multinuclear cluster(s)	Water oxidation	4	Interface 33 and D-1 or D-2

A discussion of controversies and uncertainties, as well as references, for the information summarized is given in the text.

monomer Chl. Davis et al., arguing from redox-potential considerations and axial ligation effects on Chl EPR linewidths, proposed that the unpaired electron in P-680$^+$ is localized on a single Chl macrocycle [92]. The 'hole-localized' model for the *oxidized* P-680$^+$ cation radical does not necessarily contradict the conclusion reached by den Blanken et al. that two interacting Chls contribute to the singlet and triplet properties of the *reduced* P680 species [93].

The oxidized form of the Z species was identified as an organic radical with an EPR lineshape identical to that of the well-known PS II radical which gives rise to the so-called Signal II EPR spectrum [94,95]. Its stoichiometry is one per PS II in both thylakoids and PS II particles [89]. EPR [96,97] and optical [98,99] data are consistent in suggesting that Z is a hydroquinone species, most likely plastohydroquinone, which is one-electron-oxidized to form the hydroquinone cation radical during its reaction with P-680$^+$. Such a structure, i.e. $PQH_2^{+\cdot}$, is consistent with the high redox potential (>+1.0 V) required for $Z^{+\cdot}$ in its reaction with the water-splitting redox center [100]. EPR on oriented membranes was used to assign the major, partially resolved hyperfine splittings to the 2-methyl group of the plastoquinone moiety and to suggest an orientation for the radical such that its ring plane is perpendicular to the thylakoid membrane plane [101,102].

Despite the congruence of the EPR and optical data, some results conflict with these interpretations. Neither Takahashi and Katoh [103] nor de Vitry et al. [104] have been able to find sufficient amounts of noncovalently bound plastoquinone-9 in PS II preparations. In principle there should be three: one for the Q_A acceptor, one for Z and one for the stable Signal II species usually designated $D^{+\cdot}$. The former group finds 2 PQ/PS II while the latter finds only 1.15. Several explanations are possible, including: (a) either Z or D is covalently bound PQ or noncovalently bound but modified in the isoprenoid chain; (b) the Z and D concentrations have declined during purification (there is some indication that this occurs [89]); and (c) either Z or D or both are not quinones. The assigned 2-methyl origin of the partially resolved splittings in the $Z^{+\cdot}/D^{+\cdot}$ EPR spectrum has been challenged, as this causes difficulty in simulating the EPR spectrum and in understanding the rotational properties of the methyl group. Brok et al. have reinterpreted the spectral properties by retaining the $PQH_2^{+\cdot}$ identification but assigning major splittings to the 1,4-hydroxyl protons and to the isoprenoid methylene protons. In this interpretation the radical is tilted away from a perpendicular orientation with respect to the membrane plane [105,106]. The characters of the molecular orbitals implied by this model, however, are difficult to understand: it may be that ENDOR spectroscopy simply overestimates the magnitude of the $-CH_3$ coupling [107,108]. Thus, although it has been widely accepted that plastoquinone cation radicals are involved in oxidizing-side PS II electron transfer, the situation is not as clear as one would like.

2.2.2. Manganese

A substantial body of work had associated manganese with water splitting [5,10,109] and with the development of Kok's S-state model it was generally assumed that the S-state transitions correspond to valence changes in a functional manganese cluster. This assumption has turned out to be well-founded. Cheniae and Martin established a likely stoichiometry of 4 Mn per PS II [110], Theg and Sayre developed a useful Ca^{2+} wash technique by which to distinguish functional manganese [111], and Yocum and co-workers provided procedures for preparing thylakoid membranes depleted of all but the four functionally relevant metal ions [112]. The Mn stoichiometry of four is preserved in O_2-evolving PS II membrane fragments [27] and in PS II/OEC core complexes [41,42]. Photoactivation experiments (e.g. Refs. 113,114) have provided insight as to how these manganese atoms are incorporated into the water-splitting apparatus during chloroplast development.

While the Mn stoichiometry is well-determined, the organization and valence states of these ions remain uncertain. From both spectroscopic and mechanistic considerations, one expects that the metal ions function as multinuclear cluster(s) and evidence in the literature may be used to argue for either two binuclear manganese centers or for a single tetranuclear cluster. At least four different experimental approaches are relevant to this question as well as to the related valence issue. These include X-ray, UV-Vis, and magnetic resonance spectroscopies and extraction/quantitation techniques.

X-ray absorption spectroscopy (XAS) is probably the most direct. The X-ray absorption edge spectroscopy (XANES) and XAFS results obtained by Klein, Sauer and co-workers indicate that the absorption edge in resting PS II (predominantly in the S_1 state) falls close to that of Mn^{3+} model compounds. The absorption fine structure suggests that each Mn atom is ≈2.7. Å away from one additional Mn [115–117]. Owing to complications introduced by ligand covalency on the edge position and to multiple scattering effects on the fine structure, however, these results do not rule out somewhat higher Mn valences for S_1 nor do they rule out the possibility that additional manganese atoms occur at somewhat longer ($\gtrsim 3$ Å) distances. Upon formation of S_2, a sharp K-edge increase (≈1.78 eV) is observed which is interpreted to indicate changes in Mn ligation and/or valence in the $S_1 \rightarrow S_2$ transition [116].

Velthuys and Van Gorkom and co-workers detected UV absorption changes which occur during S-state advance [99,118,119]. On the basis of model compound work, the latter group suggested that each S-state transition corresponds to an Mn(III) → Mn(IV) valency change. Renger and co-workers have also been active in applying UV optical techniques to the OEC [120,121]. They differ from the Leiden group in interpretation, preferring to assign some of the changes to associated redox active ligand(s) (Section 4.3). Dismukes and Mathis provided data which support an Mn multinuclear formulation as they have detected near-IR intervalence transitions, presumably originating from the manganese cluster, in the S_2 and S_3 states [122]. Although Dismukes and co-workers are among the first to suggest a tetranuclear Mn cluster [123], it now appears as if this group favors a somewhat modified dimer of binuclear centers formulation [12,124].

The most extensively characterized spectroscopic signature of the Mn ensemble is the S_2 state multiline EPR signal initially observed by Dismukes and Siderer [125]. They suggested either Mn_2(III,IV) or Mn_4((III)$_3$,IV) formulations for this species from their spectral simulations [123]. Hansson et al. confirmed the experimental observation, but find better agreement with the data by assuming an Mn_2(II,III) binuclear cluster [126,127]. Brudvig, de Paula and Beck [128–130] have shown that the EPR characteristics of the multiline are sensitive functions of the manner in which the sample is prepared. A significant aspect of this work is that it shows that, under certain conditions, the $S = \frac{1}{2}$ multiline EPR signal arises from an excited state in a manifold of states of different spin multiplicities. Similar observations have also been made by Rutherford et al. [131–133] and by Dismukes [12]. The implication of this result is that magnetic exchange interactions must occur between at least three and more likely all four of the Mn ions in the water-splitting site, if one excludes the possibility that other paramagnets (e.g., iron) are involved in the OEC. The latter seems unlikely, as there is no evidence for metal requirements in oxygen evolution other than Mn and Ca^{2+}. These data and their interpretation provide an explanation for the second EPR signal which has been detected from the OEC, the g 4.1 signal [131,134], as the spectral properties of this species are characteristic of an $S = \frac{3}{2}$ state. This could function as either the ground state for the $S = \frac{1}{2}$ multiline or as the ground state in a parallel ladder of states which is created by temperature or solvent perturbation of the Mn ensemble. Brudvig and co-workers have carried out a magnetic resonance analysis for such a situation [129,130] and Rutherford et al. have provided experimental support to indicate that both the multiline and the g 4.1 EPR signals arise from different configurations of the S_2 state [132,133]. There is precedent in the iron-sulfur protein associated with nitrogenase for similar perturbations to spin-state orderings in a multinuclear, exchange-coupled system [135,136]. Brudvig and Crabtree [137] coupled these ideas on the origin of the multiline and g 4.1 signals with considerations of the kinds of structures which would support the pattern of ferro- and antiferromagnetic coupling required to account for the EPR data. They propose a cubane-like, Mn_4O_4 cluster as the structure of the manganese ensemble in the lower S states (Section 4.3). In agreement with the optical data above [118], they suggest that the S-state transitions involve primarily Mn(III) to Mn(IV) valence changes in the cluster. This interpretation of the multiline and g 4.1 EPR signals is appealing as it does not require invoking additional electron carriers in the PS II/OEC. More detailed tests of the model, for example, the observation of an excited state $S = \frac{1}{2}$ signal developing out of the $S = \frac{3}{2}$, g 4.1 signal at higher temperature, are most likely in progress.

The fourth area of activity in which Mn valence and organization are addressed involves quantitation of the amount of Mn released from the PS II/OEC following perturbation. This method led to the early estimates of manganese stoichiometry [138,139] and has been refined and used to study the effects of a number of PS II inhibitory treatments [112]. With the realization of the role of the peripheral polypeptides in maintaining PS II/OEC integrity, these studies have continued and a clearer picture of the factors which control Mn binding is emerging.

Several treatments remove the peripheral polypeptides (Table 3). If these are carried out at high chloride concentration, manganese extraction is minimal; if chloride is low, two of the four manganese ions are extracted. Kuwabara et al. found that the ability to reconstitute oxygen-evolution activity by polypeptide ad-

TABLE 3

Manganese solubilization under various treatments

Treatment[a]	Mn^{2+} release	Polypeptide release	Ref.
$CaCl_2$, high Cl^-	< 10%	17, 23, 33	81, 84, 140, 141
$CaCl_2$, low Cl^-	≃ 50%	17, 23, 33	84, 141
urea, high Cl^-	< 10%	17, 23, 33	142
urea, low Cl^-	40–50%	17, 23, 33	142, 143
$LaCl_3$, high Cl^-	< 5%	17, 23, 33	144
$LaCl_3$, low Cl^-	50–60%	17, 23, 33	144
$LaCl_3$, hydroquinone	> 90%	17, 23, 33	–[c]
NaCl	< 5%	17, 23	27, 71, 142, 145
NaCl, hydroquinone	80–90%	17, 23	145
NH_2OH	> 95%	(17, 23, 33)[b]	146

[a] High Cl^- means generally greater than 200 mM.

[b] Only partial removal of those polypeptides occurs during this procedure.

[c] D.F. Ghanotakis, G.T. Babcock and C.F. Yocum, unpublished.

dition declines as these two Mn are released and is completely lost when the Mn content has decreased to 50% [84]. When the perturbing treatment includes a reducing reagent, significantly greater amounts of Mn, approaching four per PS II, are extracted. After extraction of the 17 and 23 kDa polypeptides, which does not release Mn, treatment with a variety of reductants (H_2O_2, Fe^{2+}, NH_2NH_2, benzidine, hydroquinone) displaces up to 90% of the endogenous Mn [145]. The 33 kDa polypeptide is retained under these reducing conditions even though extensive Mn release occurs. Yocum et al. [112] and, in more detail, Tamura and Cheniae [146] have studied Mn release during hydroxylamine treatment. At least three, and up to four, metal atoms per PS II are solubilized. The latter workers note a close correlation between Mn release and inactivation of O_2 evolution and find that activity decreases to zero only when four Mn per PS II have been extracted. For a series of substituted hydroxylamines [147], they correlated effectiveness in Mn extraction with ability to reach and reduce the Mn ensemble, thereby implicating Mn oxidation states higher than +2 in OEC function. Reductant-induced Mn extraction is inhibited by light (e.g. Ref. 148). Ghanotakis et al. [145] conclude that photon absorption leads to higher Mn valence states as the S ensemble advances toward S_4. Increased ligand field stabilization for Mn valences above II was postulated to produce tighter binding of the metal and a more extraction-resistant Mn cluster. A similar argument has been advanced by Abramowicz and Dismukes in their work on the extraction of the 33 kDa polypeptide [149].

Taken together, the extraction work suggests that the lower S states are predominantly in the Mn(III) state and that oxidation to Mn(IV) or higher accompanies advancement of the OEC. Because the approach detects only solubilized Mn, however, it does not provide deep insight into the functional organization of the metal atoms. For example, incomplete solubilization of Mn does not necessarily imply heterogeneous Mn pools. Some attempts have been made to address the status of the remaining, bound Mn under these conditions by monitoring the magnetic interaction between Mn and Z [112,150] or the ability to form the multiline EPR [141]. Unfortunately, these techniques are indirect at best. XAFS or the Q-band EPR technique recently introduced by Bricker and co-workers [151] may provide more information.

To summarize the current situation with respect to the organization and valence of manganese in the PS II/OEC: the four metal ions are organized as (a) multinuclear cluster(s); in the low S states Mn(III) appears to be a principal valence state that increases to Mn(IV) (or higher) as the S-state ensemble is advanced.

2.3. Cofactor requirements

Two ionic species, Ca^{2+} and Cl^-, are required for oxygen evolution. The Ca^{2+} requirement is absolute, as no other cation has been found to substitute. Cl^- may be replaced with retention of O_2 evolution, most effectively by Br^- and to a lesser extent by NO_3^-, I^- and HCO_3^-. F^-, OH^-, acetate, sulfate and phosphate are ineffective. Ca^{2+} and Cl^- function at different sites, neither appears to undergo redox chemistry, and both are required for O_2 evolution (e.g. Refs. 8,82).

The Ca^{2+} effect in higher plants has been recognized only recently [152–156], although early work [157,158] had established a Ca^{2+} requirement in blue-green algae which led to suggestions of a similar involvement in higher plants [159]. The reason for the discrepancy is now clear. The higher plant systems, but not cyanobacteria [160,161], contain the 17 and 23 kDa polypeptides. These lower the external Ca^{2+} concentration requirement [162], the 23 kDa being more effective [163], and retard the depletion of the bound cation by conventional Ca^{2+} chelators [82]. With the 17 and 23 kDa polypeptides in place, the Ca^{2+} binding constant is in the 0.3 mM range; in the absence of these peptides the binding constant increases by ≈ 10 [162]. The number and homogeneity of Ca^{2+} binding sites are uncertain although it appears that less than ten Ca^{2+} are required and that as few as two may suffice (G.M. Cheniae, unpublished). The 17 and 23 kDa polypeptides are clearly not the principal sites of functionally relevant Ca^{2+} binding [156,162]. From sequence studies of a number of Ca^{2+}-binding polypeptides, the general characteristics of a Ca^{2+} binding site have emerged [164,165] and it will be interesting to see if intrinsic polypeptides in PS II have homologies to these Ca^{2+} binding sequences. The suspicion that these may be found is reinforced by the fact that lanthanides compete effectively with Ca^{2+} for these sites but inhibit function [144]. Since Ca^{2+} does not appear to undergo redox chemistry associated with water-splitting reactions, its role is likely to be organizational/structural; the functional manifestations of its depletion are considered in more detail below.

The status of research and early work on the chloride requirement in the PS II/OEC has been reviewed recently [15,19]. Cl^-, like Ca^{2+}, exerts its function in the electron transfer reactions which lead to O_2 evolution and is not apparently required for PS II primary photochemistry. Ionic volume appears to be the key in determining effectiveness, with Cl^- being optimal [166–168]. Larger ions (e.g. phosphate and sulfate) are unlikely to have access to the sites(s), while smaller, harder ions (F^-, OH^-) are competitive with Cl^- [169,170] and inhibit. This suggests that strength of binding is important and that dissociation/reassociation of Cl^- at its binding site(s) during S-state turnover is critical to proper function. The OH^- results also corroborate the pH dependence of Cl^- binding [168] and the fact that Cl^- depletion occurs more readily at higher pH [171,172]. Interestingly, and somewhat suprisingly, the neutral, free-base forms of primary amines are competitive inhibitors of Cl^- function [170,173,174].

There appear to be at least two sites of Cl^- function in the PS II/OEC which may be distinguished by their different binding constants as well as by their different physiological manifestations [124,175,176] (Section 3.2). The number of functionally relevant chloride ions per PS II is uncertain; a determination is complicated by the occurrence of 'non-specific' Cl^- binding, which is estimated at 1 per 16 chlorophylls in thylakoids [177]. As with Ca^{2+}, the peripheral 17 and 23 kDa polypeptides facilitate Cl^- binding [8,178]. Here a clear role of the 17 kDa polypeptide has been established by Akabori et al., who showed that even with the 23 kDa polypeptide present, the 17 kDa was required for maximal oxygen-evolution activity at Cl^- concentrations less than 3 mM [179]. Also similar to Ca^{2+}, the 17 and 23 kDa polypeptides do not provide the principal Cl^- binding sites, as Cl^- is required for O_2-evolution activity when these polypeptides are removed. Other than this negative conclusion, the identity of the Cl^- binding site(s) remains controversial, as are ideas as to its function. Some workers favor positively charged amino acid residues and suggest a role in proton release/charge neutralization in water splitting (Refs. 169,176; see also Ref. 15); others have argued that Cl^- serves as a ligand to the manganese ions in the water-splitting site and facilitates electron transfer [19,170].

3. *Electron transfer in the oxygen-evolving unit*

The sequence of electron transfers which shuttles holes from P-680$^+$ to the site of water oxidation has been studied extensively in both oxygen-evolving and inhibited preparations and was reviewed in detail by Van Gorkom [14]. In general, there is a reasonably good understanding of these processes in untreated preparations (Table 4). The development of inside-out thylakoids and O_2-evolving PS II particles has provided a number of new ways by which to inhibit the system. Electron transfer following these newer treatments is less well understood, but the now wider array of inhibition methods is providing a finer understanding of the role of the various PS II components in electron transfer and water splitting.

3.1. *Electron transfer in the untreated PS II/OEC*

The photosynthetically wasteful electron/hole recombination between P-680$^+$ and Q_A^- occurs with a half time of $\approx$100 μs [180]. Thus efficient photosynthesis re-

TABLE 4

Kinetic parameters associated with several events in the water-splitting process

	$S_0 \rightarrow S_1$	$S_1 \rightarrow S_2$	$S_2 \rightarrow S_3$	$S_3 \rightarrow S_4$	Refs.
P-680$^+$ → P-680	23 ns	23 ns	50, 250 ns	50, 250 ns	186
Z → Z$^+$	< 3 μs	< 3 μs	< 3 μs	< 3 μs	193, 195
Z$^+$ → Z	~50 μs	~50 μs	~400 μs	≃ 1 μs	193, 198
$S_n \rightarrow S_{n+1}$	30 μs	110 μs	350 μs	1.3 ms[a]	121, 201
H^+ release	250 μs	–	200 μs	1.2 ms	272
(amount)	(1)	(0)	(1)	(2)	
O_2 release	–	–	–	≃ 1 ms	199

[a] Oxygen release is somewhat slower in PS II membrane fragments than in chloroplasts or in algae; see Ref. 14.

quires that either P-680$^+$ is reduced or Q_A^- is oxidized in considerably shorter times. The observation that Q_A^- transfers electrons to subsequent acceptors in the 100–500 μs time range [181] indicates that the reduction of P-680$^+$ must be in the sub-μs range. Owing to this fast reduction time and to the fact that electron donation to P-680$^+$ is slowed in preparations which do not evolve O_2 (Section 3.2), establishing the functionally relevant time course of P-680$^+$ reduction proved difficult. Mauzerall [182] and Van Best and Mathis [183] showed that this occurs in the ns range and Witt and co-workers extended this approach substantially [184–186]. The reduction of P-680$^+$ is multiphasic and dependent on the number of oxidizing equivalents stored in the water-splitting ensemble. For the states S_0 and S_1, P-680$^+$ is reduced with a major phase of 20 ns and a minor phase of 35 μs; for S_2 and S_3 the prevalent phases are 50 and 260 ns, with an additional minor phase of 35 μs [187]. The amplitude of the 35 μs phase is S-state-dependent and is maximal for S_2 and S_3. In the interpretation of Witt and co-workers, the 35 μs component is a consequence of rapid equilibria on the donor side of PS II and corresponds to forward, charge separation-conserving electron transfer. Van Gorkom [14] suggests that the 35μs component arises from back reaction in PS II β centers (see Chapter 4), a view which is consistent with Eckert and Renger's conclusion that charge storage does not occur as a result of the 35 μs P-680$^+$ decay phase [188].

To explain the S-state dependence of the sub-μs decay phases, Witt and co-workers postulated two donors operating in series between the OEC and P-680 and an equilibrium between the first and P-680$^+$ that depends on the net charge on the OEC. In this model, the second donor corresponds to the EPR-detectable Z species [185]. The question of the number of endogenous donors between P-680$^+$ and the OEC was reviewed lucidly by Bouges-Bocquet in 1980 [189]. Thus far the only donor for which there exists direct spectroscopic evidence is the Z species discussed above, although under nonphysiological conditions Cyt *b*-559 [190], carotenoid [191] and chlorophyll [192] may donate to P-680$^+$ (see Chapter 4). In O_2-evolving preparations, the rise of Z$^+$ has been measured by EPR with 3 μs time resolution and found to be instrument-limited [193]. This observation may be rationalized by either one or two intermediate donor models. In inhibited preparations, however, the data clearly favor only a single intermediate. The rise of Z$^+$

parallels the decay of P-680$^+$ in inhibited PS II particles [194,195] and in Cl^--depleted preparations only two reducing equivalents, one from the $S_1 \rightarrow S_2$ transition, the other from the oxidation of Z, can be readily extracted from the donor side [196,197]. Van Gorkom has suggested an interesting rationalization for some of these observations within the two-intermediate framework [14]. However, given the data in inhibited systems and the lack of direct, physical evidence for donors other than Z, it appears at present that the one-intermediate model is minimally sufficient to explain the bulk of the existing data.

While there is uncertainty as to whether Z is the immediate donor to P-680$^+$ in O_2-evolving preparations, there is general agreement that Z^+ is the immediate oxidant of the OEC manganese. Babcock et al. used time-resolved EPR to show that the reduction of $Z^{\dot{+}}$ is S-state dependent [198]. During the $S_2 \rightarrow S_3$ and $S_3 \rightarrow S_4 \rightarrow S_0$ transitions, $Z^{\dot{+}}$ is reduced in 400 μs and 1 ms, respectively. The latter time corresponds to the overall rate-limitation in O_2 evolution [199]. This indicates that electron transfer, and not H_2O/O_2 chemistry, is rate-limiting in water splitting, consistent with Sinclair and Arnason's observation of a lack of a significant deuterium isotope effect in O_2 evolution [200]. Z^+ reduction during the $S_0 \rightarrow S_1$ and $S_1 \rightarrow S_2$ transitions was suggested to occur in times fast relative to their $\approx$100 μs time resolution. Boska and Sauer provided support for this by showing fast ($\approx$50 μs) decay transients in the Z^+ reduction kinetics under steady-state conditions [193]. These results argue against parallel donor models in which $Z^{\dot{+}}$ functions only with S_2 and S_3 and an alternate intermediate serves S_1 and S_0 [189]. Dekker et al. [201] have monitored the S-state transitions directly in their time-resolved optical work and have found complementary times for the S-state transitions (Table 4; the slight slowing down of the final step is due to the fact that these measurements were made in PS II particles [14]). These results are significant not only for their kinetic importance but also because they provide support for the clever but complicated deconvolution procedures used to obtain optical difference spectra for the various S states. The interpretation of these data continues to be controversial, however. Results from Witt's group [202] support the pattern of charge accumulation in the Leiden model but disagree with those of Dekker et al. [201] in the assignment of absorption spectra, particularly on $S_0 \rightarrow S_1$ [186]. Lavergne [203] and Renger and Weiss [121] have conflicting proposals as well (Section 4.2). The latter group has also determined the $Z^{\dot{+}}S_1$ reaction kinetics and finds good agreement with the other optical and EPR data [121].

With the exception of the S_4 state, which reacts in $\approx$ 1 ms to produce O_2, the higher S states are remarkably stable. The early literature on this topic has been reviewed [204]. Both S_2 and S_3 survive for several seconds in the dark following their formation. These times were originally measured in electrode oxygen-evolution experiments (e.g. Refs. 2–4). Recently, thermoluminescence has been used as a probe of higher S-state decay [205–208], and Inoue, Rutherford and co-workers extended early work implicating the redox state of the PQ pool in S-state deactivation by determining the $S_2Q_B^-$ and $S_3Q_B^-$ and the $S_2Q_A^-$ and $S_3Q_A^-$ recombination times [205,211]. These results correlate with earlier O_2 electrode and delayed luminescence measurements [204,209,210]. This group has also used the technique to study protonation of both donors and acceptors to PS II [211].

A remarkable aspect of the higher S-state stability is that their redox potentials are necessarily high (at least +0.6 V and more likely in the +0.9 V range), which requires that stabilization is achieved by kinetic, not thermodynamic, means. This apparently involves simple limitation of access of endogenous and most exogenous reductants to the manganese ensemble, with the 17, 23 and 33 kDa peripheral polypeptides playing a role in this shielding mechanism. Velthuys has developed a useful assay of this phenomenon by monitoring optical changes which result from TMPD reduction of donor side components; a significant decrease in the second-order rate constant is observed in oxygen-competent vs. oxygen-inhibited PS II particles [212,213]. Similar access limitation has been shown for the exogenous donor, benzidine, in salt-washed preparations and the 33 kDa polypeptide has been implicated in this process [27]. The importance of the 17 and 23 kDa polypeptides in the shielding mechanism has been demonstrated by their role in limiting access of H_2O_2 to the OEC in inside-out thylakoids [214]. One class of compounds, lipophilic anions (also termed ADRY reagents by Renger, who has characterized these most extensively (e.g. Ref. 215)) appears to be able to overcome the accessibility barrier and deactivate higher S states effectively. Models in which the lipophilic anion acts as a direct reductant [216] or as a redox-inactive catalyst [215] have been proposed.

Chloride depletion provides additional stabilization of the higher S states [172,217] and Velthuys showed that NH_3 and methyl amines similarly increase the stability of S_2 and S_3 [218,219]. Homann et al. interpreted the altered thermoluminescence properties of Cl^--depleted samples to indicate that Cl^- displacement provides a thermodynamically stabilized higher S state i.e., the redox potential of the state is lowered [217]. Ghanotakis et al. had invoked a similar thermodynamic stabilization argument for the amine case to rationalize their finding that NH_3 blocks benzidine electron donation to the higher S states in NaCl-washed PS II particles [27]. Thus, a plausible explanation for inhibition of O_2 evolution by Cl^- depletion or by amine binding is that in the treated systems the higher S states are structurally modified and no longer thermodynamically able to oxidize water. Ono et al. have reported an experiment which shows that the S_2 state is indeed modified by Cl^- depletion, but which suggests that the relationship between structure and thermodynamics may be more complicated than indicated above [220]. In the absence of the 17 and 23 kDa polypeptides, the S_2 multiline signal may only be generated if sufficient Cl^- is present [221,222]. However, if Cl^- is added in the dark *after* single flash excitation of Cl^--depleted, 17 kDa and 23 kDa polypeptide-depleted membranes, the multiline signal spontaneously regenerates. Signal regeneration by dark Cl^- addition shows the prior formation of a modified S_2; the spontaneity of multiline formation suggests that the modified and regenerated S_2 states are not thermodynamically far apart. Several factors influence this result, however, including assay temperature and Cl^- concentration, and a more detailed consideration of these seems appropriate.

In the original formulation of the S-state model, Kok et al. postulated that the dark resting state of the OEC was a 3:1 mixture of S_1 and S_0, respectively [3]. This unusual distribution has provoked considerable speculation as to its origin [204] although no definitive model has emerged. Several developments have sparked renewed interest in this phenomenon. Beck et al. have shown that the characteristics of the multiline EPR signal are dependent upon the dark incubation time prior to its generation [128]. This suggests that S_0 and S_1 states are subject to long-term reorganization, either in the manganese ensemble directly or in the protein and lipid matrix. Plijter et al. found that the dark S_0/S_1 distribution is pH sensitive, at least in PS II particles, and that at high pH almost 100% S_0 is favored in dark-adapted material [223]. The observed pK_a for this phenomenon, 7.6, correlated with a similar pK_a for a structural modification in PS II which allows access of trypsin to the OEC [224]. Alternatively, at lower pH the S_1 state appears to occur in nearly 100% of the centers [225]. The apparent 3:1 ratio, which is determined from the oxygen yield pattern, was attributed to the interception of oxidizing equivalents by the reduced form of the stable EPR Signal II species, D, during the initial flashes in the sequence. Earlier EPR data had shown that this species reacts with the state S_2 or S_3 to form $D^{\dot{+}}$ and S_1 or S_2, respectively [226]. Rutherford has correlated these observations to suggest that D^+ functions to oxidize S_0 to S_1 in the dark so as to prevent accumulation of the lowest S state [17]. Brudvig has made a similar suggestion and, as noted above, points out that the lower S states may be expected to be more labile to metal loss (personal communication).

3.2. *Electron transfer in the PS II/OEC following inhibition*

With the progress made recently in understanding the interplay between the peripheral polypeptides and the required small ions, chloride, calcium and manganese, the effects of PS II oxidizing side inhibition have become both more clear and more complicated. A good deal of effort is currently being spent defining the loci of inhibitory treatments, particularly when manganese is not released by the inhibition. The situation when manganese is released is more clear and is dealt with first.

A number of PS II inhibitor treatments , such as Tris-washing and NH_2OH extraction, had been well established as releasing Mn from the OEC [109]. Åkerlund, Andersson and co-workers showed that these treatments also released the peripheral polypeptides [71]. Mathis and co-workers showed that the ns phases in P-680$^+$ reduction that occur in untreated preparations are replaced by μs components in inhibited samples [180]. The predominant decay phase is pH dependent, $\approx$2 μs at pH 8 and $\approx$45 μs at pH 5, and is attributed to P-680$^+$ reduction by a secondary donor, usually designated D_1 in optical work, which is identical to the EPR-detectable Z species [180,194,195]. The bulk pH and salt concentration dependencies of this phase implicate local membrane pH in influencing its time course [227]. The decreased rate of electron transfer to P-680$^+$ following Tris inhibition has been explained as indicating a shift of $\approx$120 mV in the Z^+/Z redox potential [228]. Slower reduction phases, $\approx$100 μs and pH independent, may also be detected after inhibition and correspond to P-680$^+$ Q_A^- recombination. Ford and Evans have pointed out that the rate of the recombination depends on the redox state of Z [229]. Warden and Goldbeck found that another donor competes with Z in reducing P-680$^+$ in these preparations. Its redox potential (240 mV) precludes a direct role for this species in charge accumulation and the authors tentatively identify it as Cyt *b*-559 [230]. Only one treatment so far, NH_2OH at modest concentration, is known to inhibit electron transfer specifically between Z and P-680 [29,231]; interestingly, this inhibition is reversible.

With the 17, 23, and 33 kDa polypeptides and Mn removed, reduction of $Z^{\dagger}$ occurs at the expense of endogenous reductants. Exogenous donors, when added, accelerate this reduction in a straightforward second-order kinetic process that provides the principal electron entry site to PS II in inhibited preparations. The efficiency of the donor increases as its hydrophobicity increases, reflecting the hydrophobic environment of the $Z^{\dagger}$ species [232].

A variety of treatments have been developed which allow more delicate manipulation of the polypeptide/ionic cofactor composition of the PS II/OEC. With care, manganese is usually retained in its binding site(s) following these treatments, at least some of the S-state transitions occur, and partial or full reactivation of O_2 evolution may be achieved. Thus it is now possible to remove either the 17 kDa polypeptide selectively or the 17 and 23 kDa polypeptides with salt-washing, with loss of O_2-evolution activity, provided Cl^- and Ca^{2+} concentrations are sufficiently low [7,8]. The latter preparations have been characterized most extensively. The reduction time of P-680$^+$ shifts into the μs time range under 'multiple turnover' conditions [233], although evidence from optical [32,234], EPR [27,222,235], luminescence [236], thermoluminescence [237] and TMPD oxidation [213] indicates that the lower S-state transitions proceed. The Cl^- and Ca^{2+} requirements for the S-state transitions are likely to be different, however. In the absence of both Cl^- and Ca^{2+}, the system apparently proceeds only to a modified S_2 state upon illumination [220]. The multiline EPR is absent [238,239], but may be induced by Cl^- addition either before [221,222] or after [220] photon absorption without reactivation of O_2. In Cl^--sufficient, Ca^{2+}-deficient preparations the system appears unable to advance beyond Z^+S_3 to split water (e.g. Ref. 236). An understanding of electron transfer in these preparations is evolving but controversies continue. The origin of the disputes lies, no doubt, in the complexity of the interactions. Two factors may be particularly bothersome. (1) Even in the absence of the 17 and 23 kDa polypeptides, there is evidence of tight and/or residual Ca^{2+} binding [82,156,234]. Care must be taken to deplete this site prior to assay. Recent reports indicate, for example, that multiline EPR formation may require both Ca^{2+} and Cl^- [43,222], contrary to earlier conclusions that only Cl^- is necessary. (2) There are likely to be multiple binding sites, particularly for Cl^-, in the OEC and these may have different pH and Cl^- concentration dependencies (see below). The conflicting reports [217,220] on the possibility of forming a modified S_3 state in Cl^--deficient preparations may be a manifestation of this phenomenon.

Addition of high concentrations of both Ca^{2+} and Cl^- to 17, 23 kDa depleted preparations restores water splitting (Section 2.3). In the reactivated samples, however, there are changes in the behavior of the system: (a) the miss parameter, α, is about double its value in untreated controls [213,234] (but see Ref. 236), (b) the O_2-release reaction is slowed by a factor of about 4, although electron transfer remains limiting [234], and (c) Cyt *b*-559 [240] remains in its low-potential form [241,242]. Readdition of the polypeptides in addition to Ca^{2+} and Cl^- restores high-potential *b*-559 (D.F. Ghanotakis, unpublished), but does not affect either α or the slowed O_2 release [234].

Characterization of electron transfer in preparations depleted of all three peripheral polypeptides but which retain manganese has begun [213,237]. The data thus far are contradictory, but indicate that in these preparations some of the S-state transitions occur as well.

Several PS II inhibitory treatments exert their effects only when the OEC is in its higher S states. These include Cl^- depletion, F^- treatment, incubation at alkaline pH and amine inhibition. It appears now that all four of these treatments are related to the chloride requirement for O_2 evolution, as F^-, OH^- and amines are competitive inhibitors (with respect to Cl^-) of water splitting (Section 2.3). Recent data suggest two different sites for Cl^- in the OEC — one at the S_2 state, the other in facilitating the water-splitting $S_3^* \rightarrow S_0$ transition. The S_2 site is demonstrated by the fact that only two reducing equivalents may be extracted from the donor side of PS II following Cl^- depletion under certain conditions [196,197,220]. Chloride titration of the restoration of water-splitting activity [176], oxygen-release phase lag measurements [175] and EPR and O_2-evolution measurements [124] provide evidence for the second Cl^--sensitive site at the $S_3 \rightarrow S_4 \rightarrow S_0$ transition. Damoder et al. have found that the Cl^- binding constant for this site is roughly an order of magnitude less than for the site controlling the advance beyond the modified S_2 state; moreover, they find a slightly different effectiveness for the various anions which substitute for Cl^- at these two sites. Baianu et al. [177] may have provided an estimate of the binding constant for the higher-affinity site ($\approx$0.5 mM). The concept of two sites of Cl^- action does not necessarily imply distinct physical structures, however; the S_2 and the $S_3 \rightarrow S_4 \rightarrow S_0$ loci may be the same molecular site but with binding constant parameters altered by, for example, a valence change nearby.

Although the picture which is emerging for Cl^- in PS II is beginning to clear, ambiguities and disagreements remain (e.g., compare Ref. 124 with Ref. 220). Part of the confusion may arise from the competition between OH^- and Cl^- for the sites involved. Discrepancies in the literature regarding the effect of pH on the inhibition of electron transfer in PS II may also reflect this competition. Briantais et al [243] had implicated the S_2 state as the inhibition site at high pH. More recent work on the S_2 multiline EPR signal, almost all of it in PS II membrane fragments, has yielded contradictory results. Cole et al. note reversible inhibition of multiline formation at pH 7.5 [244] and a pH titration of O_2-evolution inhibition similar to that reported by Sandusky et al. [78]. Damoder and Dismukes [245] and Beck et al. [246], however, see normal multiline formation at this pH. Cole et al. suggest that the reversible inhibition reflects OH^-/Cl^- competition, which implies that the extent of alkaline inhibition will be a function of Cl^- concentration. Such Cl^- effects have been reported (e.g., Refs 247,248) and recently Vass et al. [249] have shown such a Cl^- protecting effect at high pH. They associated the OH^--sensitive step with the $S_3 \rightarrow S_4 \rightarrow S_0$ transition and identify this as the Cl^--requiring step. It appears, therefore, that the recent OH^- data may be reconciled within the context of a two-site Cl^--binding model in which OH^- is a competitive inhibitor at both sites. In this regard Preston and Pace have noted recently that Cl^- binds at both the S_2 and the S_3 transitions in their NMR work on this question [250].

The amine situation appears to be similar. Both S_2 and S_3 are amine sensitive [218,251] and the inhibitory effectiveness of a series of amines scales with their basicity [96,170]. These observations imply a Lewis acid binding site in the OEC and were interpreted to indicate that amines inhibit by binding to Mn as substrate analogues for H_2O. Power saturation measurements on $Z^{\dagger}$ support this suggestion as they show minimal perturbation to the $Z^{\dagger}$/Mn magnetic interaction upon NH_3 binding and a disruption of this interaction as the size of the amine increases [96,150]. Sandusky and Yocum established the link between amines and Cl^- by showing that the former are competitive inhibitors of the latter [170]. The only exception is NH_3, which occupies an additional, noncompetitive site. They suggested that the NH_3-specific site was sterically hindered and may be identical to the site inferred by Radmer and Ollinger from their hydrazine and hydroxylamine studies [147].

The results summarized above show that the Cl^- requirement for O_2 evolution is subject to interference by a variety of treatments. One interpretation of these data is that Cl^- acts by ligating the manganese of the OEC directly and that this is the functional basis for its effects [19,124,170]. This model accounts well for amine inhibition, especially the correlation between basicity and effectiveness, and for F^- and OH^- inhibition, if it is postulated that ligation/deligation of the anion must occur during the water-splitting process [124]. There is, however, no direct evidence that Cl^- ligates manganese. XAFS data do not show a chloride ion in the first coordination sphere of manganese in either S_1 or S_2 [115–117], no ^{35}Cl or ^{37}Cl superhyperfine structure on the multiline EPR is demonstrably present [252], and among the amine inhibitors only NH_3, presumably occupying the noncompetitive site [246], has been shown to alter the S_2 EPR signal. Alternative models for Cl^- function [15,19,176,217] invoke the anion in a charge-neutralization function to maintain proper protein association and conformation or to facilitate deprotonation of the water molecules as they are metabolized to oxygen in the OEC. Coleman and Govindjee [176] have recently summarized the status of this view of Cl^- function and note that it is able to account for a good fraction of the information currently available on Cl^- function. The conflicting models reflect the importance of obtaining a clear resolution of the site(s) and mode(s) of Cl^- action in the PS II/OEC.

4. *Water oxidation in the oxygen-evolving unit*

A variety of models for the sequence of events which occur in the OEC during water splitting has been reviewed recently [15,16]. In general, these models are based on a dimeric Mn cluster and feature water oxidation to produce bound, partially oxidized species in the intermediate S states. Recent observations relevant to the mechanism of water oxidation, not all consistent with this view, are discussed below and three representative models are considered.

4.1. Substrate and substrate analogue binding

Early attempts to assess water binding and exchange at the manganese atoms of the OEC relied on monitoring water proton relaxation in chloroplast suspensions by NMR [253,254]. Once artifactual effects caused by adventitiously bound metal ions were excluded, little effect on bulk proton T_1 values by OEC manganese was apparent [255–257], indicating slow water exchange between the OEC and the solution phase. Sharp has returned to these measurements recently and observes small ($\approx$1%) effects which oscillate with S state and which are sensitive to the integrity of the OEC [258].

Hansson et al. have used $H_2{}^{17}O$ line-broadening of the S_2 state EPR multiline signal to conclude that oxygen from water is ligated to manganese in this intermediate state [259]. From the magnitude of the coupling observed, they rule out superoxide and hydroxyl radicals as the bound species, but not peroxide, hydroxide or water itself, and conclude that binding occurs in the S_0, S_1 or S_2 states. The binding stoichiometry is ambiguous and the broadening can be reasonably accounted for by assuming 1, 2 or 3 water oxygens bound. These results are interesting in view of the observation that NH_3 apparently ligates Mn in the S_2 state (presumably at the site non-competitive with Cl^-), but not in S_0 or S_1. If NH_3 binds as a substrate analogue, as is generally thought [170,219,246,251], this implies that H_2O does not ligate until S_2 is reached. To test this conclusion and the underlying assumption regarding H_2O/NH_3 competition, it would be useful to know whether the $H_2{}^{17}O$ broadening of the multiline EPR is still detected with NH_3 bound.

The conclusions from the ammonia-binding work contrast with several studies which have been carried out with redox-active substrate analogue, notably the hydroxylamines, the hydrazines and peroxide. Bouges showed two-flash retardation in the O_2 yield when chloroplasts were incubated with low concentrations of NH_2OH [260] and similar observations were made for H_2O_2 [261]. Radmer and Ollinger made a detailed study of this reaction with a series of substituted hydroxylamines [147,262,263] and concluded that N_2 is the only stable, volatile oxidation product. Whether the hydroxylamine site is actually a molecularly well-defined binding site or simply reflects outer-sphere electron transfer is uncertain, although Radmer and Ollinger suggested the former and identified it as OEC manganese. The mechanism envisioned by these authors invokes reduction of S_1 to S_0 by the hydroxylamines or by hydrazine followed by displacement of water and binding of the nitrogen compound in S_0. On the basis of molecular geometry/efficiency considerations, a manganese binuclear cluster model for the S_0 substrate binding site was proposed. Förster and Junge examined the proton release pattern for NH_2OH oxidation by PS II and note two protons liberated on the first flash [264]. They also note cooperative NH_2OH binding and suggest a model in which two or four NH_2OH molecules interact with S_1 to produce an unperturbed

S_0 after one flash [265]. Hanssum and Renger carried out analyses of both NH_2OH and NH_2NH_2 as water substrate analogues [266]. While they conclude that the interaction with NH_2OH is complex, they note that NH_2NH_2 is more well-behaved and exhibits cooperative binding to two sites in the lower S states. A significant aspect of this work is that the authors were able to measure second-order rate constants for the interaction of the nitrogen compounds with the water-splitting ensemble; these should be useful in future work. That this will be necessary is clear from the above: one set of water analogues – hydroxylamines, hydrazine and peroxide – apparently bind in the S_0 and S_1 states; a second presumed water analogue, ammonia, does not appear to bind in S_0 or S_1 but interacts with S_2 and S_3 avidly. This paradox requires clarification. It is clear, however, that water or its oxidation products are bound in the S_2 state although binding in S_0 or S_1 is not ruled out.

4.2. The occurrence of water chemistry

In Kok's original formulation, the S ensemble accumulated charge linearly with protons absorbed. The water-splitting reaction was envisioned as a concerted event which occurred once four oxidixing equivalents had been stored. This view came into question when proton release which accompanies water oxidation was measured. Fowler showed that proton liberation occurred in the lower S states [267] and Saphon and Crofts indicated that the pattern was 1,0,1,2 for the successive S-state transitions beginning with S_0 [268]. These measurements are difficult to make, which led to some controversy as to the correct pattern, but there is now general agreement that it is 1,0,1,2 [269,270]. Junge and co-workers have been able to time-resolve the various protolytic events [271,272] (Table 4). Only for the release of the last two protons does the proton kinetics match the electron transfer kinetics; the proton on the $S_0 \rightarrow S_1$ step lags and that on the $S_2 \rightarrow S_3$ transition leads the transfer of oxidizing equivalents into the OEC. They attribute the latter behavior to proton release and rebinding accompanying Z redox chemistry, in agreement with data from Tris-washed samples [273,274]. The possibility that the proton reactions which accompany electron transfer through Z arise from membrane Bohr effects has been discussed [97] and data supporting this interpretation have been presented [275].

The findings that proton release in the intermediate S states precedes O_2 generation led to the view that water chemistry was already occurring in the lower S states. This feature has been incorporated into most of the mechanistic models for O_2 evolution, i.e., the production and Mn ligation of partially oxidized H_2O species, usually a peroxide for energetic reasons [16], in the lower S states. The structures of these intermediates are generally based on an electron stoichiometry which reflects the one photon absorbed/one electron transferred sequence and on the 1,0,1,2 pattern for proton release. Data on electrochromic shifts clearly confirm both the charge and the proton changes with S-state transition [276,277], but data demonstrating that H^+ release actually reflects direct water chemistry, rather than either membrane Bohr effects or H_2O deprotonation without H_2O redox chemistry, have been difficult to obtain. Renger considered this situation clearly in an early discussion of molecular models for O_2 evolution [278].

Results from two different experimental approaches argue, in fact, that water redox chemistry does not occur until the S_4 state has been achieved. Dekker et al. (Sections 2.2.2 and 3.1) concluded that the absorption changes which accompany the successive S-state transitions are independent of S state and that each one most likely reflects the oxidation of Mn(III) to Mn(IV) [118]. Although the model compound data upon which the latter conclusion is based are weak, it is consistent with data discussed above which argue for high-valency Mn in the OEC. Regardless of the details of the valence changes, however, these data indicate Mn-centered oxidation rather than ligand (i.e., water)-centered oxidation and stress the charge-accumulating role of the Mn ions in the OEC. As noted in Section 3.1, however, this interpretation is controversial and Renger and Weiss have recently concluded that the optical changes for $S_0 \rightarrow S_1$ and $S_2 \rightarrow S_3$ are distinct from those on $S_1 \rightarrow S_2$. These results led them to favor a model in which bound peroxide is formed as a bridging ligand in a binuclear Mn center in the S_2 state [203].

Radmer and Ollinger used mass spectrometry to show that only $^{16}O_2$ was produced from chloroplasts flashed to the S_3 state in the presence of $H_2{}^{18}O$ which was then exchanged with $H_2{}^{16}O$ before the final O_2-producing flash. From this they conclude that no non-exchangeable, partially oxidized water intermediates are formed in the lower S states and argue that this implies concerted water-to-oxygen chemistry upon the $S_4 \rightarrow S_0$ transition [279]. Although the demonstration that the oxygen atom in the $Fe^{IV}{=}O$ unit in the horseradish peroxidase intermediate, Compound II, exchanges under certain conditions [280,281] weakens the Radmer/Ollinger argument, it nonetheless remains the most straightforward interpretation of the data. Finally, Krishtalik, from a thermodynamic analysis [282], concludes that concerted electron transfer, rather than stepwise transfer, is more likely within the energetic constraints of O_2 evolution.

The present situation with respect to water binding and redox chemistry may be summarized as follows. Water exchange is slow but the substrate is bound in the S_2 state. Formation of the oxygen–oxygen bond is usually thought to occur at the peroxide level in S_2 or S_3 in models in which water chemistry occurs before O_2 release but there are now data which argue that water/oxygen chemistry is concerted. The final oxygen-liberating step, however, is limited by electron transfer, not water chemistry.

4.3. Representative models of oxygen evolution

Three models which have appeared recently are discussed in this section (Fig. 1). They are representative of a variety of models which have been proposed [15,16,121,124,283] and are intended to provide an illustration of current thinking on O_2-evolution mechanisms. Only the essential structural features of these models are summarized in Fig. 1; the details of the structural transitions involved in the S-state changes are discussed in the original publications but have been omitted for clarity.

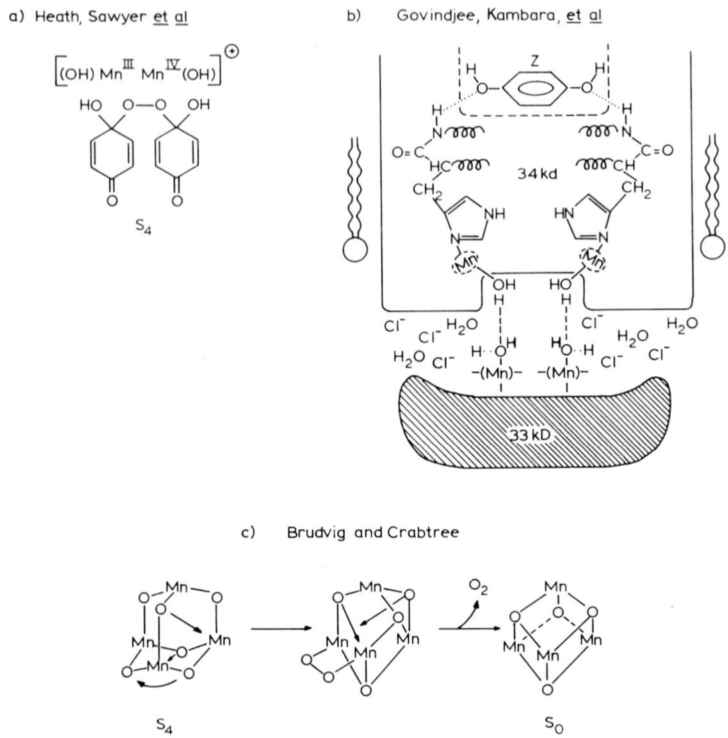

Fig. 1. Representative models for the molecular structure of the OEC active site. See text for details.

The model proposed by Sawyer, Heath and co-workers is summarized in Fig. 1a and represents the state of the OEC in S_4 just prior to O_2 release [284,285]. A specific charge-accumulating role for only 2 of the 4 Mn associated with the OEC is postulated. The unique feature of this model, involvement of two quinone species in the actual water-splitting chemistry, is based on the observation that hydroxide can be oxidized in the presence of quinone to give a peroxide-bridged biquinone structure. In this model the binuclear Mn center plays only a charge-accumulating role. It does not, as envisioned in most other models, serve as the template for oxygen–oxygen bond formation and is used only to oxidize the quinone-bound peroxide in the final O_2-releasing step.

Govindjee, Kambara and co-workers have proposed a model (Fig. 1b) which invokes two functionally distinct binuclear manganese clusters and a pair of redox-active ligands which shuttles electrons to Z [286,287]. The manganese pool associated with the extrinsic 33 kDa polypeptide functions to move protons out of the water-splitting Mn cluster associated with the intrinsic 34 kDa peptide. The concept of redox-active ligands is apparently an outgrowth of inorganic chemical work carried out by Hendrickson [288] and Meyer [289] and has been incorporated into a model of O_2 evolution by Sauer and co-workers as well [116]. In the Kambara/Govindjee formulation the redox-active ligands are capable of undergoing one-electron oxidation and, in concert with the intrinsic Mn cluster, are responsible for charge accumulation. In the model in Fig. 1b, histidine imidazole serves this function although other oxidizable amino acids could be proposed. For example, there is a stable tyrosine free radical in ribonucleotide reductase [290]. Alternatively, Pistorius has suggested the occurrence of flavin in oxidizing side reactions and this could also plausibly fulfill such a role [291]. Govindjee and co-workers have discussed their model extensively in terms of data available on water splitting and show consistency with this information.

The third model, Fig. 1c, has been developed by Brudvig and Crabtree and is based on a tetranuclear Mn cluster [137]. In the lower S states (S_0, S_1 and S_2) a Mn_4O_4 cubane structure is envisioned to rationalize the EPR properties of the various S_2 configurations which have been observed. The transition to S_3 is accompanied by the addition of two more oxygen ligands (either O^{2-} or OH^-) to form the Mn_4O_6 adamantane structure. Upon formation of S_4, in which the four most likely Mn valences are considered to be IV, IV, IV and V, the adamantane structure collapses to the cubane arrangement with release of O_2. Of the models proposed this is probably the most novel, but it has precedence, as pointed out by Brudvig and Crabtree, in the Fe_4S_4 cluster in iron-sulfur proteins and in the inorganic literature.

5. *Conclusions*

The water-splitting apparatus in photosynthesis, until recently thought to be one of the most mysterious and delicate assemblies in nature, is turning out to be biochemically tractable. The picture emerging is that the PS II/OEC core is a multicomponent membrane protein that incorporates both photochemical and water-splitting activities into one unit. The number of spectroscopic techniques which provide information on the state of the system at intermediate oxidation levels is increasing. With further improvements in the biochemistry, such as the core preparations currently being developed, new techniques, for example, magnetic susceptibility, should become useful. The electron transfer cofactors have been identified to a large extent and their reactions characterized in some detail. Their protein binding sites and geometric arrangement are currently the subject of intense scrutiny. The involvement of the ionic cofactors, Ca^{2+} and Cl^-, has been discovered, although their modes of action are only now coming into view. In this regard, researchers in oxygen evolution have realized for some time that the states S_2 and S_3 have unusual properties. Part of this, no doubt, is due to the positive charge they carry as the result of the proton release pattern, but a part may also arise from the likelihood that these two states are uniquely subject to ligand exchange reactions. The mechanism of water oxidation and the organization of the active-site manganese atoms have lost a good deal of their nebulousness, and models based on solid metallobiochemical and inorganic chemical principles are being postulated. The uniqueness of the photosynthetic reaction whereby water is split is likely to be paralleled by uniqueness in the catalytic structures.

Acknowledgements

Research in photosynthesis continues to be a running conversation among the participants. This review owes much to this conversation of which I'm glad to be a part. The work done in my lab was due largely to the skills of Drs. C.T. Yerkes, W.J. Buttner, P.J. O'Malley, D.F. Ghanotakis and T.K. Chandrashekar, encouraged by a continuing collaboration with Professor C.F. Yocum and supported by the McKnight Foundation, by the Photosynthesis Program of the Competitive Research Grants Office of the U.S. Department of Agriculture and by NIH (GM 37300).

References

1 Klimov, V.V., Allakhverdiev, S.I., Demeter, S. and Krasnovskii, A.A. (1980) Dokl. Akad. Nauk. SSSR 249, 227–230.
2 Joliot, P., Barbieri, G. and Chabaud, R. (1969) Photochem. Photobiol. 10, 309–329.
3 Kok, B., Forbush, B. and McGloin, M. (1970) Photochem. Photobiol. 11, 457–475.
4 Weiss, C.,Jr. and Sauer, K. (1970) Photochem. Photobiol. 11, 495–503.
5 Cheniae, G. (1980) Methods Enzymol. 69, 349–363.
6 Capaldi, R.A., Malatesta, F. and Darley-Usmar, V.M. (1983) Biochim. Biophys. Acta 726, 135–148.
7 Ghanotakis, D.F. and Yocum, C.F. (1985) Photosynth. Res. 7, 97–114.
8 Murata, N. and Miyao, M. (1985) Trends Biochem. Sci. 10, 122–124.
9 Andersson, B. (1986) in Encyclopedia for Plant Physiology: New Series, Vol. 19. Photosynthetic Membrane (Staehelin, A. and Arntzen, C.J., eds.) pp. 447–456, Springer-Verlag, Berlin.
10 Amesz, J. (1983) Biochim. Biophys. Acta 726, 1–12.
11 Dismukes, G.C. (1986) Photochem. Photobiol. 43, 99–115.
12 Dismukes, G.C. (1986) in Manganese in Metabolism and Enzyme Function, in press.
13 Arntzen, C.J. and Prakasi, H.B. (1986) in Encyclopedia for Plant Physiology: New Series, Vol. 19. Photosynthetic Membrane (Staehelin, A. and Arntzen, C.J., eds.) pp. 457–467, Springer-Verlag, Berlin.
14 Van Gorkom, H.J. (1985) Photosynth. Res. 6, 97–112.
15 Govindjee, Kambara, T. and Coleman, W. (1985) Photochem. Photobiol. 42, 187–210.
16 Renger, G. and Govindjee (1985) Photosynth. Res. 6, 33–55.
17 Zimmermann, J.-L. and Rutherford, A.W. (1985) Physiol. Vég. 23, 425–434.
18 Diner, B.A. (1986) in Encyclopedia for Plant Physiology: New Series, Vol. 19. Photosynthetic Membrane (Staehelin, A. and Arntzen, C.J., eds.) pp. 423–436, Springer-Verlag, Berlin.
19 Critchley, C. (1985) Biochim. Biophys. Acta 811, 33–46.
20 Huzisige, H., Usiyama, H., Kikuti, T. and Azi, T. (1969) Plant Cell Physiol. 10, 441–455.
21 Briantais, J.M. (1969) Physiol. Vég. 7, 135–180.
22 Stewart, A.C. and Bendall, D.S. (1979) FEBS Lett. 107, 308–312.
23 Yamamoto, Y., Doi, M., Tamura, N. and Nishimura, M. (1981) FEBS Lett. 133, 265–268.
24 Berthold, D.A., Babcock, G.T. and Yocum, C.F. (1981) FEBS Lett. 134, 231–234.
25 Kuwabara, T. and Murata, N. (1982) Plant Cell Physiol. 23, 533–539.
26 Murata, N., Miyao, M., Omata, T., Matsunami, H. and Kuwabara, T. (1984) Biochim. Biophys. Acta 765, 363–369.
27 Ghanotakis, D.F., Babcock, G.T. and Yocum, C.F. (1984) Biochim. Biophys. Acta 765, 388–398.
28 Dunahay, T.G., Staehlin, L.A., Seibert, M., Ogilvie, P.D. and Berg, S.P. (1983) Biochim. Biophys. Acta 764, 179–193.
29 Ghanotakis, D.F. and Babcock, G.T. (1983) FEBS Lett. 153, 231–234.
30 Ford, R.C. and Evans, M.C.W. (1983) FEBS Lett. 160, 159–163.

31 Critchley, C., Andersson, B., Ryrie, I.J. and Anderson, J.M. (1984) Biochim. Biophys. Acta 767, 532–539.
32 Wensink, J., Dekker, J.P. and Van Gorkom, H.J. (1984) Biochim. Biophys. Acta 765, 147–155.
33 Agalidis, I. and Velthuys, B.R. (1986) FEBS Lett. 197, 263–266.
34 Petrouleas, V. and Diner, B.A. (1986) Biochim. Biophys. Acta 849, 264–275.
35 Zimmermann, J.-L. and Rutherford, A.W. (1986) Biochim. Biophys. Acta, submitted.
36 Vernon, L.P. and Shaw, E.R. (1971) Methods Enzymol. 23, 277–289.
37 Diner, B.A. and Wollman, F.-A. (1980) Eur. J. Biochem. 110, 521–526.
38 Satoh, K., Nakatani, H.Y., Steinback, K.E., Watson, J. and Arntzen, C.J. (1983) Biochim. Biophys. Acta 723, 142–150.
39 Tang, X.-S. and Satoh, K. (1985) FEBS Lett. 179, 60–64.
40 Ikeuchi, M., Yuasa, M. and Inoue, Y. (1985) FEBS Lett. 185, 316–322.
41 Ghanotakis, D.F. and Yocum, C.F. (1986) FEBS Lett. 197, 244–248.
42 Ghanotakis, D.F., Demetriou, D.M. and Yocum, C.F. (1987) Biochim. Biophys. Acta, in press.
43a Schatz, G.H. and Witt, H.T. (1984) Photobiochem. Photobiophys. 7, 77–89.
43b Satoh, K., Ohno, T. and Katoh, S. (1985) FEBS Lett. 180, 326–330.
44 Nakatani, H.Y., Ke, B., Dolan, E. and Arntzen, C.J. (1984) Biochim. Biophys. Acta 765, 347–352.
45 Camm, E.L. and Green, B.R. (1984) J. Cell Biochem. 23, 171–179.
46 Inoue, Y. (1983) in The Oxygen-Evolving System of Photosynthesis (Inoue, Y., et al., eds.) pp. 439–450, Academic Press, Tokyo.
47 Omata, T., Murata, N. and Satoh, K. (1984) Biochim. Biophys. Acta 765, 403–405.
48 de Vitry, C., Wollman, F.-A. and Delepelaire, P. (1984) Biochim. Biophys. Acta 767, 415–422.
49 Bricker, T.M., Pakrasi, H.B. and Sherman, L.A. (1985) Arch. Biochem. Biophys. 237, 170–176.
50 Yamagishi, A. and Katoh, S. (1985) Biochim. Biophys. Acta 807, 74–80.
51 Pfister, K., Steinback, K.E., Gardner, G. and Arntzen, C.J. (1981) Proc. Natl. Acad. Sci. USA 78, 981–985.
52 Metz, J.G., Wong, J. and Bishop, N.I. (1980) FEBS Lett. 114, 61–64.
53 Metz, J.G. and Seibert, M. (1984) Plant Physiol. 76, 829–832.
54 Rochaix, J.D., Droh, M., Rahire, M. and Malnoe, P. (1984) Plant Mol. Biol. 3, 363–370.
55 Youvan, D.C., Bylina, E.J., Alberti, M., Begusch, H. and Hearst, J.E. (1984) Cell 37, 949–957.
56 Deisenhofer, J., Epp, O., Miki, K., Huber, R. and Michel, H. (1985) Nature 318, 618–624.
57 Trebst, A. and Depka, B. (1985) in Antennas and Reaction Centers of Photosynthetic Bacteria – Structure, Interactions and Dynamics (Michel-Beyerle, M.E., ed.), pp. 216–224, Springer-Verlag, Berlin.
58 Trebst, A. (1986) Z. Naturforsch. 41c, 240–245.
59 Rao, J.K.M., Hargrave, P.A. and Argos, P. (1983) FEBS Lett. 156, 165–169.
60 Rutherford, A.W. (1986) Biochem. Soc. Trans., in press.
61 Rutherford, A.W. and Acker, S. (1986) Biophys. J., in press.
62 Tapie, P., Choquet, Y., Wollman, F.A., Diner, B. and Breton, J. (1986) Biochim. Biophys. Acta 850, 156–161.
63 Debus, R.J., Feher, G. and Okamura, M.Y. (1985) Biochemistry 24, 2488–2500.
64 Metz, J.G., Pakrasi, H.B., Seibert, M. and Arntzen, C.J. (1986) FEBS Lett., in press.
65 Metz, J.G., Bricker, T.M. and Seibert, M. (1985) FEBS Lett. 185, 191–196.
66 Marder, J.B., Goloubinoff, P. and Edelman, M. (1984) J. Biol. Chem. 25, 3900–3908.
67 Ljundberg, U., Henrysson, T., Rochester, C.P., Åkerlund, H.-E. and Andersson, B. (1986) Biochim. Biophys. Acta 849, 112–120.
68 Widger, W.R., Cramer, W.A., Hermodson, M., Meyer, D. and Gullifor, M. (1984) J. Biol. Chem. 259, 3870–3876.
69 Westhoff, P., Alt, J., Widger, W.R., Cramer, W.A. and Herrmann, R.G. (1985) Plant Mol. Biol. 4, 103–110.
70 Babcock, G.T., Widger, W.R., Cramer, W.A., Oertling, W.A. and Metz, J.G. (1985) Biochemistry 24, 3638–3645.
71 Åkerlund, H.-E., Jansson, C. and Andersson, B. (1982) Biochim. Biophys. Acta 681, 1–10.
72 Sayre, R.T. and Cheniae, G.M. (1982) Plant Physiol. 69, 1084–1095.

73 Kuwabara, T. and Murata, N. (1982) Biochim. Biophys. Acta 680, 210–215.

74 Yamamoto, Y., Hermodson, M.A. and Krogmann, D.W. (1986) FEBS Lett. 195, 155–159.

75 Oh-oka, H., Tanaka, S., Wada, K., Kuwabara, T. and Murata, N. (1986) FEBS Lett. 197, 63–66.

76 Kuwabara, T., Murata, T., Miyao, M. and Murata, N. (1986) Biochim. Biophys. Acta 850, 146–155.

77 Moller, B.L. and Hoj, P. (1983) Carlsberg. Res. Commun. 48, 161–185.

78 Sandusky, P.D., DeRoo, C.L.S., Hicks, D.B., Yocum, C.F., Ghanotakis, D.F. and Babcock, G.T. (1983) in The Oxygen-Evolving System of Photosynthesis (Inoue, Y., et al., eds.) pp. 189–199 Academic Press, Tokyo.

79 Murata, N., Miyao, M. and Kuwabara, T. (1983) ibid. pp. 213–222.

80 Miyao, M. and Murata, N. (1983) Biochim. Biophys. Acta 725, 87–93.

81 Ono, T.A. and Inoue, Y. (1984) FEBS Lett. 168, 281–286.

82 Ghanotakis, D.F., Babcock, G.T. and Yocum, C.F. (1985) FEBS Lett. 192, 1–3.

83 Ono, T. and Inoue, Y. (1986) Biochim. Biophys. Acta 850, 380–389.

84 Kuwabara, T., Miyao, M., Murata, T. and Murata, N. (1985) Biochim. Biophys. Acta 806, 283–289.

85 Andersson, B., Larsson, C., Jansson, C., Ljundberg, U. and Åkerlund, H.-E. (1984) Biochim. Biophys. Acta 766, 21–28.

86 Ljundberg, U., Åkerlund, H.-E., Larsson, C. and Andersson, B. (1984) Biochim. Biophys. Acta 767, 145–152.

87 Bowlby, N.R. and Frasch, W.D. (1986) Biochemistry 25, 1402–1407.

88 Ljundberg, U., Åkerlund, H.-E., Larsson, C. and Andersson, B. (1984) Biochim. Biophys. Acta 767, 145–152.

89 Babcock, G.T., Ghanotakis, D.F., Ke, B. and Diner, B.A. (1983) Biochim. Biophys. Acta 723, 276–286.

90 Mathis, P. (1977) in Primary Processes of Photosynthesis (Barber, J., ed.) pp. 269–302, Elsevier, Amsterdam.

91 Rutherford, A.W., Paterson, D.R. and Mullet, J.E. (1981) Biochim. Biophys. Acta 635, 205–214.

92 Davis, M.S., Forman, A. and Fajer, J. (1979) Proc. Natl. Acad. Sci. USA 76, 4170–4174.

93 den Blanken, H.J., Hoff, A.J., Jongenelis, A.P.J.M. and Diner, B.A. (1983) FEBS Lett. 157, 21–27.

94 Babcock, G.T. and Sauer, K. (1975) Biochim. Biophys. Acta 376, 329–344.

95 Blankenship, R.E., Babcock, G.T., Warden, J.T. and Sauer, K. (1975) FEBS Lett. 51, 287–293.

96 Ghanotakis, D.F., O'Malley, P.J., Babcock, G.T. and Yocum, C.F. (1983) in The Oxygen-Evolving System of Photosynthesis (Inoue, Y., et al., eds.) pp. 91–101, Academic Press, Tokyo.

97 O'Malley, P.J. and Babcock, G.T. (1984) Biochim. Biophys. Acta 765, 370–379.

98 Diner, B.A. and De Vitry, C. (1984) in Advances in Photosynthesis Research (Sybesma, C., ed.) Vol. I, pp. 407–412, Martinus Nijhoff/Dr. W. Junk, The Hague.

99 Dekker, J.P., Van Gorkom, H.J., Brok, M. and Ouwehand, L. (1984) Biochim. Biophys. Acta 764, 301–309.

100 Wood, P.M. and Bendall, D.S. (1976) Eur. J. Biochem. 61, 337–344.

101 O'Malley, P.J., Babcock, G.T. and Prince, R.C. (1984) Biochim. Biophys. Acta 766, 283–288.

102 Rutherford, A.W. (1985) Biochim. Biophys. Acta 807, 189–201.

103 Takahashi, Y. and Katoh, S. (1986) Biochim. Biophys. Acta 848, 183–192.

104 de Vitry, C., Carles, C. and Diner, B.A. (1986) FEBS Lett. 196, 203–206.

105 Brok, M., Ebskamp, F.C.R. and Hoff, A.J. (1985) Biochim. Biophys. Acta 809, 418–428.

106 Brok, M., Horikx, J.T.G. and Hoff, A.J. (1986) FEBS Lett. 194, 322–326.

107 Hurst, G.C., Henderson, T.A. and Kreilick, R.W. (1985) J. Am. Chem. Soc. 107, 7294–7299.

108 Henderson, T.A., Hurst, G.C. and Kreilick, R.W. (1985) J. Am. Chem. Soc. 107, 7299–7306.

109 Radmer, R. and Cheniae, G.M. (1977) in Primary Processes of Photosynthesis (Barber, J., ed.) pp. 303–348, Elsevier, Amsterdam.

110 Cheniae, G.M. and Martin, I.F. (1970) Biochim. Biophys. Acta 197, 219–239.

111 Theg, S.M. and Sayre, R.T. (1979) Plant Sci. Lett. 16, 321–326.

112 Yocum, C.F., Yerkes, C.T., Blankenship, R.E., Sharp, R.R. and Babcock, G.T. (1981) Proc. Natl. Acad. Sci. USA 78, 7507–7511.

113 Ono, T. and Inoue, Y. (1983) Biochim. Biophys. Acta 723, 191–201.

114 Callahan, F.E. and Cheniae, G.M. (1985) Plant Physiol. 79, 777–786.
115 Kirby, J.A., Goodin, D.B., Wydrzynski, T., Roberston, A.S. and Klein, M.P. (1981) J. Am. Chem. Soc. 103, 5537–5542.
116 Goodin, D.B., Yachandra, V.K., Britt, R.D., Sauer, K. and Klein, K. (1984) Biochim. Biophys. Acta 767, 209–216.
117 Yachandra, V.K., Guiles, R.D., McDermott, A., Britt, R.D., Dexheimer, S.L., Sauer, K. and Klein, M.P. (1986) Biochim. Biophys. Acta 850, 324–332.
118 Dekker, J.P., Van Gorkom, H.J., Wensink, J. and Ouwehand, L. (1984) Biochim. Biophys. Acta 767, 1–9.
119 Velthuys, B.R. (1980) Annu. Rev. Plant Physiol. 31, 545–567.
120 Renger, G. and Weiss, W. (1983) Biochim. Biophys. Acta 722, 1–11.
121 Renger, G. and Weiss, W. (1986) Biochim. Biophys. Acta 850, 184–196.
122 Dismukes, G.C. and Mathis, P. (1984) FEBS Lett. 178, 51–54.
123 Dismukes, G.C., Ferris, K. and Watnick, P. (1982) Photobiochem. Photobiophys. 3, 243–256.
124 Damoder, R., Klimov, V.V. and Dismukes, G.C. (1986) Biochim. Biophys. Acta 848, 378–391.
125 Dismukes, G.C. and Siderer, Y. (1981) Proc. Natl. Acad. Sci. USA 78, 274–278.
126 Hansson, O. and Andréasson, L.-E. (1982) Biochim. Biophys. Acta 679, 261–268.
127 Andréasson, L.-E., Hansson, Ö. and Vänngård, T. (1983) Chem. Scripta 21, 71–74.
128 Beck, W.F., de Paula, J.C. and Brudvig, G.W. (1985) Biochemistry 24, 3035–3043.
129 de Paula, J.C. and Brudvig, G.W. (1985) J. Am. Chem. Soc. 107, 2643–2648.
130 de Paula, J.C., Beck, W.F. and Brudvig, G.W. (1986) J. Am. Chem. Soc. 108, 4002–4009.
131 Zimmermann, J.-L. and Rutherford, A.W. (1984) Biochim. Biophys. Acta 767, 160–167.
132 Zimmermann, J.-L. and Rutherford, A.W. (1986) Biochemistry, in press.
133 Völker, M., Renger, G. and Rutherford, A.W. (1986) Biochim. Biophys. Acta 851, 424–430.
134 Casey, J.L. and Sauer, K. (1984) Biochim. Biophys. Acta 767, 21–28.
135 Lindahl, P.A., Day, E.A., Kent, T.A., Orme-Johnson, W.H. and Münck, E. (1985) J. Biol. Chem. 260, 11160–11173.
136 Hagen, W.R., Eady, R.R., Dunham, W.R. and Haaker, H. (1985) FEBS Lett. 189, 250–254.
137 Brudvig, G.W. and Crabtree, R.H. (1986) Proc. Natl. Acad. Sci. USA 83, 4586–4588.
138 Lozier, R., Baginsky, M. and Butler, W.L. (1971) Photochem. Photobiol. 14, 323–328.
139 Blankenship, R.E. and Sauer, K. (1974) Biochim. Biophys. Acta 357, 252–260.
140 Ono, T. and Inoue, Y. (1983) FEBS Lett. 164, 255–260.
141 Imaoka, A., Akabori, K., Yanagi, M., Izumi, K., Toyoshima, Y., Kawamori, A., Nakayama, H. and Sato, J. (1986) Biochim. Biophys. Acta 848, 201–211.
142 Miyao, M. and Murata, N. (1984) FEBS Lett. 170, 350–354.
143 Miyao, M. and Murata, N. (1984) Biochim. Biophys. Acta 765, 253–257.
144 Ghanotakis, D.F., Babcock, G.T. and Yocum, C.F. (1985) Biochim. Biophys. Acta 809, 173–180.
145 Ghanotakis, D.F., Topper, J.N. and Yocum, C.F. (1984) Biochim. Biophys. Acta 767, 524–531.
146 Tamura, N. and Cheniae, G. (1985) Biochim. Biophys. Acta 809, 245–259.
147 Radmer, R. and Ollinger, O. (1982) FEBS Lett. 144, 162–166.
148 Sharp, R.R. and Yocum, C.F. (1981) Biochim. Biophys. Acta 635, 90–104.
149 Abramowicz, D.A. and Dismukes, G.C. (1984) Biochim. Biophys. Acta 765, 318–328.
150 Yocum, C.F. and Babcock, G.T. (1981) FEBS Lett. 130, 99–102.
151 Mavankal, G., McCain, D.L. and Bricker, T.M. (1986) FEBS Lett. 202, 235–239.
152 Ghanotakis, D.F., Babcock, G.T. and Yocum, C.F. (1984) FEBS Lett. 167, 127–130.
153 Yerkes, C.T. and Babcock, G.T. (1981) Biochim. Biophys. Acta 634, 19–29.
154 Nakatani, H.Y. (1984) Biochem. Biophys. Res. Commun. 120, 299–304.
155 Miyao, M. and Murata, N. (1984) FEBS Lett. 168, 118–120.
156 Boussac, A., Maison-Peteri, B., Etienne, A.-L. and Vernotte, C. (1985) Biochim. Biophys. Acta 808, 231–234.
157 Piccioni, R. and Mauzerall, E. (1976) Biochim. Biophys. Acta 423, 605–609.
158 Brand, J.J., Mohanty, P. and Ford, D.C. (1983) FEBS Lett. 155, 120–124.
159 Barr, R., Troxel, K.S. and Crane, F.L. (1983) Plant Physiol. 72, 309–315.
160 Koike, H. and Inoue, Y. (1985) Biochim. Biophys. Acta 807, 64–73.

161 Stewart, A.C., Ljundberg, U., Åkerlund, H.E. and Andersson, B. (1985) Biochim. Biophys. Acta 808, 353–362.
162 Ghanotakis, D.F., Topper, J.N., Babcock, G.T. and Yocum, C.F. (1984) FEBS Lett. 170, 169–173.
163 Boussac, A. and Etienne, R.L. (1985) C. R. Acad. Sci., in press.
164 Kretsinger, R.H. and Barry, C.D. (1975) Biochim. Biophys. Acta 405, 40–50.
165 Babu, Y.S., Sack, J.S., Greenhough, T.J., Bugg, C.E., Means, A.R. and Cook, W.J. (1985) Nature 315, 37–40.
166 Hind, G., Nakatani, H.Y. and Izawa, S. (1969) Biochim. Biophys. Acta 172, 277–289.
167 Kelley, P.M. and Izawa, S. (1978) Biochim. Biophys. Acta 502, 198–210.
168 Critchley, C., Baianu, I.C., Govindjee and Gutowsky, H.S. (1982) Biochim. Biophys. Acta 682, 436–445.
169 Homann, P. (1985) Biochim. Biophys. Acta 809, 311–319.
170 Sandusky, P.O. and Yocum, C.F. (1986) Biochim. Biophys. Acta 849, 85–93.
171 Theg, S.M. and Homann, P.H. (1982) Biochim. Biophys. Acta 679, 221–234.
172 Izawa, S., Muallem, A. and Ramaswamy, N.K. (1983) in The Oxygen-Evolving System of Photosynthesis (Inoue, Y., et al., eds.) pp. 293–302, Academic Press, Tokyo.
173 Sandusky, P.O. and Yocum, C.F. (1983) FEBS Lett. 162, 339–343.
174 Sandusky, P.O. and Yocum, C.F. (1984) Biochim. Biophys. Acta 766, 603–611.
175 Sinclair, J. (1984) Biochim. Biophys. Acta 764, 247–252.
176 Coleman, W. and Govindjee (1985) Proc. 16th FEBS Conference (Ovchinnikov, Yu.A., ed.) Part B, 21–28, VNU Science Press, Utrecht.
177 Baianu, I.C., Critchley, C., Govindjee and Gutowsky, H.S. (1984) Proc. Natl. Acad. Sci. USA 81, 3713–3717.
178 Andersson, B., Critchley, C., Ryrie, I.J., Jansson, C., Larsson, C. and Anderson, J.M. (1984) FEBS Lett. 168, 113–117.
179 Akabori, K., Imaoka, A. and Toyoshima, Y. (1984) FEBS Lett. 173, 36–40.
180 Conjeaud, H., Mathis, P. and Paillotin, G. (1979) Biochim. Biophys. Acta 546, 280–291.
181 Crofts, A.R. and Wraight, C.A. (1983) Biochim. Biophys. Acta 726, 149–185.
182 Mauzerall, D. (1972) Proc. Natl. Acad. Sci. USA 69, 1358–1362.
183 Van Best, J.A. and Mathis, P. (1978) Biochim. Biophys. Acta 505, 178–188.
184 Schlodder, E., Brettel, K., Schatz, G.H. and Witt, H.T. (1984) Biochim. Biophys. Acta 765, 178–185.
185 Brettel, K., Schlodder, E. and Witt, H.T. (1984) Biochim. Biophys. Acta 766, 403–415.
186 Witt, H.T., Schlodder, E., Brettel, K. and Saygin, Ö. (1986) Photosynth. Res. 10, 453–472.
187 Schlodder, E., Brettel, K. and Witt, H.T. (1985) Biochim. Biophys. Acta 808, 123–131.
188 Eckert, H.J. and Renger, G. (1980) Photochem. Photobiol. 31, 501–511.
189 Bouges-Bocquet, B. (1980) Biochim. Biophys. Acta 594, 85–103.
190 Knaff, D.B. and Arnon, D.I. (1969) Proc. Natl. Acad. Sci. USA 63, 956–962.
191 Schenck, C.C., Diner, B.A., Mathis, P. and Satoh, K. (1982) Biochim. Biophys. Acta 680, 216–227.
192 de Paula, J.C., Innes, J.B. and Brudvig, G.W. (1985) Biochemistry 24, 8114–8120.
193 Boska, M. and Sauer, K. (1984) Biochim. Biophys. Acta 765, 84–87.
194 Boska, M., Sauer, K., Buttner, W. and Babcock, G.T. (1983) Biochim. Biophys. Acta 722, 327–330.
195 Weiss, W. and Renger, G. (1986) Biochim. Biophys. Acta 850, 173–183.
196 Itoh, S., Yerkes, C.T., Koike, H., Robinson, H.H. and Crofts, A.R. (1984) Biochim. Biophys. Acta 766, 612–622.
197 Theg, S.M., Jursinic, P. and Homann, P.H. (1984) Biochim. Biophys. Acta 766, 636–646.
198 Babcock, G.T., Blankenship, R.E. and Sauer, K. (1976) FEBS Lett. 61, 286–289.
199 Joliot, P., Hoffnung, M. and Chabaud, R. (1966) J. Chem. Phys. 63, 1423–1441.
200 Sinclair, J. and Arnason, T. (1974) Biochim. Biophys. Acta 368, 393–400.
201 Dekker, J.P., Plijter, J.J., Ouwehand, L. and Van Gorkom, H.J. (1984) Biochim. Biophys. Acta 767, 176–179.
202 Saygin, Ö. and Witt, H.T. (1986) Photobiochem. Photobiophys. 10, 71–82.
203 Lavergne, J. (1986) Photochem. Photobiol. 43, 411–423.
204 Wydrzynski, T.J. (1982) in Photosynthesis: Energy Conversion by Plants and Bacteria (Govindjee, ed.) pp. 469–506, Academic Press, New York.

205 Rutherford, A.W., Crofts, A.R. and Inoue, Y. (1984) Biochim. Biophys. Acta 682, 457–465.
206 DeVault, D., Govindjee and Arnold, W. (1983) Proc. Natl. Acad. Sci. USA 80, 983–987.
207 Demeter, S. and Vass, I. (1984) Biochim. Biophys. Acta 764, 24–32.
208 Koike, H., Siderer, Y., Ono, T. and Inoue, Y. (1986) Biochim. Biophys. Acta 850, 80–89.
209 Lavergne, J. and Etienne, A.-L. (1980) Biochim. Biophys. Acta 593, 136–148.
210 Rutherford, A.W. and Inoue, Y. (1983) FEBS Lett. 165, 163–170.
211 Rutherford, A.W., Renger, G., Koike, H. and Inoue, Y. (1984) Biochim. Biophys. Acta 767, 548–556.
212 Velthuys, B. (1983) in The Oxygen-Evolving System of Photosynthesis (Inoue, Y., et al., eds.) pp. 83–90, Academic Press, Tokyo.
213 Tamura, N., Radmer, R., Lantz, S., Cammarata, K. and Cheniae, G. (1986) Biochim. Biophys. Acta 850, 369–379.
214 Schröder, W.P. and Åkerlund, H.-E. (1986) Biochim. Biophys. Acta 848, 359–363.
215 Hanssum, B., Dohnt, G. and Renger, G. (1985) Biochim. Biophys. Acta 806, 210–220.
216 Ghanotakis, D.F., Yerkes, C.T. and Babcock, G.T. (1982) Biochim. Biophys. Acta 682, 21–31.
217 Homann, P.H., Gleiter, H., Ono, T. and Inoue, Y. (1986) Biochim. Biophys. Acta 850, 10–20.
218 Velthuys, B.R. (1975) Biochim. Biophys. Acta 396, 392–401.
219 Velthuys, B.R. (1976) Ph.D. thesis, Univ. Leiden, Leiden, The Netherlands.
220 Ono, T., Zimmermann, J.L., Inoue, Y. and Rutherford, A.W. (1986) Biochim. Biophys. Acta 851, 193–201.
221 Toyoshima, Y., Akabori, K., Imaoka, A., Nakayama, H., Ohkouchi, N. and Kawamori, A. (1984) FEBS Lett. 176, 346–350.
222 de Paula, J.C., Li, P.M., Miller, A.-F., Wu, B.W. and Brudvig, G.W. (1986) Biochemistry 25, 6487–6494.
223 Plijter, J.J., de Groot, A., van Dijk, M.A. and Van Gorkom, H.J. (1986) FEBS Lett. 195, 313–318.
224 Völker, M., Ono, T., Inoue, Y. and Renger, G. (1985) Biochim. Biophys. Acta 806, 25–34.
225 Vermaas, W.F.J., Renger, G. and Dohnt, G. (1984) Biochim. Biophys. Acta 764, 194–202.
226 Babcock, G.T. and Sauer, K. (1973) Biochim. Biophys. Acta 325, 483–503.
227 Conjeaud, H. and Mathis, P. (1986) Biophys. J., in press.
228 Yerkes, C.T., Babcock, G.T. and Crofts, A.R. (1983) FEBS Lett. 158, 359–363.
229 Ford, R.C. and Evans, M.C.W. (1985) Biochim. Biophys. Acta 807, 1–9.
230 Golbeck, J.H. and Warden, J.T. (1986) Biochim. Biophys. Acta, in press.
231 Den Haan, G.A., de Vries, H.G. and Duysens, L.N.M. (1976) Biochim. Biophys. Acta 430, 265–281.
232 Yerkes, C.T. and Babcock, G.T. (1980) Biochim. Biophys. Acta 590, 360–372.
233 Åkerlund, H.-E., Brettel, K. and Witt, H.T. (1984) Biochim. Biophys. Acta 765, 7–11.
234 Dekker, J.P., Ghanotakis, D.F., Plijter, J.T., Van Gorkom, H. and Babcock, G.T. (1984) Biochim. Biophys. Acta 767, 515–523.
235 Boska, M., Blough, N.V. and Sauer, K. (1985) Biochim. Biophys. Acta 808, 132–139.
236 Boussac, A., Maison-Peteri, B., Vernotte, C. and Etienne, A.-L. (1985) Biochim. Biophys. Acta 808, 225–230.
237 Ono, T.-A. and Inoue, Y. (1985) Biochim. Biophys. Acta 806, 331–340.
238 Blough, N.V. and Sauer, K. (1984) Biochim. Biophys. Acta 767, 377–381.
239 Franzén, L.-G., Hansson, Ö. and Andréasson, L.-E. (1985) Biochim. Biophys. Acta 808, 171–179.
240 Butler, W.L. (1978) FEBS Lett. 95, 19–23.
241 Briantais, J.-M., Vernotte, C., Miyao, M., Murata, N. and Picaud, M. (1985) Biochim. Biophys. Acta 808, 348–351.
242 Ghanotakis, D.F., Yocum, C.F. and Babcock, G.T. (1986) Photosynth. Res. 9, 125–134.
243 Briantais, J.-M., Vernotte, C., Lavergne, J. and Arntzen, C.J. (1977) Biochim. Biophys. Acta 461, 61–74.
244 Cole, J., Boska, M., Blough, N.V. and Sauer, K. (1986) Biochim. Biophys. Acta 848, 41–47.
245 Damoder, R. and Dismukes, G.C. (1984) FEBS Lett. 174, 157–161.
246 Beck, W.F., de Paula, J.C. and Brudvig, G.W. (1986) J. Am. Chem. Soc. 108, 4018–4022.
247 Critchley, C. (1983) Biochim. Biophys. Acta 724, 1–5.

248 Nash, D., Miyao, M. and Murata, N. (1985) Biochim. Biophys. Acta 807, 127–133.
249 Vass, I., Koike, H. and Inoue, Y. (1985) Biochim. Biophys. Acta 810, 302–309.
250 Preston, C. and Pace, R.J. (1985) Biochim. Biophys. Acta 810, 388–391.
251 Frasch, W.D. and Cheniae, G.M. (1980) Plant Physiol. 65, 735–745.
252 Yachandra, V.K., Guiles, R.D., Sauer, K. and Klein, M.P. (1986) Biochim. Biophys. Acta 850, 333–342.
253 Wydrzynski, T., Zumbulyadis, N., Schmidt, P.G. and Govindjee (1975) Biochim. Biophys. Acta 408, 349–354.
254 Wydrzynski, T., Zumbulyadis, N., Schmidt, P.G., Gutowsky, H.S. and Govindjee (1976) Proc. Natl. Acad. Sci. USA 73, 1196–1198.
255 Robinson, H.H., Sharp, R.R. and Yocum, C.F. (1980) Biochim. Biophys. Acta 593, 414–426.
256 Govindjee and Wydrzynski, T. (1981) in Photosynthesis II. Electron Transport and Photophosphorylation (Akoyunoglou, G., ed.) pp. 293–303, Balaban Intl. Science Service, Philadelphia, PA.
257 Sharp, R.R. and Yocum, C.F. (1983) Photobiochem. Photobiophys. 5, 193–199.
258 Srinivasan, A.N. and Sharp, R.R. (1986) Biochim. Biophys. Acta 850, 211–217.
259 Hansson, Ö., Andréasson, L.-E. and Vänngård, T. (1986) FEBS Lett. 195, 151–154.
260 Bouges, B. (1971) Biochim. Biophys. Acta 234, 102–112.
261 Velthuys, B.R. and Kok, B. (1978) Biochim. Biophys. Acta 502, 211–221.
262 Radmer, R. (1979) Biochim. Biophys. Acta 546, 418–425.
263 Radmer, R. and Ollinger, O. (1981) Biochim. Biophys. Acta 637, 80–87.
264 Förster, V. and Junge, W. (1985) Photochem. Photobiol. 41, 191–194.
265 Förster, V. and Junge, W. (1985) FEBS Lett. 186, 153–157.
266 Hanssum, B. and Renger, G. (1985) Biochim. Biophys. Acta 810, 225–234.
267 Fowler, C.F. (1977) Biochim. Biophys. Acta 462, 414–421.
268 Saphon, S. and Crofts, A.R. (1977) Z. Naturforsch. 32c, 617–626.
269 Velthuys, B.R. (1980) FEBS Lett. 115, 167–170.
270 Wille, B. and Lavergne, J. (1982) Photobiochem. Photobiophys. 4, 131–144.
271 Förster, V., Hong, Y-Q. and Junge, W. (1981) Biochim. Biophys. Acta 638, 141–152.
272 Förster, V. and Junge, W. (1985) Photochem. Photobiol. 41, 183–190.
273 Renger, G. and Voelker, M. (1982) FEBS Lett. 149, 203–207.
274 Förster, V. and Junge, W. (1984) in Advances in Photosynthesis Research (Sybesma, C., ed.) Vol. 1, pp. 305–308, Martinus Nijhoff/Dr. Junk, The Hague.
275 Chandrashekar, T.K., Rodriguez, I.D., O'Malley, P.J. and Babcock, G.T. (1986) Photosynth. Res. 10, 423–430.
276 Saygin, Ö. and Witt, H.T. (1984) FEBS Lett. 176, 83–87.
277 Saygin, Ö. and Witt, H.T. (1985) FEBS Lett. 187, 224–226.
278 Renger, G. (1978) in Photosynthetic Oxygen Evolution (Metzner, H., ed.) pp. 239–248, Academic Press, London.
279 Radmer, R. and Ollinger, O. (1986) FEBS Lett. 195, 285–289.
280 Sitter, A.J., Reczek, C.M. and Terner, J. (1985) J. Biol. Chem. 260, 7515–7522.
281 Hashimoto, S., Tatsuno, Y. and Kitagawa, T. (1986) Proc. Natl. Acad. Sci. USA 83, 2417–2421.
282 Krishtalik, L.I. (1986) Biochim. Biophys. Acta 849, 162–171.
283 Critchley, C. and Sargeson, A.M. (1984) FEBS Lett. 177, 2–5.
284 Webber, A.N., Spencer, L., Sawyer, D.T. and Heath, R.L. (1985) FEBS Lett. 189, 258–262.
285 Spencer, L., Sawyer, D.T., Webber, A.N. and Heath, R.L. (1986) Science, submitted.
286 Kambara, T. and Govindjee (1985) Proc. Natl. Acad. Sci. USA 82, 6119–6123.
287 Padhye, S., Kambara, T., Hendrickson, D.N. and Govindjee (1986) Photosynth. Res. 9, 103–112.
288 Lynch, M.W., Hendrickson, D.N., Fitzgerald, B.J. and Pierpont, C.G. (1984) J. Am. Chem. Soc. 106, 2041–2049.
289 Gilber, J.A., Eggleston, D.S., Murphy, W.R., Geselowitz, D.A., Gersten, S.W., Hodgson, D.J. and Meyer, T.J. (1985) J. Am. Chem. Soc. 107, 3855–3864.
290 Reichard, P. and Ehrenberg, A. (1983) Science 221, 514–519.
291 Pistorius, E.K. and Gau, A.E. (1986) Biochim. Biophys. Acta 849, 203–210.

J. Amesz (ed.) *Photosynthesis*

CHAPTER 7

Photophosphorylation in chloroplasts

MORDHAY AVRON*

Biochemistry Department, Weizmann Institute of Science, Rehovot 76100, Israel

1. History

A major role in ATP synthesis for the photosynthetic machinery was suggested as early as 1943 by Rubens [1] and Emerson et al. [2], but was hotly debated thereafter. Not until 1954, when Arnon and collaborators [3] presented clear evidence for light-dependent ATP formation by isolated thylakoid preparations, and Frenkel [4] by isolated chromatophores of purple bacteria, could photophosphorylation be regarded as an established new biochemical reaction, even though still raising many doubts among the more skeptical. Further work in the late '50s by Jagendorf, Avron and Arnon and collaborators [5–9] clearly established the conditions under which vigorous photophosphorylation could be observed, and removed any lingering criticisms within the scientific community. These investigations also established the obligatory coupling of ATP synthesis to the photoinduced electron transport reactions, and delineated the two basic photophosphorylative reactions that can be observed: one coupled to cyclic electron flow (unfortunately termed 'cyclic photophosphorylation') which is accompanied by no net electron transport, and a second coupled to net electron flow, thus requiring the addition of at least an electron acceptor in stoichiometric amounts (water serving as the electron donor).

During the '60s and '70s and early '80s emphasis has slowly changed from unraveling the components of the system to understanding the mechanism of the overall reaction and its measurable partial reactions and to fractionation and purification of the essential components and their reconstitution into active complexes [10–21].

---

* This review was written during the tenure of the author as a Distinguished Visiting Investigator at the Biochemistry Department of the Roche Institute of Molecular Biology, Nutley, NJ 07110, U.S.A.

2. *General characteristics*

2.1. *Relation to electron transport*

The coupling of the phosphorylation reaction to electron transport has generally been quantitatively evaluated by measurements of two parameters: the ATP/e_2 ratio, and the dependence of the rate of electron transport on concomitant phosphorylation. Both measurements were subjects of major experimental controversies, but can be said to have reached a measure of general consensus in recent years.

The P/e_2 ratio measures the number of ATP molecules synthesized per 2 electrons traversing the electron transfer system studied. For the complete electron transport system from water to $NADP^+$, most workers agree today that this number can exceed 1, and is most probably maximally 1.33 [22]. In terms of the generally accepted chemiosmotic (electrochemical-potential) interpretation of the bioenergetic events in photophosphorylation the observed ratio in any one experiment would depend on the maximal intrinsic ratio of the system, which can be deduced from its H^+/e^- ratio and H^+/ATP ratio, decreased by proton 'leaks' in the imperfectly sealed thylakoid vesicle system. Such leaks have been analysed as being due to at least two major sources [23]. (a) An unspecific leak through the thylakoid membrane; this leak seems to be dependent mainly and linearly on the ΔpH across the vesicular system studied and is pH independent. It is therefore (see below) of a lesser significance at the higher ΔpH values, where ATP formation proceeds at its maximal rate. (b) A leak through part of the ATP synthase complex; this leak is strongly pH dependent, and was already indicated by the early observations of the sharp pH dependence of non-phosphorylative electron flow, and its inhibition by ATP synthase inhibitors such as Dio-9 and DCCD [24]. It was also shown to be partially inhibited by micromolar concentrations of ATP [22] and an antibody to the ATP synthase [25], and was quantitatively evaluated [25]. By decreasing the leak pathway, these inhibitors increase the steady-state ΔpH developed across the thylakoids, particularly in the high pH range [23]. A maximal ATP/e_2 ratio of 1.3, as obtained by extrapolative techniques which minimize the effect of dissipative proton fluxes [22], implies an H^+/e^- ratio of 2, and an H^+/ATP ratio of 3. Indeed, both ratios are in agreement with most of the experimentally determined flux ratios, and the thermodynamically determined energetics in the system under a variety of experimental conditions.

The stimulation of the rate of electron flow by concomitant phosphorylation is again a strongly pH-dependent phenomenon, with maximal stimulation of 3–4-fold observed around pH 7.5. The high and low rates of electron flow are interpreted, in chemiosmotic terms, as being limited by the proton flux sustained in the thylakoid preparations under optimal light intensity (i.e., maximal pH gradient across the thylakoid vesicles), in the presence and absence of turnover of the ATP synthase. Thus, under optimal conditions, about 1/3 to 1/4 of the maximal proton flux which is sustained during active photophosphorylation can flow via non-productive pathways. As indicated above, this non-productive proton flow has been analysed as being due to at least two processes: a non-specific leak through the membrane and a pH-dependent leak through the ATP synthase. The latter accounts for about half of the total proton flux at pH 8 and about 2/3 at pH 8.5.

Equal stimulations are observed when turnover of the ATP synthase is induced by the presence of the phosphorylation reagents leading to ATP synthesis, or by a dissipative turnover, in the presence of arsenate in place of inorganic phosphate, for example.

2.2. Coupling sites

The areas in the electron transport pathway where energy conservation is observed are termed coupling sites. One site, between plastoquinone and Cyt *f*, was originally identified in thylakoid preparations by the cross-over phenomenon [14]: when ADP, for example, is added to illuminated chloroplasts when electron flow is severely limited by the phosphorylation reaction, all electron carriers which precede the coupling site will be oxidized, while all carriers which follow the coupling site will be reduced.

At least two sites can be identified by chemiosmotic principles [12]; that is, sites at which transthylakoid proton movement is coupled to electron transport. One is in the reduction of the non-heme iron protein by plastoquinone (i.e. between plastoquinone and Cyt *f*, in agreement with the former technique), and a second at the water oxidation reaction. Since water oxidation has been shown to occur on the inside of the thylakoid vesicles, each water molecule oxidized leaves two protons intravesicularly, resulting, by chemiosmotic principles, in the creation of a high-energy state. Two coupling sites should result in a maximal H^+/e_2^- of 4, in agreement with the above discussed conclusions from ATP/e_2^- measurements.

Several observations complicate the above simple conclusions. First, in several systems a so-called 'Q-cycle' has been shown to operate, which results in an H^+/e^- ratio per site which exceeds one [12,26]. However, under high illumination and in the steady state such a 'Q-cycle', if it exists, seems to contribute only a minor component to the observed ATP synthesis. Secondly, since thylakoid preparations can catalyse significant cyclic photophosphorylation where no net electron flow can be observed, the simultaneous operation of a cyclic and a linear electron flow can result in apparent high H^+/e^- ratios and/or ATP/e_2^- ratios. Again, methods are available which minimize such effects to the point where they should not constitute a significant error in the determination of such ratios [10,22]. Finally, artificial coupling sites can be induced by the addition of exogenous electron carriers. Thus, in the most efficient photophosphorylative reaction, that catalysed by phenazine-methosulfate or pyocyanine, it was clearly demonstrated that the intermediate high-energy state, in the form of a transmembrane proton concentration gradient, was created by the cyclic transport of the protonated electron-carrier to the intravesicular space followed by the exit of the non-protonated carrier. This pitfall in determining intrinsic coupling sites can also be avoided by carefully choosing the experimental conditions.

2.3. *Uncouplers and energy-transfer inhibitors*

Uncouplers are compounds which release the limitation of the electron transport rate imposed by the phosphorylation machinery, simultaneously inhibiting ATP synthesis. In view of the above discussion, uncouplers may be classified into two general catagories: those which increase the general membrane permeability to protons and those that interact specifically with the ATP synthase, increasing proton leakage through it. Most uncouplers, such as NH_3, FCCP, SF-6847, nigericin in the presence of K^+, belong to the former group. However, in recent years several reagents, particularly sulfhydryl-modifying compounds such as *N*-ethylmaleimide derivatives, mercuric ion and derivatives and silver ions [27], were shown to uncouple by inducing a proton leak through the ATP synthase. Also, an ATP synthase which lacks its ε subunit (see Section 5.1) is proton-leaking and thus uncoupled [28].

A special case is the 'uncoupling', as evidenced by the high rates of electron transport, which are induced by high pH (8.5 – 9.5). In this case, the 'uncoupling', which is again via the ATP synthase since it is fully inhibited by ATP synthase inhibitors, does not result in inhibition of ATP synthesis. This seems to be due to the fact that the phosphorylating reagents themselves (mostly ATP) seem capable of resealing the proton leak induced by high pH. Resealing of the leak in the ATP synthase by phosphorylating reagents was also observed with mercuric ions and *N*-ethylmaleimide, but in this case with no restoration of ATP synthesis [27].

Energy-transfer inhibitors block ATP synthesis by inhibiting the ATP synthase in a manner that does not lead to proton leakage. Therefore, they do not accelerate the rate of electron transport, but their inhibition of the phosphorylation-accelerated electron transport, for example, is fully reversed by the further addition of membrane-leakage-causing uncouplers. Among the common energy-transfer inhibitors are Dio-9, phlorizin, tentoxin, DCCD and triphenyltin. The former three are thought to interact with the medium-facing part of the ATP synthase (CF_1), the latter two with the membrane embedded part (CF_0).

3. *Partial reactions*

Our present understanding of the phosphorylation reaction has been greatly aided by the ability to study several partial reactions of the overall process, and follow them during fractionation of the intact thylakoid vesicle.

3.1. *ATPase*

Thylakoid vesicles as normally isolated possess little or no ATPase activity, despite their ability to catalyse vigorous photophosphorylation. Several treatments elicit an ATPase activity which is catalysed by the membrane-bound ATP synthase. Such treatments involve both the imposition of a transmembrane proton electrochemical gradient ($\Delta\bar{\mu}_{H^+}$) and the reduction of a disulfide group on the γ subunit of the enzyme [29,30].

The mechanism whereby this activation occurs is rather complex but is suggested to involve the following steps [13,17,31,32,33]: (a) $\Delta\bar{\mu}_{H^+}$-induced conformational change in the enzyme, which (b) exposes a disulfide group to reduction by external thiol reductants, and results in (c) removal of bound ADP from the enzyme, producing (d) an enzyme which can respond to relatively low $\Delta\bar{\mu}_{H^+}$ in its ATP synthetic activity, and which hydrolyses ATP.

After activation, and in the presence of ATP, the enzyme hydrolyses ATP in the dark at a steady rate for long periods. However, if the activated membrane is allowed to stay in the dark in the absence of catalytic hydrolysis of ATP, its ability to act as an ATPase slowly decays. The mechanism of this deactivation is again rather complex but clearly involves both reoxidation of the reduced enzyme, dissipation of $\Delta\bar{\mu}_{H^+}$ and rebinding of ADP [13,34,35].

The activation and inactivation of the membrane-bound ATPase occur also in vivo and can be demonstrated in intact chloroplasts. Here, a thiol reductant need not be added, since the photochemically reduced protein, thioredoxin, seems to fulfill this function [36].

The activated membrane-bound ATPase is functionally coupled to proton movements. Thus, a transmembrane pH gradient (acid inside) of a magnitude similar to that observed during light-induced coupled electron flow is developed during ATP hydrolysis. ATP hydrolysis is stimulated, while the coupled proton transport is inhibited, by the addition of uncouplers, indicating that the rate of ATP hydrolysis is also partially limited by the electrochemical gradient which it creates. Nevertheless, attempts to measure H^+/ATP ratios in this system yielded numbers much below the expected ratio of 3.

3.2. ATP-P_i exchange

The same treatments which elicit the ATPase reaction in the membrane-bound ATP synthase induce simultaneously a dark ATP-P_i exchange reaction. It was recently demonstrated that this exchange reaction is due to the simultaneous occurrence of phosphorylation and ATPase activity and therefore the use of the term 'exchange reaction' may be a misnomer [37,38]. It is suggested that, as noted above, the induced ATP hydrolysis produces a transmembrane electrochemical proton gradient which in turn drives the ATP synthetic reaction.

As expected, the exchange, but not the ATPase, is strongly inhibited by uncouplers.

The ATP-P_i exchange reaction has been very useful in demonstrating the activity of the complete ATP synthase (CF_0-CF_1) complex, since it requires an intact system properly incorporated within a vesicular system [39]. Thus, isolated CF_1 or CF_0-CF_1, though possessing ATPase activity, do not show ATP-P_i exchange until properly incorporated into a liposomal system.

3.3. ^{18}O exchange

During ATP synthesis from ADP and phosphate (at neutral pH or above), one hydroxyl ion is released per ATP molecule synthesized. That oxygen originates from the inorganic phosphate, and so the bridge oxygen between the β and γ phosphates of ATP is always provided by the ADP [10]. However, studies with ^{18}O-labeled reagents indicated that the membrane-bound ATP synthase catalyses a vigorous ATP-H_2O exchange reaction in the absence of net ATP synthesis, in which all the oxygens of the terminal phosphate exchange with water oxygens [40]. This has been interpreted as indicating a reversible hydrolysis of ATP in its enzyme-bound form, with no release of the products (ADP and P_i) to the medium. In addition, about one oxygen from water is incorporated into ATP during the normal synthetic reaction in the light.

3.4. Post-illumination phosphorylation

When chloroplasts are pre-illuminated in the absence of a phosphorylation component under conditions in which massive proton uptake occurs, ATP synthesis can be observed in the dark in the post-illumination period when the missing component is added [17]. The amount of ATP synthesized is a function of the number of protons stored in the inner thylakoid compartment during the pre-illumination period, which is in turn a function of the intrathylakoid buffer capacity, and the pH gradient sustained [14].

As was shown for photophosphorylation, ATP synthesis during the post-illumination phase will proceed only when the pH gradient across the thylakoid vesicle exceeds a value of about 2.5. Post-illumination phosphorylation seems to be driven essentially by the pH gradient with little or no contribution from a transmembrane electrical gradient [41,42]. Nevertheless, an artificial superimposition of a transmembrane electrical gradient during the transition from the pre-illumination phase to the dark ATP-synthesizing phase greatly enhances ATP formation.

3.5. Acid-base phosphorylation

A rapid pH change of acid (pH 4–5) preincubated thylakoids to a higher pH (8–9) elicits ATP formation in the dark. The amount of ATP synthesized is again a function of the pH gradient and the number of protons (i.e., buffer capacity) stored in the intrathylakoid compartment [17]. If a transmembrane electrical gradient is imposed on the system, simultaneously with the pH gradient, the amount of ATP synthesized is a function of the total electrochemical gradient, with a threshold requirement of about 150 mV [42,43]. The fact that the relative contribution of the ΔpH and $\Delta\psi$ can be varied over a wide range without altering the magnitude of the threshold required has been interpreted as indicating that the threshold represents an essential thermodynamic limitation [43,44]. However, the fact that thiol modulation does shift the threshold value [45] suggests that the threshold, as observed in untreated thylakoids, represents more than just a thermodynamic limitation for ATP synthesis.

Recently the kinetics of acid base phosphorylation was studied using rapid mixing and quenching techniques [44,46]. In contrast with post-illumination phosphorylation, where the transmembrane pH gradient is slowly created in the light [41], no time lag is observed here, since the transmembrane electrochemical potential already exists when the reaction is commenced. The decay of the 'high-energy state', in the absence of phosphorylation, is a function of the decay kinetics of the transmembrane ΔpH and $\Delta\psi$, which decay with markedly different half-times [44].

3.6. Electric-field phosphorylation

ATP synthesis can also be induced in isolated thylakoids by subjecting them to a brief electric field pulse [12,47,48]. The driving force is presumably an induced transthylakoid electric potential of about 200 mV which is induced by field pulses of around 1000 V per cm. The amount of ATP synthesized is linear with the number of pulses, with no apparent lag, and with the duration of each pulse. Less than 1 molecule of ATP is synthesized per pulse per ATP synthase, but the 'rate' of ATP synthesis compares favorably with the rate observed during light phosphorylation if one assumes that ATP is synthesized only during the pulse.

4. Mechanism

4.1. The electrochemical potential hypothesis

The electrochemical potential hypothesis (chemiosmotic hypothesis, Mitchell's hypothesis) has formed the basis for interpretation of most of the available data on the mechanism of photophosphorylation [14]. The hypothesis suggests that proton transport into the inner thylakoid space, coupled to light-induced electron transport, is the source of the high-energy state which drives ATP synthesis. The available energy can be quantitatively estimated by summing the energy available from the transmembrane pH gradient (ΔpH) (around 60 mV per pH unit) and that available from the transmembrane electrical gradient ($\Delta\psi$). The latter is formed when proton movement is not fully balanced by countermovement of other charged ions (negatively charged ions inwards or positively charged ions outwards).

4.2. ΔpH generation and utilization

As already mentioned, there are two established sites of proton uptake coupled to electron transfer in the thylakoid membrane in the absence of added electron carriers: one at the water oxidation reaction and the second at the plastoquinone to iron-sulfur protein reaction. This would predict that the H^+/e^- stoichiometry measured during electron transport should be 2. However, designing an unambiguous experiment to determine this exact ratio in isolated thylakoids turned out to be more difficult than it seemed at first. The literature contains, therefore, numerous values for this ratio [12], some of which are indeed close to 2.

The transthylakoid pH gradient which coupled electron transport can maintain has been measured by a variety of techniques [14]. Despite the limitations of all these techniques [49], which are in all cases indirect, it can be said that thylakoids can develop an electron-transport-driven pH gradient of at least 3 pH units, and possibly as high as 4. It has been shown by several laboratories that in the steady state in the light the transthylakoid pH gradient is by far the major energy storage form, $\Delta\psi$ providing a small contribution, if at all. The magnitude of the ΔpH generated in the light is similar to that required to drive maximal ATP synthesis in the dark by acid-base phosphorylation.

ATP synthesis utilizes the proton concentration gradient created by electron transport, and therefore the magnitude of the gradient is indeed smaller by about 0.5 pH unit during ATP synthesis than in its absence. The number of protons which need to traverse the ATP synthase for the synthesis of one ATP molecule was estimated both from direct measurements and from thermodynamic analysis [22]. Most workers agree that this ratio is very close to 3. When combined with an H^+/e^- ratio of 2, this predicts a maximal ATP/e_2^- ratio of 1.3, in agreement with this ratio as determined by extrapolative techniques [22].

4.3. $\Delta\psi$ generation and utilization

As already mentioned, little or no $\Delta\psi$ is maintained by thylakoids in the steady state in the light. This seems to be due to the rather high non-specific permeability of the thylakoid membrane to the many ions which are always present in reaction mixtures [23]. Nevertheless, highly significant transmembrane electrical gradients can be demonstrated across thylakoid membranes under special circumstances [50,51]. The most important one seems to be within a second or two following a dark-to-light transition. The formed $\Delta\psi$ decays rapidly so that after a few seconds in the light the major energy storage device is the ΔpH.

The $\Delta\psi$ formed following a dark-to-light transition has been shown to be the major driving force for the ATP synthesized during the first second or so after turning the light on, and is therefore the major energy source for ATP synthesis driven by a light flash or a sequence of light flashes [12,52]. ΔpH and $\Delta\psi$ have been shown to be energetically equivalent as driving forces for ATP formation [43], but their rate of decay is markedly different [23,44].

4.4. The threshold

When the rate of ATP synthesis is plotted versus the total electrochemical gradient which serves as its driving force, a non-linear curve is obtained with little or no ATP synthesis until a threshold value of about 120–150 mV is reached, and a close to linear dependence thereafter. This behavior has been demonstrated in all of the ATP-synthesizing systems: photophosphorylation, post-illumination phosphorylation and acid-base phosphorylation, and under a variety of energy-limiting conditions such as light intensity, inhibitors, uncouplers, etc.

The mechanism underlying the threshold requirement is not clear. It could reflect a simple thermodynamic requirement for a transmembrane electrochemical gradient of a sufficient magnitude so that three protons traversing the ATP synthase provide a sufficient driving force for the synthesis of an ATP molecule under the prevailing conditions [43]. Alternatively, it could just be a reflection of the fact that ATP synthesis may depend on the 3rd power of the proton concentration while proton leakage through the membrane would be expected to depend only on the first power [12,16]. Thus, at low ΔpH most of the protons would efflux through the non-productive leak pathway while at high ΔpH the opposite will be true. It may also reflect a requirement for a conformational change, possibly accompanied by the release of bound ADP, before the ATP synthase can function in its ATP-synthetic role [53–56].

4.5. *Bulk vs. local electrochemical potentials [57]*

The original electrochemical potential hypothesis postulated that the energy provided by the coupled proton transport is stored in the form of a transmembrane bulk electrochemical gradient of protons, and is used therefrom by the membrane-bound ATP synthase [33]. Unquestionably, the data briefly discussed above support the contention that (a) bulk electrochemical gradients are formed by coupled proton flow, and (b) such bulk electrochemical gradients can serve as the driving force for ATP synthesis.

Nevertheless, an accumulating body of evidence suggests that in the intact thylakoid (and in similar energy-transducing membrane systems) ATP synthesis can be observed under conditions where it cannot be rationalized as being driven solely by the bulk electrochemical gradient [58]. These observations have led several workers to suggest that in vivo there may be a more direct path from the electron transport system to the ATP synthase. Such paths generally involve either compartmentation of the coupling system into small coupling units, each with its own 'bulk' electrochemical gradient, or a direct innermembrane path of proton transport from the electron carrier to the ATP synthase in a manner which does not fully equilibrate, within the required time domain, with the bulk electrochemical gradient. In these cases the bulk electrochemical gradient can be looked upon as an 'energy-buffer' system with a relatively slow equilibration with the direct driving system.

5. *The ATP synthase*

The ATP synthase is composed of two distinct complexes [15,16]. CF_1 refers to the water-soluble complex which extends from the thylakoid membrane to the stroma, and CF_0 to the membrane embedded complex to which it is attached.

5.1. *CF_1 – isolation, properties and reconstitution*

CF_1 was originally isolated as a coupling factor, that is, a protein which when removed from the thylakoid membrane leaves a membrane unable to catalyse photophosphorylation, and which when reconstituted into it restores this ability. The original technique for removing the CF_1 from thylakoid membranes was by treatment with EDTA under low-salt conditions. This technique removes only a part of the CF_1 but renders the CF_1-less membranes fully inactive in catalysing photophosphorylation. Activity can be essentially fully restored by reincubating the isolated and purified CF_1 with the CF_1-free membranes in the presence of Mg^{2+}. A method for fully removing the CF_1 from thylakoids, while permitting good reconstitution, has since been developed and involves treating the thylakoids with 2 M NaBr under carefully controlled conditions [59].

CF_1-less membranes are very leaky to protons, but this leakiness is eliminated by the recombination of the membranes with the isolated CF_1, or with any of several inhibitors of the CF_0 portion of the ATP synthase that is still membrane-bound in the CF_1-less membrane. It was thus concluded that CF_0 is a proton channel whose function is to deliver energetic protons to CF_1, where ATP synthesis takes place.

CF_1 is a multicomponent complex with a molecular weight of about 400000 [60,61]. It is composed of five subunits termed α, β, γ, δ, ε in order of decreasing molecular weight (Chapter 10). The ratio of these units per molecule has been a matter of controversy, but recent evidence would seem to support a structure composed of $\alpha_3\beta_3\gamma\delta\varepsilon$ [60,62]. Each subunit seems to carry a specific function: the α subunit contains tight nucleotide-binding sites which may function in regulation and undergoes major conformational changes during catalysis, the β subunit contains the catalytic site for ATP synthesis and hydrolysis, the γ subunit seems to be involved in proton transport through the ATP synthase and is essential for observing ATPase activity in isolated subunit complexes, the δ subunit seems to be involved in channelling protons from the membrane-bound CF_0 to CF_1 [63], and the ε subunit is required to observe ATP synthesis in reconstituted systems and inhibits the ATPase activity of the ε-less complex [64].

CF_1 as isolated is totally inactive. However, it can be 'activated' to catalyse ATP hydrolysis by a variety of techniques, including heat, incubation with a sulfhydryl reagent such as DTT, and treatment with proteases, organic solvents or detergents. The mechanisms of activation by these treatments are not identical and have been clarified in some cases as being due to loosening of the interaction between the ε subunit and the remaining complex (heat, solvents, detergents), reduction of a specific disulfide in the γ subunit (DTT), or partial cleavage of the α subunit (proteases) [15,16,65–67]. Although the membrane-bound activated ATPase activity is always magnesium dependent, the ATPase activity of the activated isolated CF_1 may be either calcium or magnesium dependent, depending on the activation procedure employed. Lack of magnesium dependence is related to potent inhibition of the calcium-dependent complexes by added magnesium, but the exact mechanism involved in this cation specificity is not clear at present. Since converting a latent complex such as CF_1 to an active ATPase may, in principle, involve nothing more than providing a mechanism for release into the medium of the normally inwardly pumped protons, it should not be surprising that a variety of unrelated treatments can 'nick' the intact CF_1 so that it becomes 'leaky' at different positions.

5.2. CF_0-CF_1 – *isolation, properties and reconstitution*

The development of an isolation and purification procedure for the complete CF_0-CF_1 complex from thylakoids [68] made it possible to study the properties of the active proton-translocating ATP synthase complex. The isolated CF_0-CF_1 contains three subunits, in addition to the five attributed to CF_1, termed I, II and III in order of decreasing molecular weight. Subunit III has been clearly identified as the DCCD-binding protein within the membrane-bound CF_0 [69,70]. When incorporated into artificial liposomes, in its isolated and purified form, it catalyses a DCCD-sensitive proton movement through the liposome membrane. It is therefore generally assumed that subunit III provides the backbone of the proton channel which functions within the membrane-bound CF_0 [71]. The functions of the other two subunits have not been clarified, but they are generally assumed to play a role in structuring the proton channel and in providing the binding linkage of CF_1. The stoichiometry of the subunits per CF_0 has also not been fully elucidated, but it seems to involve multiple copies of all three subunits, with subunit III existing in as many as 6 copies per CF_0.

The isolated CF_0-CF_1 has been incorporated into phospholipid liposomes and shown to carry in this form most of the energy-transducing functions which it catalyses within the thylakoid membranes. Thus, the reconstituted ATP synthase carries out ATP-dependent proton translocation resulting in both a ΔpH and a $\Delta\psi$ developing across the reconstituted liposomes [72,73]; an uncoupler-sensitive ATP-P_i exchange reaction [39]; and ATP formation driven by artificially imposed ΔpH and $\Delta\psi$ [39,74,75], or by electric field pulses [56]. The ATP synthase proteoliposomes provide the simplest system available today for the study of electrochemical-gradient-driven phosphorylation.

6. *Reverse reactions*

A unique aproach to studying in the dark the energy-transducing reactions in the process of photophosphorylation is via investigations of energy-dependent reverse reactions. Reverse reactions can be driven either by ATP or by artificially imposed electrochemical gradients [76].

6.1. *ATP-driven reactions*

The simplest ATP-driven reaction, ATP-driven proton uptake, has already been discussed. After activation the membrane-bound ATP synthase pumps protons into the inner thylakoid space, coupled to ATP hydrolysis. Both ΔpH and $\Delta\psi$ are produced with magnitudes similar to those produced by light-driven proton transport [51,77].

ATP, under similar experimental conditions, has also been shown to drive reverse electron flow, leading to the oxidation of an external electron donor, such as DTT or hydroquinone, and the reduction of Q_A. The reaction seems to involve, as an obligatory intermediate in its major kinetic phase, the ATP-driven proton uptake. Under some conditions, a second rapid kinetic phase can be observed in this reaction, which is uncoupler insensitive [78,79]. The origin of this rapid phase is not clear, but it seems to reflect a more direct route of coupling of the ATP synthase to the electron transport system. ATP-driven Q_A reduction has been demonstrated to occur in intact chloroplasts [80], and therefore may have a regulatory function in vivo.

ATP has also been shown to drive a reversal of the photochemical reaction itself, leading to light emission, or luminescence, from PS II [81]. To demonstrate the reaction, the photosystem must have an available positively charged partner on the water side of PS II, so that it is limited by the electrons delivered to Q_A by the reverse reaction. The positively charged partner is produced by a brief flash of light, with the subsequent addition of ATP in the dark producing the reverse electron flow luminescence. The mechanism by which ATP enhances luminescence has been analysed as being due to two factors: increase in the concentration of reduced Q_A, and promotion by the ATPase-produced transmembrane electrical gradient of the recombination of the reduced Q_A and its oxidized partner, which results in luminescence.

ATP-induced luminescence was also demonstrated to occur in intact chloroplasts [80].

6.2. *Reactions driven by an electrochemical potential*

If ATP reversal operates via the intermediate formation of an electrochemical proton gradient, as the chemiosmatic hypothesis predicts, it should be possible to show the same reverse reaction by the direct imposition of a transmembrane electrochemical proton gradient of the proper polarity.

Indeed, acid-base transition induces reduction of Q_A [76]. The reduction is transient, does not require addition of an external electron donor, but is dependent upon the presence of reduced carriers between the two photosystems which act as electron donors for the transient reduction [82,83].

An acid-base transition also drives reversal of the PS II-induced charge separation indicated by luminescence [84]. The reaction again requires preprovision of an oxidized partner by preillumination, and is stimulated by a simultaneously imposed electric potential gradient of the proper polarity [85].

7. *Conclusion*

During the last thirty years, intensive investigations by numerous laboratories converted photophosphorylation from a highly debatable and marginally detectable process to a well-established and well-dissected reaction. We have today a wealth of information about the overall photochemical steps, the electron transport reactions driven by it, the electrochemical gradient driven by the electron transport, and the overall reaction responsible for ATP synthesis by the enzyme-bound ATP synthase. All these reactions have been dissected into smaller sub-thylakoidal active complexes. Reactions of isolated reaction centers, electron transport complexes such as the Cyt b_6-f complex, and the isolated complete ATP synthase and its subunits can be studied individually. Nevertheless, many problems, like the manner whereby three protons traversing a membrane-bound ATP synthase can drive ATP synthesis, remain a puzzle and a challenge to future investigators.

References

1 Rubens, S. (1943) J. Am. Chem. Soc. 65, 279–282.
2 Emerson, R.L., Stauffer, J.F. and Umbreit, W.W. (1944) Am. J. Bot. 11, 51–56.
3 Arnon, D.I., Allen, M.B. and Whatley, F.R. (1954) Nature 174, 394–396.
4 Frenkel, A.W. (1954) J. Am. Chem. Soc. 76, 5568–5569.
5 Avron, M., Jagendorf, A.T. and Evans, M. (1957) Biochim. Biophys. Acta 26, 262–269.
6 Allen, M.B., Whatley, F.R. and Arnon, D.I. (1958) Biochim. Biophys. Acta 27, 16–23.
7 Jagendorf, A.T. and Avron, M. (1958) J. Biol. Chem. 231, 277–290.
8 Arnon, D.I., Whatley, F.R. and Allen, M.B. (1958) Science 127, 1026–1034.
9 Avron, M. and Jagendorf, A.T. (1959) J. Biol. Chem. 234, 967–972.
10 Avron, M. (1977) Annu. Rev. Biochem. 46, 143–155.
11 Reeves, S.G. and Hall, D.O. (1978) Biochim. Biophys. Acta 463, 275–297.
12 Witt, H.T. (1979) Biochim. Biophys. Acta 505, 355–427.
13 Shavit, N. (1980) Annu. Rev. Biochem. 49, 111–138.
14 Avron, M. (1981) in Biochemistry of Plants (Stumpf, P. and Conn, E.E., eds.) Vol. VIII, pp. 163–191, Academic Press, New York.
15 Nelson, N. (1982) Curr. Top. Bioenerg. 11, 1–33.
16 McCarty, R.E. and Moroney, J.V. (1984) in The Enzymes of Biological Membranes (Martonosi, A., ed.) Vol. 4, 2nd Edn., pp. 383–413, Plenum Press, New York.
17 Trebst, A. and Avron, M. (eds.) (1977) Photosynthetic Electron Transport and Photophosphorylation, Encyclopedia of Plant Physiology, New Series, Vol. 5, 820 pp., Springer-Verlag, Heidelberg.
18 Akoyunoglon, G. (ed.) (1981) Photosynthesis: Electron Transport and Photophosphorylation, Vol. 2, Balaban Int. Sciences, Philadelphia, PA.
19 Selman, B.R. and Selman-Reimer, S. (1981) Energy Coupling in Photosynthesis, Elsevier/North Holland, New York.
20 Nicholls, D.P. (1982) Bioenergetics, Academic Press, New York.
21 Sybesma, C. (ed.) (1984) Advances in Photosynthesis, Vol. II, Martinus-Nijhoff/Dr. W. Junk, The Hague.
22 Davenport, J.W. and McCarty, R.E. (1984) Biochim. Biophys. Acta 766, 363–374.
23 Avron, M. (1981) in Photosynthesis (Akoyunoglon, G., ed.) Vol. 2, pp. 917–928, Balaban Int. Sciences, Philadelphia, PA.
24 Avron, M. (1972) in Proc. 2nd International Congress in Photosynthesis (Forti, G., Avron, M. and Melandri, A., eds.) Vol. II, pp. 861–871, N.V. Junk, The Hague.
25 Abbott, M.S. and Dilley, R.A. (1983) Arch. Biochem. Biophys. 222, 95–104.
26 Moss, D.A. and Bendall, D.S. (1984) Biochim. Biophys. Acta 767, 389–395.
27 Underwood, C. and Gould, J.M. (1980) Biochim. Biophys. Acta 589, 287–298.
28 Richter, M.L., Patrie, W.J. and McCarty, R.E. (1984) J. Biol. Chem. 259, 7371–7373.
29 Jagendorf, A.T. (1981) in Photosynthesis (Akoyunoglou, G., ed.) Vol. 2, pp. 719–730, Balaban Int. Sciences, Phildelphia.
30 Ketcham, S.R., Davenport, J.W., Warncke, K. and McCarty, R.E. (1984) J. Biol. Chem. 259, 7286–7293.
31 Junesch, U. and Graber, P. (1984) in Advances in Photosynthesis (Sybesma, C., ed.) Vol. 2, pp. 5431–5436, Nijhoff/Junk, The Hague.

32 Sherman, P.A. and Wimmer, M.J. (1984) Eur. J. Biochem. 139, 267–371.
33 Mills, J.D. (1984) in H^+ ATPase (Papa, S., Altendorf, K. Ernster, L. and Packer, L., eds.) pp. 349–358, Adriatica Editrice, Bari.
34 Shahak, Y. (1982) Plant Physiol. 70, 87–91.
35 Shahak, Y. (1985) J. Biol. Chem. 260, 1459–1464.
36 Mills, J.D., Mitchell, P. and Schurmann, P. (1980) FEBS Lett. 112, 173–177.
37 Davenport, J.W. and McCarty, R.E. (1981) J. Biol. Chem. 256, 8947–8954.
38 Sherman, P.A. and Wimmer, M.J. (1982) J. Biol. Chem. 257, 7012–7017.
39 Pick, U. and Racker, E. (1979) J. Biol. Chem. 254, 2793–2799
40 Wimmer, M.J. and Rose, J.A. (1977) J. Biol. Chem. 252, 6769–6775.
41 Vinkler, C., Avron, M. and Boyer, P.D. (1980) J. Biol. Chem. 255, 2263–2266.
42 Hangarter, R.P. and Good. N.E. (1984) Biochemistry 23, 122–130.
43 Hangarter, R.P. and Good, N.E. (1982) Biochim. Biophys. Acta 681, 397–405.
44 Horner, R.D. and Moudrianakis, E.N. (1985) J. Biol. Chem. 260, 6153–6159.
45 Mills, J.D. and Mitchell, P. (1982) FEBS Lett. 144, 63–67.
46 Graber, P., Junesch, U. and Schatz, G.H. (1984) Ber. Bunsenges. Phys. Chem. 88, 599–608.
47 Witt, H.T. and Schlodder, E. (1981) Biochim. Biophys. Acta 635, 571–584.
48 Vinkler, C. and Kornstsein, R. (1982) Proc. Natl. Acad. Sci. USA 79, 3183–3187.
49 Casadio, R. and Melandri, B.A. (1985) Arch. Biochem. Biophys. 238, 219–228.
50 Schuurmans, J.J., Casey, R.P. and Kraayenhof, R. (1978) FABS Lett. 94, 405–409.
51 Admon, A., Shahak, Y. and Avron, M. (1982) Biochim. Biophys. Acta 681, 405–411.
52 Vinkler, C., Avron, M. and Boyer, P.D. (1978) FEBS Lett. 96, 129–134.
53 Schumann, J. and Strottman, H. (1981) in Photosynthesis (Akoyunoglu, G., ed.) Vol. 2, pp. 881–892, Balaban Int. Sciences, Philadelphia, PA.
54 Rumberg, B. and Becher, U. (1984) in H^+ ATPase (Papa, S., Altendorf, K., Ernster, L. and Packer, L., eds.) pp. 421–430, Adriatica Editrice, Bari.
55 Mills, J.D. and Mitchell, P. (1984) Biochim. Biophys. Acta 764, 93–104. 66.
56 Graber, P., Schlodder, E. and Witt, H.T. (1984) in H^+ ATPase (Papa, S., Altendorf, K., Ernster, L. and Packer, L., eds.) pp. 431–440, Adriatica Editrice, Bari.
57 Ferguson, S.J. (1985) Biochim. Biophys. Acta 811, 47–96.
58 Westerhoff, H.V., Melandri, B.A., Venturoli, G., Azzone, G.F. and Kell, D.B. (1984) Biochim. Biophys. Acta 768, 257–292.
59 Nelson, N. and Eitan, E. (1979) in Cation Fluxes Across Biomembranes (Mukohata, Y. and Packer, L., eds.) pp. 409–415, Academic Press, New York.
60 Moroney, J.V., Lopresti, L., McEwen, B.F., McCarty, R.E. and Hammes, G.G. (1983) FEBS Lett. 158, 58–62.
61 Merchant, S., Shaner, S.I. and Selman, B.R. (1983) J. Biol. Chem. 258, 1026–1031.
62 Suss, K.H. and Schmidt, O. (1982) FEBS Lett. 144, 213–218.
63 Patrie, W.J. and McCarty, R.E. (1984) J. Biol. Chem. 259, 11121–11128.
64 Richter, M.I., Patrie, W.J. and McCarty, R.E. (1984) J. Biol. Chem. 259, 7371–7373.
65 Pick, U. and Bassilian, S. (1982) Biochemistry 21, 6144–6152.
66 Finel, M., Rubenstsein, M. and Pick, U. (1984) FEBS Lett. 166, 85–89.
67 Yu, F. and McCarty, R.E. (1985) Arch. Biochem. Biophys. 238, 61–68.
68 Pick, U. and Racker, E. (1979) J. Biol. Chem. 254, 2793–2799.
69 Sigrist-Nelson, K. and Azzi, A. (1980) J. Biol. Chem. 255, 10638–10643.
70 Nelson, N. (1980) Ann. N.Y. Acad. Sci. 358, 25–35.
71 Junge, W., Hong, Y.Q., Qian, L.P. and Viale, A. (1984) Proc. Natl. Acad. Sci. USA 81, 3078–3082.
72 Admon, A., Pick, U. and Avron, M. (1985) J. Membrane Biol. 86, 45–50.
73 Shahak, Y., Admon, A. and Avron, M. (1982) FEBS Lett. 150, 27–31.
74 Gräber, P., Rogner, M., Samoray, D. and Hauska, G. (1984) in Advances in Photosynthesis Research (Sybesma, C., ed.) Vol. II, pp. 427–430, Nijhoff/Junk, The Hague.
75 Admon, A. and Avron, M. (1985) Physiol. Veg. 23, 687–695.
76 Shahak, Y. and Avron, M. (1980) Methods Enzymol. 69C, 630–641.
77 Schreiber, U. and Rienits, K.G. (1982) FEBS Lett. 141, 287–291.

78 Schreiber, U. (1984) Biochim. Biophys. Acta 767, 70–79.
79 Schreiber, U. (1984) Biochim. Biophys. Acta 767, 80–86.
80 Schreiber, U. (1980) FEBS Lett. 122, 121–124.
81 Avron, M. and Schreiber, U. (1979) Biochim. Biophys. Acta 546, 448–454.
82 Shahak, Y., Pick, U. and Avron, M. (1976) in Enzymes, Electron Transport Systems (Densuelle, P. and Michelson, A.M., eds.) Vol. 40, pp. 305–314, Elsevier, Amsterdam.
83 Hardt, H. (1981) Biochim. Biophys. Acta 635, 631–644.
84 Shahak, Y., Siderer, Y. and Avron, M. (1977) in Photosynthetic Organelles, Structure and Function (Miyachi, S., Katoh, S., Fujita, Y. and Shibata, K., eds.) pp. 115–127, Japan Soc. Pl. Physiol., Tokyo.
85 Avron, M., Admon, A. and Shahak, Y. (1981) in Energy Coupling in Photosynthesis (Selman, B.R. and Selman-Reimer, S., eds.) pp. 15–23, Elsevier, New York.

J. Amesz (ed.) *Photosynthesis*

CHAPTER 8

Carbon dioxide assimilation

FRASER D. MACDONALD and BOB B. BUCHANAN

Division of Molecular Plant Biology, Hilgard Hall, University of California, Berkeley, CA 94720, U.S.A.

1. Introduction

Life on our planet obtains its substance and energy through the process of photosynthesis, a grand device by which green plants use the electromagnetic energy of sunlight to synthesize carbohydrates (CH_2O) (Eqn. 1) and other cellular constituents from carbon dioxide and water.

$$CO_2 + 2H_2O^* \xrightarrow{\text{light}} (CH_2O) + O_2^* + H_2O \qquad (1)$$

Photosynthesis may be broadly divided into two phases: a light phase, in which the electromagnetic energy of sunlight is trapped and converted into ATP and NADPH, and a synthetic phase, in which the ATP and NADPH generated by the light phase are used for biosynthetic carbon reduction. As described below, light also functions in the regulation of the synthetic or carbon reduction phase of photosynthesis and in related biochemical processes of chloroplasts.

In most plants, the major products of photosynthesis are starch (formed in chloroplasts and sucrose (formed in the cytosol). Both of these products (collectively called photosynthate) are formed from photosynthetically generated dihydroxyacetone phosphate (DHAP) via pathways that in some respects are similar to the gluconeogenic pathway of animal cells. In the first case, DHAP is converted to hexose phosphates, which, in turn, are converted to starch within the chloroplast. In sucrose synthesis, DHAP (or a derivative) is transported to the cytosol and there it is converted to sucrose.

All oxygenic (oxygen-evolving) organisms from the simplest prokaryotic cyanobacteria to the most complicated land plants have a common pathway for the reduction of CO_2 to sugar phosphates. This pathway is known as the reductive pentose phosphate (RPP), Calvin-Benson or C_3 cycle.

Although the RPP cycle is the fundamental carboxylating mechanism, a number of plants have evolved adaptations in which CO_2 is first fixed by a supplementary pathway and then released in the cells in which the RPP cycle operates. One of these supplementary pathways, the C_4 pathway, involves special leaf anatomy and a division of biochemical labor between cell types. Plants endowed with this pathway, through greater efficiency, are able to flourish under conditions of high light intensity and elevated temperatures. A second supplementary pathway was first found in species of the Crassulaceae and is called Crassulacean acid metabolism (CAM). These plants are often found in dry areas and fix CO_2 at night into C_4 acids. During the day, the leaves can close their stomata to conserve water while CO_2 released from the C_4 acids is converted to sugar phosphates by the RPP cycle using absorbed light energy.

CO_2 fixation is also found in many bacteria, both photosynthetic and non-photosynthetic. The purple sulfur and purple nonsulfur bacteria employ the RPP cycle as do plants. The photosynthetic green bacteria, however, use a group of ferredoxin-linked carboxylases in a pathway known as the reductive carboxylic acid cycle [1].

In the following sections, we will first describe the RPP cycle, the C_4 pathway and CAM. We will then discuss what is known of the regulation of these pathways and the way in which the activity of the RPP cycle is coordinated with the utilization of photosynthate.

2. *The reductive pentose phosphate cycle*

The reductive pentose phosphate cycle is the only fundamental carboxylating mechanism in plants. In C_3 plants the entire process of photosynthesis (the light reactions and the RPP cycle) occurs within chloroplasts. The enzymes catalysing steps in the RPP cycle are water-soluble and are located in the soluble portion (chloroplast stroma or extract).

Elucidation of the pathway was chiefly the work of Calvin, Benson, Bassham and co-workers, although there were important contributions by others. In their experiments they used green algae, *Chlorella* and *Scenedesmus*, but since that time their results have been confirmed many times in a wide variety of higher plants.

The crux of the pathway (Fig. 1) is the carboxylation of ribulose 1,5-bisphosphate (Rbu-1,5-P_2) at the C-2 carbon, giving rise to a short-lived six-carbon intermediate which is cleaved to produce two molecules of 3-phosphoglycerate (3-PGA) (Eqn. 2). This reaction is catalysed by ribulose-1,5-bisphosphate carboxylase oxygenase (rubisco), one of the most abundant proteins on earth.

$$\text{Rbu-1,5-P}_2 + CO_2 + H_2O \rightarrow 2\ \text{3-PGA} \qquad (2)$$

The first two reactions involved in the further metabolism of 3-PGA utilize ATP and NADPH generated by the light reactions of photosynthesis. 3-PGA is first phosphorylated by ATP to give 1,3-diphosphoglycerate (DiPGA), which is then reduced by NADPH to give glyceraldehyde 3-phosphate (G3P). The enzymes involved are 3-PGA kinase and NADP-G3P dehydrogenase (NADP-G3PDH) respectively (Eqn. 3).

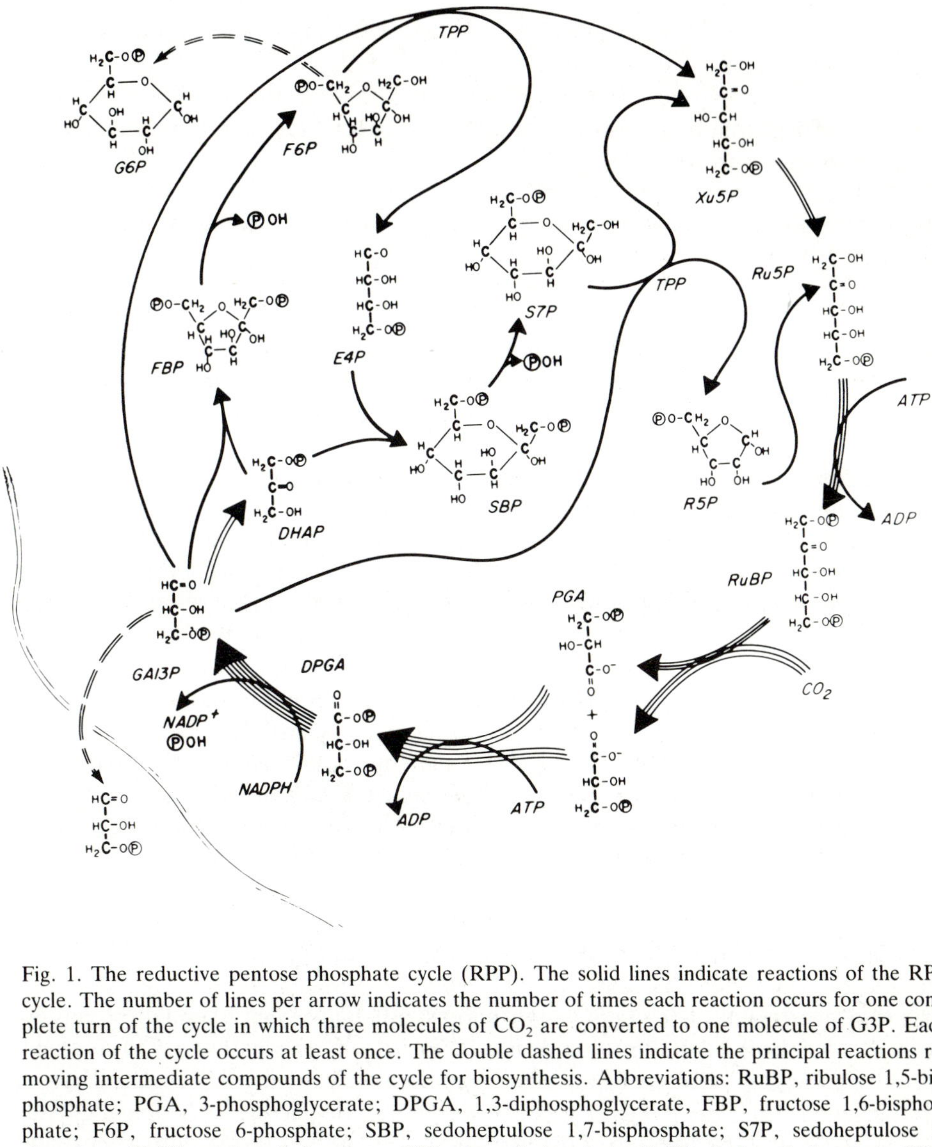

Fig. 1. The reductive pentose phosphate cycle (RPP). The solid lines indicate reactions of the RPP cycle. The number of lines per arrow indicates the number of times each reaction occurs for one complete turn of the cycle in which three molecules of CO_2 are converted to one molecule of G3P. Each reaction of the cycle occurs at least once. The double dashed lines indicate the principal reactions removing intermediate compounds of the cycle for biosynthesis. Abbreviations: RuBP, ribulose 1,5-bisphosphate; PGA, 3-phosphoglycerate; DPGA, 1,3-diphosphoglycerate, FBP, fructose 1,6-bisphosphate; F6P, fructose 6-phosphate; SBP, sedoheptulose 1,7-bisphosphate; S7P, sedoheptulose 7-phosphate; Xu5P, xylulose 5-phosphate; R5P, ribose 5-phosphate; Ru5P, ribulose 5-phosphate; TPP, thiamine pyrophosphate. From Ref. 1.

$$\text{3-PGA} \overset{\text{ATP} \quad \text{ADP}}{\rightleftharpoons} \text{DiPGA} \overset{\text{NADPH} \quad \text{NADP}}{\rightleftharpoons} \text{G3P} + \text{P}_i \tag{3}$$

Intermediates formed from G3P are utilized (Fig. 1) via a series of isomerizations, condensations and rearrangements resulting in the conversion of five molecules of triose phosphate to three of pentose phosphate, eventually ribulose 5-phosphate (Rbu-5-P). Phosphorylation of Rbu-5-P with ATP regenerates the original carbon acceptor Rbu-1,5-P_2, thus completing the cycle.

The RPP cycle displays four features which are necessary for its role as a fundamental carboxylating system [2].

(i) The rubisco reaction has a highly favorable equilibrium ($\triangle G' = -35.1$ kJ);

(ii) The carboxylating enzyme has a high affinity for CO_2, which occurs at a relatively low concentration in air;

(iii) There is a cyclic regeneration of the CO_2 acceptor Rbu-1,5-P_2 from the products of the carboxylation reaction, thus allowing the continued operation of the cycle;

(iv) The cycle is capable of the net production of fixed carbon in the form of triose phosphate. For every three turns of the cycle during which six molecules of 3-PGA are formed, five molecules must be utilized to reform three molecules of Rbu-1,5-P_2 while the sixth 3-PGA molecule is available as an end product (photosynthate) for biosynthetic reactions (predominantly starch and sucrose synthesis).

In addition to its carboxylase activity, rubisco can act as an oxygenase. In this reaction, molecular O_2 is bound and reacts with Rbu-1,5-P_2 to give 3-PGA and 2-phosphoglycolate [3]. 2-Phosphoglycolate cannot be utilized in the RPP cycle and thus represents a loss of fixed carbon. This loss is partly compensated by the process of photorespiration during which three-quarters of the lost carbon is returned to the chloroplast as 3-PGA [4]. The oxygenase reaction is greatly reduced by lowered O_2 or raised CO_2 pressure (compared to air levels) and hence photorespiration is greatly reduced in C_4 plants, CAM plants, algae and cyanobacteria which, as discussed below, possess CO_2-concentrating mechanisms. The oxygenase activity of rubisco may be necessary to protect the chloroplast against photooxidation damage when CO_2 is limiting [5]. Alternatively, it has been suggested that Rbu-1,5-P_2 oxygenation is an inevitable consequence of the carboxylation mechanism of rubisco [6,7].

3. The C_4 pathway

The C_4 (dicarboxylic acid) pathway of photosynthetic carbon assimilation may be seen as a biochemical elaboration of the RPP cycle. In this pathway CO_2 is transferred via the C-4 carboxyl of C_4 acids to the reactions of the RPP cycle. Discovered in sugar cane, the pathway was first thought to be peculiar to tropical grasses but was later found in species of dicotyledons, *Amaranthus* (Amaranthaceae), and *Atriplex* (Chenopodiaceae).

Unlike the RPP cycle in which carboxylation and carbon reduction are restricted to the chloroplast, the C_4 pathway involves the interaction of two cell types and several different compartments within these cells. C_4 plants are characterized by a radial leaf anatomy (Kranz anatomy) in which one cell type, the mesophyll cells, surrounds the other type, bundle sheath cells. This arrangement of the cell types and the division of labor between them is central to the functioning of the C_4 pathway. Carbon dioxide is first captured in the outer tissues (mesophyll) and then transported to the inner tissues (bundle sheath) where CO_2 and reducing power (as NADPH) are released. The bundle sheath chloroplasts are the exclusive site of the RPP cycle, and the CO_2 pressure therein is raised, allowing CO_2 to compete effectively with O_2 at the catalytic site of rubisco. This in turn minimizes phosphoglycolate production, the principal substrate for photorespiration in leaves. Furthermore, any CO_2 produced by photorespiration in C_4 plants would have to find its way out past the mesophyll cells where it would be recaptured by the C_4 carboxylation reaction. The ability of C_4 plants to restrict photorespiration in this way appears to be the principal factor in allowing them to flourish in conditions of bright light and warm temperatures.

The initial step in the C_4 pathway is the carboxylation of phospho*enol*pyruvate (PEP) in the cytoplasm of the mesophyll cells. The reaction is catalysed by PEP carboxylase (Eqn. 4).

$$\text{PEP} + HCO_3^- \rightarrow \text{oxaloacetate} + P_i \qquad (4)$$

As for the rubisco step of the RPP cycle, the PEP carboxylase reaction is virtually irreversible ($\Delta G' = -35.6$ kJ) and the enzyme has a very high affinity for CO_2 (as bicarbonate).

Subsequent metabolism of oxaloacetate (OAA) varies according to species. Three main types of C_4 pathway are recognized, of which the most extensively studied is that shown by plants such as *Zea mays* (corn) (Fig. 2). In these plants (here called type-1 C_4 plants) OAA is reduced to malate via NADP-malate dehydrogenase in mesophyll chloroplasts. Malate is then transported to bundle sheath chloroplasts and oxidatively decarboxylated by NADP-malic enzyme to produce pyruvate, CO_2 and NADPH. Pyruvate is recycled to the mesophyll cells while the CO_2 and NADPH are used in the RPP cycle in the bundle sheath chloroplast. The original C_3 carbon acceptor (PEP) is regenerated from pyruvate in the mesophyll chloroplast by the activity of pyruvate, P_i dikinase [8] (Eq. 5).

$$\text{Pyruvate} + \text{ATP} + P_i \rightarrow \text{PEP} + \text{AMP} + \text{pyrophosphate} \qquad (5)$$

Although the bundle sheath chloroplasts contain all the enzymes of the RPP cycle, there is now evidence that some of the 3-PGA formed by the activity of rubisco is exported to the mesophyll cells [9]. Bundle sheath chloroplasts of maize are deficient in photosystem II activity and apparently cannot produce sufficient NADPH to reduce all of the 3-PGA formed to triose phosphate. Responsibility for this step is thus shared with the mesophyll chloroplasts which recycle triose phosphate to the bundle sheath as DHAP. This transport of 3-PGA from bundle sheath to mesophyll permits the synthesis of sucrose in the mesophyll cell cytoplasm. The evidence suggests that the mesophyll cells are the major site of sucrose synthesis [10–13]. Sucrose phosphate synthetase, one of the regulatory enzymes of sucrose synthesis and fructose 6-phosphate,2-kinase (Fru-6-P,2K), the enzyme synthesizing fructose 2,6-bisphosphate — a potent regulator of cytoplasmic sucrose synthesis (see Section 5.4.1) — are both almost completely confined to the mesophyll cells.

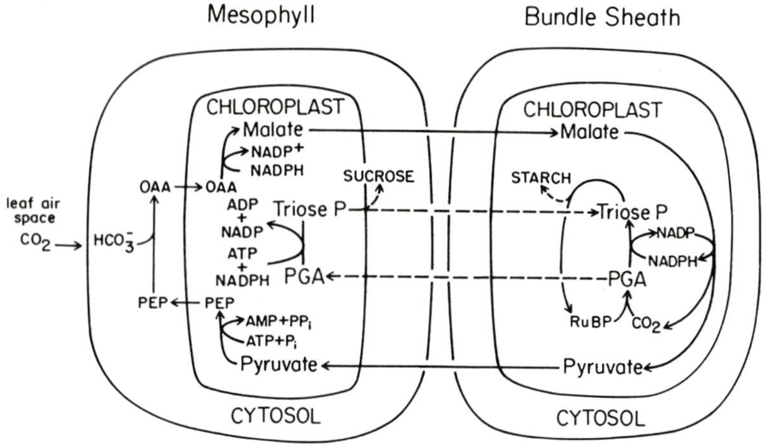

Fig. 2. The C_4 cycle of CO_2 fixation in photosynthesis. The pathway shown is that occurring in Type-1 C_4 plants such as *Zea mays*. Abbreviations: RuBP, ribulose 1,5-bisphosphate; PGA, 3-phosphoglycerate; PEP, phospho*enol*pyruvate; OAA, oxaloacetate. The partial triose-P/PGA shuttle is based primarily on evidence demonstrating concentration gradients that would support metabolite flux between the two cell types.

Two other types of C_4 pathways are recognized. In type-2 plants, (*Atriplex spongiosa)* and type-3 (*Panicum maximum*) plants, malate is replaced by aspartate as the major C_4 acid transported to the bundle sheath cells. After transport, aspartate is converted to OAA by transamination. In type-2 plants, OAA is reduced to malate, which in turn is decarboxylated by NAD-malic enzyme in the bundle sheath cell mitochondria to give NADH, CO_2 and pyruvate. In type-3 plants, OAA is decarboxylated in the cytosol by PEP carboxykinase in the presence of ATP, yielding PEP, CO_2 and ADP. The return of carbon to the mesophyll cells for regeneration of the CO_2 acceptor occurs as pyruvate (or alanine to maintain nitrogen balance) in type-2 and as PEP (or again perhaps as alanine) in type-3. These variations in the C_4 pathway are summarized in Table 1 (see also Ref. 14).

Although in each type of C_4 pathway there is an initial carboxylation catalyzed by PEP carboxylase, the plant's ability to produce a net increase in fixed carbon depends on subsequent release of CO_2 and refixation by the RPP cycle. In this sense, the RPP cycle is still the fundamental carboxylating mechanism of these plants. It should be noted that C_3 plants also contain a cytosolic PEP carboxylase which is capable of fixing CO_2. However, C_3 plants lack the biochemical and structural specialization as well as the division of labor between cell types that make possible the classical C_4 type of photosynthesis.

4. *Crassulacean acid metabolism*

Although first discovered in species of the Crassulaceae, the presence of CAM is now well established in various families of higher plants with succulent stems or leaves [15].

TABLE 1

Decarboxylation of C_4 acids in representative C_4 species

Type	Type of bundle sheath decarboxylase	Major substrate moving from	
		mesophyll to bundle sheath cells	bundle sheath to mesophyll cells
1. *Zea mays*	NADP-malic enzyme[a]	Malate	Pyruvate
2. *Atriplex spongiosa*	NAD-malic enzyme[b]	Aspartate	Alanine/pyruvate
3. *Panicum maximum*	PEP carboxykinase[c]	Aspartate	Alanine/PEP

[a]Chloroplastic; [b]mitochondrial; [c]cytosolic.

CAM employs a biochemical strategy similar to C_4 plants in that CO_2 is first fixed by carboxylation of PEP to produce malate. The malate is later decarboxylated, and the resulting CO_2 is refixed by the RPP cycle. The difference between the CAM and C_4 strategies lies in the separation of PEP carboxylation from the RPP cycle. In C_4 plants the two processes are separated spatially (mesophyll cells and bundle sheath cells), whereas CAM plants separate the PEP carboxylation from the RPP cycle temporally (night and day).

As discussed above, the spatial separation of these processes in C_4 plants necessitates a degreee of structural organization in the form of Kranz anatomy. CAM plants do not show such anatomy but have other specializations because the temporal separation of the synthesis and decarboxylation of C_4 acids requires storage of large amounts of C_4 acids in the vacuole.

The diurnal cycle of CAM (Fig. 3) can be considered to begin with CO_2 fixed at night by PEP carboxylase, which, as in C_4 plants, is located in the cytoplasm. The bulk of the OAA resulting from this carboxylation is reduced to malate by NAD-malate dehydrogenase. Malate is considered the end product of nocturnal CO_2 fixation and is largely stored in the vacuole, so as to provide a substrate for decarboxylation during the day. Under normal conditions, most of the CO_2 consumed in nocturnal CO_2 fixation is derived from the air and is taken up through open stomata. Respired CO_2 is also, however, taken up and under stress conditions, which cause stomatal closure, may be significant [17]. It is now believed that the major source of PEP, the substrate for dark CO_2 fixation, is the glycolytic breakdown of reserve glucan (mainly starch).

The mode of malate utilization during the day varies according to species and is similar to the variants of C_4 photosynthesis [15]. In the majority of CAM plants, malate released from the vacuole is decarboxylated by NAD(P)-specific malic enzyme to yield CO_2, NAD(P)H and pyruvate. In members of the Liliaceae, Bromeliaceae, Asclepiadaceae and some CAM species of other families, malate is oxidized to OAA which undergoes decarboxylation by PEP carboxykinase, probably in the cytosol [15].

The further metabolism of pyruvate or PEP is still under investigation. Increasing support is available for the view that the C_3 residue from decarboxylation is converted, by a reversal of glycolysis, into a storage carbohydrate which later serves

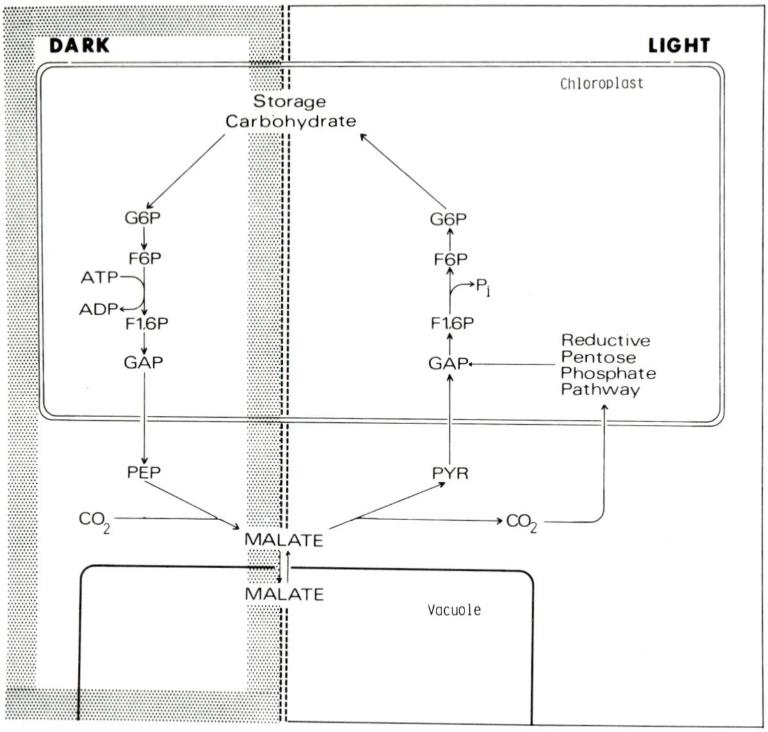

Fig. 3. Carbon flow during Crassulacean acid metabolism (CAM). The simplified pathway shown is that occurring in malic enzyme type plants. The location of the decarboxylation reaction is believed to be the mitochondria (NAD-malic enzyme type) or the cytosol [16] or chloroplast (NADP-malic enzyme type) [15]. Abbreviations: G6P, glucose 6-phosphate; F6P, fructose 6-phosphate; F16P, fructose 1,6-bisphosphate; GAP, glyceraldehyde 3-phosphate; PEP, phospho*enol*pyruvate; PYR, pyruvate.

as the carbon source for nocturnal production of PEP for carboxylation [17]. In some plants, however, at least part of the C_3 pool may be broken down to CO_2 by glycolysis and the tricarboxylic acid cycle.

The CO_2 formed by C_4 acid decarboxylation and by oxidation of the resultant C_3 residue is refixed in the chloroplast by the RPP cycle which, as in C_3 plants, is driven by ATP and NADPH produced by the light reactions of photosynthesis. In some species reserve glucan formation in the chloroplasts, from triose phosphate produced by the RPP cycle, contributes to the carbohydrate pool which later is used in the synthesis of PEP, the carbon acceptor for nocturnal CO_2 fixation. Since the CO_2 source for fixation by the RPP cycle is endogenous, typical CAM plants are able to close their stomata during the day. This confers two advantages. Firstly, the loss of water is severely restricted during the heat of the day. Secondly, the internal leaf CO_2 concentration may rise to as much as 4% [17], strongly favoring the carboxylation reaction of rubisco at the expense of oxygenation which would otherwise lead to photorespiratory losses.

Certain features of CAM such as the diurnal rhythm in gas exchange and C_4 acid

formation may undergo considerable altera
to varying photoperiod and water stress. S
shift completely from normal C_3 pathways to
stress. Thus, like C_4 photosynthesis, CAM is
ultimately depends on the RPP cycle for t
suitable for growth and respiration.

5. *Regulation of the reductive pentose*

The principal and ultimate regulator of chl
light. In fulfilling its regulatory role, light is
converted to regulatory signals that modulat[illegible] enzymes. Such regulation is essential because enzymes for degrading carbohydrates coexist in chloroplasts with enzymes of carbohydrate synthesis. Selected biosynthetic enzymes are light-activated, whereas degradative enzymes are light-deactivated. In this way chloroplasts minimize the concurrent functioning of enzymes or pathways that operate in opposing directions ('futile cycling') and thereby maximize the efficiency of temporally disparate metabolic processes. The regulatory function of light thus maintains 'enzyme order' by ensuring that carbon dioxide assimilation takes place during the day and carbohydrate degradation occurs primarily at night [18,19]. Through the provision of DHAP, formed either from newly fixed carbon or the breakdown of stored starch, chloroplasts are able to supply carbon for the cytosolic synthesis of sucrose — the transport carbohydrate in most plants — and thereby meet the energy needs of nonphotosynthetic (heterotrophic) tissues at all times.

5.1. Identification of the sites of regulation

It is believed that the sensitivity of a metabolic pathway to regulation resides principally in only a small number of the total steps in the pathway [20]. Such regulatory steps characteristically have large, negative free-energy changes (ΔG) and thus are essentially irreversible. The physiological free energy changes (ΔG^s) for the reactions of the RPP cycle were calculated by Bassham and Krause [21] from measurements of the steady-state levels of radioactive compounds in photosynthesizing *Chlorella*. The reactions shown to be substantially displaced from equilibrium and therefore potential sites of metabolic regulation were those catalysed by rubisco, fructose 1,6-bisphosphatase (FBPase), sedoheptulose 1,7-bisphosphatase (SBPase) and phosphoribulokinase (PRK).

Further evidence as to the importance of these sites in the regulation of the pathway comes from the analysis of light–dark and dark–light transient changes in levels of metabolites. It would be expected that increasing flux through a regulated step would lead to depletion of the pathway substrate for that step and that decreasing flux would lead to a rise in the concentration of that substrate. The kinetic analysis of such experiments is complicated by the cyclic nature of the path-

.ction of substrate for one reaction may be affected by the sequent step. Nevertheless, analysis confirms that the reactions .isco, FBPase, SBPase and PRK are of greatest significance in con- .lux through the RPP pathway (see Ref. 1 for review).

.rast, the enzymes involved in the reduction of 3-PGA to triose phosphate .er catalyse a freely reversible oxidation/reduction, the direction of which, in .o, is largely determined by the levels of ATP and ADP, NADPH and NADP. In the light, with high levels of ATP and of NADPH the reactions proceed in the direction of triose phosphate driven by the production of 3-PGA and consumption of triose phosphate. In steady-state photosynthesis this provides for a coordination of the activity of parts of the cycle. Any component tending to increase the activity of PRK, for example, will cause the consumption of ATP and production of ADP. This in turn will slow the rate of 3-PGA reduction, leading to decreased synthesis of Rbu-5-P, bringing the cycle back into balance.

5.2. *Mechanisms of regulation*

5.2.1. *Regulation of ribulose-1,5-bisphosphate carboxylase oxygenase*

The capacity of rubisco to carboxylate Rbu-1,5-P_2 is determined by the concentration of substrates (Rbu-1,5-P_2, CO_2 and O_2), and the amount and activity of enzyme. Under conditions of low CO_2 and high light, it is possible to show a direct correlation between the rubisco content and CO_2 fixation of spinach leaves [22]. During short-term changes in the rate of photosynthesis, however, modulation of the activity of the enzyme occurs [23,24]. Activation of the enzyme involves the formation of a complex with CO_2 and the subsequent addition of a divalent metal ion (Mg^{2+}, in vivo) to form the activated ternary complex [25]. The equilibrium of this reaction is sensitive to pH, and low pH in the stroma would be expected to lead to deactivation. Upon illumination, protons move rapidly from the stromal compartment into the thylakoids, causing an increase in stromal pH from about 7.0 to 8.0. The efflux of H^+ is countered by an influx of other cations, possibly including Mg^{2+} [26] and thus both the pH and Mg^{2+} concentration in the stroma have been proposed as being favorable for rubisco activation in the presence of CO_2. Two other mechanisms for the activation/deactivation of rubisco have recently been reported. Work with a mutant of *Arabidopsis* that requires a high CO_2 concentration for growth has led to the proposal of a mechanism whereby rubisco is activated by light [27]. The mechanism, which presently appears to be unrelated to other systems of enzyme regulation, involves a newly identified protein, rubisco activase, that links light to enzyme activity. While details of the activation mechanism are yet to be established, it presently appears that light-induced changes in the electrochemical potential of thylakoid membranes are involved. Such a mechanism for the regulation of rubisco by light could explain results obtained over the years by a number of different laboratories [18,28,29].

A second novel mechanism of rubisco regulation involves a phosphorylated inhibitor of catalysis which can occupy the catalytic site of the enzyme. The discovery of this inhibitor [30,31] followed the observation that rubisco extracted from formation may undergo considerable alterations during the seasons or in response to varying photoperiod and water stress. Some plants, 'facultative CAM plants', shift completely from normal C_3 pathways to CAM in the response to salt and water stress. Thus, like C_4 photosynthesis, CAM is a secondary process in which the plant ultimately depends on the RPP cycle for the net production of reduced carbon suitable for growth and respiration.

5. *Regulation of the reductive pentose phosphate cycle*

The principal and ultimate regulator of chloroplast carbohydrate metabolism is light. In fulfilling its regulatory role, light is absorbed by chlorophyll and is then converted to regulatory signals that modulate selected enzymes. Such regulation is essential because enzymes for degrading carbohydrates coexist in chloroplasts with enzymes of carbohydrate synthesis. Selected biosynthetic enzymes are light-activated, whereas degradative enzymes are light-deactivated. In this way chloroplasts minimize the concurrent functioning of enzymes or pathways that operate in opposing directions ('futile cycling') and thereby maximize the efficiency of temporally disparate metabolic processes. The regulatory function of light thus maintains 'enzyme order' by ensuring that carbon dioxide assimilation takes place during the day and carbohydrate degradation occurs primarily at night [18,19]. Through the provision of DHAP, formed either from newly fixed carbon or the breakdown of stored starch, chloroplasts are able to supply carbon for the cytosolic synthesis of sucrose — the transport carbohydrate in most plants — and thereby meet the energy needs of nonphotosynthetic (heterotrophic) tissues at all times.

5.1. *Identification of the sites of regulation*

It is believed that the sensitivity of a metabolic pathway to regulation resides principally in only a small number of the total steps in the pathway [20]. Such regulatory steps characteristically have large, negative free-energy changes (ΔG) and thus are essentially irreversible. The physiological free energy changes (ΔG^s) for the reactions of the RPP cycle were calculated by Bassham and Krause [21] from measurements of the steady-state levels of radioactive compounds in photosynthesizing *Chlorella*. The reactions shown to be substantially displaced from equilibrium and therefore potential sites of metabolic regulation were those catalysed by rubisco, fructose 1,6-bisphosphatase (FBPase), sedoheptulose 1,7-bisphosphatase (SBPase) and phosphoribulokinase (PRK).

Further evidence as to the importance of these sites in the regulation of the pathway comes from the analysis of light–dark and dark–light transient changes in levels of metabolites. It would be expected that increasing flux through a regulated step would lead to depletion of the pathway substrate for that step and that decreasing flux would lead to a rise in the concentration of that substrate. The kinetic analysis of such experiments is complicated by the cyclic nature of the pathway, since the production of substrate for one reaction may be affected by the regulation of a subsequent step. Nevertheless, analysis confirms that the reactions catalysed by rubisco, FBPase, SBPase and PRK are of greatest significance in controlling the flux through the RPP pathway (see Ref. 1 for review).

In contrast, the enzymes involved in the reduction of 3-PGA to triose phosphate together catalyse a freely reversible oxidation/reduction, the direction of which, in vivo, is largely determined by the levels of ATP and ADP, NADPH and NADP. In the light, with high levels of ATP and of NADPH the reactions proceed in the direction of triose phosphate driven by the production of 3-PGA and consumption of triose phosphate. In steady-state photosynthesis this provides for a coordination of the activity of parts of the cycle. Any component tending to increase the activity of PRK, for example, will cause the consumption of ATP and production of ADP. This in turn will slow the rate of 3-PGA reduction, leading to decreased synthesis of Rbu-5-P, bringing the cycle back into balance.

5.2. Mechanisms of regulation

5.2.1. Regulation of ribulose-1,5-bisphosphate carboxylase oxygenase

The capacity of rubisco to carboxylate Rbu-1,5-P_2 is determined by the concentration of substrates (Rbu-1,5-P_2, CO_2 and O_2), and the amount and activity of enzyme. Under conditions of low CO_2 and high light, it is possible to show a direct correlation between the rubisco content and CO_2 fixation of spinach leaves [22]. During short-term changes in the rate of photosynthesis, however, modulation of the activity of the enzyme occurs [23,24]. Activation of the enzyme involves the formation of a complex with CO_2 and the subsequent addition of a divalent metal ion (Mg^{2+}, in vivo) to form the activated ternary complex [25]. The equilibrium of this reaction is sensitive to pH, and low pH in the stroma would be expected to lead to deactivation. Upon illumination, protons move rapidly from the stromal compartment into the thylakoids, causing an increase in stromal pH from about 7.0 to 8.0. The efflux of H^+ is countered by an influx of other cations, possibly including Mg^{2+} [26] and thus both the pH and Mg^{2+} concentration in the stroma have been proposed as being favorable for rubisco activation in the presence of CO_2. Two other mechanisms for the activation/deactivation of rubisco have recently been reported. Work with a mutant of *Arabidopsis* that requires a high CO_2 concentration for growth has led to the proposal of a mechanism whereby rubisco is activated by light [27]. The mechanism, which presently appears to be unrelated to other systems of enzyme regulation, involves a newly identified protein, rubisco activase, that links light to enzyme activity. While details of the activation mechanism are yet to be established, it presently appears that light-induced changes in the electrochemical potential of thylakoid membranes are involved. Such a mechanism for the regulation of rubisco by light could explain results obtained over the years by a number of different laboratories [18,28,29].

A second novel mechanism of rubisco regulation involves a phosphorylated inhibitor of catalysis which can occupy the catalytic site of the enzyme. The discovery of this inhibitor [30,31] followed the observation that rubisco extracted from

Phaseolus leaves in the light was significantly more active than from darkened leaves, despite optimal in vitro activation of the enzyme with CO_2 and Mg^{2+}. Several studies have shown that phosphorylated compounds can be effective inhibitors of rubisco in vitro. The results with *Phaseolus*, however, are the first to document the importance in vivo of a compound which is light-modulated and present in sufficient amounts to reduce dark enzyme activity to close to zero [22].

5.2.2. *The ferredoxin/thioredoxin system*

Light regulates specific enzymes via a number of mechanisms [18,19,28,29,32,33]. Important among these is the ferredoxin/thioredoxin system, involving ferredoxin, ferredoxin-thioredoxin reductase (FTR), and a thioredoxin. Thioredoxins are proteins typically with a molecular weight of 12000 that are widely, if not universally, distributed in the animal, plant and bacterial kingdoms. Thioredoxins undergo reversible reduction and oxidation through changes in a disulfide group (S–S → 2 SH). In the ferredoxin/thioredoxin system, a thioredoxin (Td) is reduced via the iron-sulfur protein FTR, by ferredoxin (Fd), which itself is reduced by the chlorophyll system of illuminated chloroplast thylakoid membranes (Eqns. 6 and 7).

$$4\,Fd_{ox} + 2\,H_2O \xrightarrow{\text{light}} 4\,Fd_{red} + O_2 + 4H^+ \tag{6}$$

$$2\,Fd_{red} + Td_{ox} \xrightarrow{\text{FTR}} 2\,Fd_{ox} + Td_{red} \tag{7}$$

Two different thioredoxins, designated thioredoxin *f* and thioredoxin *m*, are a part of the ferredoxin/thioredoxin system in oxygenic photosynthetic organisms [18,34–36]. In the reduced state, the two thioredoxins selectively activate enzymes of carbohydrate biosynthesis, including FBPase, SBPase and PRK. In addition, thioredoxins have been shown to activate NADP-G3PDH – and deactivate glucose-6-phosphate dehydrogenase (G6PDH) [18,19], a key enzyme of the oxidative pentose phosphate pathway, the alternative route of carbohydrate degradation besides glycolysis. The ferredoxin/thioredoxin system also functions in chloroplasts in regulating other enzymes such as NADP-malate dehydrogenase (NADP-MDH) [18,19,28,37] and the 'coupling factor' (CF_1-ATPase) [37,38]. The type of thioredoxin interacting with each of these chloroplast enzymes is shown in Fig. 4. Cyanobacteria, C_3, C_4 and Crassulacean acid metabolism (CAM) plants have been shown to utilize the ferredoxin/thioredoxin system in enzyme regulation for these processes.

The ferredoxin/thioredoxin system functions by changing the sulfhydryl status of target enzymes. NADP-MDH, which catalyses the synthesis of malate in chloroplasts of C_3 and (especially) C_4 plants, is activated by a net transfer of reducing equivalents (hydrogen) from reduced thioredoxin to enzyme disulfide (S–S) groups, thereby yielding oxidized thioredoxin *m* and reduced (SH) enzyme [32,37]. It is thought that deactivation of NADP-MDH takes place through the oxidation (in the dark) of SH groups on reduced thioredoxin and the reduced (activated) enzyme. There is evidence that this light-dependent reduction mechanism also per-

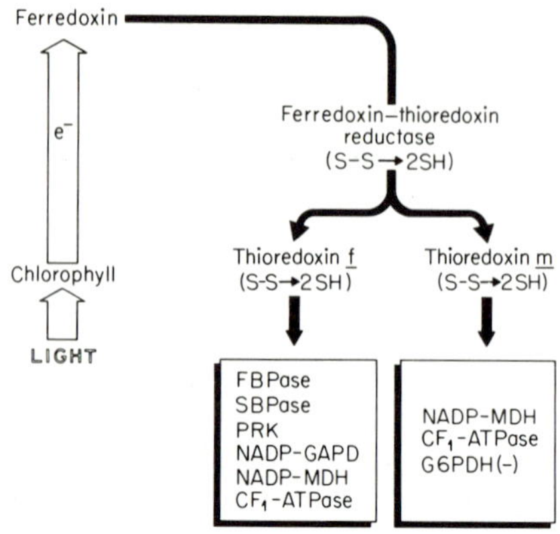

Fig. 4. Enzymes regulated by the ferredoxin/thioredoxin system. The role of an FTR S–S group in the reduction of thioredoxins is based on unpublished findings of Droux, Miginiac-Maslow, Jacquot, Gadal and Buchanan. The role of thioredoxins in regulating phosphoglycerate kinase of C_4 mesophyll cells is not indicated.

tains to the activation of FBPase, although disulfide–sulhydryl exchange may be involved in this case.

Another mechanism of light-dependent enzyme activation has been proposed in which a membrane-bound dithiol-containing factor (light-effect mediator or LEM) reduced by the photosynthetic electron transport system reductively activates regulated enzymes in the chloroplast [28]. Certain facets of this mechanism may be identical to the ferredoxin/thioredoxin system while other aspects are still the subject of debate [18,33].

In summary, current evidence [39–41] is thus consistent with the view that the ferredoxin/thioredoxin system functions in photosynthetically diverse types of plants as a master switch to restrict the activity of degradatory enzymes and activate biosynthetic enzymes in the light. It is significant that enzymes controlled by the ferredoxin/thioredoxin system (FBPase, SBPase, NADP-G3PDH, and PRK) function in the regenerative phase of the reductive pentose phosphate cycle that is needed to sustain its continued operation – i.e., to regenerate the carbon dioxide acceptor, Rbu-1,5-P_2, from newly formed 3-PGA. It seems likely that one of these thioredoxin-linked enzymes limits the regeneration of Rbu-1,5-P_2.

5.2.3. Coordinate regulation of photosynthetic enzymes

Biochemical processes are generally regulated not by one but by several interacting systems of regulation. From early work, it was concluded that the ferredoxin/thioredoxin system acts jointly with other light-actuated systems in achieving a particular regulatory effect – e.g., light-induced shifts in concentration of metabolite effectors and pH [18,19].

Since those early studies, results from a number of laboratories support such a coordinate function of the different regulatory systems. Noteworthy among the metabolite effector studies are: demonstration of the inhibition of thioredoxin-linked NADP-MDH activation by NADP [42,43], the inhibition of PRK by compounds such as 6-phosphogluconate [44], and the enhancement of thioredoxin-linked FBPase and SBPase activation by substrate (sugar bisphosphate) and divalent cations (Ca^{2+}, Mn^{2+}) [45–47]. Results pertinent to the role of other cations (Mg^{2+}) on the thioredoxin-linked activation of FBPase have also been presented [48]. In short, it appears that the ferredoxin/thioredoxin system functions jointly with mechanisms promoting light-dependent shifts in pH and metabolites in the regulation of a number of chloroplast enzymes.

Considerable debate has centered on the question of whether it is the activity of rubisco or the rate of regeneration of Rbu-1,5-P_2 (governed by the regulatory steps of the rest of the cycle, FBPase, SBPase and PRK) that primarily sets the rate of the RPP cycle and CO_2 fixation in vivo. While this question is still very much an open one [49], recent results suggest that during rapid changes from high to low light, the rate of photosynthesis at subsaturating light intensity in certain plants is determined by the rate of Rbu-1,5-P_2 regeneration and not by the activity of rubisco [50]. Subsequent variation in the activation state of rubisco does, however, occur so as to match the Rbu-1,5-P_2 saturated rate of carboxylase activity to the rate of Rbu-1,5-P_2 regeneration.

6. *Compartmentation and triose phosphate transport*

Because the RPP cycle is an exporter of fixed carbon, regulation of the cycle at several points may be insufficient to prevent the intermediates from being consumed by other metabolic processes. In addition to the biochemical controls discussed above there is also compartmentation.

The chloroplast is encircled by a double membrane called the envelope. Of the two membranes, the inner is practically impermeable to hydrophylic compounds, such as P_i, phosphate esters, dicarboxylates, glucose and sucrose. Transport of certain of these metabolites is accomplished by carrier proteins, specific for groups of compounds. Individual carriers have been shown to facilitate the transport of P_i and phosphate esters, dicarboxylates, ATP and ADP, and glucose.

The carrier protein facilitating P_i and phosphate ester transport is of particular interest in leaves in connection with carbon processing – i.e., the synthesis, transport and degradation of carbohydrate, all of which occur in the cytosol [51]. This metabolite carrier, called the phosphate translocator, is a polypeptide with a molecular mass of 29 kDa and is a major component of the inner envelope membrane [52,53]. The phosphate translocator mediates the counter-transport of 3-PGA, DHAP and P_i. The rate of P_i transport alone is three orders of magnitude lower than with simultaneous DHAP or 3-PGA counter-transport [54]. Consequently operation of the phosphate translocator keeps the total amount of esterified phosphate and P_i constant inside the chloroplast. Significantly, the carrier is specific for the divalent anion of phosphate.

The principal physiological function of the phosphate translocator is to supply the cytosol with fixed carbon in the form of DHAP or 3-PGA. It has been demonstrated that DHAP is the main metabolite released to the cytosol in the light, even if the stromal concentration of 3-PGA is significantly higher than that of DHAP. This light-dependent restriction of 3-PGA transport is seemingly a consequence of the alkaline pH of the stroma which renders 3-PGA trivalent (and immobile) but does not change DHAP, which remains divalent (and freely transportable) during illumination. In the dark, when stromal pH returns to neutrality, the transport of 3-PGA (which is then largely divalent) is significantly increased. Interestingly, the translocator seems not to be under regulation but to respond to concentration gradients by mass action [53]. In addition to 3-PGA and DHAP, glucose, formed in the nocturnal breakdown of starch and transported by the glucose translocator [55], also contributes to the carbohydrate transported from chloroplasts to cytosol. The relative importance of these C_3 and C_6 transport systems in supplying photosynthate to the cytosol, especially in the dark, remains to be determined.

During photosynthesis, chloroplasts convert CO_2, water and P_i to triose phosphates that are exported to the cytosol. Phosphate is therefore a substrate of this process and the continued operation of the RPP cycle is dependent on the utilization of triose phosphate for the synthesis of starch (in the chloroplast) and sucrose (in the cytosol). These synthetic processes release P_i, preventing the level of free P_i in the cell from falling to a concentration where photosynthesis may be limited by its availability. Such a limitation of photosynthesis is observed during O_2-insensitive CO_2 assimilation [56] and is suggested by the increase in CO_2 fixation detected on feeding P_i via the transpiration stream to a cut leaf [57]. It has long been known that isolated chloroplasts require a continuous supply of P_i in order to sustain photosynthesis.

A further ramification of the translocator-mediated exchange of exported triose phosphate and imported P_i pertains to starch synthesis. When cytosolic metabolism and P_i availability are limited, leading to a high 3-PGA/P_i ratio in the chloroplast, starch synthesis will be stimulated. This occurs because ADP-glucose pyrophosphorylase, the major regulatory enzyme in starch synthesis, is strongly activated by 3-PGA and inhibited by P_i [29]. As mentioned above, starch synthesis from triose phosphate will release P_i, relieving to some extent the P_i limitation of CO_2 fixation.

7. Coordination of CO_2 fixation and sucrose synthesis

The requirement of chloroplast photosynthesis for P_i and the release of P_i by sucrose synthesis in the cytosol require that these two processes be closely coordinated. Part of this coordination, as explained above, lies in the characteristics of the triose phosphate translocator. Results obtained in the last few years have led to the identification of a second component serving this function. Fructose 2,6-bisphosphate (Fru-2,6-P_2) coordinates the metabolism of sucrose, starch and CO_2 fixation and, in so doing, links metabolic processes of the chloroplast with those of the cytosol.

7.1. Fructose-2,6-bisphosphate

Fru-2,6-P_2, discovered as a phosphofructokinase 'activation factor' in liver [58], is now accorded a central role in the hormonal regulation of glycolysis and gluconeogenesis in mammalian tissues. Soon after elucidation of its function in animal cells, Fru-2,6-P_2 was reported to occur in plant tissues — etiolated mung beans and spinach leaves [51]. Fru-2,6-P_2 was also shown to activate pyrophosphate,fructose 6-phosphate,1-phosphotransferase (PFP) from these sources. PFP catalyses the reversible phosphorylation of fructose 6-phosphate (Fru-6-P) by pyrophosphate and is believed to be important in the regulation of glycolysis and gluconeogenesis (sucrose synthesis) in plant tissues.

Studies with spinach revealed: (1) Fru-2,6-P_2 is present in the cytosolic fraction of photosynthetic (leaf parenchyma) cells; (2) a PFP that is strongly activated by Fru-2,6-P_2 is present in the cytosol; (3) Fru-2,6-P_2 strongly inhibits cytosolic FBPase, an important regulatory enzyme of sucrose synthesis; and (4) Fru-2,6-P_2 is not present in chloroplasts in significant amounts. The results thus demonstrated that in leaves, as well as in nonphotosynthetic tissues, Fru-2,6-P_2 can affect sucrose metabolism by inhibiting cytosolic FBPase, a key enzyme of sucrose synthesis, and by activating PFP, an enzyme that, because of the reversibility of the reaction it catalyses, can potentially function in both sucrose synthesis and breakdown. Significantly, in contrast to animal systems, there was no large effect of Fru-2,6-P_2 on plant phosphofructokinase (PFK).

Fru-2,6-P_2 may act at points of carbohydrate processing other than FBPase and PFP. A Fru-2,6-P_2-activated UDP-glucose phosphorylase (a newly found enzyme activity) was detected in potato tubers [59], and 6-phosphogluconate dehydrogenase of castor beans was reported to be inhibited by Fru-2,6-P_2 [60]. The physiological significance of these regulatory responses, as well as the recently reported Fru-2,6-P_2-linked activation of phosphoglucomutase [61], remains to be clarified.

In the initial studies on plants (for review see Ref. 51), a substrate-specific fructose-6-phosphate,2-kinase (Fru-6-P,2K) was identified in leaves, specifically in the cytosol fraction. Experiments revealed that leaf Fru-6-P,2K was regulated by metabolite effectors: P_i and Fru-6-P served as activators and 3-phosphoglycerate (3-PGA) and DHAP as inhibitors. Also, an enzyme was partially purified from spinach leaves that selectively hydrolyzed Fru-2,6-P_2 to Fru-6-P and P_i. The enzyme, designated fructose-2,6-bisphosphatase (Fru-2,6-P_2ase), was strongly inhibited by its products, Fru-6-P and P_i. Thus the regulation of Fru-2,6-P_2ase by metabolites was found to be opposite to the regulation of Fru-6-P,2K which, as noted above, is activated by the same metabolites (Fig. 5). An activator metabolite of the leaf Fru-2,6-P_2ase has not yet been found. Fru-6-P,2K and Fru-2,6-P_2ase activities in leaves could not be separated by the purification procedure used. This may indicate that, as in animal tissues, the two enzyme activities are carried by a single protein.

It should be noted that Fru-6-P,2K and Fru-2,6-P_2ase of animal tissues are regulated by phosphorylation via a cAMP-dependent protein kinase that, in turn, is regulated hormonally. So far, there is no evidence that plant Fru-6-P,2K and Fru-

Effector	Fructose-6-P, 2Kinase	Fructose-2,6-bisphosphatase
Fructose-6-P	Activator	Inhibitor
Phosphate	Activator	Inhibitor
Dihydroxyacetone P	Inhibitor	No effect
3-Phosphoglycerate	Inhibitor	No effect

Fig. 5. Metabolite effectors regulating the synthesis and degradation of Fru-2,6-P_2 in leaves.

2,6-P_2ase are regulated by phosphorylation physiologically. However, recent evidence suggests that plant Fru-6-P,2K is regulated covalently in addition to its regulation by metabolites [62,63] though the nature of this covalent mechanism remains to be determined.

As discussed above, the P_i released in sucrose synthesis is recycled to the chloroplast, via the phosphate translocator in strict counter-exchange for triose phosphate [53]. 3-PGA can also be transported by this same carrier but, as also noted above, its export from the chloroplast in the light is restricted.

It is thus obvious that the metabolites modulating Fru-6-P,2K and Fru-2,6-P_2ase occupy strategic positions in the pathway of sucrose synthesis in leaves. Extensive export of triose phosphates by chloroplasts into the cytosolic C_3 pool would lower the Fru-2,6-P_2 concentration (by inhibiting Fru-6-P,2K) and thereby promote the use of photosynthate for sucrose synthesis by relieving the Fru-2,6-P_2-linked inhibition of cytosolic FBPase. On the other hand, elevated levels of P_i (e.g., in the dark) or Fru-6-P (e.g., as sucrose accumulated in the leaf) would tend to raise the Fru-2,6-P_2 concentration and thus restrict sucrose synthesis or favor sucrose degradation. These ideas were substantiated by measuring the levels of Fru-2,6-P_2 and of effector metabolites in spinach leaves in a range of conditions. When the rate of photosynthesis was decreased by lowering the light intensity or the carbon dioxide concentration, there was a 2- to 4-fold increase in the Fru-2,6-P_2 concentration which could be accounted for by the decreasing concentration of triose phosphate in the leaves [51]. On the other hand, when a variety of treatments was used so that sucrose accumulated in the leaves, a 3- to 6-fold increase of Fru-2,6-P_2 resulted (see also Ref. 64). This increase was attributed to an increased hexose phosphate content in the leaves [51], or more specifically to an increase in the Fru-6-P concentration in the cytosol. The role of chloroplasts in controlling the content of Fru-2,6-P_2 in the cytosol via export and import of central metabolites is illustrated diagrammatically in Fig. 6.

7.1.1. Relationship to carbon partitioning

The results described above suggested that Fru-2,6-P_2 could integrate the carbon metabolism of leaves by serving as a regulatory link between chloroplasts and cytosol. As such, it could function to control carbon partitioning — i.e., the conversion to and accumulation of newly fixed carbon as sucrose or starch. Hence, as more photosynthate is made available and the triose phosphate pools (especially

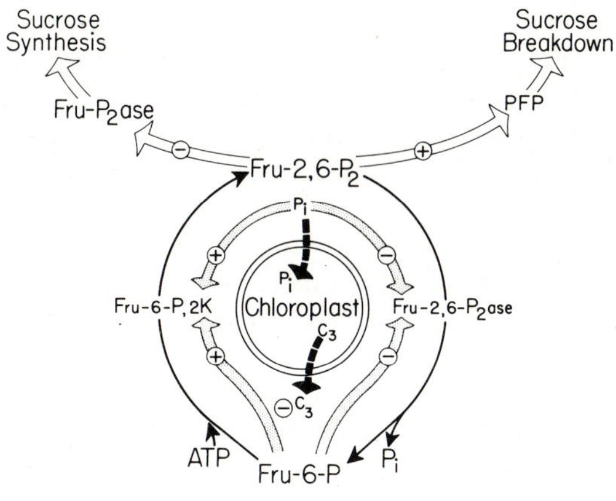

Fig. 6. Role of chloroplasts and effector metabolites of Fru-2,6-P_2-linked control of cytosolic sucrose transformations in spinach leaves. Fru-P_2ase is equivalent to FBPase. Regulation of Fru-6-P,2K and Fru-2,6-P_2ase is indicated by '+' for activation and '−' for inhibition.

DHAP) rise in the cytosol, the cytosolic FBPase is stimulated by the decreased Fru-2,6-P_2 and increased Fru-1,6-P_2. It has been concluded that cytosolic FBPase of leaves is regulated by alterations of Fru-2,6-P_2, AMP and P_i via a network reflecting the energy status of chloroplasts [65]. The subsequent production of hexose phosphate should stimulate sucrose phosphate synthase as this enzyme is activated by glucose 6-phosphate (Glu-6-P) [66]. Thus, changes in DHAP concentration and the accompanying alteration of the Fru-2,6-P_2 concentration provide a feed-forward mechanism to coordinate sucrose synthesis in the cytosol with the rate of carbon dioxide fixation in chloroplasts. This coordination is essential if photosynthesis is to continue, as a large fraction of the triose phosphate produced in chloroplasts must be used to regenerate Rbu-1,5-P_2 to allow the continued function of the RPP cycle.

In addition to this feed-forward mechanism, Fru-2,6-P_2 functions in feedback control of sucrose synthesis. When sucrose accumulates in the leaf, the hexose phosphate concentration increases [51] (the reason for this is still unclear), leading to an activation of Fru-6-P,2K. The resultant increase of Fru-2,6-P_2 then restricts the activity of cytosolic FBPase so that less sucrose, and more starch, is synthesized. In this way, when photosynthesis exceeds the rate at which sucrose can be exported, or stored in the leaf, an increasing proportion of the photosynthate is diverted into starch, which provides a store of carbohydrate that is especially important at night (see above).

8. *Regulation of C_4 photosynthesis*

The regulation of C_4 photosynthesis presumably must satisfy all or most of the conditions that were required for C_3 metabolism: light–dark regulation and adjustment of rate-limiting steps to accommodate the changing physiological needs of the plant such as sucrose synthesis and export versus starch formation in the leaves. In addition, however, the C_4 pathway must be regulated so as to maintain the concentration gradients of metabolites between the bundle sheath and mesophyll cell and to allow for operation of the carboxylating and decarboxylating steps of C_4 metabolism in the light.

A detailed kinetic analysis of the C_4 pathway of metabolism has not been accomplished. Thus it is not possible to identify conclusively the regulatory steps. Recently, however, the development of new methods for the fractionation of maize leaves has allowed the estimation of the concentrations of metabolites in the mesophyll and bundle sheath cells [9]. A number of important conclusions can be drawn from the data. Firstly, the concentration of PEP (3 mM) in the mesophyll cells is well in excess of the estimated K_m of the carboxylation enzyme PEP carboxylase (0.35 mM [67]). As this reaction is believed to be considerably displaced from equilibrium ($\Delta G' = -35.6$ kJ) PEP carboxylase fulfills the criteria for a flux-generating step [68]. There are a number of reports of regulation of PEP carboxylase by metabolites: activation by sugar phosphates and inhibition by organic acids [67]. The physiological significance of these effects on the light-activation of PEP carboxylase is uncertain, though one of the most consistently effective activators, Glu-6-P, is present in the mesophyll cells at 3.0 mM [9], a concentration sufficient to give considerable activation in vitro of PEP carboxylase from other C_4 plants [67].

The enzyme catalysing the synthesis of PEP from pyruvate, ATP and P_i in the mesophyll chloroplast — pyruvate, P_i dikinase (PPDK) — has also been implicated in regulation. The reaction may not be far displaced from equilibrium unless the pyrophosphate produced is rapidly hydrolysed (chloroplasts of maize leaves are thought to contain an active inorganic pyrophosphatase [69]). The extractable activity of PPDK, however, declines rapidly in darkened leaves and recovers upon illumination [70]. The nature of this light-linked activation has been recently clarified [71]. The results show that ADP functions as donor for the phosphorylation of PPDK. Interestingly, a single regulatory protein catalyses both the phosphorylation and dephosphorylation of the dikinase in a reaction rendering the phosphorylated enzyme inactive and the dephosphorylated enzyme active. It is proposed that on darkening there will be a rise in the ADP concentration due to inhibition of photophosphorylation which will stimulate the phosphorylation, and hence inactivation, of pyruvate P_i dikinase.

The decarboxylases used by different C_4 plants (malic enzyme, PEP carboxykinase) are believed to catalyse a non-equilibrium reaction. Little is known, however, about the regulatory characteristics of these enzymes.

It is likely that the regulatory mechanisms discussed in Section 5 apply to the regulation of CO_2 fixation in C_4 plants. In particular it is known that the ferredoxin/thioredoxin system of light-linked enzyme activation (see Section 5.2.2) is present in C_4 plants. NADP-malate dehydrogenase, FBPase and SBPase from maize leaves are regulated in this way [33].

In addition to work with C_3 species, Fru-2,6-P_2 has been studied in maize leaves.

In the initial study, activities catalysing the synthesis and breakdown of Fru-2,6-P_2 were identified and localized in cells isolated from corn leaves. Fru-6-P,2K and Fru-2,6-P_2ase were localized mainly, if not entirely, in the leaf mesophyll cells [13]. Such a distribution of these activities — together with the later finding that Fru-2,6-P_2 shows a similar distribution [11] — aligned cytosolic functions of C_4 mesophyll cells with processes taking place largely in the cytosol of parenchyma cells of C_3 plants, in particular sucrose synthesis [9,10].

The C_4 Fru-6-P,2K and Fru-2,6-P_2ase activities were regulated by metabolite effectors in a manner generally similar to their counterparts in C_3 species [13]. One significant difference was the high concentration of DHAP required for the inhibition of corn Fru-6-P,2K relative to is spinach counterpart ($I_{0.5}$ = 1.0 vs. 0.3 mM). It now seems that this difference in DHAP sensitivity reflects an adaptation of the corn Fru-6-P,2K to its in vivo biochemical environment in which, based on current evidence, DHAP can attain concentrations of 5 mM or even higher [11]. It remains to be seen whether such adaptations are related to the high rates of photosynthesis that C_4 plants are typically capable of achieving.

In summary, it is generally assumed that regulation of C_4 photosynthesis involves most of the mechanisms discussed earlier for C_3 photosynthesis. There are, however, a number of specializations in the light-dark activation and deactivation of the enzymes involved in the initial fixation of CO_2 in the mesophyll and its release in the bundle sheath. Additional controls are required for the enzymes metabolizing compounds which travel down diffusion gradients between the cell types.

9. *Regulation of Crassulacean acid metabolism*

During the light period, when CO_2 is being fixed in the chloroplasts by the RPP pathway, it is likely that the mechanisms discussed above for C_3 photosynthesis are also functional in CAM plants [19]. Additional control mechanisms are expected to provide for the efficient functioning of the diurnal cycle of CO_2 fixation. The functioning of CAM cannot, however, be interpreted solely in terms of enzymology, but rather will involve cellular compartmentation of enzymes and metabolites together with intracellular transport processes [17].

The carboxylation of PEP during nocturnal CO_2 fixation is believed to be considerably displaced from equilibrium and therefore a potential site of regulation. It would be expected that PEP carboxylase activity is high during the night and strongly inhibited by day so as to avoid competition with the RPP pathway. Recent results have clarified previous confusion in the literature on the diurnal variation of PEP carboxylase activity and its sensitivity to allosteric regulation. The enzyme from *Crassula argentea* exists primarily in the form of a tetramer of a 100 kDa subunit at night and as a dimer of the same subunit during the day. The tetrameric enzyme from night leaves is insensitive to malate, while the dimeric form from day leaves is strongly inhibited by malate [72]. During nocturnal CO_2 fixation, malate is rapidly transported to the vacuole and the malate-insensitive form of the enzyme will be active. Glu-6-P and P_i, which are activators of PEP carboxylase, are also reported to be high in CAM-performing cells during the night. During the day it is expected that the malate concentration in the cytoplasm will rise as it is released from the vacuole. The day form of the enzyme (dimer) will be strongly inhibited by malate and thus refixation of CO_2 by PEP carboxylase will be severely restricted. Interestingly, the enzyme which during the day releases CO_2 from malate, NAD-malic enzyme, also exists as a low-activity dimer and high-activity tetramer [73]. The factors which appear to deactivate PEP carboxylase (high pH and malate [74]) are those which turn on the activity of malic enzyme [75]. As in C_4 plants, malic enzyme is likely to catalyse a non-equilibrium step in CAM plants and here this reaction may be important in controlling carbon flux during the day.

Another step in which control is likely of crucial importance is the chloroplast phosphofructokinase (PFK) reaction which functions in nocturnal glycolytic production of PEP from reserve glucan. This enzyme from C_3 plants is strongly inhibited by PEP. In CAM plants, however, PFK is two orders of magnitude less sensitive to inhibition [76,77], thus allowing continued glycolysis in the face of high PEP concentrations. It remains to be seen whether PFK from CAM chloroplasts is inhibited by NADPH, a metabolite seemingly important in light–dark regulation of the enzyme in C_3 species [78].

Although these regulatory mechanisms give a clue as to how the diurnal rhythm of CAM is maintained, many aspects of the process are still poorly understood and insufficient metabolite data are available to pinpoint definitively the regulatory steps in the pathway.

10. Concluding comments

The pathway of carbon dioxide assimilation by the RPP cycle has been known for three decades. During this interval, it has been established that light functions not only to produce ATP and NADPH to drive the cycle, but also to regulate selected enzymes. In oxygen-evolving systems (chloroplasts and cyanobacteria), light absorbed by chlorophyll is converted to several different regulatory signals – changes in pH, metabolite effectors, and sulfhydryl groups – that collectively interact to 'inform' selected enzymes that the light is on (or off) and that their activities should be altered accordingly. In the case of the sulfhydryl changes, the light signal is carried from chlorophyll-containing thylakoid membranes via ferredoxin to thioredoxins, which, through redox changes in their own sulfhydryl groups, bring about changes in the sulfhydryl status of target enzymes, thereby altering key activities and directing major biosynthetic and degradatory pathways in the appropriate direction. With certain enzymes, the light-produced alkalization of the chloroplast stroma and increase in the concentration of selected metabolite effectors enhance the sulfhydryl effects. By linking these regulatory changes to light, the cell is able to be in command of its biosynthetic and degradatory capabilities at all times and to direct available resources to increase growth and survival under a wide range of environmental conditions. It is significant that photosynthetic bacteria (anaer-

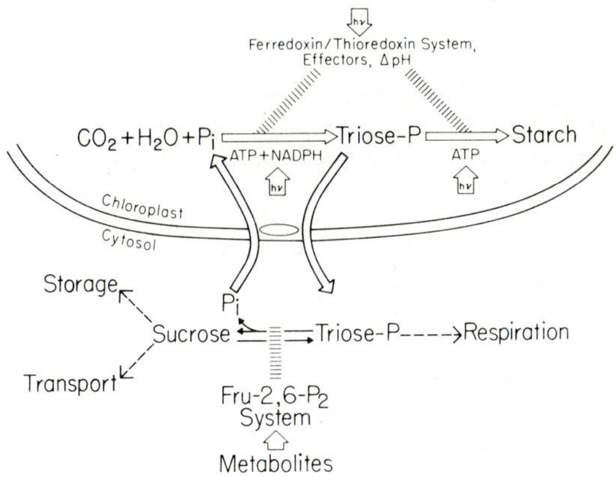

Fig. 7. Relationship of carbon processing in the cytosol to photosynthetic carbon dioxide assimilation in chloroplasts. The dual function of light in supplying ATP and NADPH and in regulation is shown.

obic photosynthetic organisms that lack the ability to evolve oxygen) seemingly do not regulate their metabolic processes in this manner.

In higher plants, which utilize photosynthetically fixed carbon to form transportable sugars such as sucrose, the photomodulation systems of chloroplasts interact with a newly discovered metabolite-directed system of enzyme regulation of the cytosol (Fig.7). Here, Fru-2,6-P_2 plays a key role. In leaves, Fru-2,6-P_2 acts as a regulatory link between chloroplasts and the cytosol, thus (i) allowing metabolic communication between these compartments, and (ii) signalling changes in environmental conditions so that carbon processing – i.e., the synthesis, degradation and transport of carbohydrate in the cytosol – can be adjusted in accord with the plants' needs. In performing its function, Fru-2,6-P_2 acts at several levels, i.e., sucrose synthesis (FBPase), sucrose degradation (PFP regulation), and the related process of carbon partitioning (accumulation of photosynthetically fixed carbon as sucrose versus starch).

Thus, the evidence at hand is in accord with the view that the Fru-2,6-P_2 system connects cytosolic carbohydrate metabolism with the light-directed regulatory mechanisms of chloroplasts, and with other regulatory signals significantly altering cytosolic metabolite status. This role of Fru-2,6-P_2 as an environmental sensor enables plants to make effective use of available energy for processes taking place either in leaves or in distal sink tissues.

Acknowledgements

Work from the authors' laboratory was supported by grants from the National Science Foundation, Competitive Research Grants office of the U.S. Department of Agriculture, National Space and Aeronautics Administration, and Chevron Chemical Company.

References

1 Bassham, J.A. and Buchanan, B.B. (1982) in Photosynthesis (Govindjee, ed.) Vol. II, pp. 141–189, Academic Press, New York.
2 Walker, D.A. (1974) in Plant Carbohydrate Biochemistry (Pridham, J.B., ed.) pp. 7–26, Academic Press, New York.
3 Lorimer, G.H., Andrews, T.J. and Tolbert, N.E. (1973) Biochemistry 12, 18–23.
4 Ogren, W.L. and Chollet, R. (1982) in Photosynthesis (Govindjee, ed.) Vol. II, pp. 191–230, Academic Press, New York.
5 Heber, U. and Krause, G.H. (1980) Trends Biochem. Sci. 5, 32–34.
6 Andrews, T.J. and Lorimer, G.H. (1978) FEBS Lett. 90, 1–9.
7 Pierce, J. and Reddy, G.S. (1986) Arch. Biochem. Biophys. 245, 483–493.
8 Hatch, M.D. and Slack, C.R. (1970) Annu. Rev. Plant Physiol. 21, 141–162.
9 Leegood, R.C. (1985) Planta 164, 163–171.
10 Furbank, R.T., Stitt, M. and Foyer, C.M. (1985) Planta 164, 172–178.
11 Stitt, M. and Heldt, H.W. (1985) Planta 164, 179–188.
12 Usuda, H. and Edwards, G.E. (1980) Plant Physiol. 65, 1017–1022.
13 Soll, J., Wötzel, C. and Buchanan, B.B. (1985) Plant Physiol. 77, 999–1003.
14 Edwards, G.E. and Walker, D.A. (1983) C_3, C_4: Mechanisms and Cellular and Environmental Regulation of Photosynthesis, University of California Press, Berkeley.
15 Kluge, M. and Ting, I. (1978) Crassulacean Acid Metabolism, Springer Verlag, New York.
16 Spalding, M.H., Schmitt, M.R., Ku, S.B. and Edwards, G.E. (1979) Plant Physiol. 63, 738–743.
17 Kluge, M. (1982) in Photosynthesis (Govindjee, ed.) Vol. II, pp. 231–262, Academic Press, New York.
18 Buchanan, B.B. (1980) Annu. Rev. Plant Physiol. 31, 341–374.
19 Buchanan, B.B. (1984) BioScience 34, 378–383.
20 ap Rees, T. (1980) in The Biochemistry of Plants, Vol. III, (Preiss, J., ed.) pp. 1–42, Academic Press, New York.
21 Bassham, J.A. and Krause, G.H. (1969) Biochim. Biophys. Acta 189, 207–221.
22 Seeman, J.R. (1986) in Biological Control of Photosynthesis (Marcelle, R., Clijsters, H. and Van Poucke, M., eds.) pp. 71–82, Martinus Nijhoff, Dordrecht.
23 Perchorowicz, J.T., Raynes, D.A. and Jensen, R.G. (1981) Proc. Natl. Acad. Sci. USA 78, 2985–2989.
24 Taylor, S.E. and Terry, N. (1984) Plant Physiol. 75, 82–86.
25 Miziorko, H.M. and Lorimer, G.H. (1983) Annu. Rev. Biochem. 52, 507–535.
26 Barber, J. (1976) Trends Biochem. Sci. 1, 33–36.
27 Portis, A., Salvucci, M. and Ogren, W. (1985) Photosynth. Res. 7, 193–201.
28 Anderson, L.E. (1979) in Photosynthesis II, Photosynthetic Carbon Metabolism and Related Processes (Gibbs, M. and Latzko, E., eds.) Vol. 6, pp. 271–281, Springer Verlag, Berlin.
29 Preiss, J. (1982) Annu. Rev. Plant Physiol. 33, 431–454.
30 Seemann, J.R., Berry, J.A., Freas, S.M. and Krump, M.A. (1985) Proc. Natl. Acad. Sci. USA 82, 8024–8028.
31 Servaites, J.C., Torisky, R.S. and Chao, S.F. (1984) Plant Sci. Lett. 35, 115–121.
32 Scheibe, R. (1984) Biochim. Biophys. Acta 788, 241–247.
33 Edwards, G.E., Nakamoto, H., Burnell, J.N. and Hatch, M.D. (1985) Annu. Rev. Plant Physiol. 36, 255–286.
34 Schürmann, P., Maeda, K. and Tsugita, A. (1981) Eur. J. Biochem. 116, 37–45.
35 Yee, B.C., de la Torre, A., Crawford, N.A., Lara, C., Carlson, D.E. and Buchanan, B.B. (1981) Arch. Microbiol. 130, 14–18.
36 Tsugita, A., Maeda, K. and Schürmann, P. (1983) Biochem. Biophys. Res. Commun. 115, 1–7.
37 Jacquot, J.-P., Gadal, P., Nishizawa, A.N., Yee, B.C., Crawford, N.A. and Buchanan, B.B. (1984) Arch. Biochem. Biophys. 228, 170–178.
38 Mills, J.D. and Mitchell, P. (1982) FEBS Lett. 144, 63–67.
39 Scheibe, R. (1981) FEBS Lett. 133, 301–304.

of greatest importance, reduced sulfur compounds) characterizes the purple sulfur bacteria (Ectothiorhodospiraceae and Chromatiaceae) and the green sulfur bacteria (Chlorobiaceae). Sulfide and elemental sulfur can be utilized as electron donors by all species of these three families of bacteria, while thiosulfate oxidation is confined to a limited number of Chlorobiaceae species, to the Ectothiorhodospiraceae and to small cell forms of the Chromatiaceae [4]. During the process of sulfide oxidation, elemental sulfur is accumulated either intracellularly (in the Chromatiaceae) or extracellularly (in the Ectothiorhodospiraceae and Chlorobiaceae) [15]. In the former case, sulfide is completely oxidized to sulfate, while in the latter sulfide oxidation is often incomplete.

The division of the five families of phototrophic bacteria, described above on the basis of organotrophic versus lithotrophic growth, is not without some exceptions. Thus, some purple non-sulfur bacteria [16–19] and the green gliding bacterium *Chloroflexus aurantiacus* [20] can use sulfide and/or thiosulfate as electron donors. Similarly, some of the organisms described above as lithotrophic are able to couple the oxidation of organic compounds to the reduction of NAD^+ (vide infra). However, these purple sulfur bacteria more commonly utilize organic compounds, present simultaneously in the growth medium with reduced sulfur compounds, as carbon sources for photolithoheterotrophic growth rather than as electron donors [4,19]. Completely organotrophic growth is of relatively minor importance in the Chromatiaceae [21]. All Chlorobiaceae species [2,6] and *Chlorohepteron thalassium* [22] are obligately lithotrophic organisms.

Although this volume is devoted to the study of photosynthesis, it should be mentioned briefly that a large number of phototrophic bacteria are facultative chemotrophs and can supply their energy needs by the oxidation of substrates via respiratory pathways in the absence of light. Respiration with O_2 as the terminal electron acceptor is well documented for both the purple non-sulfur bacteria [23] and the green gliding bacteria [7,24] and can occur even in the presence of low light intensities. Recently, chemotrophic growth has been reported for a number of purple sulfur bacteria [25,26]. Aerobic growth in the dark in some cases may be either strictly heterotrophic (e.g., *Ectothiorhodospira shaposhnikovii,* Ref. 27) or strictly lithotropic [4]. While chemolithotrophic growth usually involves reduced sulfur compounds as the respiratory electron donors, chemolithoautotrophy with H_2 as the electron donor has been reported for some purple non-sulfur bacteria [28]. Growth by anaerobic respiration with nitrate replacing O_2 as the terminal electron acceptor has been observed in two species of purple non-sulfur bacteria [29,30] and, although *Rhodopseudomonas capsulata* (recently renamed *Rhodobacter capsulatus*) apparently cannot grow anaerobically with nitrate as a terminal electron acceptor, this purple non-sulfur bacterium does contain a dissimilatory nitrate reductase which can serve as a terminal acceptor during energy-yielding respiration [31,32].

Space limitations preclude a detailed treatment of all known pathways of substrate oxidation and the reader is referred to recent reviews [4,6,19] for such information. Instead of striving for a comprehensive treatment, we have thus opted to discuss only a few such pathways, focusing on those pathways for which some

J. Amesz (ed.) *Photosynthesis*

CHAPTER 9

Substrate oxidation and NAD^+ reduction by phototrophic bacteria

DAVID B. KNAFF and CHARLOTTE KÄMPF

Department of Chemistry and Biochemistry, Texas Tech University, Lubbock, Texas 79409, U.S.A.

1. Introduction

Anoxygenic phototrophic bacteria (anoxyphotobacteria) cannot use water as an electron donor during photosynthesis [1,2]. This is in contrast to the oxygenic photosynthesis carried out by higher plants, algae and cyanobacteria, in which water is oxidized to O_2 (See Chapter 6). The phototrophic bacteria rely instead on the oxidation of reduced inorganic and simple organic compounds to supply the reducing equivalents needed for the biosynthesis of cellular material via such processes as CO_2 fixation (See Chapter 8) and N_2 fixation [3]. The five families of phototrophic bacteria differ widely with respect to their energy and carbon metabolism [4] and can be divided into two major groups on the basis of their principal electron donors. In the organotrophs, organic compounds are the favored electron donors, while the lithotrophs are characterized by the oxidation of reduced sulfur compounds.

Photoorganotrophic growth with a variety of organic compounds is common for the purple non-sulfur bacteria [5,6] and the green gliding bacteria [7]. In these bacteria, oxidation of substrates that are more reduced than cell material requires the presence of a terminal electron acceptor that is more oxidized than the cell material synthesized. Thus, for example, butyrate oxidation occurs only in the presence of CO_2 [8]. The catabolism of several classes of aliphatic organic compounds such as hexoses [9], intermediates of the tricarboxylic acid cycle (e.g., citrate [10]) and amino acids [11] has been thoroughly investigated. Growth with aromatic compounds serving as both electron donors and carbon sources has also been observed in purple non-sulfur bacteria [12,13]. The oxidation of aromatic compounds such as benzoate poses considerable problems for phototrophs since phototrophic growth under anaerobic conditions precludes the use of O_2 to initiate oxidation of aromatic substrates [13]. Some purple non-sulfur bacteria require the presence of reduced sulfur compounds, in addition to organic compounds, for growth but in such cases the reduced sulfur compounds are needed for the biosynthesis of sulfur-containing compounds rather than as electron donors [14].

Photolithotrophic growth [6] with reduced inorganic compounds (e.g., H_2 and,

40 Schürmann, P. and Kobayashi, Y. (1983) in Advances in Photosynthesis Research (Sybesma, C., ed.) Vol. VII, pp. 629–632, Nijhoff/Junk, The Hague.
41 Droux, M., Jacquot, J.-P., Suzuki, A. and Gadal, P. (1983) in Advances in Photosynthesis Research (Sybesma, C., ed.) Vol. VII, pp. 533–536, Nijhoff/Junk, The Hague.
42 Ashton, A.R. and Hatch, M.D. (1983) Arch. Biochem. Biophys. 227, 406–415.
43 Scheibe, R. and Jacquot, J.P. (1983) Planta 157, 548–553.
44 Gardemann, A., Stitt, M. and Heldt, H.W. (1983) Biochim. Biophys. Acta 722, 51–60.
45 Hertig, C.M. and Wolosiuk, R.A. (1980) Biochem. Biophys. Res. Commun. 97, 325–333.
46 Wolosiuk, R.A., Hertig, C.M., Nishizawa, A.N. and Buchanan, B.B. (1982) FEBS Lett. 140, 31–35.
47 Hertig, C.M. and Wolosiuk, R.A. (1983) J. Biol. Chem. 258, 984–989.
48 Rosa, L. and Whatley, F.R. (1984) Plant Physiol. 75, 131–137.
49 Farquhar, G.D., von Caemmerer, S. and Berry, J.A. (1980) Planta 149, 78–90.
50 Mott, K.A., Jensen, R.G. and Berry, J.A. (1986) in Biological Control of Photosynthesis (Marcelle, R., Clijsters, H. and Van Poucke, M., eds.) pp. 33–43, Martinus Nijhoff, Dordrecht.
51 Cséke, C., Balogh, A., Wong, J.H., Buchanan, B.B., Stitt, M., Herzog, B. and Heldt, H.W. (1984) Trends Biochem. Sci. 9, 533–535.
52 Heber, U. and Heldt, H.W. (1981) Annu. Rev. Plant Physiol. 32, 139–168.
53 Flügge, U.I. and Heldt, H.W. (1984) Trends Biochem. Sci. 9, 530–533.
54 Fliege, R., Flügge, U.I., Werdan, K. and Heldt, H.W. (1978) Biochim. Biophys. Acta 502, 232–247.
55 Schafer, G., Heber, U. and Heldt, H.W. (1977) Plant Physiol. 60, 286–289.
56 Sharkey, T.D. (1986) in Biological Control of Photosynthesis (Marcelle, R., Clijsters, H. and Van Poucke, M., eds.) pp. 115–125, Martinus Nijhoff, Dordrecht.
57 Sivak, M.N. and Walker, D.A. (1986) in Biological Control of Photosynthesis (Marcelle, R., Clijsters, H. and Van Poucke, M., eds.) pp. 1–31, Martinus Nijhoff, Dordrecht.
58 Furuya, E. and Uyeda, K. (1980) Proc. Natl. Acad. Sci. USA 77, 5861–5864.
59 Gibson, D.M. and Shine, W.E. (1983) Proc. Natl. Acad. Sci. USA 80, 2491–2494.
60 Miernyk, J.A., MacDougall, P.S. and Dennis, D.T. (1984) Plant Physiol. 76, 1093–1094.
61 Galloway, C.M., Dugger, W.M. and Black, C.C. (1985) Plant Physiol. 79, 920–922.
62 Stitt, M., Cséke, C. and Buchanan, B.B. (1986) Plant Physiol. 80, 246–248.
63 Stitt, M. and Heldt, H.W. (1986) Z. Pflanzenphysiol. 41, 291–296.
64 Huber, S.C. and Bickett, D.M. (1984) Plant Physiol. 74, 445–447.
65 Stitt, M. and Heldt, H.W. (1985) Plant Physiol. 79, 599–608.
66 Doehlert, D.C. and Huber, S.C. (1983) FEBS Lett. 153, 293–297.
67 O'Leary, M.H. (1982) Annu. Rev. Plant Physiol. 33, 297–315.
68 Newsholme, E.A. and Crabtree, B. (1981) Trends Biochem. Sci. 6, 53–56.
69 Simmons, S. and Butler, L.G. (1969) Biochim. Biophys. Acta 172, 150–157.
70 Hatch, M.D. (1981) in Photosynthesis, Vol. IV, Regulation of Carbon Metabolism (Akoyunoglou, G., ed.) pp. 227–236, Balaban International Science Services, Philadelphia, PA.
71 Burnell, J.N. and Hatch, M.D. (1986) Arch. Biochem. Biophys. 245, 297–304.
72 Wu, M.-X. and Wedding, R.T. (1985) Arch. Biochem. Biophys. 240, 655–662.
73 Grover, S.D. and Wedding, R.T. (1984) Arch. Biochem. Biophys. 234, 418–425.
74 Wu, M.-X. and Wedding, R.T. (1985) Plant Physiol. 77, 667–675.
75 Willeford, K.O. and Wedding, R.T. (1986) Plant Physiol. 80, 792–795.
76 Sutton, B.G. (1975) Aust. J. Plant Physiol. 2, 377–387.
77 Sutton, B.G. (1975) Aust. J. Plant Physiol. 2, 389–402.
78 Cséke, C., Nishizawa, A.N. and Buchanan, B.B. (1982) Plant Physiol. 70, 658–661.

Note added in proof. While this article was being typeset, our laboratory obtained evidence that the activities synthesizing and degrading fructose-2,6-bisphosphate in spinach leaves reside on different proteins (Macdonald, F.D., Cséke, C., Chou, Q. and Buchanan, B.B. (1987) Proc. Natl. Acad. Sci. USA, in press).

details are available at the molecular level. Particular emphasis has been placed on comparative aspects of electron transfer pathways in photosynthetic and non-photosynthetic systems.

2. Energy-dependent vs. direct reduction of $NAD(P)^+$

2.1. Purple bacteria

Under conditions of phototrophic growth, where light-dependent cyclic electron flow is the dominant electron transfer pathway and provides the energy required for ATP production [33–35], the role of substrate oxidation in phototrophic bacteria is to provide the reducing equivalents (in the form of NAD(P)H and reduced ferredoxin) required for biosynthesis. Although it had been known since 1960 that cells of various purple bacteria could reduce $NAD(P)^+$ at fairly high rates during illumination [36], a considerable time elapsed before any mechanistic information on the pathway(s) of $NAD(P)^+$ reduction became available. A particular problem in formulating a plausible scheme for NAD^+ reduction in purple bacteria arose from thermodynamic considerations. The ubiquinone and menaquinone primary acceptors that serve as the first reduced species stable for times greater than 1 ns in these bacteria [35,37–39] have E_m (pH 7.0) values between 0 and −100 mV [35,37,38,40,41], considerably more positive than that of the NAD^+/NADH couple (E_m = −320 mV at pH 7.0). Even if an unprotonated semiquinone anion, a stronger reductant than the neutral, protonated semiquinone, were formed at the primary acceptor site [40], direct reduction of $NAD(P)^+$ by the reduced primary acceptor would still be thermodynamically unfavorable. Similar considerations apply to the green gliding bacterium *Cfl. aurantiacus,* where the menaquinone primary acceptor [42,43] has E_m (pH 8.1) = −50 mV [44,45]. It was thus proposed [41] that $NAD(P)^+$ photoreduction in these bacteria occurred via an energy-dependent, 'reverse' electron flow with a high energy state (the protonmotive force, $\Delta\bar{\mu}_{H^+}$, according to the chemiosmotic hypothesis – Refs. 33,34) providing the energy needed to 'pump' electrons uphill from weak reductants (e.g., succinate, E'_m = +30 mV) to NAD^+. As shown in Fig. 1, $\Delta\bar{\mu}_{H^+}$ can be generated either during cyclic electron flow in the light [33–35] or by light-independent ATP hydrolysis catalysed by the reversible, protonmotive F_0F_1-ATPase [41,46]. Early evidence for this proposal came from experiments in which 'chromatophores' (subcellular vesicles derived from the intracytoplasmic membranes of phototrophic bacteria) prepared from the purple non-sulfur bacterium *Rhodospirillum rubrum* were shown to transfer electrons from succinate to NAD^+ in the dark if ATP or pyrophosphate hydrolysis was available to supply energy [47]. ATP-dependent reduction of NAD^+ in the dark was subsequently demonstrated in chromatophores from several other species of purple non-sulfur bacteria [41]. NAD^+ photoreduction in the absence of ATP, as predicted from the scheme in Fig. 1, was shown to be dependent on cyclic electron flow to supply the energy needed to pump electrons uphill from succinate. A number of specific inhibitors of cyclic electron flow blocked

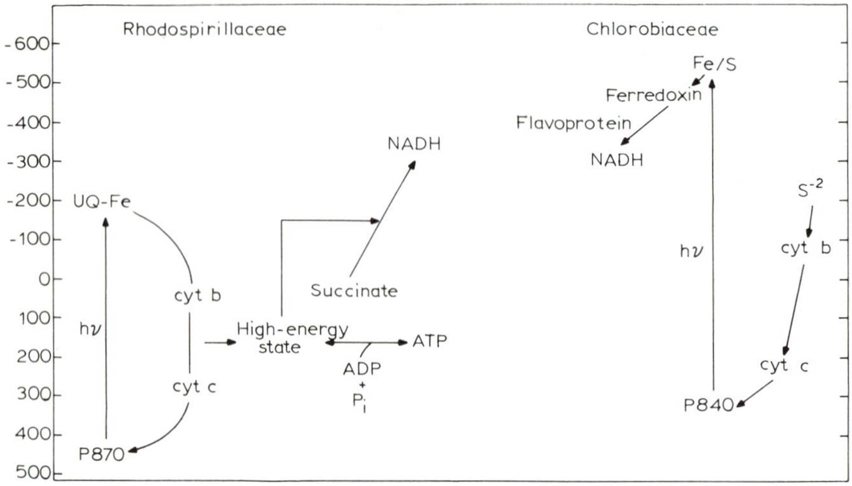

Fig. 1. Mechanism of NAD^+ photoreduction in purple and green sulfur bacteria. UQ-Fe represents the Fe^{2+}-quinone complex present at the primary quinone site of purple bacteria, although in some species menaquinone replaces ubiquinone. Fe/S represents the iron-sulfur center that functions as an early acceptor in green sulfur bacteria. The earliest electron acceptors have been omitted in the case of both green and purple bacteria. The involvement of Cyt *b* in S^{2-} oxidation in green bacteria is speculative but is based on inhibition by antimycin A (Ref. 67).

NAD^+ photoreduction in chromatophores isolated from several purple non-sulfur bacteria [41,48–50] and from the purple sulfur bacterium *Chromatium vinosum* [51] but did not inhibit ATP-driven NAD^+ reduction in the dark.

The most convincing evidence for the energy-dependent nature of NAD^+ reduction in phototrophic purple bacteria comes from the demonstration that uncouplers of oxidative or photosynthetic phosphorylation (agents that collapse $\Delta\bar{\mu}_{H^+}$ by rendering biological membranes permeable to protons but do not affect electron flow directly – Ref. 52) completely eliminate both light-dependent NAD^+ reduction and the ATP-dependent NAD^+ reduction observed in the dark [41,48–51]. This uncoupler sensitivity of $NAD(P)^+$ reduction has been observed not only in chromatophores but also in intact cells [53,54] and with a variety of electron donors, including reduced sulfur compounds (i.e., thiosulfate [54] and sulfide [55]). Another feature shared by both ATP-driven and light-dependent NAD^+ reduction in these phototrophic purple bacteria is the sensitivity of the NAD^+ reduction to inhibitors of mitochondrial NADH dehydrogenase (Complex I) such as rotenone [47,48–51,56], amytal [47] and piericidin A [49]. As inhibition was observed regardless of the electron donor used and these compounds inhibit neither cyclic electron flow nor ATP hydrolysis nor act as uncouplers, it could be concluded that the inhibitors acted on the enzyme actually involved in NAD^+ reduction [41]. The inhibition of NAD^+ reduction in phototrophic purple bacteria by these highly specific inhibitors of mitochondrial NADH dehydrogenase suggests that the mitochondrial and bacterial enzymes may be chemically similar despite the fact that, in vivo, they function predominantly in opposite directions. An additional similarity between the two systems is the presence in membranes of phototrophically grown purple non-sulfur [57–59] and purple sulfur bacteria [51,60] of iron-sulfur clusters with E_m values and EPR spectra similar to those reported for some of the iron-sulfur clusters present in mitochondrial Complex I [61]. An NADH-reducible iron-sulfur cluster has recently been observed [62], using EPR spectroscopy, in membranes isolated from the green gliding bacterium *Cfl. aurantiacus,* suggesting the possibility that this bacterium may also contain an NADH dehydrogenase-like enzyme. The possible presence, in phototrophic bacteria, of an enzyme related to mitochondrial Complex I is of particular interest in the light of recent evidence that these bacteria also contain a cytochrome bc_1 complex [35,59,63,64] similar to that (Complex III) found in mitochondria (Refs. 63, 65; see also Chapter 8).

2.2. *Green sulfur bacteria*

An early indication that a different situation prevailed in the green sulfur bacteria came from the observation that NAD^+ photoreduction, coupled to the oxidation of electron donors such as sulfide, catalysed by a cell-free preparation from *Chlorobium limicola* was not inhibited by uncouplers [66,67] or by rotenone [66]. Furthermore, unlike the case in purple phototrophic bacteria, NAD^+ photoreduction in *Chl. limicola* requires the presence of ferredoxin and a flavin-containing ferredoxin:$NAD(P)^+$ oxidoreductase [66–68]. These results point to a direct (i.e., independent of energy from $\Delta\tilde{\mu}_{H^+}$) mechanism for $NAD(P)^+$ photoreduction in green sulfur bacteria (see Fig. 1). Support for such a direct mechanism comes from the observation that among the early electron acceptors in the reaction centers of green sulfur bacteria is at least one photoreducible iron-sulfur center with $E_m \leqslant -550$ mV [37,69–72]. Photoreduction of this iron-sulfur center during the initial light-dependent reaction of photosynthesis [69] would allow a subsequent, thermodynamically favorable reduction of $NAD(P)^+$ [37,41] in a manner similar to that carried out by PS I of oxygenic photosynthesis [38,39].

3. *Succinate oxidation*

As indicated in Sections 1 and 2, succinate is an electron donor widely utilized for $NAD(P)^+$ reduction by phototrophic purple bacteria. The membrane-bound enzyme responsible for succinate oxidation has been solubilized and partially characterized in the purple non-sulfur bacteria *R. rubrum* [73,74] and *Rhodopseudomonas sphaeroides* (recently renamed *Rhodobacter sphaeroides*) [57]. In situ characterization of the iron-sulfur centers likely to be associated with succinate dehydrogenase has been accomplished for *Rps. capsulata* [59] and *C. vinosum* [51]. Of particular interest is the presence of a succinate-reducible [51,57,58,73] and fumarate-oxidizable [51] iron-sulfur cluster with E_m' near +50 mV that, like center S-3 [60,61,75,76] of mitochondrial succinic dehydrogenase (Complex II), is paramagnetic in the oxidized state. The enzyme in phototrophic bacteria also appears to have one or two ferredoxin-like (i.e., paramagnetic in the reduced state) iron-sulfur centers that correspond to centers S-1 (succinate-reducible, E_m' ranging from

−50 to +120 mV, depending on the bacterial species: Refs. 51,57–59,73,74) and S-2 (not reducible by succinate, with E_m' ranging from −250 to −380 mV: Refs. 57–59, 74) of mitochondrial Complex II, respectively. Additional similarities between the membrane-bound, succinate oxidizing complexes of purple phototrophic bacteria and of mitochondria are the similar peptide subunit compositions [73], the presence of covalently-bound FAD [73] and the sensitivity to the inhibitor 2-theonyltrifluoroacetone [51,58]. It thus appears that the membranes of purple phototrophic bacteria resemble those of eukaryotic mitochondria not only in containing electron transfer complexes similar in structure and function to mitochondrial Complexes III (the cytochrome bc_1 complex) and I (vide supra) but also in utilizing a multi-subunit enzyme similar to mitochondrial Complex II to catalyse the oxidation of succinate to fumarate. As the electron acceptor for mitochondrial Complex II is known to be ubiquinone [61,76,77], it is likely that in phototrophic purple bacteria the electrons originating from the oxidation of succinate are conveyed to ubiquinone. It is known [35,78–81] that the membranes of such bacteria contain large pools of ubiquinone and it is likely that ubiquinol produced by succinate dehydrogenase becomes a part of this pool. As mitochondrial Complex I utilizes ubiquinone as an acceptor of electrons from NADH oxidation [61,82], analogy suggests that this pool ubiquinol serves as the electron donor for the energy-dependent, reverse electron transfer to NAD^+ catalysed by the bacterial analogue of mitochondrial Complex I.

4. *Sulfide oxidation*

Although there is still some uncertainty concerning the pathway by which sulfide is oxidized to sulfate, a likely pathway has been proposed by Trüper and co-workers [6,19]. The proposed sequence in bacteria such as *C. vinosum* includes an initial oxidation of sulfide to sulfite followed by the oxidative conversion of sulfite to adenylsulfate (APS) and, finally, the phosphorylytic cleavage of APS to form sulfate plus ADP:

$$S^{2-} \xrightarrow{\nearrow 6e^-} SO_3^{2-} \xrightarrow[\nearrow \text{AMP}]{\nearrow 2e^-} \text{APS} \xrightarrow[\nearrow P_i]{} \text{ADP} + SO_4^{2-}$$

The initial six-electron oxidation of sulfide to sulfite is catalysed by a soluble, dissimilatory sulfite reductase that contains siroheme and at least one iron-sulfur center as prosthetic groups [83–85]. While similar enzymes in plants and in non-phototrophic bacteria usually function to reduce sulfite to sulfide in assimilatory pathways, the enzyme in photolithoautotrophically grown *C. vinosum* appears to function in the reverse direction, with the electrons from sulfide oxidation being delivered to an as yet unidentified acceptor. Evidence is also available for the presence of APS reductase [86] and ADP sulfurylase [19], the enzymes needed to complete the conversion of sulfite to sulfate, in *C. vinosum* and other purple sulfur bacteria. This pathway involves energy conversion coupled to substrate oxidation, resulting in the net formation of one new pyrophosphate bond in ADP.

While sulfide oxidation in purple sulfur bacteria ultimately results in the formation of sulfate, cells can also oxidize sulfide to elemental sulfur. It appears possible that the oxidation of S^{2-} to S^0 may represent a side, storage pathway rather than serving as a step in the direct oxidation of sulfide to sulfate (vide supra). Recently, considerable information has become available concerning the role of flavocytochrome *c*-552 [86] in catalysing the oxidation of S^{2-} to S^0 in *C. vinosum*. The *C. vinosum* flavocytochrome *c*-552 contains two non-equivalent subunits, one of which (M_r = 46 000) contains a single covalently bound FAD, while the other (M_r = 21 000) contains two *c*-type hemes [87]. Fukumori and Yamanaka initially demonstrated that the protein displayed sulfide:cytochrome *c* oxidoreductase activity with a number of *c*-type cytochromes, including equine mitochondrial cytochrome *c*, serving as electron acceptors and proposed, on the basis of these in vitro results, that flavocytochrome *c*-552 was responsible for the oxidation of S^{2-} to S^0 observed in vivo. Support for this hypothesis came from the demonstration [88] that, as is the case in vivo, elemental sulfur is the major product of the oxidation of S^{2-} catalysed by flavocytochrome *c*-552 in vitro. Further support for the proposal of Fukumori and Yamanaka came from the discovery that *C. vinosum*, previously thought not to contain a soluble *c*-type cytochrome similar to the mitochondrial, does indeed contain such a cytochrome [89–92]. The soluble *C. vinosum* Cyt *c*-550 (M_r = 15 000; E'_m = +240 mV), which can be replaced by equine Cyt *c* in an in vitro *C. vinosum* cyclic electron flow assay system [90], can also function as an electron acceptor in the flavocytochrome *c*-552-catalysed oxidation of S^{2-} [93].

Evidence from gel-filtration chromatography [88,94], affinity chromatography [93] and cross-linking [95] studies suggested that *C. vinosum* flavocytochrome *c*-552 forms an electrostatic complex with *C. vinosum* Cyt *c*-550 [93], equine Cyt *c* [88,93–95] and other *c*-type cytochromes that can serve as electron acceptors in the flavocytochrome *c*-552-catalysed oxidation of S^{2-}. Initial studies on complex formation between flavocytochrome *c*-552 and Cyt *c* had provided evidence that the complex was involved in the oxidation of S^{2-} [88,93] and that lysine residues on Cyt *c* [88,93] provided the positive charges for electrostatic interaction with as yet unidentified negatively charged groups on the heme subunit of flavocytochrome *c*-552 [93]. More recent cross-linking studies [95] support this conclusion and studies with Cyt *c* derivatives modified at single, specific lysine residues have identified 5 lysines on the ‘front’ side of Cyt *c* (see Fig. 2A), surrounding the cytochrome's exposed heme edge [96], that are involved in positioning the two proteins in the catalytically active complex [97]. Complex formation with *C. vinosum* flavocytochrome *c*-552 also protects these front-side lysine residues from covalent modification by acetic anhydride but, as shown in Fig. 2B, does not protect ‘back-side’ lysines [97]. Of considerable interest, in the light of other similarities (described above) between electron transfer reaction in mitochondria and in phototrophic bacteria, is the fact that essentially the same lysines involed in binding Cyt *c* to the

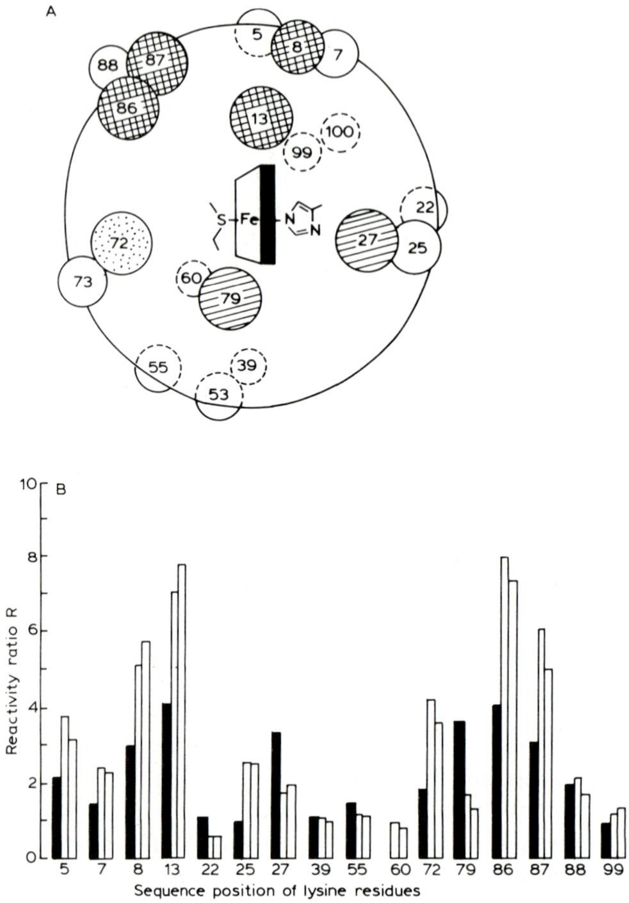

Fig. 2. (A) A schematic diagram of equine Cyt *c* from the front of the heme crevice. The approximate positions of the β-carbons of the lysine residues are indicated by closed and dashed circles for residues located toward the front and back of the molecule, respectively. Differential chemical modification indicates that some residues are protected by both flavocytochrome *c*-552 and mitochondrial redox partners (cross-hatched), or only by flavocytochrome *c*-552 (hatched), or only by mitochondrial enzymes (stippled). (B) Comparison of reactivity ratios (*R*) obtained by differential chemical modification of equine Cyt *c* in the presence and absence of flavocytochrome *c*-552 (filled bars), mitochondrial Cyt bc_1 complex (left open bar) and mitochondrial Cyt *c* oxidase (right open bar). Data for mitochondrial redox partners are from Ref. 98. In the case of the mitochondrial redox partners, *R* values for lysines 55, 72 and 99 are average values for lysines 53+55, 72+73 and 99+100. The *R* values represent, after a series of corrections, the ratio of acetylation of a specific lysine residue in free Cyt *c* to the acetylation of the same residue in the Cyt *c*:flavocytochrome *c*-552 complex. The larger the *R* value, the greater the extent of protection against acetylation.

C. vinosum flavocytochrome *c*-552 are involved in binding Cyt *c* to its mitochondrial reaction partners (see Fig. 2, and Refs. 97–99).

The green sulfur bacterium *Chl. limicola* contains a flavocytochrome *c*-553 [100] that consists of an FAD-containing subunit (M_r = 47000) and a subunit (M_r =

11000) with a single heme *c* [101-103]. Like the *C. vinosum* flavocytochrome *c*-552, the *Chl. limicola* protein exhibits sulfide:cytochrome *c* oxidoreductase activity [104–106]. While several *c*-type cytochromes can function as electron acceptors for the *Chl. limicola* flavocytochrome *c*-553-catalysed oxidation of sulfide in vitro, the acceptor in vivo is likely to be a second soluble *c*-type cytochrome, Cyt *c*-555 [104–106]. Cyt *c*-555 (M_r = 10000; E'_m = +145 mV), which has been found in all green sulfur bacteria studied so far [6], shows some similarities in tertiary structure to eukaryotic mitochondrial Cyt *c* [96,107]. Recent studies utilizing affinity chromatography [108] suggest that an electrostatic complex between *Chl. limicola* flavocytochrome *c*-553 and Cyt *c*-555 is the catalytically active species involved in sulfide oxidation. As is the case in the *C. vinosum* system described above [93,95], the heme subunit of the *Chl. limicola* flavocytochrome *c*-553 appears to contain the major site for binding the acceptor Cyt *c* [108]. Additional evidence for similarities between the *C. vinosum* and *Chl. limicola* flavocytochrome systems comes from the demonstration that the *Chl. limicola* Cyt *c*-555 can bind to the *C. vinosum* flavocytochrome *c*-552 and serve as an electron acceptor during the oxidation of sulfide catalysed by the *C. vinosum* enzyme [108]. These results and kinetic data [109,110] are consistent with a similar mechanism for the oxidation of sulfide to elemental sulfur in both bacteria, with electrons flowing from sulfide to the flavin moiety of the flavocytochrome *c*, then to the heme(s) of the flavocytochrome and ultimately to the heme group of the acceptor Cyt *c* (Cyt *c*-550 in *C. vinosum* or Cyt *c*-555 in *Chl. limicola*).

5. *Thiosulfate oxidation*

As indicated above, some forms of green sulfur bacteria are able to oxidize thiosulfate. While several pathways for thiosulfate oxidation appear to exist [6], the best-characterized [105,111] utilizes a thiosulfate:cytochrome *c* oxidoreductase to catalyse electron transfer from thiosulfate to soluble Cyt *c*-551 (M_r = 45000, E'_m = +135 mV) in *Chl. limicola* f. *thiosulfatophilum* [100]. The observation that Cyt *c*-551 is not present in non-thiosulfate-utilizing strains of green sulfur bacteria [6,9,112] supports the proposed role of this cytochrome in thiosulfate oxidation. Reduced Cyt *c*-551 reduces Cyt *c*-555 [105,111], giving the latter cytochrome a central role in accepting electrons from both sulfide and thiosulfate. Affinity chromatography studies suggest that the electron transfer from Cyt *c*-551 to Cyt *c*-555 involves an electrostatic complex between the two cytochromes [108]. Electron donation from Cyt *c*-555 to the reaction center of green sulfur bacteria can then presumably occur, with the electrons ultimately being used to reduce NAD^+ (vide supra).

The pathway(s) for thiosulfate oxidation in purple sulfur bacteria are poorly understood, with both a membrane-bound *c* cytochrome [113] and a soluble iron-sulfur protein [114] having been proposed as possible acceptors of electrons arising from thiosulfate oxidation. However, it is known that in *C. vinosum* electrons from thiosulfate can be used to reduce $NAD(P)^+$ in an uncoupler-sensitive reaction (vide supra). Thiosulfate addition to intact *C. vinosum* cells also reduces soluble Cyt *c*-550 [89]. It is thus possible that electrons from thiosulfate are transferred via Cyt *c*-550 and the cyclic electron transfer chain [89,90,115] to the quinone pool of *C. vinosum* and the quinol formed then serves as an electron donor for NAD^+ reduction via energy-dependent, reverse electron flow involving the bacterium's rotenone-sensitive enzyme.

Acknowledgements

The authors would like to thank Dr. Davide Zannoni, Robert Blankenship and R.C. Fuller for kindly providing access to manuscripts prior to publication. Work in the authors' laboratory was supported, in part, by grants (to D.B.K.) from the Robert A. Welch Foundation (D-710) and the U.S. National Science Foundation (PCM-8109635 and PCM-8408564).

References

1 Gibbons, N.E. and Murray, R.G.E. (1978) Int. J. Syst. Biol. 28, 1–6.
2 Trüper, H.G. and Pfennig, N. (1981) in The Prokaryotes: A Handbook on Habitats, Isolation and Identification of Bacteria (Starr, M.P., Stolp, H., Trüper, H.G., Ballows, A. and Schlegel, H.G., eds.) pp. 299–312, Springer, Berlin.
3 Madigan, M., Cox, S.S. and Stegeman, R.A. (1984) J. Bacteriol. 157, 73–78.
4 Kondratieva, E.N. (1979) in International Review of Microbial Biochemistry (Quayle, J.R., ed.) Vol. 21, pp. 118–175, University Park Press, Baltimore.
5 van Niel, C.B. (1944) Bacteriol. Rev. 8, 1–118.
6 Trüper, H.G. and Fischer, U. (1982) Phil. Trans. R. Soc. London B298, 529–542.
7 Castenholz, R.W. and Pierson, B.K. (1981) in The Prokaryotes: A Handbook on Habitats, Isolation and Identification of Bacteria (Starr, M.P., Stolp, H., Trüper, H.G., Ballows, A. and Schlegel, H.G., eds.) pp. 495–500, Springer, Berlin.
8 van Dijken, I.P. and Harder, W. (1975) Biotechnol. Bioeng. 17, 15–30.
9 Conrad, R. and Schlegel, H.G. (1978) J. Gen. Microbiol. 105, 305–313.
10 Giffhorn, F. (1980) FEMS Microbiol. Lett. 7, 225–228.
11 Herbert, R., Siefert, E. and Pfennig, N. (1980) Arch. Microbiol. 125, 105–110.
12 Dutton, P.L. and Evans, W.C. (1978) in The Photosynthetic Bacteria (Clayton, R.K. and Sistrom, W.R., eds.) pp. 719–726, Plenum, New York.
13 Harwood, C.S. and Gibson, J. (1986) J. Bacteriol. 165, 504–509.
14 Neutzling, O., Imhoff, J. and Trüper, H.G. (1984) Arch. Microbiol. 137, 256–261.
15 Imhoff, J. (1984) Int. J. Syst. Bacteriol. 34, 338–339.
16 Hansen, T. and Veldkamp, H. (1973) Arch. Microbiol. 92, 45–58.
17 Kepen, O.J., Pedan, L.V. and Rodova (1980) Microbiology 49, 562–566.
18 Rodova, N.A. and Pedan, L.V. (1980) Microbiology 49, 157–162.
19 Trüper, H.G. (1981) in Biology of Inorganic Nitrogen and Sulfur (Bothe, H. and Trebst, A., eds.) pp. 199–211, Springer, Berlin.
20 van Niel, C.B. (1936) Arch. Mikrobiol. 7, 323–358.
21 Thiele, H.-H. (1968) Arch. Mikrobiol. 60, 124–138.
22 Gibson, J., Pfenning, N. and Waterbury, J. (1984) Arch. Microbiol. 138, 96–101.
23 Keister, D.L. (1978) in The Photosynthetic Bacteria (Clayton, R.K. and Sistrom, W.R., eds.) pp. 849–856, Plenum Press, New York.

24 Fuller, R.C. and Redlinger, T.E. (1985) in Molecular Biology of The Photosynthetic Apparatus (Steinbeck, K.E., Bonitz, S., Arntzen, C.J. and Bogorad, L., eds.) pp. 155–162, Cold Spring Harbor Laboratory, Cold Spring Harbor, NY.
25 Kondratieva, E.N., Zhukov, V.G., Ivanovsky, R.N. and Petushkora, Y.P. (1976) Arch. Microbiol. 108, 287–292.
26 Kämpf, C. and Pfennig, N. (1980) Arch. Microbiol. 127, 125–135.
27 Kondratieva, E.N. and Krasienikova, E.N. (1979) Mikrobiologiya 48, 194–201.
28 Siefert, E. and Pfennig, N. (1979) Arch. Microbiol. 122, 177–182.
29 Satoh, T., Hoshino, Y. and Kitamura, H. (1976) Arch. Microbiol. 108, 265–269.
30 Klemme, J.-H., Chyla, I. and Preuss, M. (1980) FEMS Microbiol. Lett. 9, 137–146.
31 McEwan, A.G., George, C.L., Ferguson, S.J. and Jackson, J.B. (1982) FEBS Lett. 150, 277–280.
32 Alef, K., Jackson, J.B., McEwan, A.G. and Ferguson, S.J. (1985) Arch. Microbiol. 142, 403–408.
33 Ort, D.R. and Melandri, B.A. (1982) in Photosynthesis (Govindjee, ed.) Vol. I, pp. 537–587, Academic Press, New York.
34 Junge, W. and Jackson, J.B. (1982) in Photosynthesis (Govindjee, ed.) Vol. I, pp 589–646, Academic Press, New York.
35 Cramer, W.A. and Crofts, A.R. (1982) in Photosynthesis (Govindjee, ed.) Vol. I., pp. 389–467, Academic Press, New York.
36 Amesz, J. (1963) Biochim. Biophys. Acta 66, 22–36.
37 Amesz, J. and Knaff, D.B. (1986) in Environmental Microbiology of Anaerobes (Zehnder, A.J.B., ed.) John Wiley and Sons, New York, in press.
38 Okamura, M.Y., Feher, G. and Nelson, N. (1982) in Photosynthesis (Govindjee, ed.) Vol. I, pp. 195–272, Academic Press, New York.
39 Parson, W.W. and Ke, B. (1982) in Photosynthesis (Govindjee, ed.) Vol. I, pp. 331–385, Academic Press, New York.
40 Prince, R.C. and Dutton, P.L. (1978) in The Photosynthetic Bacteria (Clayton, R.K. and Sistrom, W.R., eds.) pp. 439–453, Plenum Press, New York.
41 Knaff, D.B. (1978) in The Photosynthetic Bacteria (Clayton, R.K. and Sistrom, W.R., eds.) pp. 629–640, Plenum Press, New York.
42 Hale, M.B., Blankenship, R.E. and Fuller, R.C. (1983) Biochim. Biophys. Acta 723, 376–382.
43 Vasmel, H. and Amesz, J. (1983) Biochim. Biophys. Acta 724, 118–122.
44 Bruce, B.D., Fuller, R.C. and Blankenship, R.E. (1982) Proc. Natl. Acad. Sci. USA 79, 6532–6536.
45 Blankenship, R.E., Feick, R., Bruce, B.D., Kirmaier, C., Holten, D. and Fuller, R.C. (1983) J. Cell. Biochem. 22, 251–261.
46 McCarty, R.E. and Carmeli, C; (1982) in Photosynthesis (Govindjee, ed.) pp. 647–695, Academic Press, New York.
47 Keister, D.L. and Yike, J.J. (1967) Arch. Biochem. Biophys. 121, 415–422.
48 Jones, C.W. and Vernon, L.P. (1969) Biochim. Biophys. Acta 180, 149–161.
49 Klemme, J.-H. (1969) Z. Naturforsch. 24b, 67–76.
50 Jones, O.T.G. and Saunders, V.A. (1972) Biochim. Biophys. Acta 275, 427–436.
51 Malkin, R., Chain, R.K., Kraichoke, S. and Knaff, D.B. (1981) Biochim. Biophys. Acta 637, 88–95.
52 Heytler, P.C. (1979) Methods Enzymol. 55, 463–472.
53 Jackson, J.B. and Crofts, A.R. (1968) Biochem. Biophys. Res. Commun. 32, 908–915.
54 Doi, M., Takemiya, K. and Nishimura, M. (1980) Plant Cell Physiol. 21, 1015–1022.
55 Brune, D.C. and Trüper, H.G. (1985) Abstr., V. International Symposium on Photosynthetic Prokaryotes, Grindelwald, Switzerland, p. 237.
56 Gromet-Elhanan, Z. (1969) Arch. Biochem. Biophys. 637, 88–95.
57 Ingledew, W.J. and Prince, R.C. (1977) Arch. Biochem. Biophys. 178, 303–307.
58 Zannoni, D. and Ingledew, W.J. (1983) Arch. Microbiol. 135, 176–181.
59 Zannoni, D. and Ingledew, W.J. (1983) FEMS Microbiol. Lett. 17, 331–334.
60 Dutton, P.L. and Leigh, J.S. (1973) Biochim. Biophys. Acta 314, 178–190.
61 Ohnishi, T. (1973) Biochim. Biophys. Acta 301, 105–128.
62 Zannoni, D. and Ingledew, W.J. (1985) FEBS Lett. 193, 93–98.
63 Hauska, G., Hurt, E., Gabellini, N. and Lockau, W. (1983) Biochim. Biophys. Acta 726, 97–133.

64 Wynn, R.M., Gaul, D.F., Choi, W.-K., Shaw, R.W. and Knaff, D.B; (1986) Photosynth. Res., in press.
65 Rieske, J.S. (1976) Biochim. Biophys. Acta 456, 195–247.
66 Evans, M.C.W. (1969) in Progress in Photosynthesis Research (Metzner, H., ed.) pp. 1474–1475, Laupp, Tübingen.
67 Knaff, D.B. and Buchanan, B.B. (1975) Biochim. Biophys. Acta 503, 524–544.
68 Evans, M.C.W. and Buchanan, B.B. (1965) Proc. Natl. Acad. Sci. USA 53, 1420–1425.
69 Swarthoff, T., Gast, P., Hoff, A.J. and Amesz, J. (1981) FEBS Lett. 130, 93–98.
70 Prince, R.C. and Olson, J.M. (1976) Biochim. Biophys. Acta 423, 357–367.
71 Knaff, D.B., Buchanan, B.B. and Malkin, R. (1973) Biochim. Biophys. Acta 325, 94–101.
72 Nuijs, A.M., Vasmel, H., Joppe, H.L.P., Duysens, L.N.M. and Amesz, J. (1985) Biochim. Biophys. Acta 807, 24–34.
73 Hatefi, Y., Davis, K.A., Baltscheffsky, H., Baltsheffsky, M. and Johansson, B.C. (1972) Arch. Biochem. Biophys. 152, 613–618.
74 Carithers, R.P., Yoch, D.C. and Arnon, D.I. (1977) J. Biol. Chem. 252, 7461–7467.
75 Beinert, H. and Albracht, S.P.J. (1982) Biochim. Biophys. Acta 683, 245–277.
76 Singer, T.P. and Johnson, Michael K. (1985) FEBS Lett. 190, 189–198.
77 Ziegler, D.M. and Doeg, K.A. (1962) Arch. Biochem. Biophys. 97, 41–50.
78 Takamiya, K. and Dutton, P.L. (1979) Biochim. Biophys. Acta 546, 1–16.
79 Crofts, A.R. and Wraight, C.A. (1983) Biochim. Biophys. Acta 726; 149–185.
80 Parson, W.W. (1978) in The Photosynthetic Bacteria (Clayton, A.K. and Sistrom, W.R., eds.) pp. 455–469, Plenum Press, New York.
81 Baccarini-Melandri, A., Gabellini, N. and Melandri, B.A. (1982) in Function of Quinones in Energy Conserving Systems (Trumpower, B.L., ed.) pp. 285–298, Academic Press, New York.
82 Ragan, C.I. (1976) Biochim. Biophys. Acta 456, 249–290.
83 Kobayashi, K., Kastura, E., Kondo, T. and Ishimoto, M. (1978) J. Biochem. 84, 1205–1215.
84 Schedel, M., Vanselow, M. and Trüper, H.G. (1979) Arch. Microbiol. 121, 29–36.
85 Seki, Y., Sogawa, N. and Ishimoto, M. (1981) J. Biochem. 90, 1487–1492.
86 Bartsch, R.G. and Kamen, M.D. (1960) J. Biol. Chem. 235, 825–832.
87 Fukumori, Y. and Yamaka, T. (1979) J. Biochem. 85, 1405–1414.
88 Gray, G.O. and Knaff, D.B. (1982) Biochim. Biophys. Acta 680, 290–296.
89 van Grondelle, R., Duysens, L.N.M., van der Wel, J.A. and van der Wal, H.N. (1977) Biochim. Biophys. Acta 461, 188–201.
90 Knaff, D.B., Whetstone, R. and Carr, J.W. (1980) Biochim. Biophys. Acta 590, 50–58.
91 Tomiyama, Y., Doi, M., Takamiya, K. and Nishimura, M. (1983) Plant Cell Physiol. 24, 11–16.
92 Gray, G.O., Gaul, D.F. and Knaff, D.B. (1983) Arch. Biochem. Biophys. 222, 78–86.
93 Davidson, M.W., Gray, G.O. and Knaff, D.B. (1985) FEBS Lett. 187, 155–159.
94 Meyer, T.C., Vorkink, W.P., Tollin, G. and Cusanovich, M.D. (1985) Arch. Biochem. Biophys. 236, 52–58.
95 Vieira, B., Davidson, M., Knaff, D. and Millett, F. (1986) Biochim. Biophys. Acta 848, 131–136.
96 Salemme, R. (1977) Annu. Rev. Biochem. 46, 299–329.
97 Bosshard, H.R., Davidson, M.W., Knaff, D.B. and Millett, F. (1986) J. Biol. Chem. 261, 190–193.
98 Rieder, R. and Bosshard, H.R. (1980) J. Biol. Chem. 255, 4732–4739.
99 Margoliash, E. and Bosshard, H.F. (1983) Trends Biochem. Sci. 8, 316–320.
100 Meyer, T.E., Bartsch, R.G., Cusanovich, M.A. and Mathewson, J.H. (1968) Biochim. Biophys. Acta 153, 854–861.
101 Yamanaka, T. (1976) J. Biochem. 79, 655–660.
102 Kenney, W.C., McIntire, W. and Yamanaka, T. (1977) Biochim. Biophys. Acta 483, 467–474.
103 Yamanaka, T., Fukumori, Y. and Okunuki, K. (1970) Anal. Biochem. 95, 209–213.
104 Kusai, A. and Yamanaka, T. (1973) FEBS Lett. 34, 235–237.
105 Kusai, A. and Yamanaka, T. (1973) Biochim. Biophys. Acta 325, 304–314.
106 Yamanaka, T. and Fukumori, Y. (1980) in Flavins and Flavoproteins (Yagi, K. and Yamano, T., eds.) pp. 631–639, Japan Scientific Societies Press, Tokyo.
107 Korszun, Z.R. and Salemme, F.R. (1977) Proc. Natl. Acad. Sci. USA 74, 5244–5247.

108 Davidson, M.W., Meyer, T.E., Cusanovich, M.A. and Knaff, D.B. (1986) Biochim. Biophys. Acta 850, 396–401.
109 Meyer, T.E., Vorkink, W.P., Tollin G. and Cusanovich, M.A. (1985) Arch. Biochem. Biophys. 236, 52–58.
110 Cusanovich, M.A., Meyer, T.E. and Tollin, G. (1985) Biochemistry 24, 1281–1287.
111 Kusai, A. and Yamanaka, T. (1973) Biochem. Biophys. Res. Commun. 51, 107–112.
112 Fischer, U. (1984) in Sulfur, Its Significance for Chemistry for the Geo-, Bio- and Cosmosphere and Technology (Müller, A. and Krebs, B., eds.) pp. 383–407, Elsevier, Amsterdam.
113 Schmitt, W., Schliefer, G. and Knobloch, K. (1981) Arch. Microbiol. 130, 328–333.
114 Fukumori, Y. and Yamanaka, T. (1979) Curr. Microbiol. 3, 119–120.
115 Coremans, J.M.C.C., van der Wal, H.N., van Grondelle, R., Amesz, J. and Knaff, D.B. (1985) Biochim. Biophys. Acta 807, 134–142.

J. Amesz (ed.) *Photosynthesis*

CHAPTER 10

Structure and function of protein complexes in the photosynthetic membrane

NATHAN NELSON

Roche Institute of Molecular Biology, Roche Research Center, Nutley, NJ 07110, U.S.A.

1. Introduction

Most of the biochemical reactions catalysed by membrane proteins are vectorial events in which structural considerations play a major role. Membranes are fluid structures that allow catalytic units to have lateral freedom while maintaining absolute polarity of the biochemical reactions. Moreover, some of these reactions require very strict distances between the reactants and well-defined spatial orientation. All of these demands are fulfilled by multisubunit protein complexes that can freely float in the membrane while keeping reactive sites in strict position. Membranes that catalyse the photosynthetic reactions are the best example of such an organization.

Photosynthetic membranes of cyanobacteria, algae and plants contain four major protein complexes which harness light energy and convert it into chemical energy. Two of these protein complexes, the cytochrome b_6-f complex and the proton-ATPase complex, contain no pigments and function in biochemical reactions. The other two, PS I reaction center and PS II reaction center, function in photochemical reactions and contain pigments involved in light-harvesting and primary energy transduction. There are a few ways to define membrane protein complexes, the best definition being based upon the biochemical activity of the complex. Thus, a protein complex can be defined as the minimal structure that catalyses a characterized biochemical reaction [1,2]. Cyt b_6-f complex can be defined as the minimal structure in the photosynthetic membrane that catalyses the reduction of plastocyanin or Cyt *c* when the electron donor is an appropriate quinone (usually plastoquinone). Upon reconstitution into a membrane this reaction must be vectorial and form a protonmotive force [3,4]. The proton-ATPase complex is the minimal structure that, upon reconstitution into lipid vesicles, can form a protonmotive force by hydrolysing ATP, and can form ATP from ADP and P_i at the expense of a protonmotive force [1,5–8]. PS I reaction center is the minimal structure that catalyses the photoreduction of ferredoxin by reduced plastocyanin or Cyt *c* [1,9–11]. PS II reaction center is the minimal structure that catalyses the photoreduction of plastoquinone, while the electron donor H_2O is oxidized to O_2 and protons [12–16].

Since a minimal structure is usually difficult to resolve, copurification of various polypeptides was taken as a primordial criterion for the integrity of those polypeptides as part of the protein complex. Sometimes we have no alternative but to use our intuition and to stop the purification when we think that we have purified the complex or simply when we are exhausted. The four protein complexes were purified from the chloroplast membranes in this manner and for some of them it was proven that all of the polypeptides that copurified were necessary for the activity of the complex [17–19]. However, many conflicting results concerning the participation of other polypeptides, such as the integrity of Cyts b_6 and *f* as part of PS I, created a lot of confusion. Today it is apparent that PS I reaction center and Cyt b_6-*f* complex are not evenly distributed in the granal and stromal membranes [20]. The chloroplast membrane contains many other polypeptides, some of which interact to various degrees with the protein complexes. If every interaction is considered to be an indication of participation in the protein complex, erroneous conclusions are inevitable. Therefore, all the ideas of supermolecular organization out of the defined protein complexes should be taken with more than a grain of salt.

In this chapter I will review the structure, function and biogenesis of these four protein complexes.

2. *Cytochrome b_6-f complex*

2.1. *Structure and function of the isolated complex*

Cyt *b-c* complex forms the evolutionary link between the respiratory and photosynthetic electron-transport pathways [3]. In both systems it oxidizes quinols and reduces metalloproteins while generating protonmotive force. In cyanobacteria, the Cyt b_6-*f* complex is shared by both respiration and photosynthesis [21].

Cyt b_6-*f* complex was identified as a separate entity of the chloroplast membrane by Wessels [22] and Boardman and Anderson [23]. Later it was purified and defined as cyt b_6-*f* complex which functions between the two photosystems [24]. The similarity between the complex isolated from chloroplasts and the mitochondrial Cyt *b*-c_1 complex was pointed out. Hauska and his colleagues conducted a thorough study of Cyt *b-c* complexes from various sources and revealed a high degree of homology among them [3]. Table 1 depicts some of the properties of the complex isolated from chloroplasts. The isolated Cyt b_6-*f* complex was shown to contain four different polypeptides. No other polypeptides were proven to be a functional part of the minimal complex [3]. The enzyme ferredoxin-NADP reductase sometimes copurifies with the Cyt b_6-*f* complex, as also occurs with other complexes such as the CF_1-ATPase [25]. This does not make it a part of the complex any more than it is part of CF_1. However, there are up to three extra polypeptides of very low molecular weight that copurify with the Cyt b_6-*f* complex, which may be proven to be an integral part of it. This phenomenon of accompanying low molecular weight proteins is common to several other protein complexes, and whether

TABLE 1

Composition of Cyt b_6-f complex

Subunit	Molecular weight (kDa)	Subunit stoichiometry[d]	Location of gene	Function
I	35.3[a,b]	1	Chloroplast	Cyt f
	31.3[a]			
II	23.7[a]	1	Chloroplast	Cyt b_6
III	19[c]	1	Nucleus	Nonheme iron
IV	15.2[a]	1	Chloroplast	C-terminal of Cyt b_6-f

[a] Obtained from nucleotide sequence of the gene coding for the subunits [31,34].
[b] Precursor form contains an N-terminal extension of 35 amino acids [31].
[c] Obtained from calibrated SDS gels [34,38].
[d] Measured by relative amounts of Coomassie blue or amido black [38].

they are an integral part of the active units remains to be shown. The b_6-f complex has been isolated in an active form [3,26]. Upon reconstitution into phospholipid vesicles vectorial electron and proton transports across the membrane were demonstrated [4].

2.2. Biogenesis of cytochrome b_6-f complex

Like all the other protein complexes of the chloroplast membrane, the genes coding for the subunits of Cyt b_6-f complex are of dual genetic origin [2,27–29]. Three of the subunits, I, II and IV, are encoded on the chloroplast DNA, while subunit III (nonheme iron protein) is coded in the nucleus, decoded as a larger precursor and imported into the chloroplast by an energy-dependent vectorial-processing mechanism [2,30]. Like most of the chloroplast products, subunits II and IV are synthesized as their mature size in the stroma and incorporated into the membrane by a vectorial translation process [2,28]. Subunit I (Cyt f) is an exception to this rule, since it is a chloroplast gene product, which is synthesized as a larger precursor in spinach, pea and wheat chloroplasts [31–33]. In spinach but not in wheat the three subunits which are synthesized in the chloroplast are encoded by the same DNA strand but they are not part of one operon [31,33,34]. The genes for subunits II and IV appear to be under common transcriptional control and they may have evolved from a common ancestoral gene coding for a b-type Cyt [29].

The biogenesis of the Cyt b_6-f complex is not controlled by light, in contrast to the biosynthesis of the two reaction centers [35,36]. Etiolated leaves contain substantial amounts of Cyt b_6-f complex and the relative amounts of the various subunits did not significantly change during the light-dependent greening of etiolated seedlings [37]. During growth, additional complexes were assembled by a concerted mechanism in which all the four subunits were synthesized and assembled at the same rate. The presence of Cyt b_6-f complex in etioplasts may suggest that the partial reactions which this complex catalyses play an important role in the development of the etioplast into fully active chloroplasts.

Since all the subunits of Cyt b_6-f complex appeared to assemble at the same rate during greening of etiolated plants, mutants of Oenothera were studied for some indication as to which one of the subunits can be synthesized independently of the others [29]. The results suggested that only subunit III (nonheme iron protein) can be synthesized in the cytoplasm and imported into the chloroplast in the absence of the other subunits. Moreover, no mutant could be found in which only subunit III is missing, which suggests that subunit III may serve as a template for the assembly of the rest of the subunits.

3. *The proton-ATPase complex*

3.1. *Structure and function*

The proton-ATPase is composed of two structures, a catalytic sector (CF_1) that is hydrophilic in nature and a membrane sector (CF_0) that is hydrophobic in nature [7,8]. The function of the membrane sector is to conduct protons across the membrane in a way that will enable the catalytic sector to harness the energy stored as a protonmotive force to form ATP from ADP and P_i. The mechanism of this reaction is largely unknown and solving it is one of the major challenges in bioenergetics. In chloroplasts, a protein complex containing at least eight polypeptides catalyses the reaction of ATP formation. The catalytic sector which has a latent ATPase activity is composed of five subunits designated α, β, γ, δ and ϵ in order of decreasing molecular weight (see Table 2). In the last decade the function of each individual subunit was studied in detail [39–42], and now these are quite established through elegant studies by McCarty and his colleagues [17–19,43]. The function of the β subunit as a catalytic subunit for the ATPase activity was first demonstrated by biochemical means [8,42], but recently we learned to appreciate signals provided to us by the genetics of the system through immunological cross-reactivity and DNA sequencing of the relevant genes and their surroundings. Thus, strict preservation of amino acid sequences in β subunits of proton-ATPases from various sources, first detected by immunological cross-reactivity [44] and later demonstrated by DNA sequencing [45,46], provided convincing evidence for the importance of the β subunit in the catalytic activity of the enzyme. The functions of the other subunits of the catalytic sector (CF_1) are listed in Table 2. Every subunit has a specific function in the catalytic activity of the enzyme, and recently it was demonstrated that the phosphorylation activity of the enzyme is dependent on the present δ and ϵ subunits [17,19], so CF_1 is an example of a complex membrane protein in which the integrity of each one of its subunits has been established. The function of the ϵ subunit as an ATPase inhibitor that renders the CF_1 into a latent ATPase was demonstrated long ago [39] and reevaluated recently by McCarty and his colleagues [18]. These studies established that the ϵ subunit of CF_1 acquired a function fulfilled in the mitochondrial enzyme by yet another peptide [8].

The proton-ATPase complex, first purified by Pick and Racker [54], was reported to contain nine different subunits, four of which may belong to the membrane sector. Later studies in our laboratory detected only three subunits in the

TABLE 2

Subunit structure and function of the proton ATPase complex

Subunit	Molecular weight (Da)	Subunit stoichiometry	Site of synthesis	Homology to *E. coli* ATPase subunits	Function	Refs.
CF_1						
α	55 446[a]	3	Chloroplast	α	Nucleotide, binding and regulation	7, 8, 40, 47, 56
β	53 874[a]	3	Chloroplast	β	Active site	8, 42, 46, 56
γ	37 000[b]	1	Cytoplasm	γ	Energy transduction from pmf to ATP; template for assembly of CF_1; binding of CF_1 to the membrane	7, 8, 40, 48–50
δ	20 000[b]	1	Cytoplasm	δ	Induction of proper binding	7, 8, 18
ε	14 702[a]	1	Chloroplast	ε	ATPase inhibitor; necessary for photophosphorylation	7, 8, 17, 18, 39, 46
CF_0						
I[a]	27 060[a]	?	Chloroplast	a	?	51
I	20 900[a]	?	Chloroplast	b	Binding of CF_1?	8, 51, 52
II	14 000[b]	?	Cytoplasm	?	Assembly of CF_0?	52
III	8 000[a,b]	6	Chloroplast	c	Proton conduction	8, 51–53

[a] From DNA sequencing. [b] From SDS gels.

membrane sector which were designated subunits I, II and III in order of decreasing molecular weight [8,55,55a]. The direct function of subunit III in proton conduction was demonstrated by a variety of methods [52,53]. It was proposed that subunit I functions in binding CF_1 and subunit II is involved in the assembly of six copies of subunit III into a functional proton channel [8,52]. Very recently sequencing of the gene cluster containing the genes coding for subunits III and I as well as the α subunit of CF_1 revealed a reading frame that is transcribed and translated into another polypeptide that copurifies from chloroplasts together with the proton-ATPase complex [53]. The fact that this gene is located within this cluster suggests that its product may have a function in the complex. Once again the strength of genetic signals was demonstrated through a vigorous study conducted by Herrmann and his colleagues.

From the mere fact that CF_1 can be released from the membrane by EDTA treatment and the enzyme stays in solution without detergents, it is apparent that the catalytic sector has minimal, if any, direct interaction with the lipids of the chloroplast membrane. It is a globular protein that is held to the surface of the membrane via interaction with the membrane sector. Recently it was shown that the γ subunit is in immediate contact with the membrane sector and the δ and ε subunits may induce proper binding for catalysis [17,18]. The enzyme contains a few well-defined sites that were used for localization experiments by the method of fluorescent energy transfer [19,56–61]. These studies revealed the position of those sites and helped to localize the various subunits of CF_1 in space relative to the chloroplast membranes (for a model of CF_1, see Refs. 61 and 62). These experiments are awaiting analysis of the amino acid sequence of the γ subunit that is now under investigation in Herrmann's laboratory [148]. Definite structural analysis could be obtained only after good crystals of the enzyme become available.

3.2. Biogenesis of the proton-ATPase complex

Only three subunits of the proton-ATPase complex are coded in the nucleus of green algae and higher plants [8,55,63,64]. Subunits γ and δ of CF_1 and subunit II of the membrane sector are synthesized on cytoplasmic ribosomes as larger precursors and are transported into the chloroplast by a vectorial processing mechanism [1,2,65,66]. The genes coding for these subunits were recently isolated in Herrmann's laboratory but their sequences are not available as yet. The other subunits are coded on the chloroplast DNA. The location and the sequences of these genes are known [45–47,51]. These genes are located in two distinct clusters, one of which contains the genes coding for β and ϵ subunits, the other containing the genes coding for subunits III and I of the membrane sector and the α subunit of CF_1. The relationship of this arrangement to the operon coding for the proton-ATPase of *E. coli* did not escape our attention [51,67]. The genes are transcribed into two sets of mRNA; however, their regulation is not clear as yet [51]. It is quite likely that all of this organization, including the coding of γ, δ and subunit II in the nucleus, was designed to synchronize the assembly of the complex in a way that will prevent the assembly of an active proton channel in the membrane sector that is not gated by CF_1 [68]. In contrast to the rest of the protein complexes, partial assembly of the proton-ATPase might lead into bioenergetic disaster by collapsing the protonmotive force. Therefore, the proton-ATPase is assembled by a concerted mechanism in which all of the subunits are synthesized and assembled at the same time. Etiolated plants contain substantial amounts of the complex and during illumination these did not increase significantly [37]. This may suggest a function for the complex irrespective of the presence of a functional photosynthetic system.

4. Photosystem I reaction center

4.1. Structure and function

The function of PS I reaction center is to oxidize plastocyanin or Cyt *c* and to reduce ferredoxin. Light provides the energy for the formation of a redox potential difference of about 0.7 V and the generation of an electric potential across the chloroplast membrane that is subsequently used for ATP formation. The catalysis of the reaction requires a strict organization of various pigments and chromophores. This organization is provided by a defined multisubunit protein complex which has Chl *a* for light harvesting, β-carotene for protection [69], P-700 for donation of electrons [70], and a series of electron acceptors for accepting the electrons and stabilizing the redox and electric potentials [71–74]. Such a protein complex was first isolated from Swiss chard chloroplasts [9,10] and subsequently demonstrated in several other higher plants [37,75,76]. It was reported to be composed of six different polypeptides [9], but with the improvement of the SDS gel techniques and studies on the genes coding for one of the alleged subunits it became apparent that PS I reaction center is composed of eight different subunits [77–79]. The subunits were designated subunits Ia, Ib, II, III, IV, V, VI and VII in order of decreasing molecular weight from 83 kDa to about 8 kDa.

Table 3 summarizes some of the properties of PS I reaction center and specific functions of its individual subunits. The purified preparation contains about 100 Chl *a* molecules per P-700 [9,10]. However, this number can be decreased to about 40 while the order of the Chl *a* molecules increases [70]. Washing out more of the Chl molecules caused a decrease in the dichroic ratio, indicating that those 40 Chls are the highly oriented primary light-harvesting antenna of the reaction center. It was also shown that the β-carotene is in very close proximity to P-700 and it is highly oriented with respect to the latter [70]. P-700 as well as the primary electron acceptor (A_1) may be composed of specialized Chl *a* molecules [80]. There are at least three more electron acceptors which are part of the reaction center and their function is to slow down the rate of the reaction and thereby stabilize the redox potential difference [72]. It was not until Malkin and Bearden [81] discovered the bound ferredoxins that this part of PS I started to be understood. Today it appears that at least four different clusters are involved in the electron-accepting site of the

TABLE 3

Subunit structure and function of PS I reaction center

Subunit	Molecular weight (kDa)	Subunit stoichiometry	Site of synthesis	Pigments and functional groups	Function
Ia	83[a]	1	Chloroplast	40 Chl *a*; 1–5 β-carotene; P_{700}; Electron acceptors A_1 and A_2 (Ia and Ib)	Primary light-harvesting antenna, primary electron donor (P-700) and acceptors (A_1 and A_2)
Ib	82[a]	1	Chloroplast		
II	25[b]	1	Cytoplasm	Nonheme iron	Secondary electron acceptor (A_3 and/or A_4)
III	20[b]	1	Cytoplasm	?	Facilitates electron transport from plastocyanin to P-700
IV	18[b]	1	Cytoplasm ?	?	?
V	16[b]	1	Chloroplast ?	?	?
VI	9[b]	?	Cytoplasm ?	?	?
VII	8[b]	?	Chloroplast ?	?	?

[a] From DNA sequencing [79]. [b] From SDS gels.

PS I reaction center. The first one (A_1) may be a specialized Chl *a* molecule [80]. The second acceptor (A_2 or X) is a special bound ferredoxin [74], and two more electron acceptors (A_3 and A_4) are bound ferredoxins A and B [72–74]. All of these clusters are present in the purified reaction center and it was found to be impossible to deplete one of them without losing the NADP-photoreduction activity of the complex [9,10].

The function of each individual subunit of the PS I reaction center is not entirely clear. Subunit I was isolated in a pure form and the preparation was active in P-700 photooxidation at room temperature [9,10]. Therefore it was defined as the P-700 reaction center. It contains about 40 Chl *a* molecules per P-700 and it should contain the primary electron acceptor A_1. Recently it was shown that a similar preparation also contains the secondary electron acceptor A_2 [82]. The purified preparation of P-700 reaction center is composed of two non-identical polypeptides that are present at a ratio of one copy of each per P-700 [9–11]. A similar preparation was isolated from green algae and cyanobacteria [64,83,84]. Immunological crossreactivity among these preparations indicates conservation of amino acid sequences in the polypeptides of P-700 reaction center from cyanobacteria though Prochloron and green algae up to higher plants [37,64,83,85]. Similarly, subunit II of PS I reaction center from spinach was immunologically crossreactive with subunit II of PS I reaction centers from various sources [37]. None of the rest of the subunits had this property. Therefore, subunits I and II of the reaction centers from various sources are homologous and most probably has similar functions in PS I from the different organisms. Studies on the reaction center of the green alga Chlamydomonas revealed that subunit II is one of the bound ferredoxins and it may contain centers A or B or even both of them [64]. Using chaotropic agents for differential extraction of the different subunits, subunit II was identified as the Fe-S center apoprotein [86]. The function of the rest of the subunits is much more illusive and largely unknown. Subunit III of the higher plant reaction centers may function in the reduction of P-700 by plastocyanin [10,11]. Depletion of this subunit caused inhibition of this reaction that could be partially overcome by high Mg^{2+} concentrations [87]. However, the binding site for plastocyanin is probably situated on subunit I and the function of subunit III may be to modulate this binding site [64,87].

Structural data on the organization of the active clusters and the various subunits can be obtained from biochemical, biophysical and genetic studies. However, the final word is in the mouth of the crystallographer, after the biochemist hands him the crystals. Only one piece of solid information is available to date, and this is the amino acid sequences of subunits Ia and Ib [78,79]. These two subunits should accommodate the P-700, the primary electron acceptor (A_1), and probably the secondary electron acceptor A_2 which is the nonheme iron cluster ‘X’. The amino acid sequences of subunits I_a and I_b make it unlikely that these subunits also contain one of the bound ferredoxins ‘A’ or ‘B’ because subunit I_a contains 4 cysteine residues and subunit I_b contains only 2 cysteines [79]. These numbers are hardly sufficient for the formation of the nonheme iron cluster ‘X’ which is supposed to be a nonheme iron center [74,88]. Therefore, it may be likely that subunit II of the reaction center contains both centers 'A' and 'B' of the bound ferredoxins. Recently we isolated a preparation of PS I reaction center from the cyanobacterium *Synechocystis* which appeared to contain only subunits I and II, while all of the bound ferredoxins were present in relatively high amounts (Reilly, P., Prince, R.C. and Nelson, N., unpublished). However, until the amino acid sequence of subunit II is available, I shall refrain from speculation. The amino acid sequences of subunits Ia and Ib also revealed large amounts of tryptophan residues (28 and 30, respectively) and 43 and 39 histidine residues, respectively. The histidines appeared in a distinct pattern of couples and it may be that each Chl *a* molecule is ligated by two histidine molecules. The calculated molecular weight of subunits Ia and Ib gave values of 83.4 and 82.4 kDa respectively [79]. Until the desired structural data are obtained from crystals that do not as yet exist, one can draw two different models for the structure of PS I reaction center which have no structural similarity [2,10].

4.2. Biogenesis of photosystem I reaction center

Two genes with a considerable sequence homology were identified on the chloroplast chromosome as candidates for the genes coding for subunit I [78,79]. The first one, having a reading frame of about 83 kDa, gave translation products that interacted with antibodies raised against polypeptides synthesized according to the DNA sequence as well as with subunit I of the reaction center [78]. The second one, with a reading frame of about 82 kDa, predicted a cyanogen bromide fragment which should give a polypeptide of about 31 kDa, while the rest of the cleavage products of this and the former gene products should give polypeptides shorter than 12 kDa. Such a cleavage product with a molecular weight of about 31 kDa was demonstrated in cyanogen bromide-treated PS I reaction center [67]. Thus, both genes are expressed and their products are assembled into the functional reaction center, and subunit I is a product of two genes designated I_a and I_b. Subunit II is coded on nuclear DNA and decoded on cytoplasmic ribosomes as the larger precurser [29,63,64,77]. It is transported into the chloroplast via a vectorial processing mechanism and assembled into the membrane following proper processing. Although attempts to localize the site of synthesis of the rest of the subunits yielded partial success, the site of synthesis of these subunits in the higher plant cells is largely unknown. The isolated reaction center from *Chlamydomonas* contains four subunits [64]. Subunits I and II are homologous to the corresponding subunits of higher plants and subunit IV was shown to be synthesized on chloroplast ribosomes. Subunit III contains no methionine or cysteine and its site of synthesis could not be determined in that study, which used ^{35}S labelling [69]. It was reported that some of the low molecular weight subunits of the spinach reaction center contain no cysteine or methionine [89]. This hampered the above-mentioned studies and now it must be done the more difficult way.

Plastids of dark-grown plants have no PS I activity [90]. Following illumination, Chl *a* is synthesized and partial reactions typical of PS I first appear in the greening leaves [90]. Using subunit-specific antibodies it was shown that subunit I is present in etiolated leaves in relatively high amounts, while the rest of the subunit could not be detected [37]. After illumination of about 2 h, subunit II was detected, and only later subunit III and the other subunits started to accumulate in the plastid. During the first 2 h of illumination the amounts of subunit I did not change significantly. From this experiment and the effect of cycloheximide treatment on the assembly of PS I in *Chlamydomonas* chloroplasts [64], it was concluded that the control of the assembly of the reaction center is in the hands of the nuclear gene and subunit II may serve as a template for the assembly of PS I reaction center. Most of the multisubunit protein complexes are assembled by a concerted mechanism in which all of the subunits are synthesized and assembled at the same time. Unassembled subunits and the complexes that are not completed are rapidly degraded. This is not the case with the PS I reaction center, which is assembled stepwise during greening of etiolated plants. Most likely some reactions catalysed by the partially assembled complex are required for the development of the etioplast into a chloroplast. This does not mean that these kinds of complexes are stable in chloroplasts. It is well-known that mutations which prevent the assembly of one subunit usually lead to the disappearance of its entire complex from the membrane [29]. Thus, the same set of rules may not apply to the biogenesis of a protein complex during greening of etiolated plants and in fully developed chloroplasts.

5. *Photosystem II*

5.1. *Structure and function*

The function of PS II is to oxidize water and to reduce plastoquinone. The oxidizing side of photosystem II operates at the unusually high redox potential of about +0.82 V to give the reaction of $2H_2O \rightarrow O_2 + 4H^+$. This reaction requires a proper environment to stabilize the reactive chemical intermediates and to prevent them from damaging the system. A very involved protein complex catalyses the reaction, and it was very difficult to resolve the minimum structure which can be defined as the PS II reaction center. The original preparations of oxygen-evolving particles contained up to 15 polypeptides, some of which were PS II-related light-harvesting Chl *a*/*b* protein complexes that eventually could be removed from the reaction center while the specific activity of the preparation increased. Table 4 depicts the most likely subunits of PS II reaction center, and includes their possible role in the functional complex. The primary antenna for light-harvesting is composed of about 50 Chl *a* molecules and 10 carotenoids. These pigments are assembled into two subunits of the system with molecular weights of 43 and 47 kDa. The genes coding for these polypeptides were located on the chloroplast DNA and were sequenced [91,92]. The predicted amino acid sequences revealed a high histidine content, which may serve as ligands for the Chl *a* molecules [29]. Using SDS gels and mild dissociation conditions which maintain most of the Chl bound to the dissociated band, it was tentatively shown that P-680, the primary electron donor of PS II, is located on the 47 kDa polypeptide [93]. However, the lack of specific an-

TABLE 4

Subunit structure and function of a predicted PS II reaction center

Subunit		Molecular weight (Da)	Subunit stoichiometry	Site of synthesis	Pigments and functional groups	Function	Refs.
I	'47 kDa'	56 246[a]	1	Chloroplast	3-10 Chl *a* carotenoids, Mn? P-680? Pheophytin?	Primary light-harvesting Primary photochemistry? Water oxidation?	16, 91, 93, 124–127
II	'43 kDa'	51 785[a]	1	Chloroplast	3-10 Chl *a*; carotenoids?	Primary light-harvesting	92, 124–127
III	'35 kDa'	35 000[b]	1	Cytoplasm	Mn? Cl^-?	Water oxidation	15, 114–122, 128, 129
IV	'34 kDa, D_2'	39 465[a]	1	Chloroplast	Plastoquinone (Q_A) Pheophytin? Fe? P-680	Primary electron acceptor Primary photochemistry?	92, 107
V	'32 kDa, D_1'	38 500[a] or 34 500[a]	1	Chloroplast	Plastoquinone (Q_B)	Secondary electron acceptor	16, 97, 123
VI	'23 kDa'	23 000[b]	1	Cytoplasm	Ca^{2+}?	Participates in water oxidation	114–122
VII	'17 kDa'	17 000[b]	1	Cytoplasm	Ca^{2+}?	Participates in water oxidation	114–122
VIII	'10 kDa' *b*-559	9 390[a]	1-2	Chloroplast	Heme	Water oxidation?	130–132

[a] From DNA sequencing. [b] From SDS gels.

tibody to the 32 kDa subunit and the fact that this polypeptide is not stained by Coomassie blue make the results less conclusive. Green bands are beautiful but may lead into a trap.

There is no evidence that the other subunits contain any pigments. Cyt *b*-559 is supposed to take part in the activity of PS II but its mode of action is not clear [94]. One of the subunits of PS II reaction center, with a molecular weight of about 32 kDa, was studied in great detail, primarily because it was shown to bind herbicides [95,96]. This polypeptide was implicated in the electron-accepting site of PS II as a plastoquinone-binding protein [96]. The gene coding for this polypeptide was cloned from a variety of sources and sequence analysis revealed very high conservation from cyanobacteria up to higher plants [97–102]. Moreover, certain homology in the amino acid sequences and a predicted structural homology between the 32 kDa polypeptide and subunit L of the reaction center of purple bacteria (Chapter 3) led to the suggestion that the former has evolved from the latter [103–107]. These findings further suggested that P-680 is situated in the 32 kDa protein in a complex with yet another subunit of PS II (34 kDa) which is partially homologous to subunit M of the reaction center of purple bacteria [103–107]. Therefore, the core of the primary photochemical reactive PS II reaction center is composed of five polypeptides. The 47 kDa polypeptide contains the primary light-harvesting antenna and perhaps P-680, the 43 kDa polypeptide functions in light-harvesting, the 34 and 32 kDa polypeptides contain the primary and secondary quinones which function as electron acceptors and perhaps also contain P-680, and a 10 kDa polypeptide which is Cyt *b*-559.

The oxidizing site of PS II functions in water oxidation via a complex of Mn, Cl^- and Ca^{2+} [108–118]. This ion cluster must be in close proximity to P-680 wherever it is. The cluster is kept together by three polypeptides with molecular weights of 35, 23 and 17 kDa, which can be released from the reaction center by washing with Tris buffer [94,119–121]. These polypeptides are located on the internal side of the thylakoid membrane. The 23 and 17 kDa subunits can be separated from the reaction center following salt washing, and the resulting preparation is active in oxygen evolution when supplemented by high concentrations of Ca^{2+} [94,122]. Thus, these two polypeptides may function in holding the calcium ions in the right position, while the 35 kDa polypeptide may function in positioning of the Mn and Cl^-. From these considerations it seems that the function of light-dependent water oxidation and plastoquinone reduction is carried out by a protein complex composed of eight different polypeptides, and any preparation that contains this number of subunits and functions in oxygen evolution deserves the definition of PS II reaction center.

5.2. *Biogenesis of photosystem II*

Five of the eight subunits of photosystem II reaction center are synthesized on chloroplast ribosomes. The three 'Tris-soluble' subunits are synthesized on cytoplasmic ribosomes as larger precursors and are transported into the chloroplast by a vectorial processing mechanism [29, 119–121,133]. The genes coding for all the five subunits which are synthesized in the chloroplasts were located on the chloroplast DNA and their sequences were analysed in detail [29]. It is assumed that the products of these genes are inserted into the membrane by a vectorial translation mechanism without processing of N-terminal signal sequences [2]. Time is ripe for a detailed study of this subject. The expression of some of these subunits is regulated by light, at both the transcriptional and the translational level. The genes for the thylakoid membrane proteins are spread over half of the chloroplast DNA, and are transcribed as either monocistronic or polycistronic mRNAs [29]. So far, only two genes have been shown to be transcribed as monocistronic mRNA, the ones which code for the large subunit of ribulose bisphosphate carboxylase and the herbicide-binding 32 kDa protein. The genes for the 47 kDa subunit and those for the 43 and 34 kDa subunits are transcribed as polycistronic mRNA of various sizes [29,91]. Maturation of these mRNA species occurs through a complicated process during which post-transcriptional control takes place.

Etiolated plants contain no PS II activity and, except for cytochrome *b*-559, none of the chloroplast-encoded subunits is present in etioplasts. Upon illumination the genes coding for these subunits are turned on and the level of mRNA is increased [29,134–136]. It was proposed that this phenomenon provides the main light control over the synthesis of various chloroplast proteins, and the genes coding for these proteins were termed photogenes [135]. However, recently it was shown that etioplasts contain considerable amounts of mRNA specific for these polypeptides [29]. The mRNA was transcribable in vitro and it was suggested that yet another step of mRNA processing may be under light control [91]. Although it was suggested that the translation of specific mRNA may be controlled by light, the mechanism of such a control is not apparent [137,138]. Therefore, the steps of mRNA synthesis and maturation seem to be the main light-controlled events. This phenomenon of light control over the maturation or even translation of an mRNA is best demonstrated by the finding that the 47 kDa subunit of PS II and Cyt b_6 and subunit IV of Cyt b_6-f complex are transcribed as a common mRNA [29,91]. However, the synthesis of the 47 kDa polypeptide is tightly controlled by light, while the synthesis of Cyt b_6 and subunit IV is not affected by light [37,91]. Recently we used subunit-specific antibodies to follow the relative amounts of PS II polypeptides during greening of etiolated spinach seedlings [139]. The three extrinsic polypeptides, which are nuclear gene products, were present in etiolated leaves in relatively high amounts. The subunits of 35, 23 and 17 kDa were present at about 55, 40 and 30% of their amounts in green leaves [139]. The other polypeptides could not be detected prior to an approx. 6 h illumination period. Further illumination induced the appearance of these subunits at a relatively similar rate. The oxygen-evolution activity was developed parallel to the increase in the amounts of these polypeptides. It was concluded that the assembly of PS II during greening is a two-step process in which some of the polypeptides are synthesized and partially assembled in the dark and the rest of them are synthesized and assembled in a light-dependent process.

Most of the membrane protein complexes are quite stable and exhibit relatively slow turnover of synthesis and decomposition. Due to the harsh reaction catalysed by PS II, an accelerated turnover was reported for this system, especially for the herbicide-binding subunit of 32 kDa [95,140–144]. Exposure of leaves or green algae to naturally occurring levels of visible light can cause loss of photosynthetic activity, a phenomenon called photoinhibition [145–147]. Abundant evidence implicated PS II as the primary site of lesion in photoinhibition. It was suggested that the 32 kDa polypeptide is selectively damaged and that PS II is thereby inactivated [146,147]. These studies employed pulse labeling as a measure of the amounts of the 32 kDa polypeptide under various illumination conditions. This approach fails to detect the total amounts of the given polypeptide, and the measured high turnover and rapid disappearance of the protein under high light intensities may be partially due to the preferential labeling of the non-assembled polypeptide (Rott, R. and Nelson, N., unpublished). The mechanism proposed for the action of the 32 kDa polypeptide as an electron acceptor of PS II, its inactivation by light, and its high turnover implies that this polypeptide must undergo a rapid exchange with the assembled reaction center. This is quite a unique phenomenon for membrane protein complexes, which usually do not readily exchange their assembled subunits.

Some general features of the biogenesis of the two photosynthetic reaction centers recently surfaced. It appears that the polypeptides involved in the primary photochemical events are coded on the chloroplast DNA. All of the chlorophyll-protein complexes that are coded in the nucleus are not part of the minimal structure of the reaction centers. The chlorophyll-protein complexes which contain the primary light-harvesting pigments, subunits I_a and I_b of PS I and the 47 and 43 kDa subunits of PS II, are chloroplast gene products. The fact that subunit I of PS I reaction center is present in etioplasts, while the 47 and 43 kDa subunits of PS II are not, may suggest that the biogenesis of PS I is controlled by the nucleus and the latter is controlled by light at the level of the chloroplast.

References

1 Nelson, N. and Cidon, S. (1984) J. Bioenerg. Biomembranes 16, 11–36.

2 Nelson, N. and Riezman, H. (1984) Biogenesis of Energy-Transducing Systems. New Comprehensive Biochemistry (Ernster, L., ed.) Vol. 9, pp. 351–376, Elsevier, Amsterdam.

3 Hauska, G., Hurt, E., Gabellini, N. and Lockau, W. (1983) Biochim. Biophys. Acta 726, 97–133.

4 Hurt, E., Gabellini, N., Shanak, Y., Lockau, W. and Hauska, G. (1983) Arch. Biochem. Biophys. 225, 879–885.

5 Racker, E. (1976) A New Look at Mechanisms in Bioenergetics, Academic Press, New York.

6 Racker, E. (1985) Reconstitution of Transporters, Receptors, and Pathological States, Academic Press, New York.

7 McCarty, R.E. (1979) Annu. Rev. Plant Physiol. 30, 79–104.

8 Nelson, N. (1981) Curr. Top. Bioenerg. 11, 1–33.

9 Bengis, C. and Nelson, N. (1975) J. Biol. Chem. 250, 2783–2788.

10 Bengis, C. and Nelson, N. (1977) J. Biol. Chem. 252, 4564–4569.

11 Nelson, N. and Notsani, B. (1977) in Bioenergetics of Membranes (Packer, L., Papageorgiou, G.C. and Trebst, A., eds.) pp. 233–244, Elsevier, Amsterdam.

12 Critchley, C. (1985) Biochim. Biophys. Acta 811, 33–46.

13 Berthold, D.A., Babcock, G.T. and Yocum, C.F. (1981) FEBS Lett. 134, 231–234.

14 Kuwabara, T. and Murata, N. (1982) Plant Cell Physiol. 23, 533–539.
15 Bricker, T.A., Metz, J.G., Miles, D. and Sherman, L.A. (1983) Biochim. Biophys. Acta 724, 447–455.
16 Satoh, K., Nakatani, H.Y., Steinback, K., Watson, J. and Arntzen, C.F. (1983) Biochim. Biophys. Acta 723, 142–150.
17 Richter, M.L., Patrie, W.J. and McCarty, R.E. (1984) J. Biol. Chem. 259, 7371–7373.
18 Patrie, W.J. and McCarty, R.E. (1984) J. Biol. Chem. 259, 11121–11128.
19 Nalin, C.M., Snyder, B. and McCarty, R.E. (1985) Biochemistry 24, 2318–2324.
20 Andersson, B. and Anderson, J.M. (1980) Biochim. Biophys. Acta 593, 427–440.
21 Binder, A., Hauser, R. and Krogmann, D. (1984) Biochim. Biophys. Acta 765, 241–246.
22 Wessels, J.S.C. (1966) Biochim. Biophys. Acta 126, 581–583.
23 Boardman, N.K. and Anderson, J.M. (1967) Biochim. Biophys. Acta 143, 187–203.
24 Nelson, N. and Neumann, J. (1972) J. Biol. Chem. 247, 1817–1824.
25 Berzborn, R.J. (1980) Methods Enzymol. 69, 492–502.
26 Wood, P.M. and Bendall, D.S. (1976) Eur. J. Biochem. 61, 337–344.
27 Schatz, G. and Mason, T.L. (1974) Annu. Rev. Biochem. 43, 51–87.
28 Chua, N.-H. and Schmidt, G.W. (1979) J. Cell Biol. 81, 461–483.
29 Herrmann, R.G., Westhoff, P., Alt, J., Tittgen, J. and Nelson, N. (1985) Thylakoid membrane proteins and their genes. In Molecular Form and Function of the Plant Genome (van Vloten-Doting, L., Groot, G.S.P., Hall, T.C., eds.) pp. 233–255, Plenum Publishing Corp., New York.
30 Alt, J., Westhoff, P., Sears, B.B., Nelson, N., Hurt, G., Hauska, G. and Herrmann, R.G. (1983) EMBO J. 2, 979–986.
31 Alt, J. and Herrmann, R.G. (1984) Curr. Genet. 8, 551–557.
32 Willey, D.L., Auffret, A.D. and Gray, J.C. (1984) Cell 36, 555–562.
33 Willey, D.L., Howe, C.J., Auffret, A.D., Bowman, L.M., Dyer, T.A. and Gray, J.C. (1984) Mol. Gen. Genet. 194, 416–422.
34 Heinemeyer, W., Alt, J. and Herrmann, R.G. (1984) Curr. Genet. 8, 543–549.
35 Baker, N.R. and Butler, W.L. (1976) Plant Physiol. 58, 526–529.
36 Bradbeer, J.W. (1981) Development of photosynthetic function during chloroplast biogenesis. In The Biochemistry of Plants (Hatch M.D., and Boardman, N.K., eds.) Vol. 8, pp. 423–472, Academic Press, New York.
37 Nechushtai, R. and Nelson, N. (1985) Plant Mol. Biol. 4, 377–384.
38 Hurt, E. and Hauska, G. (1982) J. Bioenerg. Biomembranes 14, 119–138.
39 Nelson, N., Nelson, H. and Racker, E. (1972) J. Biol. Chem. 247, 7657–7662.
40 Nelson, N., Deters, D.W., Nelson, H. and Racker, E. (1973) J. Biol. Chem. 248, 2049–2055.
41 Deters, D.W., Racker, E., Nelson, N. and Nelson, H. (1975) J. Biol. Chem. 250, 1041–1047.
42 Nelson, N., Kanner, B.J. and Gutnick, D.L. (1974) Proc. Natl. Acad. Sci. USA 71, 2720–2724.
43 Moroney, J.V. and McCarty, R.E. (1982) J. Biol. Chem. 257, 5910–5914.
44 Rott, R. and Nelson, N. (1981) J. Biol. Chem. 256, 9224–9228.
45 Walker, J.E., Saraste, M., Runswick, M.J. and Gay, N.J. (1982) EMBO J. 1, 495–551.
46 Zurawski, G., Bottomley, W. and Whitfield, P.R. (1982) Proc. Natl. Acad. Sci. USA 79, 6260–6264.
47 Deno, H., Shinozaki, K. and Sugiura, M. (1983) Nucleic Acids Res. 11, 2185–2191.
48 McCarty, R.E. and Fagan, J. (1973) Biochemistry 12, 1503–1507.
49 Weiss, M.A. and McCarty, R.E. (1977) J. Biol. Chem. 252, 8007–8012.
50 Moroney, J.V., Lopresti, L., McEwen, B.F., McCarty, R.E. and Hammes, G.G. (1983) FEBS Lett. 158, 58–62.
51 Hennig, J. and Herrmann, R.G. (1986) Mol. Gen. Genet., in press.
52 Nelson, N. (1980) Ann. N.Y. Acad. Sci. 358, 25–36.
53 Nelson, N., Eytan, E., Notsani, B., Sigrist, H., Sigrist-Nelson, K. and Gitler, C. (1977) Proc. Natl. Acad. Sci. USA 74, 2375–2378.
54 Pick, U. and Racker, E. (1979) J. Biol. Chem. 254, 2793–2799.
55 Nelson, N., Nelson, H. and Schatz, G. (1980) Proc. Natl. Acad. Sci. USA 77, 1361–1364.
55a Westhoff, P., Alt, J., Nelson, N. and Herrmann, R.G. (1985) Mol. Gen. Genet. 199, 290–299.
56 Kambouris, N.G. and Hammes, G.G. (1985) Proc. Natl. Acad. Sci. USA 82, 1950–1953.

57 Cerione, R.A. and Hammes, G.G. (1982) Biochemistry 21, 745–752.
58 Cerione, R.A., McCarty, R.E. and Hammes, G.G. (1983) Biochemistry 22, 769–776.
59 Snyder, B. and Hammes, G.G. (1984) Biochemistry 23, 5787–5795.
60 Snyder, B. and Hammes, G.G. (1985) Biochemistry 24, 2324–2331.
61 Richter, M.L., Snyder, B., McCarty, R.E. and Hammes, G.G. (1985) Biochemistry 24, 5755–5763.
62 Rott, R., Cidon, S. and Nelson, N. (1984) in H^+-ATPase (ATP synthase): Structure, Function, Biogenesis of the F_0F_1 Complex of Coupling Membranes (Papa, S., et al., eds.) pp. 247–256, Adriatica Editrice, Bari, Italy.
63 Nechushtai, R., Nelson, N., Mattoo, A. and Edelman, M. (1981) FEBS Lett. 125, 115–119.
64 Nechushtai, R. and Nelson, N. (1981) J. Biol. Chem. 256, 1624–1628.
65 Chua, N.-H. and Schmidt, G.W. (1979) J. Cell Biol. 81, 461–483.
66 Ellis, R.J. (1981) Annu. Rev. Plant Physiol. 32, 111–137.
67 Nelson, N. (1986) Biochem. Soc. Trans. 14, 5–7.
68 Nelson, N. (1981) Structure and Function of CF_0. In Energy Coupling in Photosynthesis (Selman, B.R. and Selman-Reimer, S., eds.) pp. 261–268, Elsevier, Amsterdam.
69 Junge, W., Schaffernicht, H. and Nelson, N. (1977) Biochim. Biophys. Acta 462, 73–85.
70 Kok, B. (1956) Biochim. Biophys. Acta 22, 399–401.
71 Ke, B. (1973) Biochim. Biophys. Acta 301, 1–33.
72 Sauer, K., Mathis, P., Acker, S. and van Best, J.A. (1978) Biochim. Biophys. Acta 503, 120–134.
73 Bearden, A.J. and Malkin, R. (1975) Q. Rev. Biophys. 7, 131–177.
74 Evans, M.C.W., Sihra, C.K., Bolton, J.R. and Cammack, R. (1975) Nature 256, 668–670.
75 Orlich, G. and Hauska, G. (1980) Eur. J. Biochem. 111, 525–534.
76 Hauska, G., Samoray, D., Orlich, G. and Nelson, N. (1980) Eur. J. Biochem. 11, 535–543.
77 Westhoff, P., Alt, J., Nelson, N., Bottomley, W., Bunemann, H. and Herrmann, R.G. (1983) Plant Mol. Biol. 2, 95–107.
78 Fish, L.E., Kuck, U. and Bogorad, L. (1985) J. Biol. Chem. 260, 1413–1421.
79 Kirsch, W., Seyer, P. and Herrmann, G.R. (1986) Curr. Genet., in press.
80 Shuvalov, V.A. (1976) Biochim. Biophys. Acta 430, 113–121.
81 Malkin, R. and Bearden, A.J. (1971) Proc. Natl. Acad. Sci. USA 68, 16–19.
82 Warden, J. and Golbeck, J.H. (1986) Biochim. Biophys. Acta 849, 30–33.
83 Nechushtai, R., Muster, P., Binder, A., Liveanu, V. and Nelson, N. (1983) Proc. Natl. Acad. Sci. USA 80, 1179–1183.
84 Takahashi, Y., Koike, H. and Katoh, S. (1982) Arch. Biochem. Biophys. 219, 209–218.
85 Schuster, G., Nechushtai, R., Nelson, N. and Ohad, I. (1985) FEBS Lett. 191, 29–33.
86 Bonnerjea, J., Ortiz, W. and Malkin, R. (1985) Arch. Biochem. Biophys. 240, 15–20.
87 Nechushtai, R. and Nelson, N. (1981) J. Bioenerg. Biomembranes 13, 295–306.
88 Evans, M.C.W., Sihra, C.K. and Cammack, R. (1976) Biochem. J. 158, 71–77.
89 Mullet, J.E., Grossman, A.R. and Chua, N.-H. (1982) Cold Spring Harbor Symp. 46, 979–984.
90 Bradbeer, J.W. (1981) in The Biochemistry of Plants (Hatch, M.D. and Boardman, N.K., eds.) Vol. 8, pp. 423–472, Academic Press, New York.
91 Morris, J. and Herrmann, R.G. (1984) Nucleic Acids Res. 12, 2837–2850.
92 Alt, J., Morris, J., Westhoff, P. and Herrmann, R.G. (1984) Curr. Genet. 8, 597–606.
93 Green, B.R. and Camm, E.L. (1984) in Advances in Photosynthesis Research (Sybesma, C., ed.) Vol. II, pp. 99–102, Martinus Nijhoff/W. Junk, The Hague.
94 Ghanotakis, D.F. and Yocum, C.F. (1985) Photosynth. Res. 7, 97–114.
95 Pfister, K., Steinback, K.E., Gardner, G. and Arntzen, C.J. (1981) Proc. Natl. Acad. Sci. USA 78, 981–985.
96 Satoh, K., Nakatani, H.Y., Steinbach, K.E., Watson, J. and Arntzen, C.J. (1983) Biochim. Biophys. Acta 723, 142–150.
97 Zurowski, G., Bohnert, H.J., Whitfield, P.R. and Bottomley, W. (1982) Proc. Natl. Acad. Sci. USA 79, 7699–7703.
98 Hirschberg, J. and McIntosh, L. (1983) Science 222, 1346–1349.
99 Erickson, J.M., Rahire, M., Bermann, P., Delepelaire, P., Diner, P. and Rochaix, J.D. (1984) Proc. Natl. Acad. Sci. USA 81, 3617–3621.

100 Goloubinoff, P., Edelman, M. and Hallick, R.B. (1984) Nucleic Acids Res. 12, 9489–9496.

101 Erickson, J.M., Rahire, M., Rochaix, J.D. and Metz, L. (1985) Science 228, 204–207.

102 Golden, S.S. and Haselkorn, R. (1985) Science 229, 1104–1107.

103 Okamura, M.Y., Feher, G. and Nelson, N. (1982) in Photosynthesis, Energy Conversion by Plant and Bacteria (Govindjee, ed.) Vol. I, pp. 195–272, Academic Press, New York.

104 Youvan, D.L., Bylina, E.J., Alberti, M., Begusch, H. and Hearst, J.E. (1984) Cell 37, 949–957.

105 Deisenhofer, J., Epp, O., Miki, K., Huber, R. and Michel, H. (1984) J. Mol. Biol. 180, 385–398.

106 Deisenhofer, J., Epp, O., Miki, K., Huber, R. and Michel, H. (1985) Nature 318, 618–624.

107 Trebst, A. (1986) Z. Naturforsch. 41c, 240–245.

108 Radmer, R. and Cheniae, G.M. (1977) in Primary Processes in Photosynthesis (Barber, J., ed.), pp. 303–348, Elsevier, Amsterdam.

109 Amesz, J. (1983) Biochim. Biophys. Acta 725, 1–12.

110 Yocum, C.F., Yerkes, C.T., Blankenship, R.E., Sharp, R.R. and Babcock, G.T. (1981) Proc. Natl. Acad. Sci. USA 78, 7507–7511.

111 Kelly, P.M. and Izawa, S. (1978) Biochim. Biophys. Acta 502, 192–210.

112 Baianu, I.C., Critchley, C., Govindjee and Gutowsky, H.S. (1984) Proc. Natl. Acad. Sci. USA 81, 3713–3717.

113 Sandusky, P.O. and Yocum, C.F. (1984) Biochim. Biophys. Acta 766, 603–611.

114 Piccioni, R. and Mauzerall, P. (1976) Biochim. Biophys. Acta 423, 605–609.

115 Brand, J.J., Mohanty, P. and Fork, D.C. (1983) FEBS Lett. 155, 120–124.

116 Ono, T.-A. and Inoue, Y. (1983) Biochim. Biophys. Acta 723, 191–201.

117 Ghanotakis, D.F., Topper, J.N., Babcock, G.T. and Yocum, C.F. (1984) FEBS Lett. 170, 169–173.

118 Van Gorkom, H.J. (1985) Photosynth. Res. 6, 97–113.

119 Åkerlund, H.-E., Jansson, C. and Åndersson, B. (1982) Biochim. Biophys. Acta 681, 1–10.

120 Åndersson, B., Larsson, C., Jansson, C., Ljungberg, U. and Åkerlund, H.E. (1984) Biochim. Biophys. Acta 766, 21–28.

121 Åkerlund, H.E., Renger, G., Weiss, W. and Hagemann, R. (1984) Biochim. Biophys. Acta 765, 1–6.

122 Ghanotakis, D.F. and Yocum, C.F. (1986) in Ion Interactions in Energy Transfer Biomembranes (Papgeorgiou, G.C., Barber, J. and Papa, S., eds.), pp. 291–301, Plenum Press, New York.

123 Cohen, B.N., Coleman, T.A., Schmitt, J.J. and Weissbach, H. (1984) Nucleic Acids Res. 12, 6221–6230.

124 Camm, E.L. and Green, B.R. (1983) Biochim. Biophys. Acta 724, 291–293.

125 Satoh, K. and Bulter, W.L. (1978) Plant Physiol. 61, 373–379.

126 Diner, B.A. and Wollman, F.A. (1980) Eur. J. Biochem. 110, 521–527.

127 Nakatoni, H.Y., Ke, B., Dolan, E. and Arntzen, C.J. (1984) Biochim. Biophys. Acta 765, 347–352.

128 Metz, J.G., Wong, J. and Bishop, N.I. (1980) FEBS Lett. 114, 61–66.

129 Metz, J.G. and Bishop, N.I. (1980) Biochem. Biophys. Res. Commun. 94, 560–566.

130 Bendall, D.S. (1982) Biochim. Biophys. Acta 683, 119–151.

131 Widger, W., Cramer, W.A., Hermodson, M., Meyer, D. and Gullifor, M. (1984) J. Biol. Chem. 259, 3870–3876.

132 Herrmann, R.G., Alt, J., Schiller, B., Widger, W.R. and Cramer, W.A. (1984) FEBS Lett. 176, 239–244.

133 Westhoff, P., Jansson, C., Klein-Hitpass, L., Berzborn, R., Larsson, C. and Bartlett, S.G. (1985) Plant Mol. Biol. 4, 137–146.

134 Bedbrook, J.R., Link, G., Coen, D.M., Bogorad, L. and Rich, A. (1978) Proc. Natl. Acad. Sci. USA 75, 3060–3064.

135 Rodermel, S.R. and Bogorad, L. (1985) J. Cell Biol. 100, 463–476.

136 Gallagher, T.F. and Ellis, R.J. (1982) EMBO J. 1, 1493–1498.

137 Reardon, E.M. and Price, C.A. (1983) Arch. Biochem. Biophys. 226, 433–440.

138 Miller, M.E., Jurgenson, J.E., Reardon, E.M. and Price, C.A. (1983) J. Biol. Chem. 258, 14478–14484.

139 Liveanu, V., Yocum, C.F. and Nelson, N. (1986) J. Biol Chem. 261, 5296–5300.

140 Eaglesham, A.R.J. and Ellis, R.J. (1974) Biochim. Biophys. Acta 335, 396–407.

141 Grebanier, A.E., Coen, D.M., Rich, A. and Bogorad, L. (1978) J. Cell Biol. 78, 734–746.

142 Mattoo, A.K., Pick, U., Hoffman-Falk, H. and Edelman, M. (1981) Proc. Natl. Acad. Sci. USA 78, 1572–1576.

143 Hoffman-Falk, H., Mattoo, A.K., Marder, J.B., Edelman, M. and Ellis, R.J. (1982) J. Biol. Chem. 257, 4583–4587.

144 Marder, J.B., Goloubinoff, P. and Edelman, M. (1984) J. Biol. Chem. 259, 3900–3908.

145 Powels, S.B. (1984) Annu. Rev. Plant Physiol. 35, 15–44.

146 Kyle, D.J., Ohad, I. and Arntzen, C.J. (1984) Proc. Natl. Acad. Sci. USA 81, 4070–4074.

147 Ohad, J., Kyle, D.J. and Arntzen, C.J. (1984) J. Cell Biol. 99, 481–485.

148 Tittgen, J., Hermans, J., Steppuhn, J., Jansen, T., Janssen, ~~C.,~~ Andersson, B., Nechushtai, R., Nelson, N. and Herrmann, R.G. (1986) Mol. Gen. Genet. 204, 258–265.

J. Amesz (ed.) *Photosynthesis*

CHAPTER 11

Structure and function of light-harvesting pigment-protein complexes

H. ZUBER, R. BRUNISHOLZ and W. SIDLER

Institut für Molekularbiologie und Biophysik, Eidgen. Technische Hochschule Zürich, ETH-Hönggerberg – HPM, CH-8093 Zürich, Switzerland

1. Introduction

The antenna complexes of photosynthetic organisms are multi-molecular energy transport systems whose function is to funnel excited-state energy to the photochemical reaction center (RC). These light-harvesting pigment-protein complexes and the special pair of the RC (see Chapters 3 and 4) thus form a cooperative, highly regulated energy transfer and energy trapping system. This system is the starting point for a light-induced redox reaction leading ultimately to the production of ATP and the generation of reducing power. Light-harvesting antennae of the various organisms are composed generally of a number of pigment-protein (antenna) complexes of varying absorption maxima. Such complexes form systems for heterogeneous and directed energy transfer to the RC [1–3]. The structures of antenna complexes acting within the same heterogeneous energy transfer system can differ greatly. Pigment molecules (Chl, BChl, bilins, carotenoids) are highly ordered within the antenna complexes, and their position and orientation follow defined laws of symmetry (Fig. 1) [2,4]. The antenna system of photosynthetic organisms is, therefore, very efficient [5], and energy flows with little energy dissipation to the RC. Antennae contain a large number of pigment molecules, approximately 25 – 1000 pigments per RC. It is commonly assumed that energy, in the form of excited singlet states (S_1, excitons), migrates between the individual pigment molecules within approximately 10^{-12} s by means of a random walk [5–7]. Other processes, such as non-radiative relaxation or fluorescence emission, are much slower. It is further assumed that the transfer of energy occurs as inductive resonance transfer between the pigment molecules [8]. This type of energy transfer, however, is relatively insensitive to structural details of the antennae (even distribution in two dimensions of cluster formation; Fig. 1A or B) [6]. Thus cluster formation of pigments (Fig. 1B,C,D,E) in the form of antenna complexes as observed in all antenna systems, must be due to other structural or functional reasons:

(1) The well-ordered pigment clusters of the antenna complexes having different absorption maxima are the basis of heterogeneous, directed energy transfer between the antenna complexes to the RC (Fig. 1C).

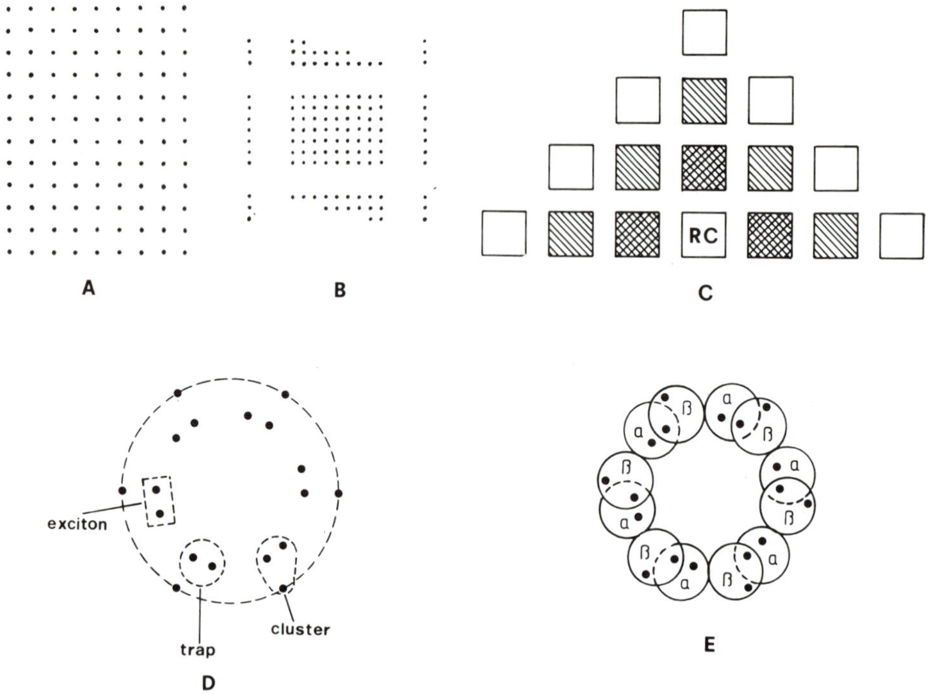

Fig. 1. Arrangement of pigments (Chl, BChl, bilins) in antenna systems (scheme). (A) Even, highly ordered distribution of pigment molecules. (B) Distribution of pigments in highly ordered clusters. (C) Arrangement of pigment clusters of the antenna complexes with different absorption maxima and spatially separated: basis for heterogeneous and directed energy transfer to the RC. (D) Pigment clusters within the antenna complex forming excitons or energy traps. (E) Regular arrangement of polypeptides which bind the pigment molecules at defined binding sites: basis for the formation of pigment clusters in the antenna complexes.

(2) This directed energy transfer is optimized by the spatial separation of the pigment clusters or antenna complexes (Fig. 1C).

In both cases random walk is minimized. Pigment clusters can also be found within the antenna complexes, for example in the form of pigment dimers (or oligomers) with a distance between the pigments of 10 – 20 Å (Fig. 1D). The reason for a strong coupling of pigments is probably the formation of localized excitons. The distances between the pigment pairs are greater (20 – 30 Å) [9]. Hypothetically, due to the symmetric (e.g. cyclic) arrangement of both dimeric and monomeric pigments, larger localized excitons, for example quasi-stationary exciton states, could be formed (Fig. 1D) [10]. In such an energy transfer system, energy traps in the form of long-wavelength-absorbing pigment molecules (possibly coupled in pigment dimers) are conceivable (Fig. 1D).

Fundamental to the formation of pigment clusters are the structure and organization of the antenna polypeptides [2,3,9]. As structural analysis has shown to date, all pigment molecules are bound at defined binding sites (not covalently bound: Chl, BChl, carotenoids; covalently bound: bilins) to relatively small polypeptides (mol. weight approx. 6 – 30 kDa). Polypeptides determine the type, number, position, orientation, distance and environment of the pigments optimal for the energy transfer. The regular and symmetrical arrangement of the pigments within the antenna complexes is based upon a regular and symmetrical arrangement of antenna polypeptides with repeating basic elements (for example, the basic elements of (α-β)-antenna polypeptide pairs in bacteria and cyanobacteria (Fig. 1E)). The regular arrangement of the antenna complexes in relation to one another within the whole antenna is likewise achieved through the specific structure and interaction of antenna polypeptides. The specific polypeptide-determined environment of the pigment (for example, in the α- or β-polypeptides; see Section 2) is the basis for their different spectral characteristics (differences in red-shift) and properties (for example, sensitizing and fluorescing pigments), which is important for the heterogeneous energy transfer. Detailed understanding of the molecular structure of antenna complexes and of polypeptides and pigment arrangement is fundamental in terms of the physical mechanism of energy transfer and in view of future theoretical and biophysical studies.

As a result of adaptation to diverse environmental conditions, photosynthetic organisms show a multitude of antenna structures (Fig. 2). An environmental factor which influences structure and function of antenna systems is light energy (spectral range, intensity). Spectral range (Fig. 2) determines the type of pigment and its polypeptide environment; light intensity determines the size of the antenna system (number and size of antenna complexes and number of pigments, or polypeptides). The size of the antennae is related to the extent of the heterogeneous energy transfer system (i.e. absorption range). The type of antenna system is dependent upon its position within the organism and upon the complexity of the photosynthetic apparatus, that is, the photosynthetic membrane. Antenna complexes are either part of the photosynthetic membrane (intramembrane complex) or they are bound to it (extramembrane complex) (Fig. 2). Extramembrane complexes are always functionally connected to intra-membrane complexes located in the proximity of the RC. Their spectral absorption range reaches downward into the blue range: this extends the range for heterogeneous energy transfer. Four main types of antenna polypeptides have been discovered in intramembrane and extramembrane complexes:

(1) hydrophobic membrane polypeptides in the intramembrane antenna complexes of Rhodospirillaceae, Chromatiaceae, Chlorobiaceae and Chloroflexaceae;
(2) mixed types of hydrophobic and globular polypeptides in antenna complexes of algae and higher plants;
(3) globular polypeptides (proteins) in extramembrane antennae (phycobilisomes, etc.) of cyanobacteria, red algae and Cryptophyceae;
(4) fibrillary polypeptides in extramembrane antennae of chlorosomes of green photosynthetic bacteria (Chlorobiaceae, Chloroflexaceae).

In all of these cases, the type of photosynthetic membrane and its inclusion in the more or less complex metabolism determines the structure of the antenna polypeptides and their organization within the antenna complexes. Significant struc-

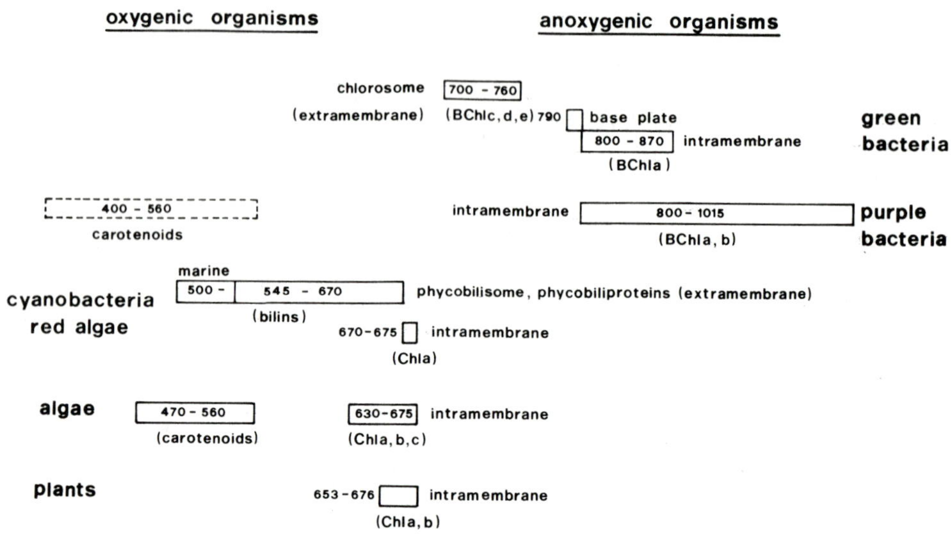

Fig. 2. Spectral range (absorption in nm) of the antenna system of oxygenic (plants, algae, cyanobacteria) and anoxygenic (green and purple bacteria) photosynthetic organisms. The antenna systems (antenna complexes) are located either within (intramembrane) or on the surface (extramembrane) of the photosynthetic membrane.

tural differences are found between intramembrane antenna polypeptides in the cytoplasmic or intracytoplasmic membrane of photosynthetic bacteria and those in the thylakoid membrane of algae and higher plants.

2. *Light-harvesting antennae of photosynthetic bacteria*

In recent years there has been considerable interest in studying the photosynthetic apparatus of purple non-sulfur bacteria, whereby special emphasis was given to spectroscopic analyses (absorption, fluorescence, fluorescence depolarization, circular and linear dichroism) of detergent-solubilized light-harvesting pigment-protein complexes (for recent reviews see Refs. 11 and 12). Within the last six years there has also been proteinchemical characterization of the antenna polypeptides, concentrating to a large extent on comparative primary structure determination (for recent reviews see Refs. 2,3,4 and 14). Among others, extensively characterized purple non-sulfur bacteria are *Rhodospirillum rubrum, Rhodopseudomonas sphaeroides* and *Rhodopseudomonas capsulata* (the latter two species were recently renamed *Rhodobacter*). With respect to structure-function relationships far less information has been obtained for members of the Chlorobiaceae and Chloroflexaceae (green bacteria) families. Here, the structural studies of antenna systems focused mainly on *Chloroflexus aurantiacus* and *Prosthecochloris aestuarii.* The main general properties of bacterial antennae are:

(1) Except for the chlorosome and the base-plate-type antenna of chlorobineae, the bacterial antenna complexes are intra-membrane bound.

(2) The photon-capturing chromophores BChl (*a, b, c,* etc.) are non-covalently bound to a polypeptide matrix.
(3) The polypeptide components are small (5–6 kDa) and in most cases span the photosynthetic membrane once (hydrophobic polypeptides, insoluble in water,

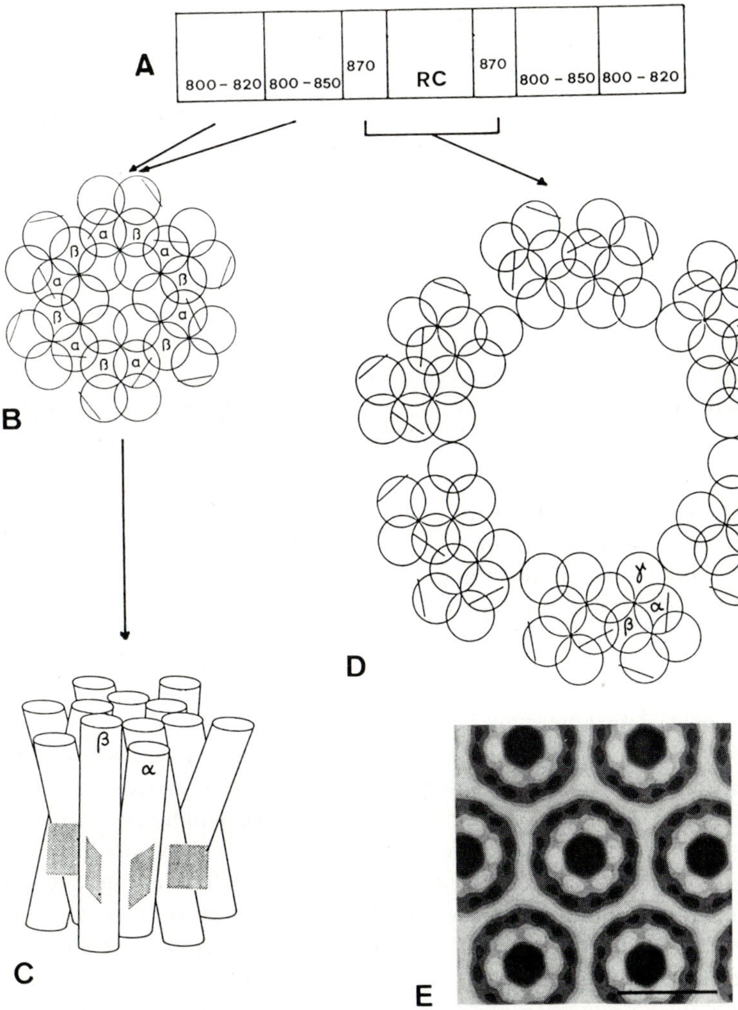

Fig. 3. Arrangement of the antenna light-harvesting complexes of purple bacteria (A) and arrangement of the antenna polypeptides in the light-harvesting complexes (B – E). (B) Possible arrangement of the B800-850 light-harvesting polypeptides $(\alpha\beta)_6$. Top view of the intramembrane α-helices of B(800-850)-α and B(800-850)-β. (C) As (B), side-view. (D) Possible arrangement of the B880 (B1015) antenna polypeptides $(\alpha\beta)_{12}$, top view of the intramembrane α-helices of α-, β-polypeptides of *Rp. viridis* surrounding the reaction center. (E) Electron micrograph and Fourier processed image of rotary-shadowed Triton X-100-treated membranes of *Rp. viridis*. Hexagonally arranged antenna polypeptides surrounding the reaction center (black center). Taken from Ref. 32.

soluble in chloroform/methanol). The antenna polypeptides build up large aggregates.

(4) In addition to BChl molecules, they may have specific carotenoids. These act as additional chromophores and protect against photo-oxidation.

2.1. Purple photosynthetic bacteria

Members of this sub-order of photosynthetic bacteria can be differentiated into (1) those of which the photosynthetic membrane appears as vesicles (inside-out, so-called chromatophores) e.g. *Rs. rubrum, Rp. sphaeroides, Rp. capsulata, Chromatium vinosum* or (2) those having stacked lamellae, e.g. *Rp. viridis, Rp. acidophila* and *Rp. palustris* or (3) those simply having their photosynthetic apparatus located within the cytoplasmic membrane without special membrane extrusion, e.g. *Rp. gelatinosa* (renamed *Rhodocyclus gelatinosus*) and *Rs. tenue.* Another usual way to differentiate purple bacteria is based on their typical near-infrared antenna absorption characteristics (Q_Y band of bacteriochlorophyll). On this basis three main antenna classes have been identified (see Table 1). (1) Purple bacteria having their bulk bacteriochlorophyll (B) organized in one antenna system with a single absorption band at 870, 880 or 1020 nm. This antenna system was designated B870, B880 or B1020 and surrounds the RC (see Fig. 3). Representatives of this class are *Rs. rubrum* with B880 and *Rp. viridis* with B1020 (BChl *b*). (2) In addition to the B880 antenna complex many purple bacteria, e.g. *Rp. sphaeroides, Rp. capsulata, Rp. gelatinosa* and *Rs. tenue,* have a second type of antenna, the B800-850 light-harvesting complex with Q_Y-absorption bands at 800 and 850 nm. (3) The presence of a third type of antenna, the B800-820 (B800-830) complex, has so far been exclusively described for *Rp. acidophila, Rp. palustris* and *Chromatium vinosum.*

2.1.1. Purple bacteria with one type of antenna system

The isolation of a spectrally intact B880 antenna complex of *Rs. rubrum* S1 (wild-type strain) was achieved by several groups [15–18]. Its basic functional unit is two strongly interacting BChl *a* molecules, as revealed by CD measurements in the near-infra red [19]. They are non-covalently bound to two small, very hydrophobic polypeptides, α and β, in an apparent ratio of 1:1 [15,17]. Amino acid sequence determinations of B880-α from the strain S1 [20] and B870-α from the carotenoidless mutant G-9$^+$ [21] revealed identical primary structures, and also for B880-β and B870-β [22] (see Fig. 4). Recently, the primary structures of the α- and β-polypeptides have been confirmed by gene sequencing [23]. The α- and the β-polypeptides show very little sequence homology (6%), indicating early separation in their evolution. Nevertheless, hydropathy profiles of the antenna polypeptides draw attention to a similar structure with respect to the arrangement in the photosynthetic membrane [22]. The profiles clearly demonstrate (identical positive/negative excursions) a structural organization into three domains: a central hydrophobic stretch region of 21–23 amino acids is flanked by polar charged N-terminal and C-terminal regions. In the hydrophobic domain a buried secondary structure, most probably an α-helix of 5–6 turns in length, is indicated by the large

TABLE 1

Comparison of the major light-harvesting complexes in purple bacteria. Taken from Ref. 12.

Antenna type	B890-protein class		B800-850-protein class		
	B890-protein	B875-protein	B800-850-protein		B800-820-protein
			Type I	Type II	
Examples of bacteria containing antenna type	*Rsp. rubrum* *C. vinosum* *Rps. acidophila* *? Rps. viridis*	*Rps. palustris* *Rps. capsulata* *Rps. sphaeroides* *Rps. gelatinosa*	*Rps. acidophila* 7750 and 7050 (high-light grown)	*C. vinosum* *Rps. acidophila* 7050 (low-light grown)	
BChl *a*: carotenoid	2:1	2:2	3:1	3:1	3:1
No. of polypeptides in isolated complex	2	2	2 or 3	2	2
No. of amino acid residues in polypeptides	52 and 54	52–58 and 47–48	53 and 42	~50–65	~50–65
Intensity of CD spectrum of long-wavelength band[a]	Strong	Weak	Strong	Strong	Strong

[a] All spectra indicate the presence of a BChl dimer.

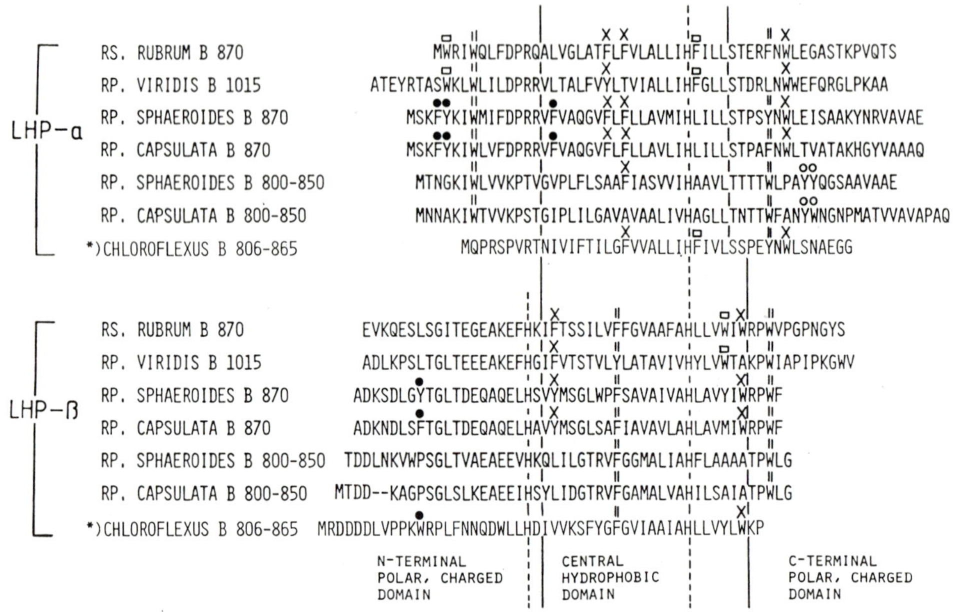

Fig. 4. Amino acid sequences of antenna polypeptides from purple and green photosynthetic bacteria. Antenna complex specific aromatic residues are marked: □ ● ×, B880 (B1015); ○, B800-850; ||, B880 and B800-850.

negative excursion of the hydropathy profiles. On the other hand, the N- and the C-terminal parts show the folding of a polypeptide chain in an aqueous environment, indicating that they are exposed to the membrane surface. Comparative primary structure analyses of a series of bacterial antenna polypeptides draw attention to overall conserved amino acid positions and typically invariant positions for specific antenna complexes. The most significant structural element present in all bacterial light-harvesting polypeptides so far sequenced is a histidine residue located within the phospholipid-bilayer half exposed to the periplasmic side (Figs. 4–6). It was proposed that this specific histidine residue offers a fifth ligand for binding the BChl molecule [21,22]. Except for the additional overall conserved Ala (Ser in the B800-850 of *Rp. sphaeroides*) in the hydrophobic stretch region (position His-4, Fig. 5), the structural elements for specific antenna complexes appear to be clustered in groups of 2–4 amino acids (Fig. 5).

Rs. rubrum, with its single antenna complex, is an ideal object to probe the membrane-surface exposed regions of the antenna polypeptides. Chromatophores (inside-out vesicles, the cytoplasmic side of the membrane outside) have been chemically modified with the hydrophilic marker diazobenzenesulfonate [24]. In addition, protease treatment of chromatophores [25–27] and of spheroplasts (periplasm outside) was carried out [27]. The surface location of the B880 complex was further investigated with antibodies raised to either the intact complex or to the individual polypeptides [27]. In addition, the secondary structures of the polypep-

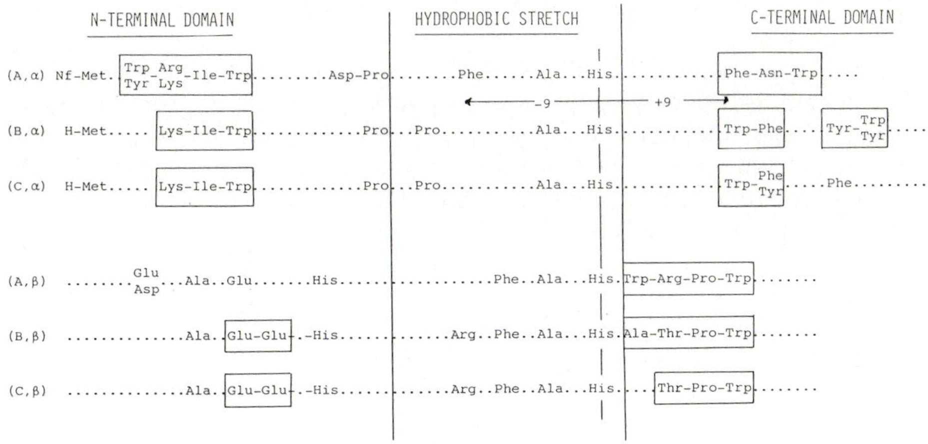

Fig. 5. Specific structural elements (amino acid residues boxed) of antenna polypeptides of B880 (A), B800-850 (B) and B800-820 (C) light-harvesting complexes from purple bacteria.

tide regions buried in the membrane or exposed on the membrane surface were investigated by far–UV CD (190–250 nm) analyses [28] and label experiments with photogenerated carbenes [29]. As a result of these experiments a tentative structural model emerged of how the antenna polypeptides and the pigments are arranged with regard to the photosynthetic membrane (Fig. 6):

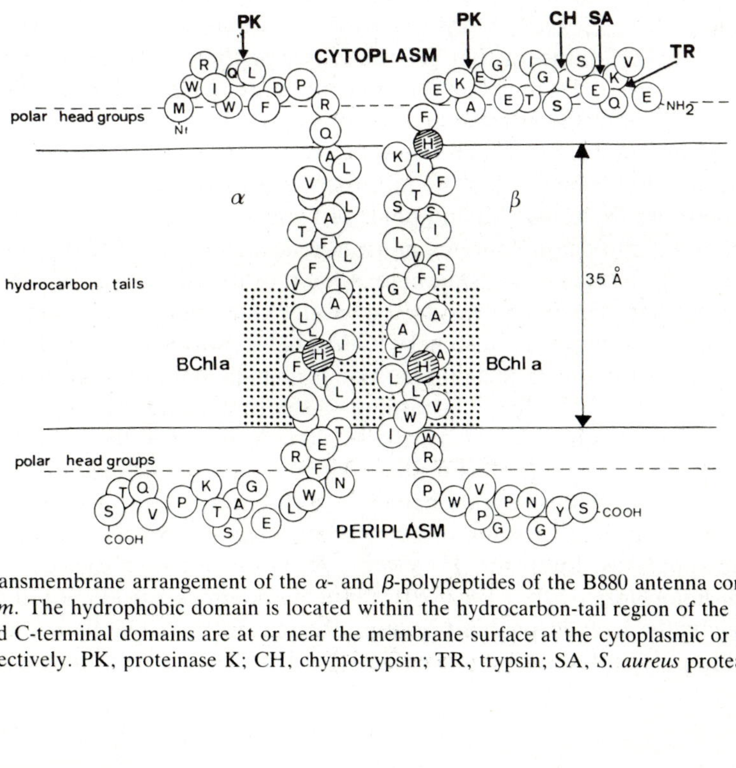

Fig. 6. Transmembrane arrangement of the α- and β-polypeptides of the B880 antenna complex from *Rs. rubrum*. The hydrophobic domain is located within the hydrocarbon-tail region of the membrane; the N- and C-terminal domains are at or near the membrane surface at the cytoplasmic or periplasmic side, respectively. PK, proteinase K; CH, chymotrypsin; TR, trypsin; SA, *S. aureus* protease.

- The N-terminal portions of both antenna polypeptides are exposed to the cytoplasmic side.
- The C-terminal portions of both polypeptides are located at the periplasmic side.
- The N-terminal domain of the α-polypeptides, partly arranged in an amphipathic α-helix, interacts with the phospholipid-bilayer/water interphase.
- The carotenoid seems to be (partly) located at the cytoplasmic surface. Its absence in *Rs. rubrum* G-9$^+$ induces a conformation of the polypeptides which is more sensitive to proteolysis.
- The α- and β-polypeptides span the membrane once as α-helices, 5–6 turns in length (21–23 amino acids).
- The exciton-coupled BChl dimer is located in that half of the lipid bilayer which is exposed to the periplasmic side.
- Possible cyclic aggregates of (α/β)-heterodimers, e.g. dodecamers $(\alpha\beta)_{12}$, are formed with 24 polypeptides and 24 BChl *a* molecules (Fig. 3).
- The interaction sites are most probably the N- and C-terminal domains.

The photosynthetic membrane of *Rp. viridis* forms stacked lamellae and is unique in showing well-ordered two-dimensional arrays of photosynthetic units in electron micrographs [30–32]. The photoreceptor complexes of *Rp. viridis* (reaction center and antenna complexes) are arranged on a hexagonal lattice with a repeat distance of approx. 130 Å, a structure which appears to be common for all BChl *b*-containing species [33]. It has been assumed that the center-core ring of about 45 Å in diameter represents the RC complex which is surrounded by a ring-like structure, roughly 20 Å wide consisting of 6 [30] or 12 [32] subunits, most likely the B1020 antenna complexes. The photoreceptor complex consisting of one RC complex surrounded by 12 pairs of α/β-heterodimers would agree well with the proposed 20–24 antenna pigments per RC, each polypeptide carrying one BChl molecule, in *Rs. rubrum*. Despite the fact that the antenna complex B1020 has not been isolated to date, the corresponding polypeptides α and β were isolated and sequenced [34]. They show a high degree of homology (α, 46%; β, 52%) to the polypeptides of *Rs. rubrum* (Fig. 4) and the same principal three-domain structure. Furthermore, by using CD and polarized infrared spectroscopy it was shown that the photoreceptor unit contains more α-helical structure (57%) than the isolated RC (47%), implying a considerable amount of α-helical structure for the antenna polypeptides [45]. In contrast to the species containing BChl *a*, a third small (36 amino acids) polypeptide, also very hydrophobic, was isolated and sequenced [34]. A particular feature of this polypeptide is its high amount of aromatic amino acids (3 Trp, 2 Tyr and 1 Phe). According to the molar ratio of this polypeptide to α and β (1:1:1) it is likely to be an additional constituent of the light-harvesting complex B1020 [34] and was thus termed B1020-γ. It was postulated that B1020-γ serves as a sort of linker-peptide between (α/β) entities and possibly represents a decisive factor in the formation of hexagonally arrayed components of the photosynthetic apparatus. Interestingly, another BChl *b*-containing species, *Ectothiorhodospira halochloris* with a B800/1015 antenna complex(es) [35,36] and analogous hexagonally arrayed pigment-protein complexes also exhibits a small, very hydrophobic polypeptide (29 amino acids), with at least 2 tryptophan residues.

Amino acid sequence analyses revealed approx. 40% homology to the B1020-γ polypeptide of *Rp. viridis* (Brunisholz et al., in preparation).

2.1.2. Purple bacteria with two types of antenna systems

Besides the B880 antenna complex, some photosynthetic bacteria have the ability to form an additional antenna complex, the B800-850, with specific structural features (Figs. 3 and 5). Among the best-characterized bacteria of this class are *Rp. sphaeroides* and *Rp. capsulata.* The 800-850 complex is more stable than the B880 antenna complex, easier to purify and thus better-characterized [37–42]. Recently, on the basis of its spectral properties it was suggested that the B800–850 antenna exists in vivo as aggregates of a minimal unit comprised of 6 BChl *a* molecules (4 BChl-850 nm, with the porphyrin ring perpendicular to the membrane, and 2 BChl-800 nm with the porphyrin ring plane more or less parallel to the membrane), 3 carotenoids (2 associated with BChl-850 nm, perpendicular to the membrane, and one associated with BChl-800 nm, parallel to the membrane) and 4 polypeptides $\alpha_2\beta_2$ (Fig. 7) [40]. The 850 nm absorption band, approx. 1.5 times larger than the 800 nm band, represents a BChl dimer, similar to that in the B880 antenna complexes, whereas a BChl monomer is responsible for the 800 nm band [41,42]. Ac-

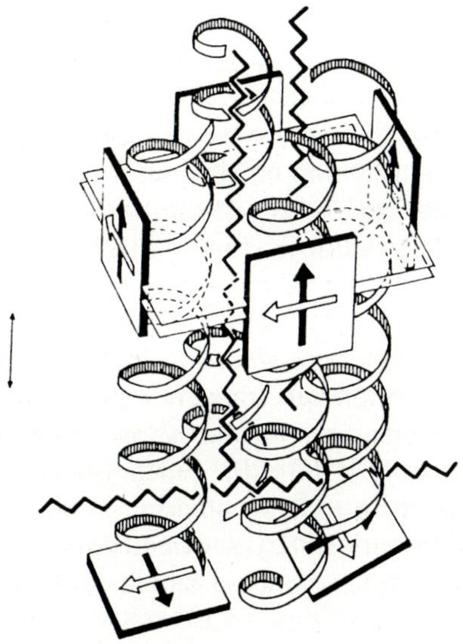

Fig. 7. Schematic picture of the proposed model for the B800-850 antenna complex. The basic unit consists of four BChl 850 molecules (upper boxes), 2 BChl 800 molecules (lower boxes), three carotenoids (zigzag lines) and two proteins, each consisting of two subunits. The helices symbolize the α-helix regions, which are supposed to be transmembrane. The Q_y transitions (open arrows) of two of the BChl 850 molecules (left front and right back) are in the same plane, while the Q_y transitions of the remaining BChl 850 molecules are in a parallel plane, which is vertically displaced by about 1 Å. The Q_x transition moments (solid black arrows) of the BChl 850 are perpendicular to these planes. The Q_y transitions of the BChl 800 molecules are both in a plane parallel to BChl 850 Q_y transitions, while the Q_x molecules are tilted out of this plane at an angle smaller than 24°. The bar represents 5 Å. Taken from Ref. 40.

cording to spectroscopic measurements [43] the BChl dimer of the B800-850 complex of *Rp. sphaeroides* shows the same strong interactive associations (exciton coupling) as found for the B880 complex of *Rs. rubrum,* whereas the B870 complexes of *Rp. sphaeroides* and *Rp. capsulata* exhibit weak CD signals (Table 1). This indicates structural variations of the B870 antenna complexes within the Rhodospirillaceae. CD measurements in the far-UV, ATR-IR (attenuated total reflected IR) spectroscopy and polarized infrared spectroscopy suggest an α-helical transmembrane organization of the antenna polypeptides of *Rp. sphaeroides* [13,28,44]. Accordingly, the α-helices are tilted at less than 40° with respect to the normal to the plane of the membrane. In the light of all this spectroscopic data revealing some structural information the interest focussed on the primary and the three-dimensional structures [46–54] of the antenna pigment-protein complexes. Proteinchemical analyses [48–52], together with DNA sequencing [53,54], yielded the primary structures of the polypeptides of the B800-850 and the B880 complexes of wild type and mutants of *Rp. sphaeroides* and *Rp. capsulata.* They are shown in Fig. 4 in comparison with antenna polypeptides of other sources. Compared to the polypeptides of the B880 complexes they exhibit the following structural characteristics: (1) similar three-domain structure; (2) similar folding of polypeptides in the two types of complexes; (3) similar aggregation tendency. On this basis, a cyclic hexamer $(\alpha\beta)_6$ structure (12 polypeptides) was proposed [2–4]. This complex contains the same basic elements of $\alpha\beta$-heterodimers (Fig. 3).

2.1.3. Purple bacteria with three or more types of antenna systems

Antenna complexes with the absorption maxima at around 800–820 nm have been reported for some purple bacteria. It is well-established that in *Chromatium vinosum, Rp. acidophila* and *Rp. palustris* [55–67] a B800-820 (B800-830) antenna complex is present along with B800-850 and B880. It appeared that its formation can be influenced by (1) light intensity, (2) temperature and (3) carbon source of the medium. In the case of *Chromatium vinosum,* for example, two extreme modifications of near-infrared absorption spectra were observed [65–67] when growing the cells heterotrophically at temperatures either above (large amount of B800-850) or below 36.5°C (large amount of B800-820). Furthermore, *Rp. acidophila* strain 7050 was shown to be very sensitive to light intensity: in dim light conditions (<200 lux) mainly B800-820 (together with B880) is formed, whereas strong light induces the formation of a B800-850 complex which is similar to those of *Rp. sphaeroides* and *Rp. capsulata* [56,58]. A second type of B800-850 complex (type II) with an 800 nm band equal in height to the 850 nm band has always been found along with the B800-820 antenna [58,62–64]. Such a variety of different antenna complexes in certain bacteria poses the intriguing question of the molecular basis for the variable spectral forms of BChl *a*. Furthermore, CD spectra in the near-IR showed that specific BChl types in different bacteria are not necessarily equally organized, even if they show absorption bands at almost the same wavelength [28,63]. The primary structures of the α- and β-polypeptides of the B800-850 complex of *Rp. acidophila* 7750 (major component of cells when grown at a temperature between 30 and 35°C) and strain 7050 (cells in strong light conditions), as well as of the

B800-820 complex of strain 7750 (cells at low temperature, 22°C; Schmidt, Brunisholz and Zuber, unpublished) and of strain 7050 (cells at low light intensity) were established [68]. Apparently, the cells of *Rp. acidophila* strain 7750 and 7050 are able to synthesize specific α/β antenna polypeptides for the B800-850 and the B800-820 light-harvesting complexes. They show the typical three-domain structure and high homology (85–90%) to the corresponding polypeptides of the B800-850 antenna complexes. Among other mutations, the change from Tyr^{44}–Trp^{45} (in B(800-850)-α) to Phe^{44}–Leu^{45} (in B(800-850)-α) is of particular interest. It may be correlated to the shift of the absorption maxima from 800–850 nm to 800–820 nm.

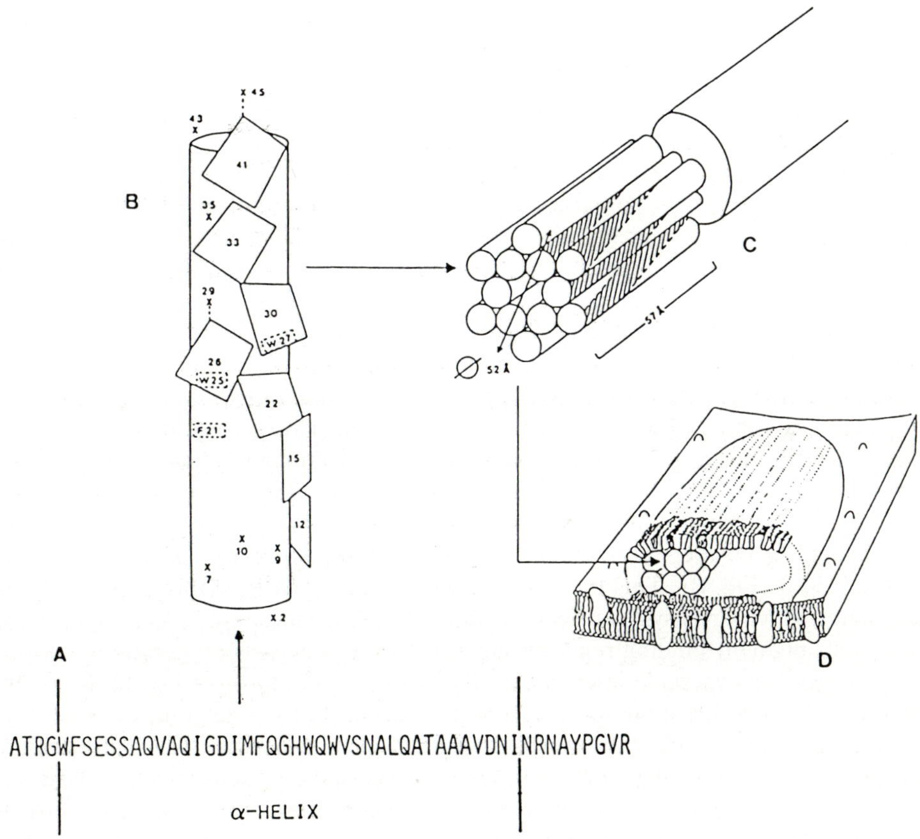

Fig. 8. Amino acid sequence of the BChl *c*-binding (antenna) polypeptide of the chlorosomes from *Chloroflexus aurantiacus* (A) and the proposed α-helix model of this antenna polypeptide (B) with possible BChl *c* binding sites (Gln 12,15,22,26,33, Asn 30,41) via the central Mg atom. (C) Chlorosome subunit ('globular subunit') (light-harvesting BChl *c*-protein complex of *C. aurantiacus*) composed of 12 polypeptide chains (α-helices) of the BChl *c*-binding polypeptide. Pairs of α-helices (dimeric basic units) are tilted to expose the areas for BChl *c* binding (hatched areas). (D) Chlorosome model (cross-section, *C. aurantiacus*) containing the chlorosome subunits of the rod-shaped elements in the chlorosome core region (taken from Refs. 69 and 72).

2.2. *Green photosynthetic bacteria: intramembrane antenna complexes, baseplate systems and the accessory antenna systems (chlorosomes)*

Within the phototrophic bacteria, the green bacteria are unique in the following features.

- Cells lack the intracellular proliferation of unit-membrane structures that bear the majority of pigments in purple bacteria.
- Besides small amounts of BChl *a*, the major chlorophylls are BChl *c, d* and *e,* pigments not found in purple bacteria.
- In all green bacteria the main antenna system is organized in small ovoid particles, the so-called chlorosomes, located adjacent to the cell membrane. The BChl *c* and *d* pigments are associated with this accessory light-harvesting system.

2.2.1. *The antenna system of Chloroflexus aurantiacus*

The photosynthetic apparatus of *Chloroflexus aurantiacus* is located mainly in two cytologically distinct compartments [69] (Fig. 8). The major antenna chlorophyll (BChl *c*) is organized in the so-called chlorosome [70] and absorbs at 740 nm. A BChl *a*-containing light-harvesting system absorbing at 808 and 866 nm, similar to that in purple bacteria, is located within the cytoplasmic membrane and represents the RC – surrounding antenna. Fluorescence spectroscopy clearly demonstrated the existence of a linker antenna (λ_{max} 790 nm) between the chlorosome BChl *c* and the B808–866 antenna [71]. It was shown that BChl *c* is associated with only one polypeptide [72]. Accordingly, approx. 14 molecules of BChl *c* are bound to a dimer of 2 copies of a 5.7 kDa polypeptide, which is consistent with spectroscopic measurements [71]. Primary structure analysis of the 5.7 kDa polypeptide revealed different structural principles (Fig. 8) as compard to the tripartite structure of purple antenna polypeptides [73]. Secondary structure predictions of the BChl *c*-binding polypeptide indicated preferentially α-helix formation between Trp^5 and Ile^{42} [73] (Fig. 8). This particular α-helix binds most probably seven BChl *c* molecules which are asymmetrically arranged on the surface of the α-helix. Opposite this pigment-binding region a segment of the α-helix is apparently involved in polypeptide–polypeptide interactions. Furthermore, as shown in Fig. 8, the BChl *c*-binding polypeptide monomers aggregate specifically to large complexes.

Organic solvent extraction of the cell membranes released two hydrophobic polypeptides. Primary structure analyses [74] revealed a structural relationship to the intra-membrane bound antenna polypeptides of purple bacteria, with a three-domain structure with extended homology around the membrane-buried His residue (Fig. 4). Accordingly, these polypeptides are assigned as B(808-866)-α and B(808-866)-β. It is interesting to note that they exhibit a higher homology to the polypeptides of the B880 complexes of purple bacteria than to the α- and β-polypeptides of the B800-850 antennae.

2.2.2. *The BChl a-protein of Prosthecochloris aestuarii*

The function of the BChl *a*-protein, located between the chlorosomes and the intramembrane antenna system (baseplate) is to accept excitation energy from the chlorosome BChl *c* or *d* and to transfer it to the reaction center BChl *a* [75]. In the case of *Prosthecochloris aestuarii* strain 2K the three-dimensional structure of the water-soluble BChl *a*-protein was determined to a nominal resolution of 2.8 Å [76]. This protein forms a trimer of tightly packed subunits of 46 kDa. Each subunit binds 7 BChl *a* molecules which occupy the space within an ellipsoid of axial dimension 45 × 35 × 15 Å. Furthermore, the electron-density maps showed for the first time specific interactions between the 7 pigments and the polypeptide matrix. Five of the BChl *a* molecules are bridged via their central Mg atom to His residues (5th ligand). The average center-to-center distance between the porphyrin rings is approximately 12 Å for nearest neighbors, whereas the closest distance between BChl molecules in adjacent subunits is 24 Å.

3. *Accessory light-harvesting antenna systems (phycobilisomes) of cyanobacteria, red algae and of cryptomonads*

Cyanobacteria, red algae and cryptophytan algae have, in addition to the antenna complexes of PS I and PS II within the thylakoid membrane, accessory extramembrane light-harvesting antennae composed of pigmented phycobiliproteins and pigment-free linker polypeptides. This antenna system enables these organisms to use additional light energy in the spectral range 450–650 nm. Electron microscopic studies of cyanobacteria (Fig. 9) and red algae show such antenna systems as large granules, the phycobilisomes (PBS), attached to the thylakoid membrane in a regular pattern. The PBS form hemidiscoidal or hemiellipsoidal structures, both with a diameter of 700–800 Å [77–83]. In the hemidiscoidal phycobilisome type [79–83], six rods (composed of 3–7 stacked discs) radiate from a core region in the centre of the PBS (composed of 2 or 3 cylinders). The PBS are coupled by this core with PS II in the thylakoid membrane (Fig. 9E), and the efficiency of the transfer of the light energy is $> 95\%$. The diameter of the discs in the rods was estimated to be 120 Å and the thickness 60 Å (reviewed in Refs. 1,77,84–86). In cryptophytan algae no supramolecular structure of phycobiliproteins could be revealed by electron microscopy [87–89].

Phycobiliproteins are chromoproteins with covalently bound open-chain tetrapyrrole pigment molecules. Phycobiliproteins account for up to 50% of the total cellular proteins and they provide the organisms with their typical blue-green, red or brown colour. A striking property of phycobiliproteins in the dissociated state is the high fluorescence in the visible range of the spectrum (Table 2). Minor protein components (10–20%) of the PBS, the so-called linker polypeptides (mainly without chromophore), are involved in the specific association of the biliproteins to the supramolecular structure of the PBS and in the modulation of light absorption properties and the energy transfer processes. Their presence in cryptophytan biliprotein antenna systems has not yet been confirmed.

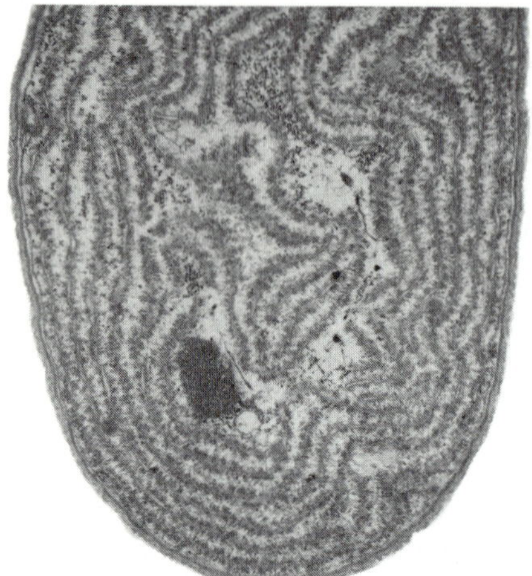

9A

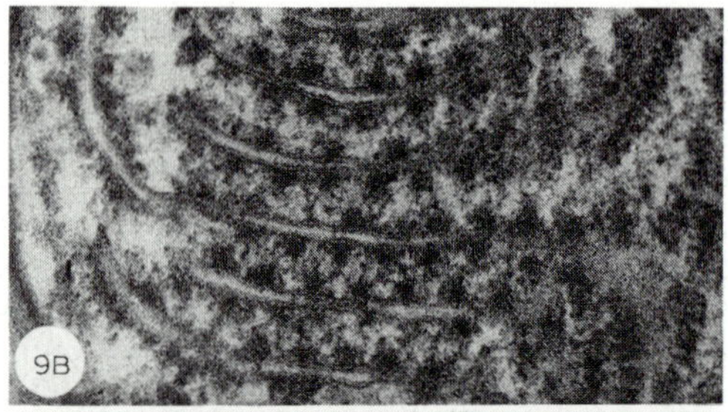

9B

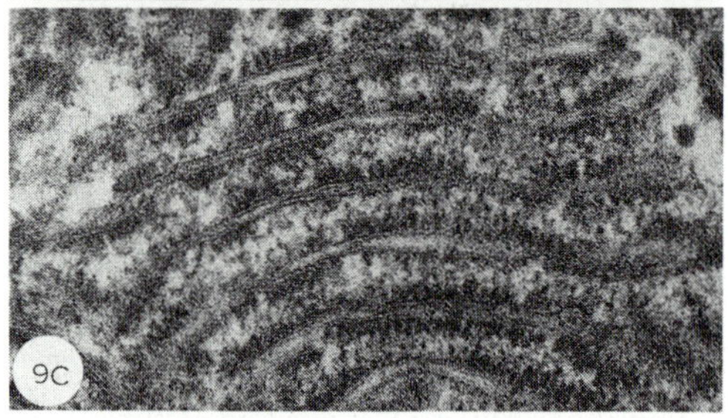

9C

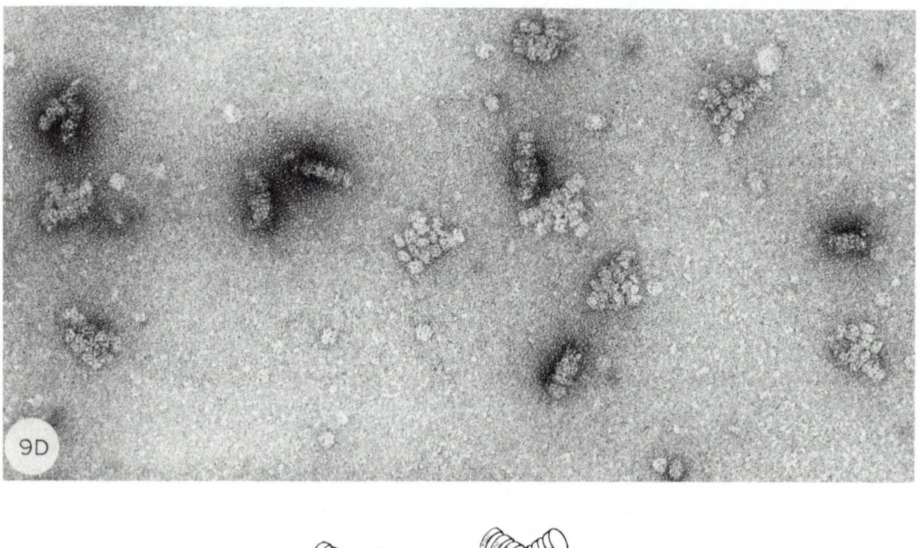

9D

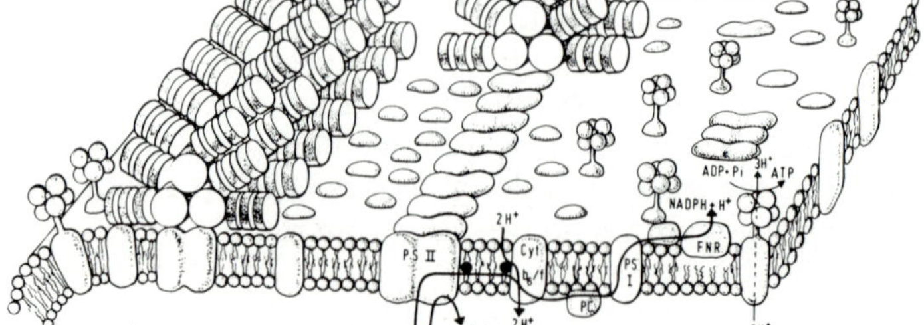
PS II
Cyt b6/f
PS I
PC
FNR
NADPH + H+
ADP • Pi
ATP
3H+
2H+
2H+
2H2O
2H+ + 1/2 O2
3H+

9E

3.1. *Pigment structure and absorption properties of phycobilin chromophores*

The exciting variations in colour and the different types of absorption spectra of the phycobiliproteins originate from five bile pigments (bilins) related to biliverdin, which is their biosynthetic precursor. Biliverdin is an oxidative degradation product of heme, the prosthetic group in hemoglobin, myoglobin and in cytochromes. The different absorption properties of the phycobilins are caused by subtle structural differences in the tetrapyrrole prosthetic group. The most frequently occurring pigments (Fig. 10) are the blue phycocyanobilin (PCB, λ_{max} 600–670 nm), the red phycoerythrobilin (PEB, λ_{max} 540–570 nm) and the red phycourobilin (PUB, λ_{max} 490–500 nm) (for reviews see Refs. 90–92). Less frequently, a purple phycobiliviolin chromophore (or cryptoviolin chromophore), PXB, λ_{max} 550–600 nm, with a typical double peak is found in cyanobacteria and cryptophytan algae [93–98]. A green, as yet unidentified biliverdin-like chromophore, PBV (λ_{max} 660–690 nm) was observed in a cryptophytan biliprotein isolated from *Chroomonas* sp. [99,100]. Phycobilin chromophores are generally singly bound to the polypeptide chain by thioether linkages via ring A or doubly bound via ring A and D [86] (Fig. 10). They are fixed in an extended conformation by their protein environment [92,101,102].

3.2. *Classification, occurrence and distribution of phycobiliproteins*

Phycobiliproteins are hydrophilic (water-soluble) polypeptides and are isolated from the cleared cell extracts by ammonium sulfate precipitation (25–60% saturation) followed by ion-exchange chromatography on DEAE-cellulose [98,103] or by dissociation of isolated PBS followed by ion-exchange chromatography [81,83,104,105]. Four biliprotein classes, isolated from cyanobacteria and red algae, can be differentiated by their absorption maxima: allophycocyanin (APC, λ_{max} 650–680 nm), phycocyanin (PC, λ_{max} 620–635 nm), phycoerythrocyanin (PEC, λ_{max} 575 nm) and the phycoerythrins (PE, λ_{max} 545–565 nm). The phycobiliproteins, although similar in the size of the polypeptides, show typical differences in their amino acid sequences, the types and numbers of chromophores (Table 2) and their location within the PBS (Fig. 11). Phycobiliproteins from cyanobacteria and red

---

Fig. 9. Electron micrographs showing phycobilisomes of *Mastigocladus laminosus* (Fischerella PCC 7603). Preparations and micrographs: W. Wehrmeyer and E. Mörschel. (A) Thin section of *M. laminosus*. The granules, representing the phycobilisomes, appear as electron-dense particles and are arranged in a regular pattern on the thylakoid membrane. (B) Enlargement of a portion of thylakoid membranes, showing phycobilisomes of *M. laminosus* in face view. (C) Enlargement of a part of the thylakoid membranes with phycobilisome rows, cut longitudinally. (D) Detailed view of a phycobilisome from *M. laminosus*. The triangular core is composed of three cylinder-shaped complexes. The rod-shaped structures, formed by double discs (two $\alpha\beta$-trimers form an $\alpha\beta$-hexamer, representing a double disc), are radiating from the sides of the core, as described by Nies and Wehrmeyer [82]. See also Fig. 11. (×200 000). (E) Schematic view of the phycobilisomes and PS II arrangement of a thylakoid membrane. The phycobilisome represents the hemidscoidal type. Drawing by E. Mörschel (unpublished results).

TABLE 2

Properties of the major biliproteins from *Mastigocladus laminosus*,[a] *Fremyella diplosiphon*[b] and *Chroomonas* sp.[c] with known primary structure

Protein	λA_{max}	λF_{max}	Quarternary structure	Bilin content per subunit	Bilin binding positions	Number of residues	% Homology between $\alpha\beta$-subunits	Refs.
APC[a]	650	660	$(\alpha\beta)_3$	α 1PCB β 1PCB	α Cys^{84} β Cys^{84}	160 161	38	103, 112, 117
C-PC[a]	620	643 647	$(\alpha\beta)_3$ $(\alpha\beta)_6L_R^{34,5}$	α 1PCB β 2PCB	α Cys^{84} β Cys^{84}, Cys^{155}	162 172	27	113, 114, 116 151, 152
PEC[a]	575	620	$(\alpha\beta)_6L_R^{34,5}$	α 1PVB β 2PCB	α Cys^{84} β Cys^{84}, Cys^{155}	162 171	23	93, 97, 111
C-PE[b]	565	575–591	$(\alpha\beta)_6$	α 2PEP β 3PEB	α Cys^{84}, Cys^{143a} β $Cys^{50\&61}$, Cys^{84}	164 184	26	115, 118
PC-645[c]	645	675	$(\alpha_1\beta\alpha_2\beta)$	α_1 1PXB α_2 1PXB β 1PVB 2PCB	α Cys^{18} α Cys^{18} β $Cys^{50\&61}$, Cys^{84}, Cys^{155}	70 85–86 177	61 27	98, 156, 157, 159

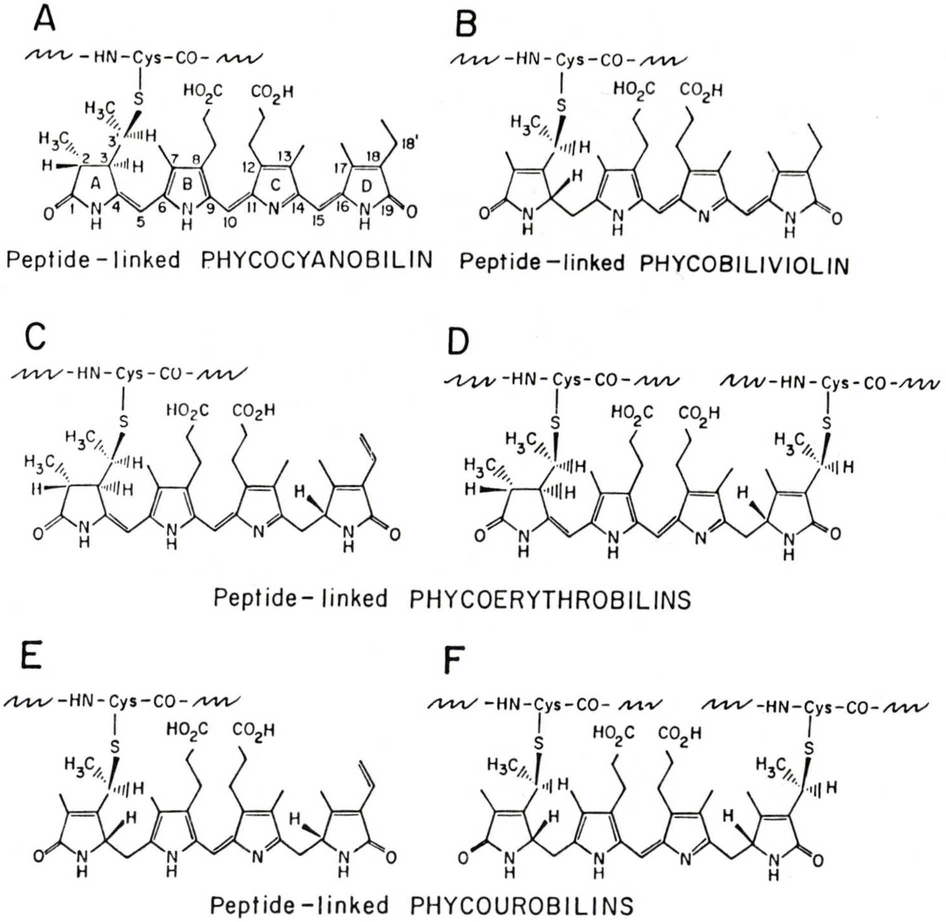

Fig. 10. Bilin-polypeptide linkages in biliproteins. (A) Phycocyanobilin linked through ring A [86]. (B) Phycobiliviolin linked through ring A [94] (J. Bishop. J.C. Lagarias, A.N. Glazer, H. Rapoport, unpublished results). (C) Phycoerythrobilin linked through ring A. (D) Doubly linked phycoerythrobilin. (E) Phycourobilin linked through ring A. (F) Doubly linked phycourobilin (A.V. Klotz, A.N. Glazer, J. Nagy, J. Bishop, H. Rapoport, unpublished results). From Glazer [86,94].

algae are composed of α- and β-subunits with an M_r of about 18–22 kDa. In addition, in red algae a third γ-subunit (M_r about 30 kDa), which has a high chromophore content and functions as a light-harvesting polypeptide, was found [106,107]. Biliproteins can be dissociated to ($\alpha\beta$)-monomers at high dilutions or in chaotropic salts, but most frequently they occur as trimers, $(\alpha\beta)_3$, or hexamers, $(\alpha\beta)_6$. Denaturing agents are required for the separation of the α- and β-subunits [108–111]. All cyanobacteria and red algae contain the blue allophycocyanin $(\alpha^{APC}\beta^{APC})_3$ and the blue phycocyanin $(\alpha^{PC}\beta^{PC})_6$ as phycobiliproteins. The facultatively occurring red biliproteins, which are at the distal end of the PBS rods, are either phycoerythrocyanin $(\alpha^{PEC}\beta^{PEC})_6$, or the phycoerythrins, C-PE $(\alpha\beta)_6$, B-PE

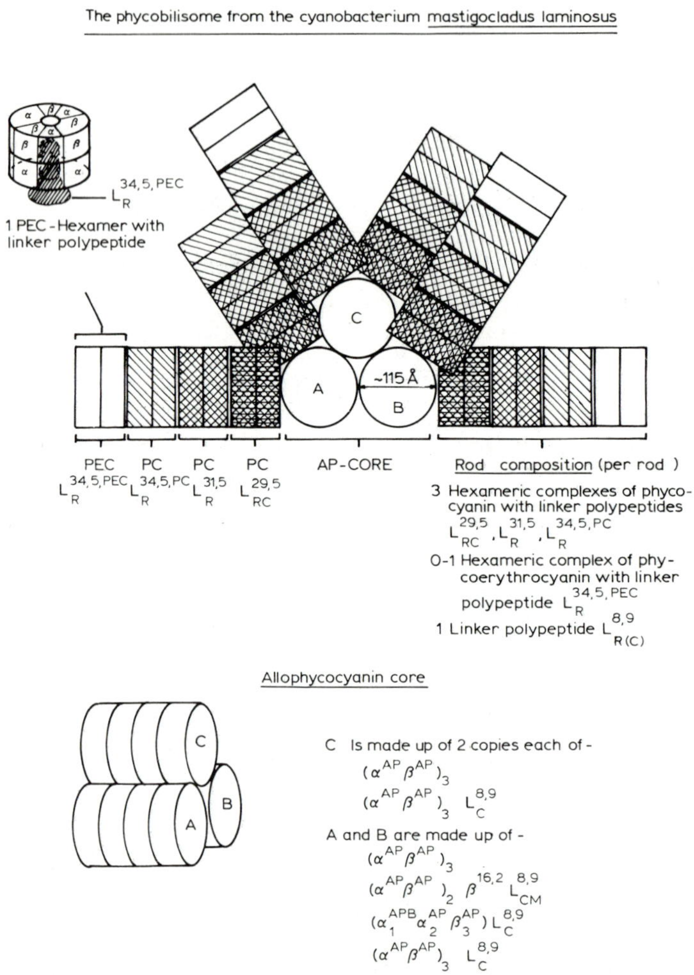

Fig. 11. Phycobilisome of *Mastigocladus laminosus* (Fischerella PCC 7603) represented in a schematic view. The structure is adapted from the PBS of Synechocystis 6701 [77] using structural and spectral data from *M. laminosus* biliproteins and linker polypeptides [22,31,82,97,105,135–137]. For the designation of the rod and core substructures, the nomenclature of Glazer [86] is applied.

or R-PE, forming an $(\alpha^{PE}\beta^{PE})_6\gamma$ complex. The primary structures (amino acid sequences) of the four types of phycobiliproteins from cyanobacteria, APC [103,112], C-PC [113,114], PEC [97] and C-PE [115], and APC [116], PE [117] and the chromopeptides of C-PE [118], R-PE [119] and B-PE [120] from red algae have been established. For example, the amino acid sequences of the whole biliprotein set from the cyanobacterium *Mastigocladus laminosus* and of $\alpha^{C\text{-}PE}\beta^{C\text{-}PE}$ from *Fremyella diplosiphon* are shown in Fig. 12. In Table 2 some physical and chemical properties of the phycobiliproteins, including the number of amino acid residues contained in the polypeptide chains, the types, numbers and binding sites of the phycobilin prosthetic groups and the amino acid sequence homologies between the α- and β-polypeptide chains, are listed.

Allophycocyanin trimers $(\alpha\beta)_3$ (λ_{max} 650–655 nm) represent the main component of the PBS core, which funnels light energy absorbed by the PBS rods to the PS II reaction centers in the thylakoid membranes. A unique form of APC, occurring in small amounts in the PBS core, is the APC-B complex ($\alpha^{APB}\alpha_2{}^{AP}\beta_3{}^{AP}$) which fluoresces at longer wavelengths (675–680 nm) and thus functions as terminal energy acceptor of the PBS together with the 75–90 kDa linker polypeptide [121–123]. APC from *M. laminosus* contains one PCB chromophore per α^{AP}- and β^{AP}-subunit bound by a thioether linkage to Cys ^{84} on each α^{APC}- and β^{APC}-subunit [103]. The sequences of APC-subunits from other cyanobacteria and red algae are 81–89% homologous to APC from *M. laminosus* [112,116].

Phycocyanin hexamers $(\alpha^{PC}\beta^{PC})_6$ (λ_{max} 620–635 nm), together with one of the linker polypeptides from the 30 kDa family, form the discs in the PBS rods attached to the APC core in the PBS [124,125]. In several cyanobacteria the type of C-PC and the number of C-PC discs in the PBS rods is regulated by light intensity and/or light quality (chromatic adaptation, reviewed in Refs. 126 and 127). C-PC from *M. laminosus* (Fig. 12) has two PCB chromophores bound to the $\beta^{C\text{-}PC}$ subunit, the second one being bound to Cys ^{155} within an insertion of 10 amino acid residues at position 151–160 [113]. The PCB chromophores are singly bound to Cys^{84} and Cys^{155} by ring A (Fig. 10).

Phycoerythrocyanins (PEC) and phycoerythrins (PE) are the facultatively occurring red biliproteins, one or the other of which may form the outer discs of the PBS rods. PEs occur in various forms as C-PE, R-PE, B-PE in cyanobacteria and red algae or as PEs in cryptomonads.

Phycoerythrocyanin (PEC, λ_{max} 575 nm) has so far been only in cyanobacteria which do not perform chromatic adaptation [93,111]. It is structurally similar to phycocyanin. The functional difference is caused by the red phycobiliviolin chromophore replacing the blue PCB chromophore bound to Cys^{84} in the α^{PEC}-subunit. The β^{PEC}-subunit contains two blue PCB chromophores bound to Cys^{84} and Cys^{155}, homologous to phycocyanin (Table 2, Fig. 12).

Phycoerythrins occur in various spectral forms. In fresh water and soil cyanobacteria they contain only the red PEB chromophores (λ_{max} 555 nm). In marine cyanobacteria and red algae the red PUB chromophores (λ_{max} 495 nm) in varying numbers and ratios are also found [86,128]. The different forms of phycoerythrins from cyanobacteria and red algae are distinguished by their spectral properties and designated C-PE (λ_{max} 560–565 nm), R-PE (λ_{max} 565–540–498 nm) and B-PE (λ_{max} 545–563–498 nm). In the primary structure of C-PE from *Fremyella diplosiphon* [115] (Fig. 12), two PEB are bound to the α-subunit. In the α-subunit the second chromophore, unique to the α-subunit of PE, is inserted together with a pentapeptide at position 143a. In the β-subunit, a doubly bound chromophore is attached to cysteine residues 50 and 61, similar to the rhodophytan phycoerythrins (B-PE and R-PE) [119,120] (Figs. 10 and 12). A unique peptide insertion of 14 amino acid residues (without chromophore) was found at position 141a–o in the β-subunit [115] (Fig. 12). B-PE of *Porphyridium cruentum* and R-PE of *Gastro-*

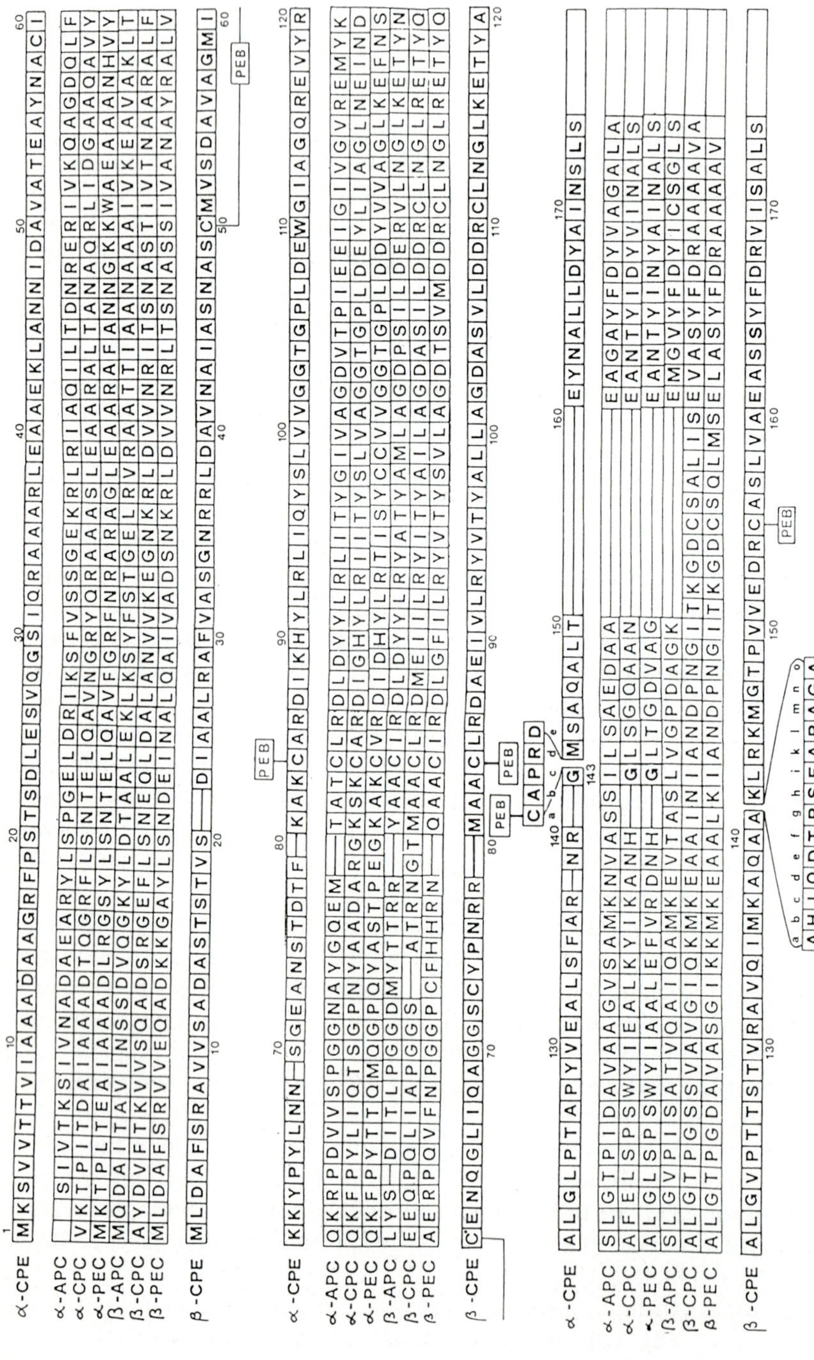

clonium coulteri contain a γ^{PE}-subunit: $(\alpha^{PE}\beta^{PE})_6\gamma^{PE}$. The amino acid sequences of the corresponding chromopeptides show strong homology, including to those of C-PE [86,118–120]. The singly bound chromophores are attached via ring A and the doubly bound chromophores via ring A and D to the polypeptide [86]. B-PE contains two red PEB in the α^{PE}-subunit and three red PEB chromophores on the β^{PE}-subunit, similar to C-PE. In the $\gamma^{B\text{-}PE}$-subunit two PEB and two PUB chromophores have been determined. The doubly bound chromophore in $\beta^{B\text{-}PE}$ at Cys^{50} and Cys^{61} is a PUB-type pigment [120]. Five chromopeptides were found in the $\gamma^{R\text{-}PE}$-subunit which contains one PEB and four PUB chromophores, singly bound by ring A to a Cys in the polypeptide [86,119]. From the amino acid sequence homologies between the α- and β-subunits from APC, CPC, CPE and PEC (Table 2), it was concluded that the phycobiliproteins developed phylogenetically from a single ancestor polypeptide chain, related to myoglobin [101,115,129].

3.3. Linker polypeptides

Sodium dodecyl sulfate/polyacrylamide gel electrophoresis (SDS-PAGE) of phycobilisome samples from different species of cyanobacteria and red algae resolved up to nine colourless polypeptides [130]. On the basis of M_r they are assigned to four classes: M_r 70–120 kD, M_r 25–35 kDa, M_r 16–22 kDa and M_r 8–10 kDa [1,85]. Two of them (the 90 kDa and 16.2 kDa) were found to contain a PCB chromophore and consequently the 'colourless' polypeptides were designated linker polypeptides [105,131,132]. Reconstitution experiments with these linker polypeptides and with dissociated biliproteins showed that they were essential for the correct arrangement of biliproteins to the PBS [133], (Fig. 11), e.g. the formation of the core and the rods, the attachment of the rods to the core, and the exact positioning of the hexameric discs within the phycobilisome rods and the PBS core [124,125]. In the PBS rods, one linker binds to a phycobiliprotein hexamer and, in the PBS core, between one and four linker polypeptides form a complex with APC. The complexes are probably formed by electrostatic interactions between the basic linker polypeptides and the acidic biliproteins [105]. By sticking to the end of the PBS rods they probably also restrict the disc stacking. The various linker polypeptides bound to the trimers or hexamers are able to modulate the spectral properties of the phycobiliprotein hexamers [125]. The absorption maximum at 620 nm obtained from the $(\alpha^{PC}\beta^{PC})_6$ is red-shifted up to 635 nm in the different $(\alpha^{PC}\beta^{PC})_6L_R^{30}$ linker polypeptide complexes. They are the bridging and directing elements for the energy transfer between the different rod and core complexes and

Fig. 12. Amino acid sequence of the three biliproteins from the phycobilisome of the cyanobacterium *M. laminosus* (Fischerella PCC 7603) $\alpha\beta$-APC, $\alpha\beta$-C-PC, $\alpha\beta$-PEC and of C-PE from the cyanobacterium *Fremyella diplosiphon* (Calothrix UTEX 481). [PEB] is the phycoerythrobilin chromophore of C-PE; the PCB-chromophores of $\alpha\beta$-APC, $\alpha\beta$-C-PC and β-PEC and the PXB-chromophore of α-PEC are bound at homologous positions (to Cys^{84} and to Cys^{115}) and are not shown. The names of the amino acids are abbreviated according to the one-letter code (Eur. J. Biochem. (1983) 183, 9–33).

PS II [134]. Primary structure information is available for linker polypeptides of *Synechococcus* 6301 [77] and *M. laminosus* [105]. The complete amino acid sequences of the $L_{RC}^{8.9}$ and the $L_{C}^{8.9\ PEC}$ have been established and large N-terminal amino acid sequences of the $L_{C}^{16.2}$, the $L_{R}^{35.5\ PC}$ and the $L_{R}^{34.5\ PEC}$ have been determined [135–137]. In *M. laminosus* parts of different linker polypeptides show homologies of up to 34% to each other and up to 23% to the biliprotein β^{PEC}. Detailed inspection of the amino acid sequence homologies led to the suggestion that the 30 kDa linker polypeptide family originates from a fusion of α- and β-biliprotein subunit genes (e.g. α^{PEC} and β^{PEC} genes). The various linker polypeptides may therefore be derived phylogenetically from a biliprotein subunit precursor and by duplications and fusions of the subunit genes [135].

3.4. The architecture of the phycobilisome

Several methods for the isolation of PBS have been established [81,104,105,138]. Principally they are based on the observation that PBS are only stable in solutions of high ionic strength, e.g. 0.75–0.9 M potassium phosphate. The PBS are detached from the membranes with 2% Triton X-100. The function of the intact PBS is tested by fluorescence emission spectra at 680 nm upon excitation at 550–650 nm. Most of the structural data describing the PBS originate from the hemidiscoidal type, as reviewed in Refs. 1,77 and 79–86. The complex architecture of the PBS rods and core is best described for the PBS of the cyanobacterium *Synechococcus* 6301 (a cyanobacterium which contains C-PC but neither PEC nor C-PE in the rods) with a bicylindrical core and for the PBS of the cyanobacterium *Synechocystis* 6701 (which contains C-PC and C-PE in the rods) with a tricylindrical core (reviewed in Refs. 1, 139 and 140). Each cylinder in the core is formed by four complexes of APC trimers with linker polypeptides: cylinder A, $(\alpha^{AP}\beta^{AP})_3$; B, $(\alpha^{AP}\beta^{AP})_3\cdot$10K; C, $(\alpha_1^{APB}\alpha_2^{AP}\beta_3^{AP})\cdot$10K ; D, $(\alpha^{AP}\beta^{AP})_2\cdot$18.5K$\cdot$99K. The six rods contain four biliprotein hexamers, each hexamer associated with a different linker polypeptide of the 30 kDa family. A rod-core-PS II connecting 18 S subassembly particle was isolated from *Synechococcus* 6301 with the polypeptide composition $((\alpha^{PC}\beta^{PC})_3 L_{RC}{}^{27})_2 - ((\alpha^{AP}\beta^{AP})_3) - ((\alpha^{AP}\beta^{AP})_2\cdot\beta^{18.3} L_{CM}{}^{75})$ [141]. The tricylindrical PBS model may also be adapted to the PBS of *M. laminosus* [82,83,105,135,142], as shown in detail in Fig. 11, and to other numerous cyanobacteria and red algae, whereas the bicylindrical core type PBS was found only in *Synechococcus* 6301.

3.5. The three-dimensional structure and the function of phycobiliproteins

In the PBS rods the phycobiliprotein hexamers can be identified by high-resolution electron microscopy as discs, subdivided into two halfs ($\alpha\beta$-trimers) of 30 Å thickness [80,143]. Deeper insight into the molecular structure of the trimers and hexamers was achieved by X-ray crystallographic analyses of biliproteins. In the last century, strikingly coloured phycocyanin and phycoerythrin crystals had already been observed by Molish [144]. Recently, several C-phycocyanins [145–147], B-phycoerythrin [147,148] and phycoerythrocyanin [149] have been crystallized successfully. Using the vapor diffusion method, C-PC isolated from the cyanobacterium *M. laminosus* formed crystals allowing X-ray diffraction analysis at 3 Å resolution [101,145]. As shown in Fig. 13C, three identical ($\alpha\beta$)-subunits (Fig. 13A,B) are arranged around a three-fold symmetry axis and form a disc with dimensions 110 × 30 Å resembling a water wheel with a central hole of 35 Å diameter. This trimer may correspond in its dimensions to a biliprotein half disc in the PBS rods. The C-PC ($\alpha\beta$)-monomer (Fig. 14A) possesses a local two-fold rotational symmetry axis and the α- and β-subunits of the C-PC monomer are thus very similar. In the α- and β-subunits, six of eight α-helices (A,B,E,F,G,H) forming the globular domain (Fig. 14A) are structurally related to the globin folding of myoglobin [101]. This reveals an interesting phylogenetic relationship between an antenna polypeptide of an oxygen-producing cyanobacterium (prokaryote) and an oxygen-binding polypeptide of higher organisms (vertebrates). The other two helices (X,Y) provide a strong contact between the C-PC α- and β-subunits (Fig. 14A). In addition to the close structural relationship between C-PC and myoglobin, the chromophore binding sites of the α^{84} and the β^{84} PCB chromophores are at positions similar to the histidine-heme binding site of myoglobin (E,e-loop). Using structural information from the *M. laminosus* C-PC structure [101] the three-dimensional structure of an $(\alpha\beta)_6$ hexameric C-PC from the cyanobacterium *Agmenellum quadruplicatum* was determined [102]. The $(\alpha\beta)_6$ hexamer is formed by two head-to-head aggregated trimers (Fig. 14B). The different extended chromophore configurations are mainly determined by the interaction of the propionic side chains with arginine residues and of the pyrrole nitrogens with aspartates [102]. The $(\alpha^{PC}\beta^{PC})_6$ hexamers form a central cave where a linker polypeptide from the 30 kDa family is probably buried in the $(\alpha^{PC}\beta^{PC})_6L_R$ complex. These linkers are presumed to link the discs and to be in tight contact with the β^{84} chromophores. They determine the configuration of the chromophores, which is correlated with the direction of the excitation energy flow from disc to disc, along the PBS rods to the PBS core by modulating the optical properties of the different biliprotein hexamers.

The combination of spectral data with the knowledge of the three-dimensional structure of C-phycocyanin and of the phycobilisome architecture resulted in the first insight into the possible pathway and the kinetics of energy transfer in this complex light-harvesting antenna. The different functions of the three PCB chromophores in the C-PC molecules, as postulated by Teale and Dale [150], could be determined within the C-PC trimers isolated from *M. laminosus* [151]: a sensitizing (s) function was assigned to the α^{84} and the β^{155} PCB (s-chromophores) and a fluorescing function to the β^{84} PCB (f-chromophore), which points into the central hole of the trimer (Fig. 13). In the C-PC trimer, prepared from *M. laminosus,* the energy transfer occurs from the α_s-chromophore in one monomer to the β_f-chromophore in the adjacent monomer, and from the β_s-chromophore to the β_f-chromophore in the same monomer [151,152]. In the C-PC trimer and hexamer, the direction of energy flow is from the outside to the inside of the hexamer, where the 30 kDa-linker polypeptide is located. A specific role for aromatic amino acid residues in this energy transfer between the chromophores is suggested [152].

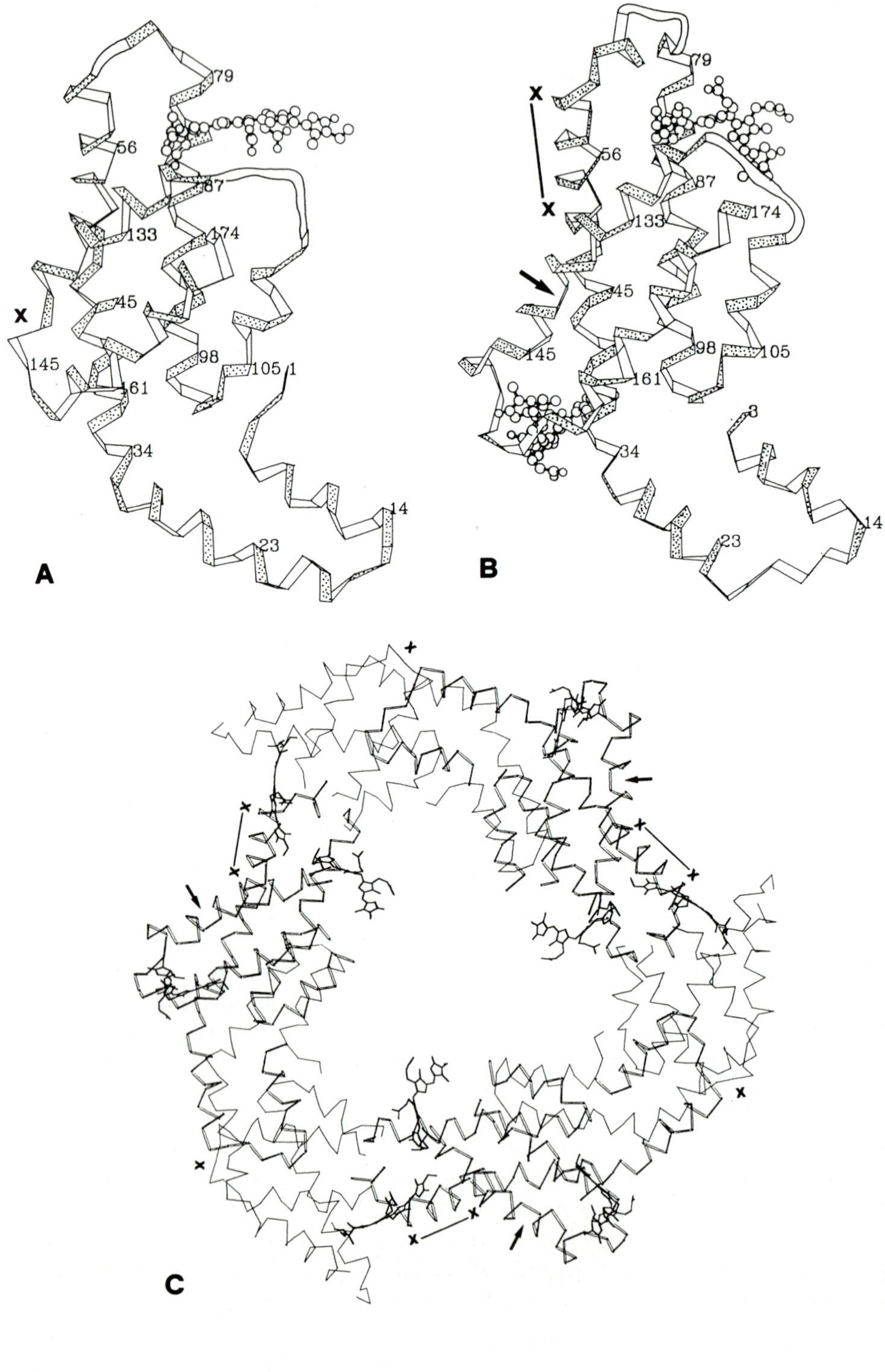
79
56
87
133
174
X
45
145
161
98
105
1
34
14
23
A
X
X
79
56
87
133
174
45
145
98
105
161
3
34
14
23
B
C

Further energy transfer within the rods may be performed by the β_f-chromophores of the PBS rods, modulated by the various 30 kDa-linker polypeptides, and directed along the central rod channel towards the APC complexes of the PBS core. Excitation energy transfer in the PBS was found to occur in up to five steps with different energy transfer rates [134]: within the discs (hexamers) from (s)- to (f)-chromophores and from (f)- to (f)-chromophores, from disc to disc, from the rods to the APC complexes in the core, to the APB-L_{CM} complex and to Chl *a* in the PS II complex. The disc-to-disc excitation energy transfer (20 ps) was supposed to be the rate-limiting step.

3.6. Cryptomonad phycobiliproteins

In contrast to cyanobacteria and red algae, the cryptomonad phycobiliproteins are located in the intra-thylakoidal lumen of the chloroplasts [87–89,153]. Cryptomonad phycobiliproteins do not aggregate to phycobilisomes but absorb light energy in the same spectral range from 450 to 650 nm as the PBS from cyanobacteria and red algae. According to their absorption maxima, the different cryptomonad biliproteins isolated from various cryptomonads are designated phycocyanins (PC) or phycoerythrins (PE). Cryptomonad biliproteins are tetrameric complexes ($\alpha_2\beta_2$) and two cryptoviolin chromophores [156] and phycoerythrin-545 (PE-545) from subunits. Phycocyanin-645 (PC-645) from *Chroomonas* sp. is built up of four subunits [154], two different α-subunits, α_1 and α_1 with a molecular mass of 9 kDa and 10 kDa, respectively, and two β-subunits with an estimated molecular mass of 15 kDa. Three different types of chromophores were located in the PC-645 molecule, which fulfills the same function as the much larger PBS [156,157]: two blue PCB chromophores and one cryptoviolin (or phycobiliviolin-like) chromophore are attached to the β-subunit and one green biliverdin-like chromophore on each of the different α-subunits. PC-612 from *Hemiselmis virescens* ($\alpha_2\beta_2$) contains six PCB and two cryptoviolin chromophores [156] and phycoerythrin-545 (PE-545) from teins from *M. laminosus.* The β-subunits of PC-645 and PE-545 showed 56% and and cryptoviolin chromophores in the α-subunits [158]. Cryptomonad biliproteins have been separated by ion-exchange chromatography on CM-cellulose in 8 M urea at low pH [159,160].

N-terminal amino acid analyses [159] of PC-645 and PE-545 revealed that the α-subunits have lost about 60 amino acid residues at their N-terminus compared with the cyanobacterial α-subunits. The chromophores are bound to Cys^{18} or Cys^{19}, respectively, homologous to Cys^{84} in cyanobacterial and rhodophytan biliproteins

Fig. 13. Three-dimensional structure of C-phycocyanin (C-PC): α-subunit and β-subunit (A and B) and $\alpha\beta$-trimer (C) derived from X-ray diffraction analysis of crystals of C-PC isolated from the cyanobacterium *Mastigocladus laminosus* (Fischerella PCC 7603) [101,145]. The possible positions of the additional PEB chromophores in C-PE from *Fremyella diplosiphon* (Calothrix UTEX 481) are adapted to the models of C-PC, using data from the amino acid sequence of C-PE [115]. Positions of the PEB chromophores at cysteine 50/61:X--X and 143a:X. The beginning of the insertion 141a–o in β-C-PE is indicated by an arrow (→).

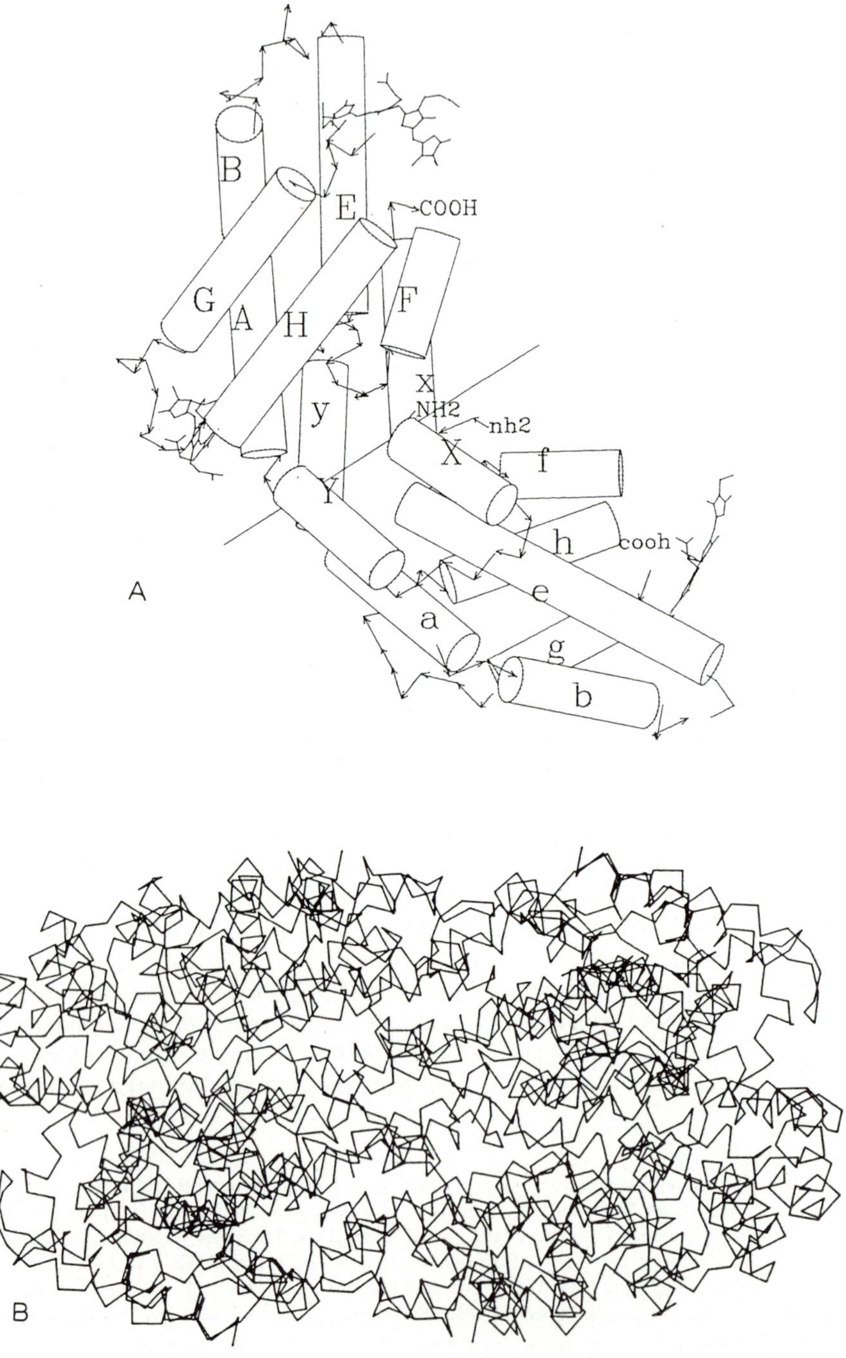
B
E
COOH
G
A
H
F
X
NH2
nh2
y
X
f
Y
h
cooh
e
a
g
b
A
B

[159]. The α-subunits are 30–34% sequence homologous to cyanobacterial biliproteins from *M. laminosus.* The β-subunits of PC-645 and PE-545 showed 56% and 53% homology to β-PEC from *M. laminosus,* similar to the *Hemiselmis virescens* (Millport 64) phycocyanin-β-subunit [161]. In the primary structure of the PC-645 β-subunit from *Chroomonas,* the cryptoviolin-chromophore is doubly bound to Cys^{50} and Cys^{61} [98], similar to the doubly bound PEB and PUB chromophores in the β-subunit of phycoerythrins. The blue PCB chromophores are at homologous positions (Cys^{84} and Cys^{155}) to cyanobacterial and rhodophytan biliproteins. The 60 amino acid residues deleted at the N-terminus of the cryptomonad biliprotein α-subunits concern the x-, y- and a-helices, which provide the α-β-subunit contact of cyanobacterial and red algal phycobiliproteins. The determination of the three-dimensional structure by X-ray crystallographic analysis was initiated by crystallization of *Chroomonas* PC-645, which forms trigonal and orthorhombic crystals and reflects beyond 3.3 Å [162].

4. *Light-harvesting antennae of algae and higher plants*

4.1. *General features*

The thylakoid membrane in algae and higher plants is highly differentiated for regulatory reasons. With the photosystems I (PS I) and II (PS II), it contains considerably more complex antenna systems than those in bacteria [163,164]. This is the case both for antenna complexes in direct proximity to the RC (core complexes CC I, CC II) and for those further away (for example LHC I and II) [1,165–169]. The structural complexity of the antenna complexes is based in part on the heterogeneity of the antenna polypeptide structure. Antenna polypeptides are partly located within the membrane, and as such have the character of a membrane protein. In exposed regions on the surface of the membrane, however, they show the characteristics of a globular protein. This heterogeneity of polypeptide structure also results in greater heterogeneity in the binding sites of the pigment molecules (Chl, carotenoids). Moreover, it has become more and more evident (in LHC II, for example) that there is a genetic heterogeneity of antenna polypeptides (a multitude of structurally similar polypeptides), probably for regulatory reasons. Either the antenna complexes contain more types of very similar polypeptides, or there exist several antenna complexes which differ only minimally in the structure of the antenna polypeptides. Heterogeneity in Chl *a* and Chl *b* structures among

Fig. 14. (A) Cylinder model representing the eight α-helices from the C-phycocyanin α- and β-subunits (α-β monomer) [101] of *Mastigocladus laminosus* (Fischerella PCC 7603). The helices A (a), B (b), E (e), F (f), G (g) and H (h) are designated according to the structure and the nomenclature of myoglobin [101]. The helices Y and X form a tight contact between the subunits and have no counterpart in the myoglobin structure. (B) Side-view (along x-axis, i.e. crystallographic 2-fold symmetry axis) of the C_α-backbone of a C-phycocyanin ($\alpha\beta$)-hexamer derived by X-ray diffraction analysis of C-PC crystals from *A. quadruplicatum* [102]. In this side-view, the two $\alpha\beta$-trimers, aggregated head to head, correspond to the two half-discs, visible by electron microscopy in the PBS-rods [80,82,83].

the various antenna complexes has also been found [170–173]. The functional significance of these diverse heterogeneities of antenna complexes most probably resides in the fact that this would allow for the formation of a highly regulated, that is, variable antenna system for heterogeneous, directed energy transfer. For this reason it is not surprising that, during the course of evolution, a multitude of antenna structures of the various species (cyanobacteria, red algae, green algae, brown algae, dinoflagellates, diatoms, etc.) arose in this highly developed photosynthetic apparatus composed of PS I and PS II. The complexity of antenna systems and antenna polypeptides has caused considerable difficulty in the identification, characterization and structural analysis of these molecules.

An unusual feature of the thylakoid membrane in algae and higher plants is its lateral heterogeneity, that is, its organization in stacked (appressed) grana regions with PS II and unstacked (non-appressed) stroma regions mainly with PS I (see Chapter 12). Correspondingly, there is a non-uniform distribution of antenna complexes in, mainly, LHC I and CC I for PS I and LHC II and CC II for PS II. The size of the appressed grana regions, or the non-appressed stroma regions, regulates the distribution of excitation energy between PS I and PS II. The antenna complex LHC II (phosphorylation of the N-terminal region of the antenna polypeptide) plays an important role [174–176] in this regulation.

4.2. Antenna complexes of photosystem I

In PS I, as in PS II, there are a number of Chl *a*/*b* protein complexes having light-harvesting and energy-transfer functions. Such complexes most probably exist in direct contact with the RC (part of the core complex), and certainly exist as peripheral LHC I antenna complexes further removed from the RC. A native PS I complex (80–180 Chl per RC, 100 kDa) with at least 6 polypeptides was isolated by solubilization of the thylakoid membrane with nonionic detergents (for example, Triton X-100) [177,178]. With further detergent treatment, the PS I complex dissociated into the core complex (CC I with the RC) and the peripheral antenna complex (LHC I) (spinach, barley, pea, *Chlamydomonas reinhardtii* [179–183]. The peripheral antenna complex (pea, spinach; Chl *a*/*b* ratio 4.0:1, typical fluorescence at 730 nm) contains 3–4 antenna polypeptides (19–25 kDa) [181,184,185]. This complex was also dissociated into two different antenna complexes – LHC Ia (2 polypeptides of 22 and 23 kDa) and LHC Ib (1 polypeptide, 20 kDa) – which differ in their fluorescence characteristics (680 nm and 730 nm) [184]. No structural data on these polypeptides are available at present. It was postulated that in *C. reinhardtii,* in addition to the peripheral antenna complex, an antenna system (with 4 polypeptides) exists, which connects the peripheral antenna energetically with the core complex CC I [183].

The core complex (CC I, CP I) with RC P700 (10–65 BChl *a,* 1–2 β-carotenes found in various preparations of pea, spinach, barley and *C. reinhardtii*) contains two main types of polypeptides (60–70 kDa) and small amounts of three to five polypeptides of unknown function (some may also have an antenna function) [165,186]. The relatively large number of bound Chl *a* molecules in the core complex (10–65), or in the polypeptides (20 each), leads to the assumption that most of these may have an antenna function (energy transfer).

The primary structure of the two large CC I polypeptides in maize chloroplasts (A_1, A_2, 83.2 and 82.5 kDa, 752 and 736 amino acid residues) was determined on the basis of their DNA sequence [186]. These two polypeptides, which very probably evolved parallel to one another after gene duplication of a precursor polypeptide (gene family, see Section 4.3), are 45% homologous in amino acid sequence. They contain 49–52% hydrophobic amino acid residues. The hydropathy plot of the two polypeptides is very similar and shows 11 possible membrane-spanning helices with 24–25 amino acid residues each. 73% of the conserved histidine residues, which may be binding sites for the hypothetical antenna Chl *a* molecules, lie in the membrane-buried segments. The important question which arises in terms of the functional equivalence or divergence of the two large polypeptides (any combinations of which may be associated with the RC or have antenna function) cannot be resolved on the basis of presently available structural and functional data (see Section 4.3 and Chapter 10).

4.3. Antenna complexes of photosystem II

The antenna complex LHC II (Chl *a*/*b* protein), which contains approximately half of the polypeptides, and of the Chl, in green plants, can be isolated by means of Triton X-100 solubilization of the thylakoid membrane and subsequent sucrose gradient centrifugation and Mg^{2+}-induced aggregation of the complex. Its pigment ratio is 4 Chl *a*, 3–4 Chl *b*, 1–2 xanthophylls; its Chl-protein ratio is 6–7:1 or 13:1 [187–191]. The polypeptides evidently form a multi-polypeptide family [192]. They originate from a multi-gene family in differing and variable amounts. It is not yet known whether the various antenna polypeptides are built into the same LHC II complex, or whether various LHC II complexes, having only one type of polypeptide each, exist. The primary structure of the Chl *a*/*b* protein (the major Chl *a*/*b* protein of pea, polypeptide 15, 228 amino acid residues) of the LHC II complex was determined by DNA sequence analysis of the c-DNA clone (pAB 96) [193]. In addition, the amino acid sequence of the precursor of the Chl *a*/*b* binding protein was derived from a larger nuclear gene, both from pea clone (AB 80, 269 amino acid residues) [194] and from *Lemna gibba* (pLg AB 19/H5 C; 264 amino acid residues) [195]. In both cases, the polypeptide chain becomes shortened at the N-terminus with the formation of the mature Chl *a*/*b* protein (36, or 35, amino acid residues are split off). The Chl *a*/*b* proteins of pea and *Lemna gibba* are up to 85% homologous in amino acid sequence. Also homologous to these polypeptides are the Chl *a*/*b* proteins of petunia LHC II [196,197]. In the latter case, the primary structure of five very similar polypeptides (sequence homology > 90%) of the multigene family of the Chl *a*/*b* protein of this organism was determined.

The functional significance of this multitude of Chl *a*/*b* proteins is unknown. Based on primary structure data and the hydropathy plot of the polypeptide chain, a three-dimensional structure model of the LHC II complex with three transmembrane segments (40% helices) was proposed for *Lemna gibba* (Fig. 15) [195]. This

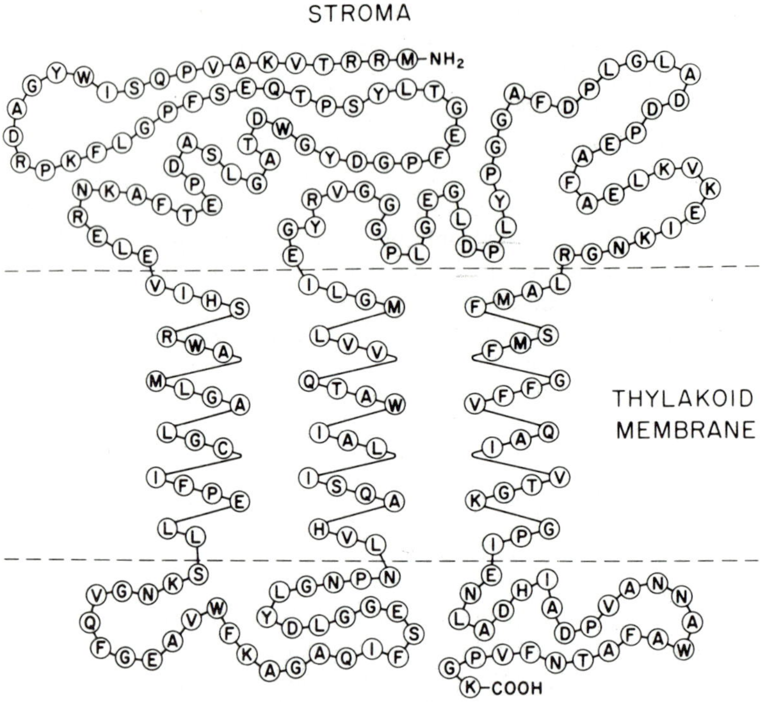

Fig. 15. Three-dimensional structure model of the LHC II (Chl *a*/*b*-protein) of *Lemna gibba* with three helical transmembrane segments in the thylakoid membrane (Karlin-Neumann et al. [195]).

model shows that a large part of the polypeptide chain (48% including the N-terminus) lies outside the lipid bilayer on the stromal side of the membrane, and a smaller part (with the C-terminus) lies on the luminal side. Its main features agree with those postulated on the basis of electron microscope studies and image reconstruction analysis of LHC II complexes arranged in two-dimensional crystals [198,199]. The model shows LHC II as trimers, with three symmetrically arranged subunits and a strong transversal asymmetry (Fig. 16). The more exposed part of the complex (10–15 Å protrusion) may lie on the stroma side of the membrane in vivo. Within the subunits, a domain structure (3–4 domains) of the polypeptide chain is also visible. A similar three-dimensional structure of LHC II complexes with transversal asymmetry was also found, using a similar method, in LHC II complexes incorporated into phospholipid vesicles [200].

One has to assume that for optimal energy transfer in the PS II, the LHC II and the core complex CC II are in close contact. In this case, there are possibly also antenna systems of some form in the direct vicinity of the RC. The PS II core complex CC II of spinach and barley was isolated by means of detergent treatment of the membrane [201–206]. It contains at least 7 polypeptides: the 47–51 kDa (CP 47, CP III, CPa-1) and 43 kDa (CP 43, CP IV, CPa-2) polypeptides with primary reaction activities (see below), two 32 kDa (the so-called D-1 and D-2 polypep-

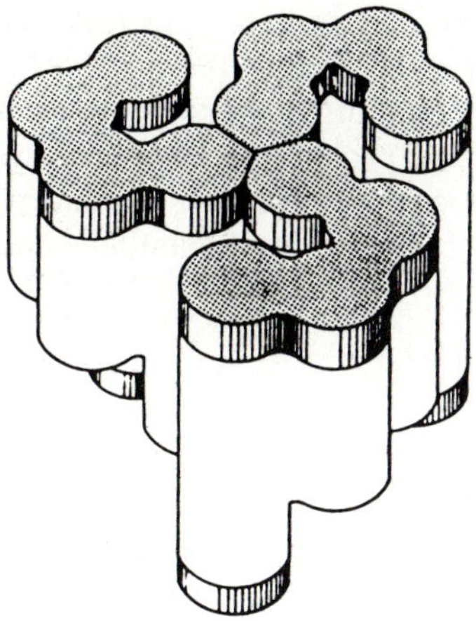

Fig. 16. Three-dimensional structure of pea LHC II determined at 16 Å resolution by electron microscopy of two-dimensional crystals and image analysis (Kühlbrandt et al. [198,199]).

tides), the 11 kDa (Cyt *b*-599) polypeptide and three hydrophilic polypeptides (34, 23 and 16 kDa, involved in water photolysis). Under conditions which do not denature the polypeptides, two functionally different Chl-binding polypeptides can be obtained from the CC II complex [207–210]. The spectral characteristics (695 nm fluorescence) of CP 47 suggest that it contains P-680 and the primary acceptor pheophytin. CP 43, with a fluorescence emission maximum at 685 nm, is probably an antenna complex which surrounds the RC (see below) [205,206].

Genes of CP 47 and CP 43 were localized in the chloroplast. The CP 47 gene is 70 kbp away from the gene of the 32 kDa herbicide-binding polypeptide (D-1), and the gene of the second 32 kDa polypeptide (D-2) overlaps the gene of the 44 kDa polypeptide. The amino acid sequences of the 47 (51) kDa polypeptides were determined by the DNA sequencing (508 and 473 amino acid residues respectively) [209,211,212]. Again, approximately 50% of the amino acid residues are hydrophobic, and for both polypeptides the hydropathy plot indicates 5–7 transmembrane segments with about 25 hydrophobic amino acid residues each (α-helices). In the CP 47 and CP 43 characteristically distributed conservative His residues and cysteine residues, which may possibly be Chl binding sites, were found. A similar situation exists with the polypeptide pairs of the PS I core complex (CC I) and the 32 kDa polypeptides. The CP 47 and CP 43 are only 20% sequence homologous; the homology is, however, 40–60% within the hydrophobic cluster region of the transmembrane segments. This seems to indicate that the polypeptide chain is similarly folded in the transmembrane segments. This structural similarity between the two polypeptides (and also the possibly identical binding site for Chl molecules) does not correspond well with the hypothesis that CP 43 has the function of an antenna polypeptide and that CP 47 contains the RC. However, recently, on the basis of sequence homology between D-1 and D-2, and the sequence homology between these two polypeptides and the L and M subunits of the reaction center of purple bacteria (see Chapter 3), it has been postulated that the D-1 and D-2 polypeptides form the core of the PS II RC [213]. In this case, CP 47 and CP43 would both represent antenna polypeptides. Whatever the case, it can be emphasized here that most probably a combined system (antenna pigments – special pair) exists in the core complex CC II (and possibly in the CC I complex) for the directed energy transfer from the antenna complexes to the RC.

References

1 Glazer, A.N. (1983) Annu. Rev. Biochem. 52, 125–157.
2 Zuber, H. (1985) Photochem. Photobiol. 42, 821–844.
3 Zuber, H. (1985) in Antennas and Reaction Centers of Photosynthetic Bacteria (Michel-Beyerle, M.E., ed.), Springer Series in Chemical Physics 42, pp. 2–14, Springer-Verlag, Berlin, New York.
4 Zuber, H., Sidler, W., Füglistaller, P., Brunisholz, R. and Theiler, R. (1985) in Molecular Biology of the Photosynthetic Apparatus (Arntzen, C., Bogorad, L., Bonitz, S. and Steinback, K., eds.), Cold Spring Harbor Symposia.
5 Pearlstein, R.M. (1982) in Photosynthesis: Energy Conversion by Plants and Bacteria (Govindjee, R., ed.) Vol. 1, pp. 293–330, Academic Press, New York.
6 Knox, R.S. (1977) in Topics in Photosynthesis (Barber, J., ed.) Vol. 2, pp. 55–97, Elsevier, Amsterdam.
7 Pearlstein, R.M. (1982) Photochem. Photobiol. 35, 835–844.
8 Förster, T. (1948) Ann. Phys. (Leipzig) 2, 55–75.
9 Zuber, H. (1985) Optical Properties and Structure of Tetrapyrroles, pp. 425–441, Walter de Gruyter, Berlin, New York.
10 Pearlstein, R.M. and Zuber, H. (1985) in Antennas and Reaction Centers of Photosynthetic Bacteria (Michel-Beyerle, M.E., ed.), Springer Series in Chemical Physics 42, pp. 53–61, Springer-Verlag, Berlin, New York.
11 van Grondelle, R. (1985) Biochim. Biophys. Acta 811, 147–195.
12 Thornber, J.P., Cogdell, R.J., Pierson, B.K. and Seftor, R.E.B. (1983) J. Cell. Biochem. 23, 159–169.
13 Nabedryk, E. and Breton, J. (1981) Biochim. Biophys. Acta 635, 515–524.
14 Cogdell, R.J. and Valentine, J. (1983) Photochem. Photobiol. 38, 769–772.
15 Picorel, R., Bélanger, G. and Gingras, G. (1983) Biochemistry 22, 2491–2497.
16 Cogdell, R.J. and Thornber, J.P. (1979) Ciba Foundation Symp. 61, Excerpta Medica, Amsterdam, 61–79.
17 Cogdell, R.J., Lindsay, J.G., Valentine, J. and Durant, I. (1982) FEBS Lett. 150, 151–154.
18 Giménez-Gallego, G., Fenoll, C. and Ramírez, J.M. (1986) Anal. Biochem. 152, 29–34.
19 Sauer, K. and Austin, L.A. (1978) Biochemistry 17, 2011–2019.
20 Gogel, G.E., Parkes, P.S., Loach, P.A., Brunisholz, R.A. and Zuber, H. (1983) Biochim. Biophys. Acta 746, 32–39.
21 Brunisholz, R.A., Cuendet, P.A., Theiler, R. and Zuber, H. (1981) FEBS Lett. 129, 150–154.
22 Brunisholz, R.A., Suter, F. and Zuber, H. (1984) Hoppe-Seyler's Z. Physiol. Chem. 365, 675–688.
23 Bérard, J., Bélanger, G., Corriveau, P. and Gingras, G. (1986) J. Biol. Chem. 261, 82–87.
24 Wiemken, V., Brunisholz, R.A., Zuber, H. and Bachofen, R. (1983) FEMS Microbiol. Lett. 16, 297–301.
25 Brunisholz, R.A., Wiemken, V., Suter, F., Bachofen, R. and Zuber, H. (1984) Hoppe-Seyler's Z. Physiol. Chem. 365, 689–701.
26 Wiemken, V. and Bachofen, R. (1982) Biochim. Biophys. Acta 681, 72–76.
27 Brunisholz, R.A., Zuber, H., Valentine, J., Lindsay, J.G., Woolley, K.J. and Cogdell, R.J. (1986) Biochim. Biophys. Acta 849, 295–303.
28 Cogdell, R.J. and Scheer, H. (1985) Photochem. Photobiol. 42, 669–678.

29 Meister, H.P., Bachofen, R., Semenza, G. and Brunner, J. (1985) J. Biol. Chem. 260, 16326–16331.
30 Miller, K.R. (1979) Proc. Natl. Acad. Sci. USA 76, 6415–6419.
31 Wehrli, E. and Kübler, O. (1980) in Proceedings of Life Science: Electron Microscopy at Molecular Dimensions (Baumeister, W. and Vogell, W., eds.) pp. 48–56, Springer Verlag, Heidelberg, New York.
32 Stark, W., Kühlbrandt, W., Wildhaber, I., Wehrli, E. and Mühlethaler, K. (1984) EMBO J. 3, 777–783.
33 Engelhardt, H., Baumeister, W. and Saxton, W.O. (1983) Arch. Microbiol. 135, 169–175.
34 Brunisholz, R.A., Jay, F., Suter, F. and Zuber, H. (1985) Biol. Chem. Hoppe-Seyler 366, 87–98.
35 Steiner, R. and Scheer, H. (1985) Biochim. Biophys. Acta 807, 278–284.
36 Steiner, R., Angerhofer, A. and Scheer, H. (1986) Z. Naturforsch. 41c, 571–592.
37 Cogdell, R.J. and Crofts, A.R. (1978) Biochim. Biophys. Acta 502, 409–416.
38 Broglie, R.M., Hunter, C.N., Delepelaire, P., Niederman, R.A., Chua, N.-H. and Clayton, R.K. (1980) Proc. Natl. Acad. Sci. USA 77, 87–91.
39 Feick, R. and Drews, G. (1978) Biochim. Biophys. Acta 501, 499–513.
40 Kramer, J.M., Van Grondelle, B., Hunter, C.N., Westerhuis, W.H.J. and Amesz, J. (1984) Biochim. Biophys. Acta 765, 156–165.
41 Feick, R. and Drews, G. (1979) Z. Naturforsch. 340, 196–199.
42 Webster, G.D., Cogdell, R.J. and Lindsay, G.J. (1980) FEBS Lett. 111, 391–394.
43 Bolt, J.D., Hunter, C.N., Niederman, R.A. and Sauer, K. (1981) Photochem. Photobiol. 34, 653–656.
44 Theiler, R. and Zuber, H. (1984) Hoppe-Seyler's Z. Physiol. Chem. 365, 721–729.
45 Nabedryk, E., Berger, G., Andrianambininboa, S. and Breton, J. (1985) Biochim. Biophys. Acta 809, 271–276.
46 Welte, W., Wacker, T., Leis, M., Kreutz, W., Shiozawa, J., Gad'on, N. and Drews, G. (1985) FEBS Lett. 182, 260–263.
47 Allen, J.P., Theiler, R. and Feher, G. (1985) Biophys. J. 47, Abstracts M-AM-A8, p. 4a.
48 Tadros, M.H., Suter, F., Drews, G. and Zuber, H. (1983) Eur. J. Biochem. 129, 533–536.
49 Tadros, M.H., Suter, F., Seydewitz, H.H., Witt, I., Zuber, H. and Drews, G. (1984) Eur. J. Biochem. 138, 209–212.
50 Theiler, R., Suter, F., Wiemken, V. and Zuber, H. (1984) Hoppe-Seyler's Z. Physiol. Chem. 365, 703–719.
51 Theiler, R., Suter, F., Zuber, H. and Cogdell, R.J. (1984) FEBS Lett. 175, 231–237.
52 Theiler, R., Suter, F., Pennoyer, J.D., Zuber, H. and Niederman, R.A. (1985) FEBS Lett. 184, 231–236.
53 Youvan, D.C. and Ismail, S. (1985) Proc. Natl. Acad. Sci. USA 82, 58–62.
54 Youvan, D.C., Bylina, E.J., Alberti, M., Begusch, H. and Hearst, J.E. (1984) Cell 37, 949–957.
55 Clayton, R.K. and Arnold, W. (1961) Biochim. Biophys. Acta 48, 319–323.
56 Pfennig, N. (1969) J. Bacteriol. 99, 597–602.
57 Thornber, J.P. (1970) Biochemistry 9, 2688–2698.
58 Cogdell, R.J., Durant, I., Valentine, J., Lindsay, J.G. and Schmidt, K. (1983) Biochim. Biophys. Acta 722, 427–435.
59 Stadnichuk, I.N., Abdurakhmanov, I.A., Kusnetsova, N. Yu. and Erokhin, Yu.E. (1983) Molekulyarnaya Biologiya 18, 685–690.
60 Fuller, R.C. and Andersen, I.A. (1958) Nature 181, 252–254.
61 Erokhin, Yu.E. and Sinegub, O.A. (1970) Molekulyarnaya Biologiya 4, 541–550.
62 Hayashi, H. and Morita, S. (1980) J. Biochem. 88, 1251–1258.
63 Miyazaki, T. and Morita, S. (1981) Photosynthetica 15, 238–243.
64 Angerhofer, A., Cogdell, R.J. and Hipkins, M.F. (1986) Biochim. Biophys. Acta 848, 333–341.
65 Mechler, B. and Oelze, J. (1978) Arch. Microbiol. 118, 91–97.
66 Mechler, B. and Oelze, J. (1978) Arch. Microbiol. 118, 99–108.
67 Mechler, B. and Oelze, J. (1978) Arch. Microbiol. 118, 109–114.
68 Brunisholz, R.A., Niederer, E., Bissig, I. and Zuber, H. (1985) V. International Symposium on Photosynthetic Prokaryotes, Grindelwald, Abstract IVH, p 295.

69 Staehelin, L.A., Golecki, J.R., Fuller, R.C. and Drews, G. (1978) Arch. Microbiol. 119, 269–277.
70 Schmidt, K. (1980) Arch. Microbiol. 124, 21–31.
71 Betti, J.A., Blankenship, R.E., Natarajan, L.A., Dickinson, L.C. and Fuller, R.C. (1982) Biochim. Biophys. Acta 680, 194–201.
72 Feick, R.G. and Fuller, R.C. (1984) Biochemistry 23, 3693–3700.
73 Wechsler, T., Suter, F., Fuller, R.C. and Zuber, H. (1984) FEBS Lett. 181, 173–178.
74 Wechsler, T., Brunisholz, R., Suter, F., Fuller, R.C. and Zuber, H. (1985) FEBS Lett. 191, 34–38.
75 Sybesma, C. and Vredenberg, W.J. (1963) Biochim. Biophys. Acta 75, 439–441.
76 Matthews, B.W., Fenna, R.E., Bolognesi, M.C., Schmid, M.F. and Olson, J.M. (1979) J. Mol. Biol. 131, 259–285.
77 Glazer, A.N. (1984) Biochim. Biophys. Acta 768, 29–51.
78 Gantt, E. and Conti, S.F. (1969) J. Bacteriol. 97, 1486–1493.
79 Koller, K.P., Wehrmeyer, W. and Mörschel, E. (1978) Eur. J. Biochem. 91, 57–63.
80 Bryant, D.A., Guglielmi, G., Tandeau de Marsac, N., Castets, A.M. and Cohen-Bazire, G. (1979) Arch. Microbiol. 123, 113–127.
81 Rosinski, J., Hainfeld, J.F., Rigbi, M. and Siegelman, H.W. (1981) Ann. Bot. 47, 1–12.
82 Nies, M. and Wehrmeyer, W. (1981) Arch. Microbiol. 129, 374–379.
83 Mörschel, E., Koller, K.P., Wehrmeyer, W. and Schneider, H. (1977) Cytobiology (Eur. J. Cell Biol.) 16, 118–129.
84 Cohen-Bazire, G. and Bryant, D.A. (1982) in The Biology of Cyanobacteria (Carr, N.G. and Whitton, B.A., eds.) pp. 143–190, Blackwell Scientific Publications, Oxford.
85 Glazer, A.N. (1982) Annu. Rev. Microbiol. 36, 173–198.
86 Glazer, A.N. (1985) Annu. Rev. Biophys. Biophys. Chem. 14, 47–77.
87 Gantt, E. (1971) J. Cell. Biol. 48, 280–290.
88 Wehrmeyer, W. (1970) Arch. Microbiol. 71, 367–383.
89 Gantt, E. (1980) in Phytoflagellates: Form and Function (Cox, E., ed.) pp. 381–405, Elsevier, Amsterdam.
90 Bennet, A. and Siegelman, H.W. (1979) in The Porphyrins, Vol. VI (Dolphin, D., ed.) pp. 493–520, Academic Press, New York.
91 Rüdiger, W. (1980) in Pigments in Plants (Czygan, F.C., ed.) pp. 314–351, Fischer Verlag, Stuttgart.
92 Scheer, H. (1981) Angew. Chem. Int. Ed. Engl. 20, 241–261.
93 Bryant, D.A. (1982) J. Gen. Microbiol. 128, 835–844.
94 Bishop, J.E., Rapoport, H., Klotz, A.V., Chan, C.F., Glazer A.N., Füglistaller, P., and Zuber, H. (1987) J. Am. Chem. Soc. 109, 875–881.
95 MacColl, R. and Guard-Friar, D. (1983) J. Biol. Chem. 258, 14327–14329.
96 Bryant, D.A., Glazer, A.N. and Eiserling, F.A. (1976) Arch. Microbiol. 110, 61–75.
97 Füglistaller, P., Suter, F. and Zuber, H. (1983) Hoppe-Seyler's Z. Physiol. Chem. 364, 691–712.
98 Sidler, W., Nutt, H.U., Kumpf, B., Wehrmeyer, W., Brenzel, A. and Zuber, H. (1986) in Progress in Photosynthesis Research (Biggins, J., ed.), Vol. II, pp. 153–156, Martinus Nijhoff, Dordrecht, Boston, Lancaster.
99 Glazer, A.N. and Cohen-Bazire, G. (1975) Arch. Microbiol. 104, 29–32.
100 MacColl, R., Habig, W. and Berns, D.S. (1973) J. Biol. Chem. 248, 7080–7086.
101 Schirmer, T., Bode, W., Huber, R., Sidler, W. and Zuber, H. (1985) J. Mol. Biol. 184, 257–277.
102 Schirmer, T., Huber, R., Schneider, M., Bode, W., Miller, M. and Hackert, M.L. (1986) J. Mol. Biol. 188, 651–676.
103 Sidler, W., Gysi, J., Isker, E. and Zuber, H. (1981) Hoppe-Seyler's Z. Physiol. Chem. 362, 611–628.
104 Gantt, E. and Lipschultz, C.A. (1972) J. Cell. Biol. 54, 313–324.
105 Füglistaller, P., Rümbeli, R., Suter, F. and Zuber, H. (1984) Hoppe-Seyler's Z. Physiol. Chem. 365, 1085–1096.
106 Gantt, E. and Lipschultz, C.A. (1974) Biochemistry 13, 2960–2966.
107 Koller, K.P. and Wehrmeyer, W. (1975) Arch. Microbiol. 104, 255–261.
108 Frank, G., Zuber, H. and Lergier, W. (1975) Experientia 31, 23–26.
109 Gysi, J. and Zuber, H. (1974) FEBS Lett. 2, 209–213.

110 Glazer, A.N. and Fang, S. (1973) J. Biol. Chem. 248, 659–662.
111 Füglistaller, P., Widmer, H., Sidler, W., Frank, G. and Zuber, H. (1981) Arch. Microbiol. 129, 268–274.
112 Minami, Y., Yamada, F., Hase, T., Matsubara, H., Murakami, A., Fujita, Y., Takao, T. and Shimonishi, Y. (1985) FEBS Lett. 191, 216–220.
113 Frank, G., Sidler, W., Widmer, H. and Zuber, H. (1978) Hoppe-Seyler's Z. Physiol. Chem. 359, 1491–1507.
114 Freidenreich, P., Apell, G.S. and Glazer, A.N. (1978) J. Biol. Chem. 253, 212–219.
115 Sidler, W., Kumpf, B., Rüdiger, W. and Zuber, H. (1986) Biol. Chem. Hoppe-Seyler 367, 627–642.
116 Offner, G.D. and Troxler, R.F. (1983) J. Biol. Chem. 258, 9931–9940.
117 Troxler, R.F., Ehrhardt, M.M., Brown-Mason, A.S. and Offner, G.D. (1981) J. Biol. Chem. 256, 12176–12184.
118 Muckle, G. and Rüdiger, W. (1977) Z. Naturforsch. 32c, 957–962.
119 Klotz, A.V. and Glazer, A.N. (1985) J. Biol. Chem. 260, 4856–4863.
120 Nagy, J.D., Bishop, J.E., Klotz, A.V., Glazer, A.N. and Rapoport, H. (1985) J. Biol. Chem. 260, 4864–4868.
121 Glazer, A.N. and Bryant, D.A. (1975) Arch. Microbiol. 104, 15–22.
122 Lundell, D.J. and Glazer, A.N. (1981) J. Biol. Chem. 256, 12600–12606.
123 Zilinskas, B.A., Greenwald, L.S., Bailey, C.L. and Kahn, P.C. (1980) Biochim. Biophys. Acta 592, 267–276.
124 Yu, M.H., Glazer, A.N. and Williams, R.C. (1981) J. Biol. Chem. 256, 13130–13136.
125 Lundell, D.J., Williams, R.C. and Glazer, A.N. (1981) J. Biol. Chem. 256, 3580–3592.
126 Bogorad, L. (1975) Annu. Rev. Plant Physiol. 26, 369–401.
127 Tandeau de Marsac, N. (1983) Bull. Inst. Pasteur 81, 201–254.
128 Rüdiger, W. (1971) Fortschr. Chem. Organ. Naturst. 29, 59–139.
129 Zuber, H. (1983) in Photosynthetic Prokaryotes: Cell Differentiation and Function (Papageorgiou, G.C. and Packer, L., eds.) pp. 23–42, Elsevier, Amsterdam.
130 Tandeau de Marsac, N. and Cohen-Bazire, G. (1977) Proc. Natl. Acad. Sci. USA 74, 1635–1639.
131 Lundell, D.J. and Glazer, A.N. (1983) J. Biol. Chem. 258, 894–901.
132 Lundell, D.J., Yamanaka, G. and Glazer, A.N. (1981) J. Cell. Biol. 91, 315–319.
133 Zilinskas, B.A. and Howell, D.A. (1983) Plant Physiol. 71, 379–387.
134 Glazer, A.N. (1985) in Abstracts of the Vth Int. Symposium on Photosynthetic Prokaryotes, Grindelwald, pp. 68.
135 Füglistaller, P., Suter, F. and Zuber, H. (1985) Biol. Chem. Hoppe-Seyler 366, 993–1001.
136 Füglistaller, P., Suter, F. and Zuber, H. (1986) Biol. Chem. Hoppe-Seyler 367, 601–614.
137 Füglistaller, P., Suter, F. and Zuber, H. (1986) Biol. Chem. Hoppe-Seyler 367, 615–626.
138 Rigbi, M., Rosinski, J., Siegelman, H.W. and Sutherland, J.C. (1980) Proc. Natl. Acad. Sci. USA 77, 1961–1965.
139 Glazer, A.N., Lundell, D.J., Yamanaka, G. and Williams, R.C. (1983) Ann. Microbiol. Inst. Pasteur 134B, 159–180.
140 Gingrich, J.C., Lundell, D.J. and Glazer, A.N. (1983) J. Cell. Biochem. 22, 1–14.
141 Yamanaka, G., Lundell, D.J., Glazer, A.N. (1982) J. Biol. Chem. 257, 4077–4086.
142 Mörschel, E. and Mühlethaler, K. (1983) Planta 158, 451–457.
143 Mörschel, E., Koller, K.-P. and Wehrmeyer, W. (1980) Arch. Microbiol. 125, 43–51.
144 Molisch, H. (1894) Bot. Z. 52, 117–186.
145 Dobler, M., Dover, S.D., Laves, K., Binder, A. and Zuber, H. (1972) J. Mol. Biol. 71, 785–789.
146 Hackert, M.L., Abad-Zapatero, C., Stevens, S.J. and Fox, J.L. (1977) J. Mol. Biol. 111, 365–369.
147 Sweet, R.M., Fuchs, H.E. and Fischer, R.G. (1977) J. Biol. Chem. 252, 8258–8260.
148 Abad-Zapatero, C., Fox, J.L. and Hackert, M.L. (1977) Biochem. Biophys. Res. Commun. 78, 266–272.
149 Rümbeli, R., Schirmer, T., Bode, W., Sidler, W. and Zuber, H. (1985) J. Mol. Biol. 186, 197–200.
150 Teale, F.W.J. and Dale, R.E. (1970) Biochem. J. 116, 161–169.
151 Mimuro, M., Füglistaller, P., Rümbeli, R. and Zuber, H. (1986) Biochim. Biophys. Acta 848, 155–166.

152 Mimuro, M., Rümbeli, R., Füglistaller, P. and Zuber, H. (1986) Biochim. Biophys. Acta 851, 447–456.

153 Gantt, E. (1979) in Biochemistry and Physiology of Protozoa (Levandowski, M. and Hufner, S.A., eds.) Vol. 1, pp. 121–137, Academic Press, New York.

154 Mörschel, E. and Wehrmeyer, W. (1975) Arch. Microbiol. 105, 153–158.

155 Mörschel, E. and Wehrmeyer, W. (1977) Arch. Microbiol. 113, 83–89.

156 MacColl, R. and Guard-Friar, D. (1983) Biochemistry 22, 14327–14329.

157 Jung, J., Song, P.-S., Paxton, R.J., Edelstein, M.S., Swanson, R. and Hazen, E.E., Jr. (1980) Biochemistry 19, 24–32.

158 MacColl, R., Guard-Friar, D. and Csatorday, K. (1983) Arch. Microbiol. 135, 194–198.

159 Sidler, W., Kumpf, B., Suter, F., Morisset, W., Wehrmeyer, W. and Zuber, H. (1985) Biol. Chem. Hoppe-Seyler 366, 233–244.

160 Guard-Friar, D. and MacColl, R. (1986) Photochem. Photobiol. 43, 81–95.

161 Glazer, A.N. and Apell, G.S. (1977) FEMS Microbiol. Lett. 1, 113–116.

162 Morisset, W., Wehrmeyer, W., Schirmer, T. and Bode, W. (1984) Arch. Microbiol. 140, 202–205.

163 Anderson, J.M. and Andersson, B. (1982) Trends Biochem. Sci. 7, 288–292.

164 Cramer, W.A., Widger, W.R., Herrmann, R.G. and Trebst, A. (1985) Trends Biochem. Sci. 10, 125–129.

165 Thornber, J.P. (1975) Annu. Rev. Plant Physiol. 26, 127–158.

166 Thornber, J.P., Alberte, R.S., Hunter, F.A., Shiozowa, J.A. and Kan, K.-S. (1977) Brookhaven Symposia Biol. 28, 132–148.

167 Thornber, J.P., Markwell, J.P. and Reinman, S. (1979) Photochem. Photobiol. 29, 1205–1216.

168 Barber, J. (1979) Photochem. Photobiol. 29, 203–207.

169 Cogdell, R.J. (1983) Annu. Rev. Plant Physiol. 34, 21–45.

170 Rebeiz, C.A., Belanger, F.C., Freyssinet, G. and Saab, D.G. (1980) Biochim. Biophys. Acta 490, 234–247.

171 Rebeiz, C.A., Belanger, F.C., Mc Carthy, S.A., Freyssinet, G. and Duggan, J.X. (1981) in Photosynthesis V. Chloroplast Development (Akoyunoglou, G., ed.) pp. 197–212, Balaban Int. Sci. Serv., Philadelphia, PA.

172 Bazzaz, M.B. and Brereton, R.G. (1982) FEBS Lett. 138, 104–108.

173 Dörnemann, O. and Senger, H. (1982) Photochem. Photobiol. 35, 821–826.

174 Bennett, J. (1983) Biochem. J. 212, 1–13.

175 Kyle, D.J., Staehelin, L.A. and Arntzen, C.J. (1983) Arch. Biochem. Biophys. 222, 527–541.

176 Kyle, D.J., Ting-Yun-Kuang, Watson, J.L. and Arntzen, C.J. (1984) Biochim. Biophys. Acta 765, 89–96.

177 Bengis, C., and Nelson, N. (1975) J. Biol. Chem. 250, 2783–2788.

178 Mullet, J.E., Burke, J.E. and Arntzen, C. (1980) Plant Physiol. 65, 814–822.

179 Bellamare, G., Bartlett, S.G. and Chua, N.-H. (1982) J. Biol. Chem. 257, 7762–7767.

180 Wollman, F.A. and Bennoun, P. (1982) Biochim. Biophys. Acta 680, 352–367.

181 Haworth, P., Watson, J.L. and Arntzen, C.J. (1983) Biochim. Biophys. Acta 724, 151–158.

182 Anderson, J.M., Brown, J.S., Lam, E. and Malkin, R. (1983) Photochem. Photobiol. 38, 205–210.

183 Ish-Shalom, D. and Ohad, J. (1983) Biochim. Biophys. Acta 722, 498–507.

184 Lam, E., Ortiz, W. and Malkin, R. (1984) FEBS Lett. 168, 10–14.

185 Lam, E., Ortiz, W., Mayfield, S. and Malkin, R. (1984) Plant Physiol. 74, 650–655.

186 Fish, L.E., Kück, U. and Bogorad, L. (1985) J. Biol. Chem. 260, 1413–1421.

187 Burke, J.J., Ditto, C.L. and Arntzen, C.J. (1978) Arch. Biochem. Biophys. 187, 252–263.

188 Bennett, J., Markwell, J.P., Skrdla, M.P. and Thornber, J.P. (1981) FEBS Lett. 131, 325–330.

189 Arntzen, C.J. (1978) Curr. Top. Bioenerg. 8, 111–160.

190 Hiller, R.G. and Goodchild, D.J. (1981) Biochem. Plants 8, 1–49.

191 Ryrie, I.J., Anderson, J.M. and Goodchild, D.J. (1980) Eur. J. Biochem. 107, 345–354.

192 Schmidt, G.W., Bartlett, S.G., Grossman, A.R., Cashmore, A.R. and Chua, N.-H. (1981) J. Cell. Biol. 91, 468–478.

193 Coruzzi, G., Broglie, R., Cashmore, A. and Chua, N.H. (1983) J. Biol. Chem. 258, 1399–1402.

194 Cashmore, A.R. (1984) Proc. Natl. Acad. Sci USA 81, 2960–2964.

195 Karlin-Neumann, G.A., Kohorn, B.D., Thornber, J.P. and Tobin, E.M. (1985) J. Mol. Appl. Genet. 3, 45–61.
196 Dunsmuir, P., Smith, S.M. and Bedbrook, J. (1983) J. Mol. Appl. Genet. 2, 285–300.
197 Dunsmuir, P. (1985) Nucleic Acid Res. 13, 2503–2519.
198 Kühlbrandt, W., Thaler, T. and Wehrli, E. (1983) J. Cell. Biol. 96, 1414–1424.
199 Kühlbrandt, W. (1984) Nature 307, 478–480.
200 Li, J. (1985) Proc. Natl. Acad. Sci. USA 82, 386–390.
201 Vernon, L.P., Ke, B. and Satoh, K. (1966) Brookhaven Symp. Biol. 19, 102–114.
202 Vernon, L.P. and Klein, S.M. (1975) Ann. N.Y. Acad. Sci. 244, 281–296.
203 Delepelaire, P. and Chua, N.-H. (1979) Proc. Natl. Acad. Sci. USA 76, 111–115.
204 Satoh, K., Nakatani, H.Y., Steinback, K.E., Watson, J. and Arntzen, C.J. (1983) Biochim. Biophys. Acta 724, 142–150.
205 Camm, E.L. and Green, B.R. (1983) J. Cell. Biochem. 23, 171–179.
206 Camm, E.L. and Green, B.R. (1983) Biochim. Biophys. Acta 774, 291–293.
207 Nakatani, H.Y., Ke, B., Dolan, E. and Arntzen, C.J. (1984) Biochim. Biophys. Acta 765, 347–352.
208 Yamagishi, A., and Katoh, S. (1983) Arch. Biochem. Biophys. 225, 836–846.
209 Westhoff, P., Alt, J. and Herrmann, R.G. (1983) EMBO J. 2, 2229–2237.
210 Bricker, T.M., Pakrasi, H.B. and Sherman, L.A. (1985) Biochem. Biophys. 237, 170–176.
211 Alt, J., Morris, J., Westhoff, P. and Herrmann, R.G. (1984) Curr. Genet. 8, 597–606.
212 Morris, J. and Herrmann, R.G. (1984) Nucleic Acids Res. 12, 2837–2850.
213 Deisenhofer, J., Epp, O., Miki, K., Huber, R. and Michel, H. (1985) Nature 318, 618–624.

J. Amesz (ed.) *Photosynthesis*

CHAPTER 12

Molecular organization of thylakoid membranes

JAN M. ANDERSON

CSIRO, Divison of Plant Industry, GPO Box 1600, Canberra, A.C.T. 2601, Australia

1. Introduction

An ultimate understanding of the molecular mechanisms of photosynthetic energy transduction requires the elucidation of the topography of the proteins and lipids of photosynthetic membranes. The secrets of the molecular logic of light-harvesting, electron flow and proton translocation lie in the specific and critical arrangement of most thylakoid proteins in multisubunit, membrane-spanning intrinsic protein complexes embedded within the fluid bilayer of the thylakoid membrane. This dynamic, asymmetric organization allows for rapid changes in both the conformation of complexes and the lateral distribution of complexes along the membrane, features which may be essential for function.

The concepts for the arrangements of proteins within membranes introduced by Singer [1] led Singer and Nicolson [2] to formulate a generalized fluid protein-lipid mosaic model for membrane structure which is still a good basis for understanding the architecture of biological membranes. Anderson [3], in 1975, first applied these concepts to thylakoid membranes. Then, with an incomplete list of all the thylakoid components, particularly the polypeptides, and no description of their molecular structure or knowledge of their location in the membrane, it was only possible to glimpse the molecular organization of thylakoid membranes, and offer instead speculations galore [3]. It is clear now that there is not only a marked asymmetry in the transverse arrangement of both proteins and lipids across the membrane, but also a marked lateral heterogeneity in the distribution of the supramolecular thylakoid complexes between the appressed and non-appressed regions of the thylakoids of higher plants and algae that contain chlorophyll *b*. This lateral heterogeneity has important consequences for light-harvesting strategies, for electron flow between the photosystems, and for the flexible photosynthetic capacities of thylakoids of plants adapted to specific habitats.

This review focusses on certain aspects of the transverse arrangement of the proteins and lipids of thylakoid membranes, and the lateral heterogeneity of these complexes between the appressed and non-appressed regions of Chl *b*-containing plants. The consequences of lateral heterogeneity will also be examined with the emphasis on areas under debate.

2. *Transverse organization of thylakoid membranes*

Asymmetry is a fundamental feature of all biological membranes, since the amphipathic lipid or intrinsic protein molecules have little chance to flip-flop across the membrane. The assembly of proteins into compact supramolecular complexes consisting of hydrophobic α-helical regions linked to more hydrophilic amino acid sequences at the membrane surfaces imparts critical order to the arrangement of pigment molecules and redox components. The vast array of fixed surface charges present, particularly in the protein domains at the outer and inner membrane surfaces, as well as part of the lipid domain, not only gives rise to an asymmetry of overall electrostatic charge, but also means that small changes in the ionic composition at either membrane surface will have a profound effect on the conformation of thylakoid proteins and lipids.

2.1. Transverse asymmetry of thylakoid lipids

Thylakoid membranes have a unique set of acyl lipids that form about 25–30% of the total thylakoid mass. The neutral galactolipids monogalactosyldiacylglycerol (MGDG) and digalactosyldiacylglycerol (DGDG) comprise about 75% of the total acyl lipids (MGDG, 50%; DGDG, 25%), together with roughly equal amounts of the negatively charged lipids phosphatidyldiacylglycerol (PG) (10%) and sulphoquinovosylglycerol (SG) (10%) and other phospholipids (5%) [4]. As found with other energy-transducing membranes, thylakoid lipids have acyl fatty acids which are highly unsaturated. The predominant fatty acid is α-linolenic acid (18:3), which accounts for 90% of the total acyl chains [4,5]. These acyl lipids form the fluid bilayer matrix in which the functional supramolecular complexes are embedded. The fluid matrix permits diffusional processes such as lateral migration of phosphorylated LHC II and electron transport by plastoquinone [4,6].

The transverse distribution of acyl lipids across the thylakoid membrane is more difficult to determine than that of the transverse protein distribution. Due to the relatively small size of the acyl lipid molecules, antibody labelling, lipolytic digestion and chemical modification may considerably modify the lipids themselves, thereby leading to lipid interchange across the membrane, and even membrane disruption. Nevertheless, partial asymmetry in the lipid distribution in the outer and inner monolayers of all biological membranes is assumed to occur [5]. Antibody labelling studies suggest that more PG is exposed at the outer membrane surface, and more MGDG and SG at the inner surface [7]. Conflicting results, however, arise from lipolytic studies by two groups [8,9]. Using chemical modification (tritium labelling of the galactose head group) of right-side-out and inside-out thylakoid vesicles, a partial asymmetry of the galactolipids was observed; 60% of both galactolipids in the outer half, and 40% in the inner half of the bilayer [10]. Since the galactolipids comprise about 75% of the lipid bilayer, this implies that the negatively charged PG and SG are preferentially located in the inner half of the thylakoid membrane [10].

2.2. *Transverse asymmetry of thylakoid proteins*

The impact of Mitchell's chemiosmotic hypothesis for biological energy conservation stimulated interest in the transbilayer arrangement of thylakoid proteins. Initial studies indicated an apparent vectorial arrangement of electron transport components, with electron donors for PS II and PS I located at the inner thylakoid surface and electron acceptor sites at the outer surface. This arrangement is consistent with charge separation across the membrane and an inward direction of proton pumping. Progress was hampered, however, by the lack of inside-out thylakoid vesicles to directly probe the inner thylakoid surface. With Albertsson's introduction of the aqueous polymer two-phase partition technique (see Ref. 11), cell or membrane fractions could be separated into fractions characterized by differences in their surface charge properties, rather than by size or density. As the asymmetry of surface charges at the outer and inner surfaces is a basic feature of thylakoids, this method of separation is extremely powerful. When thylakoids are fragmented by mechanical shearing forces in a Yeda press, both right-side-out and inside-out thylakoid vesicles are obtained [11,12]. With the aqueous polymer two-phase partition technique, the right-side-out vesicles partition to the upper dextran phase, while the inside-out vesicles partition to the lower polyethylene glycol phase [12]. The inner thylakoid surface can then be directly probed by suitable membrane-impermeable effectors and comparisons made with the outer surface.

Most of the thylakoid proteins are organized into four intrinsic protein complexes: PS II complex, Cyt *b*/*f* complex, PS I complex and ATP synthetase (Fig. 1). The electron transport complexes are linked by 'mobile' electron transport carriers, plastoquinone, plastocyanin and ferredoxin (see Chapter 10). Furthermore, chloroplasts that possess Chl *b* have the major light-harvesting Chl *a*/*b*-proteins of PS II (LHC II) that may represent over 50% of the thylakoid protein [13], as well

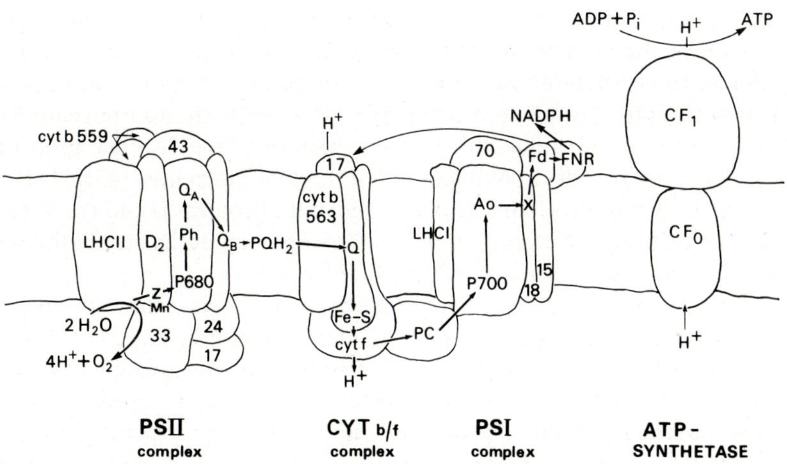

Fig. 1. Arrangement of the supramolecular protein complexes and mobile electron transport carrier in thylakoid membranes.

as the minor Chl *a*/*b*-proteins of PS I (LHC I) (see Chapter 11). The molar compositions in terms of pigments and the number of apoproteins of these heterogeneous Chl *a*/*b*-proteins that are encoded in multigene nuclear families are not yet established [13].

The biochemical strategies used to explore the topology of thylakoid membranes, especially the transverse arrangement of individual thylakoid proteins or multisubunit complexes, include a variety of non-permeable membrane probes such as proteolytic enzymes, chemical modifications and antibody labelling. It is necessary to use at least two of these complementary approaches, as negative results by an individual method do not necessarily mean that a particular peptide is inaccessible. Before discussing briefly the transverse arrangement of thylakoid proteins deduced from these biochemical techniques, the prediction of protein-folding domains in the membrane using computors to provide hydropathy index plots will be reviewed.

2.2.1. Hydropathy index plots

Recently, model building and new experimental approaches have been generated from the advances made in the molecular biology of thylakoid membranes, particularly from the nucleotide sequences of isolated thylakoid genes. Knowledge of the deduced amino acid sequences of thylakoid proteins obtained from gene sequences enables predictions to be made about their secondary and tertiary structures. Several algorithms that depend on the classification of the net hydrophobic/hydrophilic character of individual amino acids have been used to predict the putative hydrophobic, membrane-spanning regions of intrinsic proteins [14,15]. Usually hydrophobic amino acid sequences of 20 to 25 residues relatively free of charged groups or α-helix breakers (e.g. proline) may be considered as candidates for a membrane-spanning α-helix. Many thylakoid proteins, however, have extra functional groups, such as bound quinones, hemes, chlorophylls, carotenoids and iron-sulphur centres. As these functional groups are usually located within the hydrophobic region of the membrane they may modify α-helices, and allow charged groups or proline to be included in the membrane-spanning region. This might be so especially for the pigment-protein complexes which contain an unusually large number of pigment molecules per polypeptide chain (e.g. the 28 kDa apoprotein of LHC II binds up to 13 chlorophyll and 4 xanthophyll molecules [13]). It is also possible that the polypeptide chains of some thylakoid proteins could traverse the membrane as β-sheets [15]. Cramer et al. [16] have presented hydrophathy index plots and putative structures for several thylakoid proteins.

The excitement generated by these hydropathy index plots, which predict that most of the thylakoid proteins indeed span the membrane, may have led to a view that it is unnecessary to experimentally demonstrate that the proteins have segments exposed at the outer and inner surfaces. Nevertheless, biochemical evidence such as antibody labelling, protease studies and chemical modification is needed to prove all predicted structures. Already it has become evident that caution is needed in the interpretation of the hydropathy index plots. This need is demonstrated by the hydropathy index plots of the D1 (herbicide-binding) and D2 proteins of the PS II complex which were initially assigned 7 α-helices each (cf. Ref. 4). X-ray analysis of the L and M subunits of the photosynthetic bacterial reaction centre complex shows that these proteins contain only 5 α-helices [17]. In view of the close structural homology of the L and M subunits to the D1 and D2 proteins of PS II complex, it is probable that these latter proteins also have only 5 membrane-spanning segments [17].

While hydropathy index plots are very useful for building models of the transmembrane arrangement of the polypeptide chains and the functional groups, the ultimate aim of structural research is to obtain the three-dimensional structure of each protein. Following the exciting resolution to 3 Å in the X-ray structure of the purple photosynthetic bacterial reaction centre deduced by Deisenhofer et al. [17] there is a challenge to crystallize the thylakoid supramolecular complexes. Already, two-dimensional arrays of LHC II, the most abundant thylakoid protein, reconstituted into artificial membranes have been examined in the electron microscope [18,19]. LHC II appears to protrude about 20 Å on one side, and 10 Å on the other side of membrane, with each LHC II unit being 27 Å wide and 60–65 Å long. It is likely that these units exist in vivo as trimers [18].

2.2.2. *Topology of the Cyt b/f complex*

The Cyt *b/f* complex is the only electron transport complex for which the transmembrane organization of all its subunits is established. This membrane-spanning complex that functions as an intermediate electron transport complex between PS II and PS I, and translocates protons across the membrane from the stroma to the lumen, contains 4 proteins: Cyt *f* (33 kDa), Cyt *b*-563 (23 kDa), the Rieske Fe-S protein (20 kDa) and the unnamed 17 kDa protein.

Initial studies [20] with intact thylakoids, using pronase and chemical labelling, indicated that Cyt *f* and possibly the other 3 subunits were accessible at the outer membrane surface, while Cyt *f* was shown to be exposed at the inner lumenal surface [21–23]. Using the complementary approaches of proteolytic enzymes and antibody labelling in RSO and ISO vesicles, Mansfield and Anderson [24] demonstrated that all 4 subunits are exposed at both the outer and the inner surfaces. Although the use of specific carboxypeptidase had previously failed to reveal the location of the C-terminal of any of the apoproteins of the chlorophyll-proteins (Anderson, unpublished results), a specific carboxypeptidase cleaved the C-terminals of each of the subunits of Cyt *b/f* complex [24]. Significantly, it was found that all of the polypeptides had their C-terminals protruding into the stromal matrix [24]. These results support the generalization that most intrinsic proteins tend to be oriented with their N-terminals facing inwards, that is, towards the lumen for thylakoid membranes.

The transmembrane arrangement of these subunits of the Cyt *b/f* complex is also evident in the hydropathy index plots determined from the deduced amino acid sequences obtained from the sequencing of the chloroplast-encoded genes of Cyt *f*, Cyt *b*-563, and the 17 kDa protein (cf. Refs. 4 and 16). Cyt *f* is anchored in the membrane by a single α-helix located close to the C-terminal end ($\approx$ 20 amino acid residues). Most of the mainly hydrophilic polypeptide chain ($\approx$ 250 amino acid

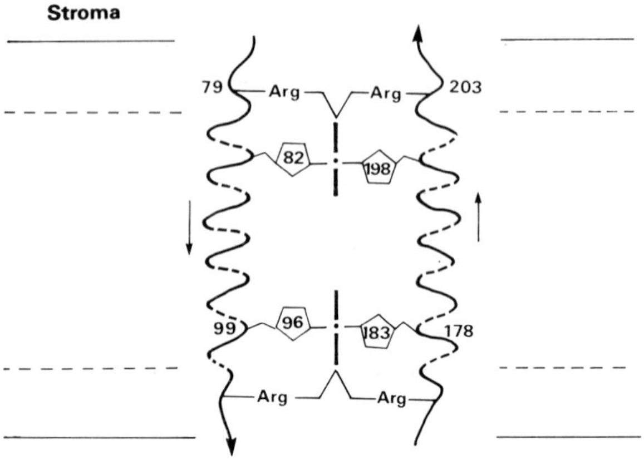

Fig. 2. A putative transmembrane arrangement of the two hemes of Cyt *b*-563 cross-linking the membrane-spanning α-helices II and V predicted by the gene sequence data and hydropathy index plots according to Cramer et al. [16].

residues) of the N-terminal portion is folded as a globular structure that encloses the heme group and projects about 4 nm into the lumen [21]. This globular structure is folded to form a negatively charged domain that is able to interact with the acidic domain of plastocyanin.

Comparison of the hydropathy index plots of Cyt *b*-563 and the 17 kDa protein, with those of mitochondrial and fungal Cyts *b* ($\approx$ 40 kDa), suggests that Cyt *b*-563 corresponds to the first half of the Cyt *b* structure, and the 17 kDa protein is associated with the second half [16]. Cramer et al. [16] propose that the two histidines of both α-helical spans II and V bind the two hemes of Cyt *b*-563. This means that the hemes are located one above each other and approximately perpendicular to the membrane plane, consistent with the established vectorial electron transport between the lumenal and stromal surfaces (Fig. 2). These 4 histidine groups (each about 4 amino acid residues in from the end of the α-helical spans) are absolutely conserved in the hydrophobic regions of seven Cyt *b* molecules (cf. Ref. 16). As stated, the thylakoid 17 kDa protein shows considerable sequence homology with the C-terminal end of mitochondrial and fungal Cyt *b*. Since the mRNA for Cyt *b*-563 and the 17 kDa protein is dicistronic, and read from Cyt *b*-563 to the 17 kDa protein, the chloroplast 17 kDa protein is the product of a split gene (cf. Ref. 4). Since the C-terminal end of the 17 kDa protein is located at the stromal surface [24], this protein, assumed from hydropathy index plots to have 3 α-helical regions, would have its N-terminal end located also at the lumenal surface (Fig. 3).

The nuclear-encoded Rieske Fe-S protein is not accessible to proteolytic enzymes in thylakoids [20] or right-side-out vesicles [24], but antibody labelling shows this peptide to be exposed at both thylakoid membrane surfaces [24]. Although this thylakoid gene has not yet been sequenced, it is likely to have a structure similar to that of *Neurospora* Fe-S protein, which has only one membrane-spanning

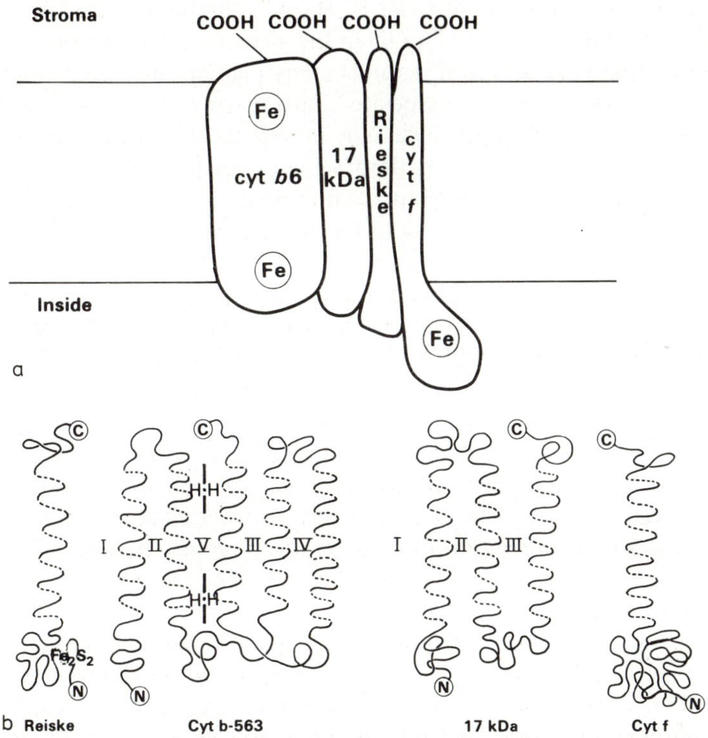

Fig. 3. A putative transmembrane arrangement of the subunits of the Cyt *b/f* complex using (a) data derived from proteases and antibody labelling [24], and (b) from gene sequences and hydropathy index plots [16].

segment. Since the C-terminal is located at the stromal surface [24], the Fe_2-S_2 centre of the Rieske protein will be located towards the lumen (Fig. 3).

In order to build up the three-dimensional structure of the Cyt *b/f* complex, near-neighbour analytical studies have to be performed. An initial cross-linking study [25] with glutaraldehyde shows cross-linking between the Rieske Fe-S protein and the 17 kDa protein, and also between Cyt *f* and the Rieske Fe-S protein. It is likely that the Cyt *b/f* complex exists as a dimer in vivo ($M_r \approx 280\,000$), as proposed for the mitochondrial Cyt *b/c* complex [26]. Electron micrographs of reconstituted Cyt *b/f* complex substantiate this idea [23], as does the finding of Cyt *f* dimers following mild cross-linking by glutaraldehyde [25].

2.2.3. Transverse organization of the Chl-proteins

The complementary techniques of protease treatment and antibody agglutination show that the Chl *a*-proteins of PS II (47 and 43 kDa apoproteins) [27] and the 68 kDa apoprotein(s) of PS I [28] all have sites exposed at both the stromal and the lumenal membrane surfaces. Similarly, both of the major apoproteins of LHC II traverse the membrane [27,28]. The gene sequences of the Chl *a*-proteins of both

PS II and PS I, and the 28 kDa protein of LHC II, show a number of possible α-helical regions in each case (see Chapter 11). Given the size of the chlorin ring (≈ 15 × 15 Å) and the fact that each apoprotein binds many pigment molecules, it is not surprising that these apoproteins are associated with the membrane α-helical regions, since protease digestions of intact thylakoids or vesicles do not release any of the chlorophyll of the PS II complex [27], the PS I complex [28,29] or LHC II [27,28]. Moreover, as neither the chlorophyll or xanthophyll molecules of LHC II are accessible to proton attack, they are likely to be located within the hydrophobic interior of the membrane [30].

Karlin-Neumann et al. [31] have presented a model for the main apoprotein of LHC II which has 3 α-helices (Chapter 11), consistent with the direct determination of the α-helical content of LHC II [32]. There is a large domain of surface-exposed protein (≈ 48%); this is also consistent with the electron microscopic pictures of reconstituted LHC II [18,19]. While most of the Chl molecules are thought to reside in the hydrophobic membrane interior, there are insufficient histidine residues present for the co-ordination of all Chl *a* molecules, and the ligand for Chl *b* has not been recognized yet. The location of the carotenoids, so often ignored, but always a constituent of all Chl-proteins, is not established [30].

2.2.4. *Intrinsic proteins of the PS II complex*

As mentioned, the Chl *a*-proteins of PS II have been shown to have domains exposed at both membrane surfaces [27,28], but biochemical studies have not established this for the other intrinsic proteins of PS II complex. Again it is evident, however, from an examination of their hydropathy index plots that each of the intrinsic subunits of PS II complex traverse the membrane (cf. Ref. 4). Both the herbicide-binding, D1 protein (32 kDa) and the D2 protein (34 kDa) probably each have 5 α-helices, as discussed above. The gene for Cyt *b*-559 has also been sequenced and a probable membrane-spanning, α-helical domain was found close to the N-terminal [33]. A second gene, coding for 39 amino acid residues located immediately distal to the Cyt *b*-559 apoprotein gene, has a putative α-helix adjacent to its C-terminus [16]. Both proteins contain histidine residues that are located 4 to 5 residues in from each end of the α-helix, suggesting that Cyt *b*-559 is a heterodimer [33]. Assuming that the N-terminals of these proteins are located at the lumen, it appears that the heme moiety of the Cyt *b*-559 heterodimer is located towards the inner half of the bilayer.

2.2.5. *Extrinsic proteins of the PS II complex*

The three extrinsic 33 kDa, 24 kDa and 18 kDa polypeptides of the oxygen-evolving complex are associated with the PS II complex at the lumenal side of the membrane [34,35] (Fig. 1). Each of the three proteins is able to re-bind stoichiometrically to depleted PS II complexes. The 33 and 24 kDa proteins bind directly to the complex, with the 33 kDa protein promoting binding of the 24 kDa protein [34,36], while the 18 kDa protein binds to the 24 kDa protein only when the latter polypeptide is bound to the PS II complex [37]. The 33 and 24 kDa proteins appear to be directly associated with two uncharacterized polypeptides of 24 kDa and 10 kDa that are also present in the PS II complex [36].

2.2.6. Transverse organization of the PS I complex

Proteolytic and immunological studies have shown that the native PS I-LHC I complex spans the membrane [28]. This applies to at least six of the 9–12 protein subunits [28,29]. Lack of positive characterization of all of the Chl *a/b*-proteins of LHC I and of limited gene sequencing studies of PS I proteins make this the least understood thylakoid complex in terms of its transmembrane arrangement. Further, there is no evidence yet for large hydrophilic domains protruding from the outer and inner membrane surfaces although they will be required for binding sites for the extrinsic proteins, plastocyanin and ferredoxin. Extensive homology between the two maize Chl *a*-proteins deduced from a photogene suggests that they are present as a P-700 heterodimer containing 11 to 13 putative α-helical regions [38].

Although it was originally suggested that plastocyanin was located in the lumen, contradictory reports followed. However, studies with vesicles of opposite sidedness verified that plastocyanin was indeed bound to the inner thylakoid surface [39], consistent with its function in the lumen [6]. The final PS I electron acceptors, ferredoxin and ferredoxin-$NADP^+$ reductase, are located on the stroma-facing side of the thylakoid membrane [40].

3. Lateral distribution of thylakoid components

3.1. Lateral asymmetry of acyl lipid distribution

Interestingly, the lateral segregation of the thylakoid supramolecular protein complexes to be described later (Section 3.2.2.) is matched by a partial lateral asymmetry of acyl lipid distribution [41,42], even though the appressed and non-appressed thylakoid regions are contiguous. Although each of the acyl lipid classes is found in both membrane regions, the ratio of MGDG to DGDG is higher in the appressed (2.4) relative to the non-appressed (1.1) membrane fractions, and there is an enrichment of anionic lipids in the grana partitions [41,42]. On a protein basis, however, the appressed regions are depleted in both galactolipids and sulpholipid, but have similar amounts of phosphatidyl-glycerol compared to whole thylakoids [41]. Murphy and Woodrow [41] estimate that the acyl lipids occupy about 14% of the appressed regions compared to 40% of the non-appressed regions, consistent with the higher amount of bilayer seen in freeze-fracture electron micrographs of non-appressed thylakoids [43]. Murphy [44] suggests that the roughly cone-shaped MGDG molecules (that are enriched in appressed membrane fractions) help to stabilize the regions of convex membrane curvature on the inner bilayer-half of the grana margins.

There is little evidence as yet for the specific association of individual acyl lipids with thylakoid complexes. However, phosphatidyldiacylglycerol, esterified with 16:3-*TRANS*-hexadecenoic (a fatty acid found only in thylakoid membranes), is associated mainly with oligomeric LHC II [45], MGDG has been implicated also in mediating interaction between the PS II complex and LHC II [46], and ATP

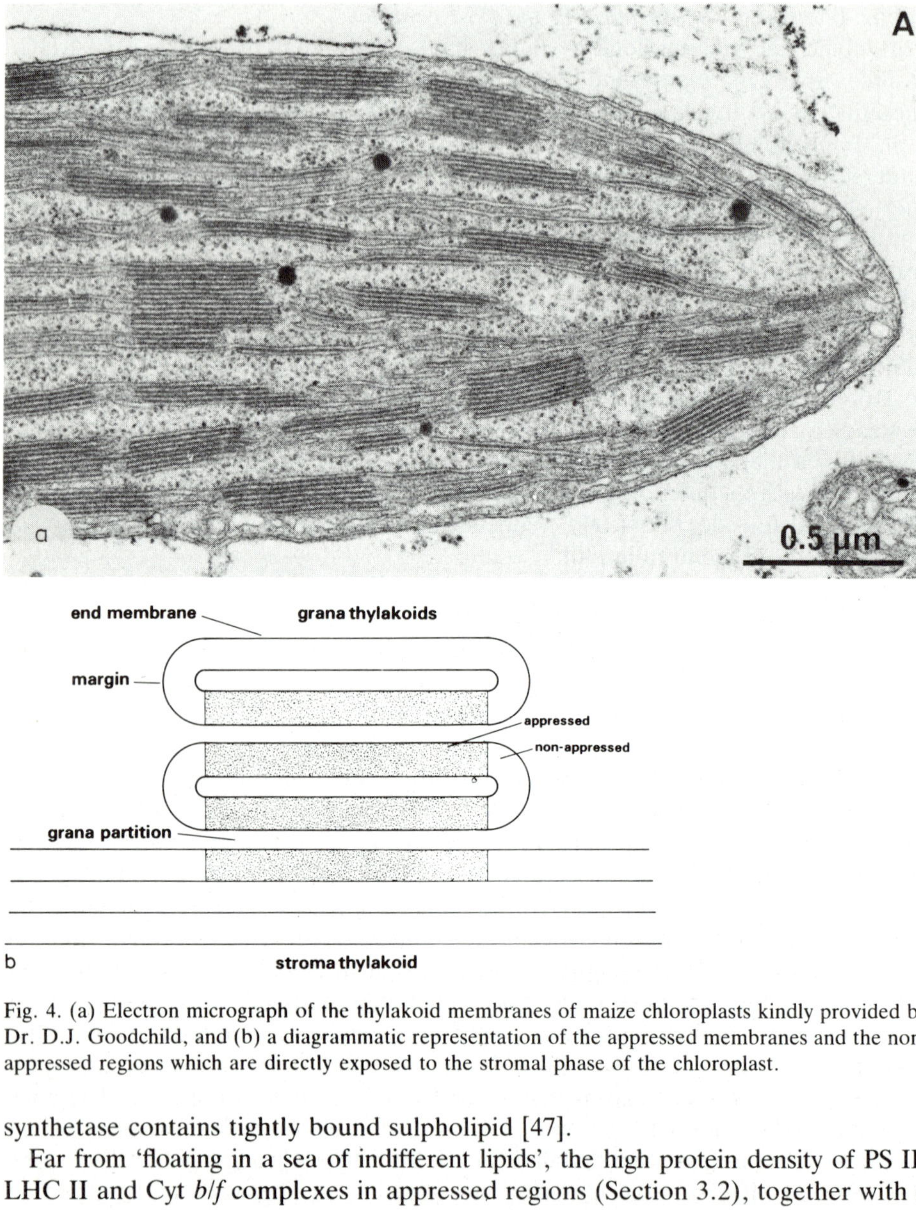

Fig. 4. (a) Electron micrograph of the thylakoid membranes of maize chloroplasts kindly provided by Dr. D.J. Goodchild, and (b) a diagrammatic representation of the appressed membranes and the non-appressed regions which are directly exposed to the stromal phase of the chloroplast.

synthetase contains tightly bound sulpholipid [47].

Far from 'floating in a sea of indifferent lipids', the high protein density of PS II-LHC II and Cyt *b*/*f* complexes in appressed regions (Section 3.2), together with a partial lateral segregation of acyl lipids, suggests that the lipids have a role in maintaining the stability of grana partitions and margins. Lipids may also contribute in the organization of photosynthetic function.

3.2. *Lateral heterogeneity in the location of thylakoid intrinsic complexes*

3.2.1. *Electron microscopic studies*

A striking, yet puzzling, feature of most plant thylakoids is their structural differentiation into the appressed regions of grana partitions, and the non-appressed regions of stroma thylakoids, grana end membranes and grana margins (Fig. 4). Freeze-fracture electron micrographs reveal distinct differences in both the size and the number of particles in the four fracture faces of appressed and non-appressed membranes [43,48,49]. This implies differences in the location of the intrinsic thylakoid protein complexes. Thylakoids readily and reversibly destack and restack depending on the ionic composition of the medium [50]. In low-salt buffers, the membranes destack and the freeze-fracture particles become evenly mixed along the entire single thylakoid membrane network; on addition of cations there is a lateral segregation of freeze-fracture particles with concomitant membrane stacking [43]. Hence, structural studies clearly show that lateral compartmentation is a basic feature of membrane stacking.

ATP synthetase was the first thylakoid complex to be positively localized by Miller and Staehelin [51] who demonstrated unequivocally by antibody labelling studies that CF_1 was present only in non-appressed membranes. The bulky CF_1 component that protrudes some 9–14 nm into the stromal matrix would prevent ATP synthetase being located in the appressed membrane regions that approach one another to $\approx$ 3 nm under illumination [6].

Comparisons of the freeze-fracture profiles of mutant and normal thylakoids of developing plastids and of isolated thylakoid protein complexes allow correlations of specific freeze-fracture particles with the major functional thylakoid complexes [6,48,49]. The correlation of PS I complex with the 10–11 nm PFu freeze-fracture particles was based on their marked decrease in size and number in P-700-deficient mutants of maize [52] and barley [53], whereas the size and number of particles in the appressed regions were not changed. These results suggest that PS I complex is present only in non-appressed regions. The large EFs particles that are proportional to the number of PS II reaction centres in mutants of barley are generally assumed to be PS II complexes, while mutants lacking LHC II have decreased numbers of PFs particles [53]. These electron microscopic studies strongly suggest a lateral heterogeneity of PS II and PS I complexes, as well as of ATP synthetase, but positive identification of the location of thylakoid complexes was lacking until recently (see Section 3.2.3).

3.2.2. *Biochemical studies*

Early fractionation studies with thylakoids that had been fragmented by detergents [54] or mechanical means [55] followed by collection of the thylakoid submembrane fractions by centrifugation provided the first evidence for lateral heterogeneity. Submembrane fractions derived from granal stacks were enriched in PS II but they also contained PS I, whereas the stroma thylakoids had mainly PS I. Sane et al. [55] proposed that appressed membranes were the site of non-cyclic electron transport, while the non-appressed membranes carried out cyclic photo-

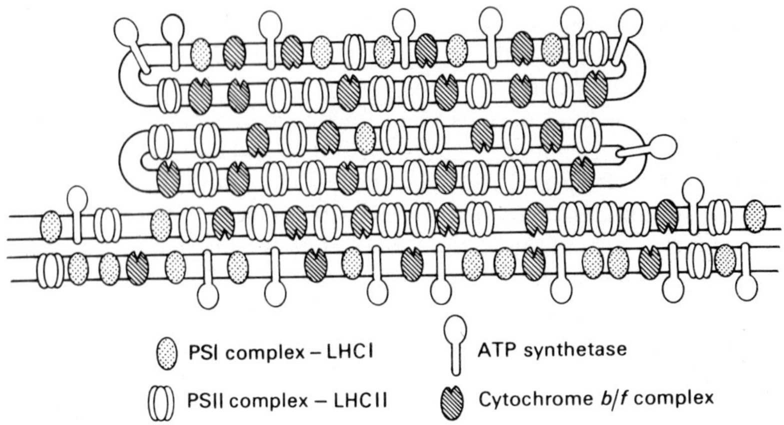

Fig. 5. Possible static representation of the lateral heterogeneity in the distribution of the supramolecular thylakoid complexes between appressed and non-appressed thylakoids [62,63].

phosphorylation. Later, stroma thylakoid fractions were shown to have some PS II activities as well [56]. Biophysical studies supported the notion of most PS II and PS I in close contact in the grana partitions, and models based on the concept of light excitation regulation between PS I and PS II being controlled by spillover depicted the pigment domains of PS II and PS I with a common pool of LHC II [57,58].

The aqueous polymer two-phase partition technique pioneered by Albertsson et al. [11] not only provides a method to separate right-side-out from inside-out vesicles (Section 2.2), but also allows the partial separation of appressed and non-appressed membrane fractions. The inside-out vesicles which partition to the lower phase were depleted in PS II activity [59]. Significantly, they were derived from the appressed membranes of the grana stacks as judged by electron microscopy [60] and their mode of formation [61]. Futhermore, analysis of the Chl-protein content revealed a substantial depletion of PS I complex, and an enrichment of PS II complex and LHC II in the appressed membrane fraction [62]. In 1980, Andersson and Anderson postulated that PS II and PS I are mainly laterally segregated, with PS I excluded from the appressed grana partitions, where most PS II-LHC II complexes are located [62,63] (Fig. 5).

The concept that photosynthesis involves PS II and PS I co-operating in series, immortalized in the Z scheme of Hill and Bendall [64], has been the corner-stone of modern photosynthesis. This scheme does not define the structural or spatial organization of the redox components of the photosystems or the mechanism of ATP synthesis. As discussed, the early fractionation studies with thylakoids fragmented by detergent [54] or mechanical shearing [55] allowed granal stacks enriched in PS II to be separated from stroma thylakoids highly enriched in PS I. The proposal of Sane et al. [55] that the appressed regions were the sites of noncyclic electron transport with PS II and PS I in close contact, and that cyclic photophosphorylation involving PS I only was located in the non-appressed regions,

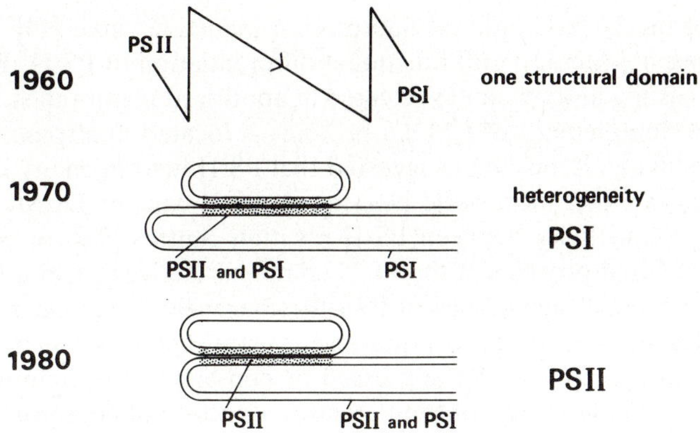

Fig. 6. Summary of the evolution of ideas of the molecular organization of the photosystems in thylakoid membranes.

dominated the ideas of molecular organization throughout the 1970s (Fig. 6). Researchers tried to demonstrate a heterogeneity in PS I, but none was found. The 1980 model of Andersson and Anderson [62,63] was initially a startling idea, since most of the pigment domains of PS II and PS I are postulated to be segregated from one another. It is the converse of the Sane et al. model [55], since there is a heterogeneity of PS II (Fig. 5). This new concept involving the spatial separation of most of the PS II domains from the PS I domains was adopted quickly.

More quantitative measurements using inside-out appressed vesicle fractions were made for the amounts of PS II and PS I reaction centres which showed a 10-fold excess of PS II over PS I [65,66], and a 3.3-fold enrichment of PS I relative to PS II in stroma thylakoids [66]. Comparable ratios were found by EPR measurements of Signal I and Signal II [67]. In contrast, analysis of the Cyt content by redox spectroscopy indicated a uniform distribution of Cyt *f* and Cyt b_6 in both appressed and non-appressed regions, with Cyt *b*-559 mostly located in appressed regions [68,69]. However, there is conflicting evidence for the location of the Cyt *b*/*f* complex. While membrane fractionation studies suggest a random distribution between stacked and unstacked membranes [68,69], the absence of Cyt *f* in the appressed membrane fraction obtained with Triton X-100 indicated to others that the Cyt *b*/*f* complex was located in stroma thylakoids only [70]. As Barber [71] had suggested that the other thylakoid complexes were laterally compartmented due to differences in their distribution of surface charges, he proposed the Cyt *b*/*f* complex might be restricted to the fret region interfacing grana and stroma thylakoids [72]. The constant stoichiometry of Cyt *f* and P-700 in maize mesophyll and bundle sheath thylakoids also led Ghirardi and Melis [73] to postulate location of the Cyt *b*/*f* complex in the fret region.

Heterogeneity of PS II is a feature of the current model for molecular organization of the thylakoid multisubunit complexes (Fig. 6). While there is no doubt that PS II is structurally and functionally heterogeneous, this heterogeneity is not yet fully defined. The idea of PS II heterogeneity was first introduced to interpret the biphasic nature of the DCMU-induced fluorescence induction curve [74]. Melis and colleagues propose a structural and functional differentiation of PS II: PS IIα has a larger light-harvesting antenna and is located in appressed membranes, while PS IIβ with a smaller complement of Chl *a*/*b*-proteins is located in stroma thylakoids [75–77]. Alternatively, it has been suggested that PS II heterogeneity is only apparent and results from either incomplete blockage of PS II units by DCMU [78], or differences in the connectivity between PS II reaction centres [79], or preferential excitation of the Chl *a*-proteins of the PS II complex relative to Chl *a*/*b*-proteins [80]. While the functional significance of PS II heterogeneity is not clear, there are undoubtedly differences in the absorption and spectral antenna unit size for PS II complexes that are located in the appressed or non-appressed membranes, and possible differences in the nature of their electron donors and acceptors. One idea is that the small amount of PS IIβ in non-appressed membranes may be required to poise cyclic photophosphorylation [65]. Developmental changes in the composition and organization of thylakoid membranes are also reflected in varying ratios of PS IIα to PS IIβ [76,81]. Perhaps PS IIβ may represent newly synthesized PS II complexes awaiting a final complement of the peripheral Chl *a*/*b*-proteins and lateral compartmentation to appressed membranes, or a specific population of LHC II-depleted PS II units following the postulated lateral migration of some phosphorylated LHC II to non-appressed membranes. Others have suggested that they represent damaged PS II units which have lost their Q_B protein upon photo-oxidation [82].

While the concept of lateral heterogeneity of thylakoid complexes with most PS II-LHC II complexes in appressed membranes, and few PS II complexes together with PS I complexes and ATP synthetase in non-appressed membranes, is generally accepted (Fig. 5), the evidence is nevertheless indirect. Membrane fractionation studies using mechanical shearing rather than detergents should minimize artefactural alterations in protein composition in the membrane fragments, and prevent lateral segregation of complexes at the point of fragmentation. While the two-phase partition technique [11] has been very useful in molecular topology studies, it is an assumption that there is no preferential extraction of any thylakoid complex into either of the aqueous phases during the phase partition steps. Further, the correlations between structure, so wondrously revealed by freeze-fracture electron microscopy, and function are indirect. Consequently, it is necessary to directly locate the distribution of individual thylakoid complexes by immunocytochemical ultrastructural studies, in order to answer the following important questions:

(1) The extent of exclusion of PS I complex from appressed membranes? While it is clear that PS I is markedly depleted in appressed regions, is there any PS I complex present in grana partitions?
(2) The location of the Cyt *b*/*f* complex?
(3) The heterogeneity of the PS II complex?

Recently, antibody labelling of ultrathin sections of embedded tissue followed by ferretin or gold provided direct visualization of the in situ distribution of thylakoid complexes. To ensure that this approach is valid, care must be taken to ensure that only specific antibodies are used.

3.2.3. 'Seeing is believing'

In immunoelectron microscopic studies with antibodies directed to two of the intrinsic subunits of the PS II complex [83] and to the extrinsic proteins of the oxygen-evolving complex of PS II [83,84], about 90% of the gold labelling occurred in the appressed regions of the thylakoid for both the intrinsic and extrinsic PS II antigens. These results demonstrate directly that most, but not all, PS II complexes are located in appressed regions [83,84], in agreement with the majority of biochemical fractionation studies. Further, Vallon et al. [83] show clearly that extrinsic peptides of the oxygen-evolving system associated with PS II complex are located in both the non-appressed and the appressed membrane regions. Antibodies to Cyt *b*-559 were localized only in appressed regions, suggesting that the high-potential form of Cyt *b*-559 may not be a component of PS IIβ [85]; however, the level of resolution is insufficient to make this assertion.

Three hypotheses have been made for the location of the Cyt *b*/*f* complex: random distribution between appressed and non-appressed regions [68,69]; location only in the fret region interfacing appressed and non-appressed membranes [72,73]; distribution in non-appressed regions only [70]. Antibodies directed towards the Cyt *b*/*f* complex [86] or Cyt *f* [87,88] of spinach thylakoids were labelled with ferritin [86] or gold [87,88] respectively; in all cases, it is seen that the Cyt *b*/*f* complex is distributed laterally in both appressed and non-appressed regions. These results argue against the suggestions of a predominant location in the fret region [72,73] or only in stroma thylakoids [70].

Fewer studies have been made as yet with the PS I complex. Vallon et al. [89] made extensive immunogold labelling studies with spinach and *Chlamydomonas reinhardtii* thylakoids. When the membranes were probed with an antibody directed against the 68 kDa apoprotein of the Chl *a*-proteins of the PS I complex, almost all of the gold label was found in the non-appressed regions [89]. Since the

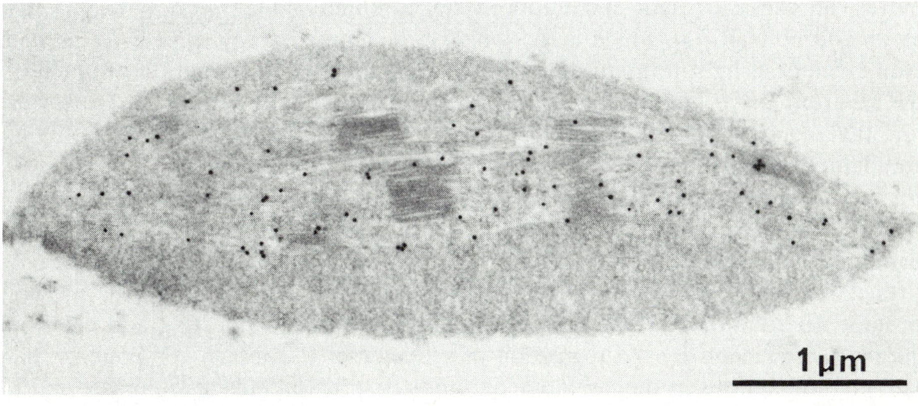

Fig. 7. Section of a spinach leaf, fixed in glutaraldehyde and embedded in K_4 M resin, which has been treated with rabbit antibody to the 68 kDa apoprotein of the P-700-Chl *a*-protein of the PS II complex followed by goat anti-rabbit antibody with 20-nm gold particles attached (see Refs. 84 and 88) (Goodchild, D.J. and Anderson, J.M., unpublished results).

localization of CF_1 at the outer surface of non-appressed thylakoids is well known [51], Vallon et al. [89] compared the gold labelling patterns of the PS I antibody with that obtained with antibodies directed against the α and β subunits of CF_1. Identical labelling patterns were obtained. Vallon et al. [89] conclude that the PS I complex is excluded from the appressed membranes of the grana partitions. In another study, antibodies directed against the 68 kDa apoprotein or the 18 and 16 kDa proteins of the PS I complex were used on ultrathin sections of spinach coupled with gold (Fig. 7). Exclusive labelling of the non-appressed membranes is evident. These results strongly suggest that few or no PS I complexes are located in grana partitions. Further immunocytochemical studies are needed to confirm this important result.

4. *Consequences of lateral heterogeneity*

Lateral heterogeneity has profound consequences for thylakoid function and structure [4,6,63] despite some uncertainty prevailing as to the absolute extent of lateral heterogeneity of the PS I complex, and the location of the Cyt *b*/*f* complex in both appressed and non-appressed membrane regions.

4.1. Light-harvesting strategies

Contrary to the concepts that prevailed in the 1960s and 1970s that the pigment systems of PS II and PS I were in close contact with each other (e.g. Refs. 57 and 58), it is now established that most of the pigment domains of PS II are widely separated from those of PS I in Chl *b*-containing chloroplasts [63]. The main light-harvesting antenna, LHC II, can transfer excitation energy very efficiently to both P-680 and P-700. Consequently, if LHC II were in contact with both reaction centres, an excess of light excitation energy to either photosystem would of necessity end up in P-700, which is the longer-wavelength energy trap; P-700 is also more efficient at light-trapping than P-680. Hence, lateral segregation of much of LHC II from P-700 limits the over-excitation of PS I relative to PS II. However, were thylakoids to possess static and fixed proportions of LHC II, subject only to modulations in light-harvesting capacity by long-term adaptation (i.e., by synthesis and breakdown of LHC II), there would be no rapid and flexible responses to regulate fluctuations in the excitation energy received by each photosystem. Furthermore, not only are the environmental light conditions continually changing for any particular chloroplast, but also the cellular demands for ATP and $NADP^+$ are not constant. In addition to the long-term regulation of light-harvesting and electron transport components by the regulation of synthesis of membrane components that leads to modulations in their relative amounts [90], thylakoids also possess short-term mechanisms which regulate the excitation energy between the photosystems, and control the extent of non-cyclic and cyclic photophosphorylation [48,91,92].

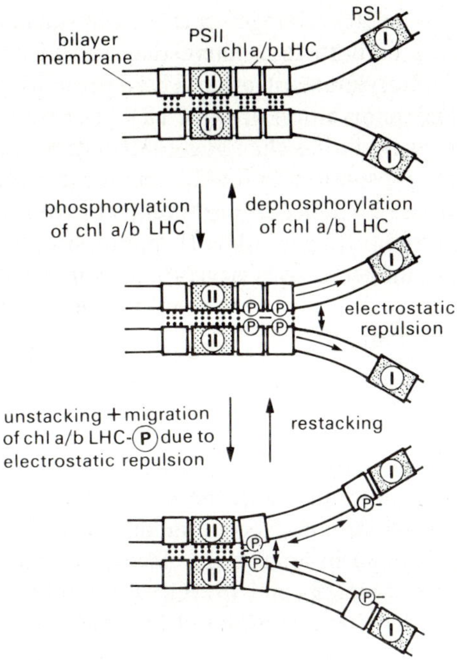

Fig. 8. Schematic diagrams adapted from Staehelin and Arntzen [48] depicting how the reversible phosphorylation of some Chl *a*/*b*-proteins of LHC II affects membrane appression and their distribution between the PS II complex-enriched appressed membranes and the non-appressed membrane regions which are enriched in PS I complex.

4.1.1. Protein phosphorylation

Photophosphorylation of key thylakoid proteins of higher plants and green algae provides a novel means of short-term regulation of excitation energy between the photosystems, and also allows thylakoid membrane organization to be dynamic and flexible, rather than static. Chloroplasts have kinase and phosphatase systems which catalyse the reversible phosphorylation of several PS II proteins, including the two major apoproteins of LHC II [91]. An excess of PS II light causes over-reduction of the PQ pool, which then causes the activation of a membrane-bound kinase which phosphorylates the threonine residue(s) at the stroma-exposed, N-terminal region of certain apoproteins of LHC II [48]. The increase of net negative charge in the phosphorylated LHC II adjacent to the edges of appressed regions results in a decrease of membrane appression [72], and the phosphorylated LHC II complexes will diffuse laterally from the appresed to non-appressed regions where they may interact with PS I (Fig. 8) [48]. A limited decrease in membrane stacking has been demonstrated following this light-mediated phosphorylation of part of LHC II in vitro (cf. Refs. 4 and 48). This strategy of detachment of some LHC II units from PS II complexes would decrease the excitation energy to PS II within a few minutes, then the plastoquinone pool would become oxidized and initiate phosphatase action, and LHC II would reassociate with PS II complexes. This mechanism, which effects no more than 20% changes in thylakoid stacking in vitro, appears to provide a fine tuning of both light-harvesting and electron flow. The significance of the photophosphorylation of other PS II proteins [91] is not yet understood. The reversible phosphorylation of some LHC II may function also in the maintenance of high efficiency during changing contributions of cyclic photophosphorylation to total photophosphorylation [92]. Increased cyclic phosphorylation by PS I would result in an imbalance in non-cyclic electron transport in favour of PS II, which could then trigger LHC II photophosphorylation [92]. Certainly electron transfer can influence light-harvesting so that the interplay between them is not static and one-way only, with light-harvesting influencing electron transport, as was orginally thought.

4.2. Electron transport strategies

Lateral heterogeneity in the distribution of the thylakoid PS II and PS I complexes raises several important questions about the long-range electron transport from the appressed regions to the stroma-exposed membranes [4,6,93]. The role of plastoquinone and/or plastocyanin in long-range electron transfer between the PS II complex and the Cyt *b*/*f* complex, and the Cyt *b*/*f* complex and the PS I complex, respectively (Fig. 5), has been considered in detail elsewhere [4,6,93]. The mobile electron transport carrier plastoquinone has an important function in acting as a redox buffer pool (10–20 molecules per P-680) between the photosystems. It is generally accepted that the diffusion of plastoquinone is rapid enough to allow for long-range electron transfer from PS II complexes to Cyt *b*/*f* complexes [4,6,93], although an accurate rate for its diffusion in the protein-rich, appressed membrane region is not available. If indeed Cyt *b*/*f* complexes are present in both membrane regions, plastoquinone will not be required to shuttle electrons between the appressed and non-appressed regions. Haehnel [6] argues from physical considerations that plastocyanin would not be able to effectively transport electrons between widely-separated Cyt *b*/*f* complexes and PS I complexes. Further, since the plastocyanin molecule resembles an ellipsoid cylinder ($4 \times 0.32 \times 2.8$ nm) its diffusion might be restricted within the narrow confines of the lumen (3–6 nm, in the light). Moreover, it is likely that the viscosity of the lumen is very high (cf. Ref. 4). Haehnel [6] considers that it is more likely that plastocyanin will either form a transient complex with Cyt *b*/*f* and PS I complexes, or there will be very short-range lateral diffusion of plastocyanin from the Cyt *b*/*f* complex to the PS I complex. On the other hand, Whitmarsh [93] considers that the rate of plastocyanin diffusion could be great enough for its role as a long-range electron shuttler. Further studies on the characterization of diffusion of electron carriers in electron transport and on the rate-limiting steps of electron transport are needed, together with direct techniques for measuring the diffusion coefficients of these electron carriers in thylakoid membranes.

To resolve this issue, it is crucial to establish whether there are some PS I complexes in the appressed grana partitions or none at all. The suggestion of extreme lateral heterogeneity with exclusion of all PS I complexes from appressed membranes [62,63] came from biochemical fractionation studies with inside-out thylakoid vesicles (obtained from phase partition [11]), which were depleted in PS I complex [62] and P-700 [65–67]. Because of the contamination of the inside-out vesicle fraction with right-side-out vesicles that are enriched in the PS I complex, it was argued that the PS I complex or P-700 observed in the inside-out vesicle fraction might be accounted for by the contaminating right-side-out vesicles [62,65–67].Recently, Atta-Asafo-Adjei and Dilley [94] found that plastocyanin markedly enhances the rates of PS I activities by the inside-out-vesicles fraction. Since plastocyanin could not stimulate PS I rates in the contaminating right-side-out vesicles, these authors [94] maintain that PS I is indeed present in the appressed membrane fraction. However, immunocytochemical studies indicate very few or no PS I complexes in grana partitions (Section 3.2.3.).

4.3. Adaptation of photosynthetic capacity

It is well known that plants grown under different environmental conditions have varying amounts of appressed and non-appressed membranes [48,95]. As Anderson [63,95] pointed out, these morphological differences in membrane stacking must be reflected in modulations in the relative stoichiometries of the supramolecular complexes according to our current concepts of lateral heterogeneity in the distribution of thylakoid complexes shown in Fig. 5. Thus, there is no a priori need to have a fixed, constant proportion of thylakoid complexes, although the actual molar composition of each of the intrinsic complexes most remain unchanged, with the exception of the Chl *a*/*b*-proteins of PS II and PS I, which are structurally heterogeneous [13]. Indeed, marked modulations in the relative distribution of Chl *a* and Chl *b* amongst the various Chl-proteins exist, as reflected in the overall Chl *a*/Chl *b* ratios of different thylakoids [90,95]. Some mutant chloroplasts, developing plastids or chloroplasts from plants receiving intermittent light have little or no Chl *a*/*b*-proteins [48]. It is well-established that thylakoids from sun plants or plants grown under high light intensity have high Chl *a*/Chl *b* ratios and high photosynthetic capacities, while chloroplasts from shade plants or plants grown under low irradiance have low Chl *a*/Chl *b* ratios and, concomitantly, lower photosynthetic capacities [90]. Low Chl *a*/Chl *b* ratios are due to more LHC II and LHC I relative to the Chl *a*-proteins of PS II and PS I complexes. In turn, more LCH II means increased amounts of appressed membranes relative to stroma-exposed membranes. Fewer stroma-exposed membranes means relatively fewer PS I complexes and ATP synthetase on a Chl basis; in turn, the electron transport capacity wil be lower, and carbon dioxide fixation will be decreased [90]. Thus, modulation in the relative amounts of thylakoid complexes and, concomitantly, the ratio of appressed and non-appressed membranes should indeed be reflected in the overall photosynthetic capacities of leaves [95]. These long-term adaptations in the modulation in the relative amounts of the thylakoid components are responses to various environmental factors such as light (irradiance, quality and duration), water loss, heat, cold, and nitrogen or mineral deficiencies. These adaptations involve a variety of strategies that include alteration in the amounts and composition of chlorophylls and carotenoids, relative numbers of PS IIα, PS IIβ and PS I units, and the amounts of Cyt *b*/*f* complex and ATP synthetase [81,90,95].

These substantial organizational, structural and functional changes that occur in response to environmental factors further suggest also that photosynthesis itself has a positive-feedback mechanism that promotes the accumulation of certain thylakoid components [90,96]. Melis et al. [96] point out that the adaptations to environmental stress are geared to ameliorate or restore the damaged system by implementing specific changes in the stoichiometries of the electron transport components and in the antenna chlorophylls of both photosystems. For example, partial inhibition of PS II by herbicides (independent of the herbicide used) results in loss of PS II activity and a significant imbalance of electron flow between PS II and PS I [97–99]. In response to this partial inhibition of PS II activity, more PS II-LHC II complexes are synthesized, resulting in lower Chl *a*/Chl *b* ratios, and higher Q_A/P-700 and Q_A/PQ ratios [97–99].

5. *Thylakoid stacking*

5.1. *Mechanisms of thylakoid stacking*

It is established that the major apoproteins of LHC II are of great importance in membrane stacking in Chl *b*-containing chloroplasts (cf. Ref. 4). Not only is LHC II mainly located in appressed membranes [62], but mutant or developing plastids lacking the 28 and 26 kDa proteins of LHC II have no stacked membranes [48]. Limited proteolysis of thylakoids or LHC II-proteoliposomes removes a short segment of the N-terminus of the apoproteins of LHC II (cf. Ref. 4) which contains the threonine residue(s) [91] that are reversibly phosphorylated during protein phosphorylation (Section 4.1.2), as well as several positively charged lysine and arginine residues. The peptide cleaved by trypsin from the N-terminal regions of LHC II is [Lys-Arg]-Ser-Thr-Thr-Lys-Lys [100]. It is thought that the local decrease in net negative surface charge in the N-terminal region of LHC II will help to decrease the overall electrostatic repulsive forces between approaching thylakoid membranes, thus promoting stacking in regions with a high density of LHC II [72,101].

The role of the N-terminal region of the apoproteins of LHC II in mediating thylakoid appression is demonstrated also in phosphorylation studies [48], where removal of this peptide sequence prevents the control of excitation energy distribution between the photosystems (Section 4.1.2). Phosphorylation of the threonine residues [91] increases the local net negative charges of LHC II located at the edges of the grana partitions, which then favours the lateral diffusion of phosphorylated LHC II into the adjacent non-appressed regions [101].

It is thought that membrane appression results from the localized decrease in the net negative surface charge of the many LHC II proteins surrounding each core PS II complex, thereby decreasing the overall electrostatic repulsive forces between adjacent membrane surfaces, and also increasing the van der Waals attractive interactions [101]. Albertsson [102] suggests that additional attractive forces between the inner thylakoid membrane pairs exposed to the lumen might also be important in keeping the thylakoids in flat sheets, and in maintaining and stabilizing the lateral segregation of PS II complex in the appressed grana thylakoids. On the other hand, PS I complexes found mainly in stroma-exposed thylakoids would not have a favourable distribution of surface stroma-exposed charges to promote membrane appression [70,72], while the CF_1 headpiece of ATP synthetase would be too bulky to allow membrane appression. If the surface charge distribution is indeed an important factor for the location of thylakoid complexes [71,101], it is more difficult to rationalize a random distribution for the Cyt *b*/*f* complex, a consideration that led Barber [72] to propose that this complex was located only in the fret region. However, if the local stroma-exposed surface charges of the Cyt *b*/*f* complex were adequately protected, this complex might be able to freely diffuse in both appressed and non-appressed regions. Alternatively, there may be some difference in the composition of the Cyt *b*/*f* complexes located in either appressed or non-appressed regions. It is important now to determine directly whether Cyt *b*/*f* complex can diffuse along the entire membrane network in vivo, and whether the extent of inter-membrane LHC II–LHC II interactions actually restricts lateral movement of PS II-LHC II complexes in the grana partitions.

5.2. *Significance of thylakoid stacking*

Many hypotheses have been evoked to account for the significance of thylakoid stacking (cf. Refs. 48 and 95). LHC II, which is essential for membrane stacking, necessarily ensures that most of the PS II-LHC II complexes are located in appressed membranes; hence the more LHC II a chloroplast has, the more stacked membranes it will have. However, it should be remembered that membrane stacking is not essential for photosynthetic electron flow, and the quantum yields of chloroplasts with different proportions of stacking are equal [48,90]. Lateral heterogeneity in the distribution of the photosystems, by the separation of most PS II pigment domains from those of PS I, limits the extent of transfer to PS I of light excitation orginally absorbed by PS II and thereby prevents the over-excitation of PS I relative to PS II. Were the bulk of the light-harvesting pigments, mainly located in LHC II, in direct contact with the reaction centres of both photosystems most of the energy would end up in P-700. Hence the lateral segregation of most PS II and PS I, together with the short-term mechanism of protein phosphorylation to finely modulate the distribution of excitation energy between PS II and PS I under fluctuating light conditions, allows Chl *b*-containing chloroplasts to maximize photosynthetic capacity under varying light conditions.

Not all plants and algae have the same light-harvesting strategies. While the three electron transport complexes and ATP synthetase appear to be structurally and functionally similar in prokaryotes and eukaryotes, there is a great diversity in accessory light-harvesting pigments and their apoproteins [102,103]. These include the water-soluble phycobiliproteins of blue-green and red algae, which have no appressed membranes, no Chl *b* and as far as is known do not possess extreme lateral segregation of PS II and PS I [102,103]. Another major group of algae possesses Chl c_1 and Chl c_2 instead of Chl *b*. These have their membranes stacked together in groups of three thylakoids that extend across the total chloroplast, and have a fixed proportion of stacked to unstacked membranes of 2:1 [103]. We know very little about either the regulation of light-harvesting or electron transport in Chl *c*-containing algae, and the extent of lateral heterogeneity in the distribution of PS II and PS I is known only for Chl *b* algae. It is likely that all of these algae have very different light-harvesting strategies and mechanisms of thylakoid appression compared to the plants and algae that possess the Chl *a*/*b*-proteins. Even with the Chl *b*-containing algae, there are a variety of Chl *a*/*b*-proteins [103]. Some marine green algae have siphonaxanthin-Chl *a*/*b*-proteins, instead of the lutein-Chl *a*/*b*-proteins of higher plants and most green algae. These ancient marine green algae have different membrane appression profiles [103], and they may not possess extreme lateral heterogeneity in the distribution of the photosystems.

6. *Epilogue*

Much of the excitement generated in the 1980s by the molecular organization studies was caused by the demonstration of lateral heterogeneity of intrinsic thylakoid complexes, and the heterogeneity of PS II. It is important now to define the extent of exclusion of the PS I complex from appressed membrane regions, to determine whether the Cyt *b*/*f* complex is indeed located in both membrane regions, and to determine the regulatory mechanisms that modulate the relative stoichiometries of the five thylakoid protein complexes, the mobile electron carriers and the acyl lipids. Future advances in the elucidation of thylakoid organization will come from biochemical studies such as immunology, chemical cross-linking and enzymic treatments, as well as from molecular genetic techniques such as site-directed mutagenesis. These methods are needed now to test the putative tertiary structures of thylakoid proteins deduced from hydropathy index plots. Crystallization of the thylakoid multi-protein complexes will allow high resolution from image analysis reconstruction, and ultimately X-ray analysis. More studies on lipid organization and lipid–protein interactions are also needed. Then having replaced the complicated yet visible thylakoid membranes with simple yet invisible concepts of molecular structure, we will need to integrate this molecular knowledge with the wider interactions of thylakoid function within the chloroplast with that of plant growth yield.

With our current level of understanding of the molecular organization of thylakoid membranes and its relation to function and structure, accumulated during the past 26 years, we are now ready to explore the mechanisms which enable plants to adapt not only to their natural habitats, but also to environmental stresses. This knowledge is vital to form the basis of rational solutions for agricultural and related industries in the production of new varieties of crop plants by either conventional breeding or genetic engineering.

References

1 Singer, S.J. (1971) in Structure and Function of Biological Membranes (Rothfield, L.I., ed.) pp. 144–222, Academic Press, New York.
2 Singer, S.J. and Nicolson, G.L. (1972) Science 175, 720–731.
3 Anderson, J.M. (1975) Biochim. Biophys. Acta 416, 191–235.
4 Murphy, D.J. (1986) Biochim. Biophys. Acta 864, 33–95.
5 Quin, P.J. and Williams, W.P. (1983) Biochim. Biophys. Acta 737, 223–266.
6 Haehnel, W. (1984) Annu. Rev. Plant Physiol. 35, 659–693.
7 Radunz, A. (1984) in Advances in Photosynthesis Research (Sybesma, C. ed.) Vol. III, pp. 151–154, M. Nijhoff/Dr. W. Junk, The Hague.
8 Rawyler, A. and Siegenthaler, P.A. (1984) In Structure and Function of Metabolism of Plant Lipids (Siegenthaler, P.A. and Eichenberger, W., eds.) pp. 409–412, Elsevier, Amsterdam.
9 Unitt, M.D. and Harwood, J.L. (1982) in Biochemistry and Metabolism of Plant Lipids (Wintermans, J.F.G.M. and Kuiper, P.J.C., eds.) pp. 331–334, Elsevier, Amsterdam.
10 Sundby, C. and Larsson, C. (1985) Biochim. Biophys. Acta 813, 61–67.
11 Albertsson, P.A., Andersson, B., Larsson, C. and Åkerlund, H.E. (1982) Methods Biochem. Anal. 28, 115–150.
12 Andersson, B., Åkerlund, H.E., and Albertsson, P.A. (1977) FEBS Lett. 77, 141–145.
13 Thornber, J.P. (1986) In Encyclopedia of Plant Physiology, New Series, Vol. 19 (Staehelin, L.A. and Arntzen, C.J., eds.) pp. 98–142, Springer-Verlag, Heidelberg.
14 Kyte, J. and Doolittle, R.F. (1982) J. Mol. Biol. 157, 105–132.
15 Eisenberg, D. (1984) Annu. Rev. Biochem. 53, 592–623.
16 Cramer, W.A., Widger, W.R., Herrmann, R.G. and Trebst, A. (1985) Trends Biochem. Sci. 10, 125–129.
17 Deisenhofer, J., Epp, O., Miki, K., Huber, R. and Michel, H. (1985) Nature 318, 618–624.
18 Kühlbrandt, W. (1984) Nature 307, 478–480.
19 Li, J. (1985) Proc. Natl. Acad. Sci. USA 82, 386–390.
20 Ortiz, W. and Malkin, R. (1985) Biochim. Biophys. Acta 808, 164–170.
21 Willey, D.L., Auffret, A.D. and Gray, J.C. (1984) Cell 36, 555–562.
22 Mansfield, R.W. and Bendall, D.S. (1984) Biochim. Biophys. Acta 766, 62–69.
23 Mörshel, E. and Staehelin, L.A. (1983) J. Cell Biol. 97, 301–310.
24 Mansfield, R.W. and Anderson, J.M. (1985) Biochim. Biophys. Acta 809, 435–444.
25 Lam, E. (1986) Biochim. Biophys. Acta 848, 324–332.
26 Leonard, K., Wingfield, P., Arad, T. and Weiss, T. (1981) J. Mol. Biol. 149, 259–274.
27 Andersson, B., Anderson, J.M. and Ryrie, I.J. (1982) Eur. J. Biochem. 123, 465–472.
28 Ryrie, I.J., Anderson, J.M., and Glare, T. (1985) Photobiochem. Photobiophys. 9, 145–157.
29 Ortiz, W., Lam, E., Chollar, S., Munt, D. and Malkin, R. (1985) Plant Physiol. 77, 389–397.
30 Sieffermann-Harms, D. (1985) Biochim. Biophys. Acta 811, 325–355.
31 Karlin-Neumann, G.A., Kohorn, B.D., Thornber, J.P. and Tobin, E.M. (1985) J. Mol. Appl. Genet. 3, 45–61.
32 Nabedryk, E., Andrianambinintsoa, S. and Breton, J. (1985) Biochim. Biophys. Acta 765, 380–387.
33 Herrmann, R.G., Alt, J., Schiller, B., Widger, W.R. and Cramer, W.A. (1985) FEBS Lett. 976, 239–244.
34 Murata, N. and Miyao, M. (1985) Trends Biochim. Sci. 10, 122–124.
35 Govindjee, Kambara, T. and Coleman, W. (1985) Photochem. Photobiol. 42, 187–210.
36 Andersson, B., Larsson, C., Jansson, C., Ljungberg, U. and Åkerlund, H.E. (1984) Biochim. Biophys. Acta 776, 21–28.
37 Ono, T. and Inoue, Y. (1984) FEBS Lett. 164, 375–378.
38 Fish, L.E., Küch, U. and Bogorad, L. (1985) J. Biol. Chem. 260, 1413–1421.
39 Haehnel, W., Berzborn, R.J. and Andersson, B. (1981) Biochim. Biophys. Acta 637, 389–399.
40 Jennings, R.C., Garlaschi, F.M., Gerola, P.D. and Forti, G. (1979) Biochim. Biophys. Acta 546, 207–219.
41 Murphy, D.J. and Woodrow, I.E. (1983) Biochim. Biophys. Acta 725, 104–112.

42 Gounaris, K., Sundby, C., Andersson, B. and Barber, J. (1983) FEBS Lett. 156, 170–173.
43 Staehelin, L.A. (1976) J. Cell Biol. 71, 136–158.
44 Murphy, D.J. (1982) FEBS Lett. 150, 19–26.
45 Remy, R., Tremolieres, A., Duval, J.C., Ambard-Bretteville, F. and Dubacq, J.P. (1982) FEBS Lett. 137, 271–275.
46 Siefermann-Harms, D., Ross, J.W., Kaneshiro, K.H. and Yamamoto, H.Y. (1982) FEBS Lett. 149, 191–196.
47 Pick, U., Gounaris, K., Weiss, M. and Barber, J. (1985) Biochim. Biophys. Acta 808, 415–420.
48 Staehelin, L.A. and Arntzen, C.J. (1983) J. Cell Biol. 97, 1327–1337.
49 Staehelin, L.A. and deWitt, M. (1984) J. Cell Biol. 24, 261–269.
50 Izawa, S. and Good, N.E. (1966) Plant Physiol. 41, 544–552.
51 Miller, K.R. and Staehelin, L.A. (1976) J. Cell Biol. 68, 30–47.
52 Miller, K.R. (1980) Biochim. Biophys. Acta 592, 143–152.
53 Simpson, D.J. (1981) In Proceedings 5th International Congress of Photosynthesis (Akoyunoglou, G., ed.) Vol. 3, pp. 15–22, Balaban International Science Services, Philadelphia, PA.
54 Anderson, J.M. and Boardman, N.K. (1966) Biochem. Biophys. Acta 112, 403–421.
55 Sane, P.V., Goodchild, D.J. and Park, R.B. (1970) Biochim. Biophys. Acta 216, 162–178.
56 Armond, P.A. and Arntzen, C.J. (1977) Plant Physiol. 59, 398–404.
57 Butler, W.L. (1978) Annu. Rev. Plant Physiol. 29, 345–378.
58 Thornber, J.P. and Barber, J. (1979) in Photosynthesis in Relation to Model Systems (Barber, J., ed.) Topics in Photosynthesis, Vol. 3, pp. 27–70, Elsevier/North Holland, Amsterdam, New York.
59 Åkerlund, H.E., Andersson, B. and Albertsson, P.A. (1976) Biochim. Biophys. Acta 449, 525–535.
60 Andersson, B., Simpson, D. and Hoyer-Hanson, G. (1978) Carlsberg Res. Commun. 43, 77–89.
61 Andersson, B., Sundby, C. and Albertsson, P.A. (1980) Biochim. Biophys. Acta 599, 391–402.
62 Andersson, B. and Anderson, J.M. (1980) Biochim. Biophys. Acta 593, 427–440.
63 Anderson, J.M. (1981) FEBS Lett. 124, 1–10.
64 Hill, R. and Bendall, F. (1960) Nature 139, 136–137.
65 Anderson, J.M. and Melis, A. (1983) Proc. Natl. Acad. Sci. USA 80, 745–749.
66 Andersson, B. and Haehnel, W. (1982) FEBS Lett. 146, 13–17.
67 Anderson, J.M. and Malkin, R. (1982) FEBS Lett. 148, 293–296.
68 Cox, R.P. and Andersson, B. (1981) Biochim. Biophys. Res. Commun. 103, 1136–1142.
69 Anderson, J.M. (1982) FEBS Lett. 138, 62–66.
70 Henry, L.E.A. and Møller, B.L. (1981) Carlsberg Res. Commun. 46, 227–242.
71 Barber, J. (1980) FEBS Lett. 118, 1–10.
72 Barber, J. (1983) Plant Cell Environ. 6, 311–322.
73 Ghirardi, M.L. and Melis, A. (1983) Arch. Biochem. Biophys. 224, 19–28.
74 Melis, A. and Homann, P.H. (1976) Photobiochem. Photobiol. 29, 373–382.
76 Thielen, A.P.G.M. and Van Gorkom, H.J. (1981) Biochim. Biophys. Acta 635, 111–120.
77 Melis, A. and Anderson, J.M. (1983) Biochim. Biophys. Acta 724, 473–484.
78 Kyle, D.J., Haworth, P. and Arntzen, C.J. (1982) Biochim. Biophys. Acta 680, 336–342.
79 Hodges, M. and Barber,J. (1983) FEBS Lett. 160, 177–181.
80 Percival, M.P., Webber, A.N. and Baker, N.R. (1984) Biochim. Biophys. Acta 767, 582–589.
81 Melis, A. (1984) J. Cell Biochem. 24, 271–285.
82 Kyle, D.J., Staehelin, L.A. and Arntzen, C.J. (1984) Proc. Natl. Acad. Sci. USA 81, 4070–4074.
83 Vallon, O., Wollman, F.A. and Olive, J. (1985) FEBS Lett. 183, 245–250.
84 Goodchild, D.J., Andersson, B. and Anderson, J.M. (1985) Eur. J. Cell Biol. 36, 294–298.
85 Rao, L.V.M., Usharani, P., Butler, W.L. and Tokuyasu, K.T. (1986) Plant Physiol. 80, 138–141.
86 Allred, D.R. and Staehelin, L.A. (1985) Plant Physiol. 78, 199–202.
87 Shaw, P.J. and Henwood, J.A. (1985) Planta 165, 333–339.
88 Goodchild, D.J., Anderson, J.M. and Andersson, B. (1985) Cell Biol. Int. Rep. 9, 715–721.
89 Vallon, O., Wollman, F.A. and Olive, J. (1986) Photobiochem. Photobiophys. 12, 203–220.
90 Anderson, J.M. (1986) Annu. Rev. Plant Physiol. 37, 93–136.
91 Bennett, J. (1984) Physiol. Plant. 60, 583–590.
92 Horton, P. (1985) in Photosynthetic Mechanisms and the Environment (Barber, J., ed.) Vol. 6, pp. 135–187, Elsevier Biomedical Press, Amsterdam.

93 Whitmarsh, J. (1986) in Encyclopedia of Plant Physiology, New Series, Vol. 19, Photosynthetic Membranes (Staehelin, L.A. and Arntzen, C.J., eds.) pp. 508–527, Springer-Verlag, Heidelberg.
94 Atta-Asafo-Adjei, E. and Dilly, R.A. (1986) Arch. Biochem. Biophys. 243, 660–667.
95 Anderson, J.M. (1982) Photobiochem. Photobiophys. 3, 225–241.
96 Melis, A., Mandori, A., Glick, R.E., Ghirardi, M.L., McCauley, S.W. and Neale, P.J. (1985) Physiol. Vég. 23, 757–765.
97 Bose, S., Mannan, R.M. and Arntzen, C.J. (1984) Z. Naturforsch 39, 510–513.
98 Mattoo, A.K., St. John, J.B. and Wergin, W.P. (1984) J. Cell Biochem. 24, 163–175.
99 Leech, R.M., Walton, C.A. and Baker, N.R. (1985) Plant Physiol. 165, 277–283.
100 Mullett, J.E. (1983) J. Biol. Chem. 258, 9941–9948.
101 Barber, J. (1982) Annu. Rev. Plant Physiol. 33, 261–295.
102 Albertsson, P.A. (1982) FEBS Lett. 149, 186–190.
103 Larkum, A.W.D. and Barrett, J. (1983) Adv. Bot. Res. 10, 1–230.
104 Anderson, J.M. and Barrett, J. (1986) in Encyclopedia of Plant Physiology, New Series, Vol. 19. Photosynthetic Membranes (Staehelin, L.A. and Arntzen, C.J. eds.) pp. 269–285, Springer-Verlag, Heidelberg.

J. Amesz (ed.) *Photosynthesis*

CHAPTER 13

Structure and exciton effects in photosynthesis

ROBERT M. PEARLSTEIN

Physics Department, Indiana-Purdue University, 1125 East 38th Street, P.O. Box 647, Indianapolis, IN 46223, U.S.A.

1. Introduction

In both antenna and reaction center (RC) pigment-protein complexes of photosynthetic organisms, pigment molecules are often close enough together to perturb each other's electronic transitions significantly. These perturbations result from resonance interactions that produce so-called *exciton effects* in optical spectra, principally absorption, circular dichroism (CD) and linear dichroism (LD). The same resonance interactions also generate electronic excited-state energy transfer through the various antenna arrays to the RCs, a very important function in photosynthesis. However, the latter topic has been extensively reviewed quite recently [1–5] and so is not further considered here. Besides, the subject of exciton effects in photosynthesis has now grown so large it requires separate treatment, and indeed is treated somewhat selectively here. For earlier reviews and some omitted topics, see Refs. 1,4,6–8.

It is especially timely to review the subject of exciton effects because, with the advent of the X-ray structural model of the *Rhodopseudomonas viridis* RC [9–12], it is becoming apparent that analyses of exciton effects exhibit a dichotomy. On the one hand there are analyses based on incomplete structural information, on the other there are those based on X-ray structural models. The former generally seem theoretically straightforward and consistent with all experimental data, while the latter tend to be theoretically involuted and inconsistent with at least some of the data. Because the underlying interactions are quite important in photosynthesis, it is worthwhile exploring this situation and trying to understand what underlies it. In Section 2 basic theoretical concepts are briefly summarized. Exciton analyses based on partial structural information are discussed in Sections 3 and 4, and those based on X-ray models are considered in Sections 5–7.

2. Theoretical concepts

The resonance interactions that give rise to the exciton effects in spectra are interactions among the electronic transition moments of closely juxtaposed pigment

TABLE 1

Formulas for calculating exciton 'stick' spectra

Spectrum Type	'Stick' spectrum formula
Exciton N-mer absorption	$A_K = \sum_{\alpha,\beta=1}^{N_L} \sum_{i,j=1}^{N_M} \mu_{\alpha i}\mu_{\beta j} (\hat{\mu}_{\alpha i}\cdot\hat{\mu}_{\beta j}) U_{\alpha iK} U_{\beta jK}$
Exciton N-mer CD	$C_K = -1.6959 \times 10^{-6} \sum_{\alpha,\beta=1}^{N_L} \sum_{i,j=1}^{N_M} \nu_{\alpha i}\mu_{\alpha i}\mu_{\beta j}R_{ij} (\hat{R}_{ij}\cdot\hat{\mu}_{\alpha i} \times \hat{\mu}_{\beta j}) U_{\alpha ik}\, U_{\beta jK}$
Exciton N-mer LD	$L_K = 1/2 \sum_{\alpha,\beta=1}^{N_L} \sum_{i,j=1}^{N_M} \mu_{\alpha i}\mu_{\beta j} [3(\hat{\mu}_{\alpha i}\cdot\hat{\varepsilon})(\hat{\mu}_{\beta j}\cdot\hat{\varepsilon}) - \hat{\mu}_{\alpha i}\cdot\hat{\mu}_{\beta j}] U_{\alpha iK}\, U_{\beta jK}$
Simple exciton dimer absorption	$A_{\pm} = \mu^2(1 \pm \cos\theta)$
Simple exciton dimer CD	$\mp C_{\pm} = 1.7 \times 10^{-6} \nu\mu^2 R\kappa_{rot},\ \kappa_{rot} = \hat{R}\cdot\hat{\mu}_1 \times \hat{\mu}_2$
Simple exciton dimer LD	$L_{\pm} = 1/4\, \mu^2[3(\cos\theta_1 \pm \cos\theta_2)^2 - 2(1 \pm \cos\theta)]$

N-mer formulas are general, and include off-resonance interactions (see Ref. 20 for inclusion of doubly-excited states.) Dimer formulas are for the simplest case only, i.e., 1-orbital interactions and zero differential environmental shifts. Subscripts α and β label excited-state energy levels; subscripts i and j label pigment molecules; N_L = number of energy levels (usually 4 or less taken for Chl or BChl), N_M = number of pigment molecules; $\mu_{\alpha i}$ = strength (in Debye) of transition of αth level in ith molecule; $\hat{\mu}_{\alpha i}$ = unit vector in the direction of that transition; $U_{\alpha iK}$ = α-ith element of the eigenvector for the Kth exciton state; $\nu_{\alpha i}$ = energy (in cm^{-1}) of the αth transition on the ith molecule; R_{ij} = distance (in Å) from the transition-moment centroid of the ith molecule to that of the jth; $\hat{R}_{ij}$ = unit vector in the direction of that distance vector; $\hat{\varepsilon}$ = unit vector in the direction of a macroscopically defined symmetry axis; θ = angle between the two dipoles of the simple exciton dimer; κ_{rot} is the geometrical factor in the CD formula for the simple exciton dimer; and θ_1 and θ_2 are the angles between each of the dipoles of the simple exciton dimer and the macroscopic symmetry axis. Units of integrated absorption (A_K) and LD (L_K) are $(\text{Debye})^2$; those of integrated CD (rotational strengths, C_K) are Debye-Bohr magneton.

molecules. The simplest theory [1,13–15] treats each transition moment (from ground to excited state) as a point dipole, and includes only interactions among otherwise degenerate transitions, i.e. exact resonances. In practice, with Chl or BChl pigments the theory is usually applied only to the lowest excited state or the lowest two excited states because higher excited states are often closely overlapping in energy.

Various corrections to the simplest theory have been used in one treatment or another. These include extended dipoles [16,17] or arrays of point monopoles [16,18] in place of point dipoles; simultaneous inclusion of off-resonance interactions [16,18,19], i.e. interactions between a transition to a given electronic state in one molecule with transitions to different states in other (identical) molecules; and a correction for so-called 'doubly excited' states [18,20]. (The last is a quantum-mechanical effect [21] which, in principle, should be quite small ($\approx$ 1% or less) for photosynthetic pigment complexes, but whose actual magnitude is somewhat controversial.) For pigment molecules in van der Waals contact, in practice only the two BChls of the RC special pair, simultaneous inclusion of charge-transfer states (see Section 5) has also been tried [22]. All of the treatments discussed here consider the possibility of nonresonant spectral effects resulting from, for example, pigment–protein interactions, by treating the zero-order transition-energy of the lowest excited state of each pigment as a parameter.

Table 1 summarizes the formulas for calculating absorption, CD, and LD 'stick' spectra (values of the integrals of the spectral bands) for the simplest theory, and with off-resonance interactions included. (See Ref. 22 for a discussion of the inclusion of charge-transfer states.) In calculating actual spectra, the treatments reviewed here, except for Ref. 16, assume Gaussian bands with assigned widths (and skews). More sophisticated methods for calculating actual spectra have been given [23–26], but so far have not proved very practical for photosynthetic complexes.

Table 2 gives formulas for calculating the interaction energies for point dipoles [1] and extended dipoles [16], which enter the equations given in Table 1. For most purposes, the formulas of both tables are included in a computer program which diagonalizes the matrix of interaction energies, and calculates spectra from the resulting matrix eigenvalues and eigenvectors.

For Chl and BChl, only the lowest 4 (neutral) excited states are considered. These are labelled, in order of ascending energy [27], Q_y, Q_x, B_x, and B_y.

3. Purple bacterial antennas

Good spectroscopic data for well-defined complexes from purple bacteria are available (Refs. 28, 29 and references therein), but general agreement is lacking on what, if anything, constitutes a minimum complex. One candidate is the so-called 'photoreceptor unit', consisting of core antenna plus reaction center [30]. This cyclic unit structure has features in common with secondary-antenna cyclic structures [31]. Some exciton analyses apply to any of these membrane antennas, others are specific. A key issue is the explanation of spectral data on the B850 part of the

TABLE 2

Exciton interaction-energy formulas

Type of transition dipole		Formula
Point		$J_{\alpha i,\ \beta j} = 5040\ \mu_{\alpha i}\mu_{\beta j}\kappa_{\alpha i,\ \beta j}/R_{ij}^3$,
	where	$\kappa_{\alpha i,\ \beta j} = \hat{\mu}_{\alpha i}\cdot\hat{\mu}_{\beta j} - 3(\hat{\mu}_{\alpha i}\cdot\hat{R}_{ij})\ (\hat{\mu}_{\beta j}\cdot\hat{R}_{ij})$
Extended		$J_{\alpha i,\ \beta j} = 5040\ \mu_{\alpha i}\mu_{\beta j}\ (L_1^{-1/2} - L_2^{-1/2} + L_3^{-1/2} - L_4^{-1/2})/d^2$,
	where	$L_1 = \sum_{m=1}^{3} [R_{ij,m} - 1/2\ d\ (\mu_{\alpha i,m} - \mu_{\beta j,m})]^2$,
		$L_2 = \sum_{m=1}^{3} [R_{ij,m} - 1/2\ d\ (\mu_{\alpha i,m} + \mu_{\beta j,m})]^2$,
		$L_3 = \sum_{m=1}^{3} [R_{ij,m} + 1/2\ d\ (\mu_{\alpha i,m} - \mu_{\beta j,m})]^2$,
	and	$L_4 = \sum_{m=1}^{3} [R_{ij,m} + 1/2\ d\ (\mu_{\alpha i,m} + \mu_{\beta j,m})]^2$

$J_{\alpha i,\beta j}$ = exciton interaction energy (in cm^{-1}) for interaction between αth transition dipole on ith molecule with βth dipole on jth molecule; $\kappa_{\alpha i,\beta j}$ = orientation factor for same interaction; d = length (in Å) of extended dipole; $R_{ij,m}$ and $\mu_{\alpha i,m}$ are the cartesian components (in any convenient cartesian coordinate system) of the unit vectors $\hat{R}_{ij}$ and $\hat{\mu}_{\alpha i}$, respectively. Other symbols are as in Table 1.

B800–850 complex (see Chapter 11). Two approaches so far have been tried. 1. First rationalize the BChl geometry in terms of spectral observations, then attempt to justify the geometric model in terms of independent structural information. 2. First introduce all available structural information into a geometric model, then attempt to deduce remaining, uncertain, geometric features from spectral observations. The first approach is considered in Section 3.1, the second in Section 3.2.

3.1. Scherz-Parson model

This is the most promising approach so far to a geometric model of a purple bacterial antenna complex [18]. Its basic ideas are simple and ingenious. If correct, it would explain much. However, the model has some significant problems.

The model is based on two assumptions. First, it is assumed that the spectral red-shift of the long-wavelength absorption band of B800-850, to 850 nm as compared with 770 nm for BChl *a* in a solution, is due mainly to an exciton dimer effect as is apparently the case [17,22] for the long-wavelength band in the *Rps. viridis* reaction center. Second, because the geometry of the two Q_y transition dipoles that gives maximal red-shift gives minimal (≈ zero) exciton contribution to CD [1,15], it is assumed that the observed rotational strength comes primarily from interactions of Q_y dipoles in one dimer with those in a second, distant, dimer. Exciton effects in absorption and CD are approximately uncoupled in this model. It should be noted that even maximal exciton contribution to the red-shift is not enough to explain the entire amount of the observed shift.

Thus, Q_y dipoles within a dimer are assumed to be nearly parallel and in line, leaving the higher-energy exciton transition with almost no dipole strength, and at a center-to-center distance of 7.5 Å producing an apparent absorption red-shift of ≈ 800 cm^{-1}. At the same time, one such dimer displays very little Q_y rotational stength. Now, a second dimer, identical in structure to the first, is placed about 28 Å from the first, and oriented so that the angular factor in the expression for rotational strength is near unity. At this much greater distance, the interactions of the dipoles in one dimer with those in the other contribute very little to the overall exciton splitting energy. The original red-shifted exciton transition is itself split into two sub-transitions of roughly equal strengths, and about 70 cm^{-1} apart. However, the excitonic CD of this 'dimer-of-dimers' is very large, with both the angular factor *and* the distance factor augmenting the rotational strength of each sub-transition.

This relatively simple BChl-tetramer model of B850 has several salutary features. With a single set of geometric assumptions, it simultaneously accounts for Q_y absorption and CD. Moreover, it also appears to explain Q_x absorption and CD. The Q_x CD bands are quite nonconservative and have small rotational strengths, while the Q_y bands are essentially conservative and of very large strengths. The Scherz-Parson model explains this as follows. In the Q_y case, the *inter*-dimer interaction is the dominant one for CD, as noted, but in the Q_x case (because molecular Q_x and Q_y transitions are perpendicular) the angular factor in the inter-dimer rotational strength expression becomes quite small so that the *in-*

tra-dimer interaction dominates. The intra-dimer excitonic CD is nonconservative because there is substantial borrowing of rotational strength by other electronic transitions for closely juxtaposed dipoles [20], whereas at a 28-Å separation this effect is negligible. This result of the model is particularly impressive because it seems to be an independent consequence of the basic structural assumptions.

On the other hand, the model is not without its drawbacks. It requires a certain structural association of 4 BChl molecules. Although a number of conceivable structures would be consistent with the tenets of the model, many otherwise plausible structures are excluded. This in itself is not a drawback; the structural restrictions may be viewed simply as predictions of the model. Indeed, the idea of a basic dimer of closely paired BChls enters other models as well (see Section 3.2). However, problems may arise if in an actual structure more than two dimers are present and interact with comparable energies. In other words, the success of the Scherz-Parson model may depend on each dimer having only one nearest neighboring dimer, a somewhat unlikely supposition on the grounds of both symmetry and energy-transfer requirements (see Section 3.2.).

A second possible difficulty of the model concerns a detail of its explanation of the BChl 850 Q_y CD bands. The model places the two exciton sub-transitions that putatively give rise to these CD bands only ≈5 nm apart, while the observed wavelength separation of positive and negative peaks is ≈30 nm [32]. This in itself is not necessarily a discrepancy. Individual transition bandwidths are 20–30 nm, so that considerable overlap of positive and negative lobes would be expected. The Scherz-Parson model thus predicts that most of the rotational strength (≈80%) is unobserved because of cancellation and the observed peak positions are artifacts of this situation. These are mathematically plausible conclusions. However, with this degree of overlap, both peak positions and observed lobe areas (apparent rotational strengths) must then be very sensitive functions of transition bandwidths and of inter-dimer geometry. For example, the ordinary band-narrowing which accompanies cooling from room temperature [32] should cause a marked increase of lobe areas and not just an increase of peak heights. Such a phenomenon has yet to be observed for purple bacterial antenna components.

In spite of these criticisms, the Scherz-Parson model is interesting and clever, and should be considered very seriously.

3.2. 'Structure-first' models

These models start from the most comprehensive structural information available. Since there is no X-ray diffraction structure yet, this comes principally from polypeptide linear sequence data and deductions therefrom. Two major structural proposals (see Chapter 11) are from Zuber and colleagues (B800-850, B875, B1015) [31] and from Loach and co-workers (B875) [33]. A proposal based on a mix of structural and spectroscopic information comes from Kramer et al. [34].

The Zuber model of B850 consists of a 'cyclic unit structure' of (here considering pigments only) BChl pairs in C6 symmetry or possibly groups of 4 BChl molecules in C3 symmetry. Pearlstein and Zuber [35] have considered the consequences of these symmetries for the possible exciton states of the B850 BChl. An interesting general conclusion of their analysis is that most of the optical absorption is limited by the symmetry to a small subset, possibly 3 or 4, of the 12 exciton states in the Q_y band. More specific predictions are speculative at this point, because the Zuber structural proposal is consistent with a range of possible BChl orientations. Pearlstein and Zuber have noted that both red-shift and blue-shift situations could arise.

It is of particular interest to ask whether the Scherz-Parson model is consistent with the Zuber structural proposal. With regard to a C6 arrangement of BChl pairs, the answer appears to be no (R. M. Pearlstein, unpublished results), even though the nearest-neighbor and next-nearest-neighbor distances (≈10 Å and ≈25 Å, respectively) are not too far from the Scherz-Parson requirements. The reason concerns both local and global aspects of the C6 symmetry. Locally, the symmetry constrains the geometry of a BChl pair and either one of its nearest-neighboring pairs in a way that is somewhat different from the Scherz-Parson tetramer-geometry. Globally, although this symmetry produces only two strongly absorbing and strongly chiral exciton sub-transitions, in accord with the Scherz-Parson model, the rotational strength of each sub-transition is *larger* than that of the latter model by more than an order of magnitude! (These global effects manifest themselves in the coherent superposition of local molecular states that constitute each exciton state, and that also appear, suitably weighted by geometric factors, in the expressions for dipole and rotational strengths. See Table 1.) Consistency between the Scherz-Parson model and a Zuber structure of groups of 4 BChls in C3 symmetry cannot be ruled out at this point.

The Loach model [33] has yet to be analysed in detail from the viewpoint of exciton effects, but some predictions have been made. In this model, as in that of Scherz and Parson or in the *Rps. viridis* RC, there is a closely coupled BChl dimer which could give rise to most of the observed red-shift of the Q_y absorption band. In addition, the Loach model suggests specific BChl–amino acid interactions which could explain the remaining red-shift. Loach et al. [33] speculate that the closely coupled dimer can also account for the observed CD spectrum, but do not explain how. They ascribe no role to the interactions of the BChls in one dimer with those in neighboring dimers toward producing exciton effects in the B875 complex. Interestingly, they do invoke a distant interaction, quite similar to that proposed for B800-850 by Scherz and Parson, to explain the observed CD spectrum of the product of octylglucoside treatment of B875.

Just in terms of the placement of the BChls, the Loach model consists of two rings of BChl dimers, one ring consisting of dimers each of which is near the amino terminuses, the other of dimers each near the carboxyl terminuses of the polypeptide pair. The rings consist of either 6 or 12 dimers each, dependent on whether B875 encloses one RC or a pair of RCs. Thus, each dimer has 3 nearest-neighboring dimers, one displaced along the polypeptide pair (transmembrane), the other two displaced laterally around the ring. The nearest-neighbor distance is ≈20 Å. Considering only a single polypeptide pair of the Loach model with its attached pair of BChl dimers, Scherz and Parson [18] note the similarity of this sub-structure to their proposal. However, they also point out that one of the two Loach dimers would have to be rotated through nearly a right angle about the polypeptide helix axis relative to the other dimer to explain the observed CD by their mechanism. Such a structural revision appears to be inconsistent with the tenets of the Loach model. Even if it were consistent, the question of three rather than one nearest-neighboring dimers must be addressed. As already discussed, a multiple-neighbor situation makes it more difficult to satisfy the Scherz-Parson conditions. In any event, the Loach model as put forward cannot explain exciton effects in optical spectra on the basis of the Scherz-Parson argument. Whether the Loach model can do so on any basis remains, as it does with the Zuber model, to be demonstrated.

Kramer et al. [34] base their unit-structure proposal (see Chapter 11, Fig. 7) on Zuber's deductions (Ref. 31 and references therein) regarding polypeptide structure and BChl binding sites, but for B850 assume a 'minimal' tetramer of BChls arranged in approximate C4 symmetry. Although such an arrangement implies roughly equal distances between all pairs of nearest-neighboring BChls around the tetrameric ring, Kramer et al. treat their tetramer as a pair of exciton dimers (ignoring the inter-dimer coupling) in proposing a set of pigment orientations that explains the observed CD. This model also accounts for observed LD and fluorescence polarization, but leaves unexplained the large red-shift of the Q_y transition. The symmetric tetramer of Kramer et al. is quite different from the pair-of-pairs arrangement of Scherz and Parson; the two models appear to be incompatible.

In summary, the model of Scherz and Parson is most promising, but has significant drawbacks. A Scherz-Parson type of explanation (red-shift and rotational strength having their origins in *distinct* exciton couplings) may be possible for a B800-850 model with C3 symmetry as proposed by Zuber. A marriage of these two concepts would, of course, only be useful if the difficulties of the Scherz-Parson concept can be resolved. On the other hand, one can not yet exclude the possibility that the observed rotational strength arises mainly from a closely coupled exciton dimer, while the red-shift has a basically nonexcitonic origin.

4. *Chl a/b-protein complex*

This complex, also known as Chl *a*/*b*-P2, is associated primarily with PS II of green plants [29]. In its minimal form the complex consists of one (24–27 kDa) polypeptide, one or two xanthophyll molecules, 3 molecules of Chl *b* and probably 4 molecules of Chl *a* [29]. The light-harvesting complex, LHC II, is thought to consist of aggregates of this minimal complex. Approximately half the total Chl, including virtually all of the Chl *b*, of green plants and green algae occurs in these complexes.

It was pointed out some time ago that many effects in optical spectra of this complex can be explained if it is assumed that the 3 molecules of Chl *b* form an exciton-coupled trimer with C3 symmetry, while the Chl *a* molecules are not so coupled (or are very weakly coupled) [36,37]. In this picture, which has recently been reviewed [1,4], the Chls *b* lie in relatively close mutual proximity, while the Chls *a* are more widely spaced, although still well within Förster transfer distances. This model quantitatively explains Q_y absorption and CD, as well as fluorescence and fluorescence polarization spectra of Chl *a*/*b*-P2. It is appealing for its simplicity and its reliance on symmetry considerations, as well as for its agreement with spectral data, but as yet it has almost no structural data to support it. Other exciton-coupled models are also possible, but probably not with C3 symmetry. Because red-shifts are much smaller for Chl than for BChl antennas, arguments of the sort used by Scherz and Parson are unnecessary for exciton models of Chl *a*/*b*-P2.

Recently, the Knox group has extended this model of Chl *a*/*b*-P2. In earlier work, only the Q_y transition of each Chl molecule was considered. Gülen and Knox [19], using geometrical and monomolecular transition parameters similar to those previously established for the model, treated 3 monomolecular transitions (Q_y, B_y and B_x) simultaneously. (Q_x is ignored because it is so weak in Chl *b*.) They found that this slightly improved the agreement of the model with absorption and CD data in the red, and more significantly provided a reasonable explanation of spectral data in the Soret region for the first time.

A potentially more significant extension has been put forward very recently by Gülen et al. [38]. Here, Chl *a*–Chl *b* exciton interactions have been invoked to explain spectral effects, principally CD, in the Chl *a* Q_y region ($\approx$670–690 nm) of LHC II, i.e. aggregated or 'assembled' Chl *a*/*b*-P2. The main idea is that one such interaction, presumably involving a Chl *b* of one Chl *a*/*b*-P2 with a Chl *a* of another, gives rise to a CD doublet with positive lobe at 650 nm and negative lobe at 685 nm. However, there are several problems with this suggestion. First, because Chl *a* and Chl *b* have distinct spectra, their Q_y transitions are not in resonance, so that the rotational strength of a Chl *a*–Chl *b* excitonic CD band is smaller than that of a comparable Chl *b*–Chl *b* (or Chl *a*–Chl *a*) band by a factor, J/Δ, where $2J$ is the exciton splitting energy and Δ is the energy difference between Chl *a* and Chl *b* Q_y levels (Ref. 1; also see Table 1). Since $J \approx 100\ \mathrm{cm}^{-1}$, and $\Delta \approx 600\ \mathrm{cm}^{-1}$, the Chl *a*–Chl *b* excitonic contribution to the 685-nm CD signal should be $\approx$6-times smaller (assuming distances and orientations are at least as favorable as for the Chl *b*–Chl *b* interaction) than the shorter-wavelength CD signals that are attributed to Chl *b*–Chl *b* interactions. However, the 685-nm signal in LHC II is observed to be at least as large as the shorter-wavelength signals [38]. Second, the reliability of different experimental techniques for disaggregating LHC II to form Chl *a*/*b*-P2 is debated (Á. Faludi-Dániel, personal communication). Third, other possible explanations of the 685-nm CD band of LHC II have not been ruled out. These include Chl *a*–Chl *a* exciton interactions, liquid crystal effects [39,40] and differential light scattering [41,42]. Thus, one cannot conclude that the origin of the Chl *a* part of the LHC II Q_y CD spectrum is understood as yet.

On the other hand, there seem to be no serious objections to the C3-symmetry model of the Chl *b* chromophores in Chl *a*/*b*-P2. Even so, because less structural information is available for these complexes than for the bacterial antennas, one must accept the model cautiously.

5. BChl a-protein from P. aestuarii

As indicated in Sections 3 and 4, exciton analysis of spectra is a potentially useful way of interpreting the spectra, understanding (B)Chl–(B)Chl interactions, and possibly verifying structural hypotheses. One also sees that it is not without controversial aspects, although as long as structural information is incomplete one is still free to assume structural features that suit the exciton analysis and thereby minimize controversy. The situation is quite different when detailed knowledge of structure is available, i.e., in practice when a structural model based on X-ray diffraction at or near atomic resolution has been established. This is the situation to be considered in the remainder of this chapter.

Historically the first X-ray structure [43–45] to undergo exciton analysis was that of the water-soluble BChl *a*-protein from the green photosynthetic bacterium *Prosthecochloris aestuarii.* The analysis [16] raised questions, and controversies, that remain unresolved after a decade. It is reviewed again here to emphasize these difficulties, to correct some misconceptions in the literature [4,46,47] regarding possible sources of the difficulties, and to discuss more recent developments. Exciton analysis of photosynthetic pigment-protein complexes is not likely to become a truly useful procedure until it produces results that agree with all relevant spectra of this particular complex.

The *Prosthecochloris* BChl *a*-protein is composed of 3 identical subunits arranged in C3 symmetry [43–45]. Each subunit consists of a single 39-kDa polypeptide enclosing a core of 7 molecules of BChl *a* (see Fig. 1). Crudely put, 6 of the 7 BChls are arrayed as a skewed ring with the seventh roughly at the ring's center. Thus, each BChl has one or more near neighbors (≈12 Å separation, center-to-center). The closest approach of two BChls not in the same subunit is ≈24 Å.

Standard exciton analyses (Ref. 16; also unpublished calculations by R.E. Fenna and independently by L.L. Shipman) based on the atomic coordinates published by Fenna et al. [44] produce calculated absorption and CD spectra with multiline features that resemble observed spectra in multiplicity and splitting energies. However, the observed overall spectral red-shift is not reproduced theoretically, nor is the pattern of intensity borrowing in the Q_y absorption spectrum or the magnitudes and signs of Q_y CD bands. These are the anomalies that continue to plague all exciton-analytical efforts in photosynthesis.

In their 1978 analysis, Pearlstein and Hemenger [16] presented two sets of theoretical results, one of which is remembered and the other not. The former comes from a nonstandard approach in which each Q_y dipole is assumed to be rotated 90° in the plane of its BChl macrocycle relative to the direction assigned [27] by molecular orbital theory. This single assumption virtually solves all of the anomalies just noted, except for the overall red-shift. However, this striking finding has not led anywhere, because so far no physical basis for such a rotation of electronic transition moments has been proven. Nonetheless, there may be a kernel of truth in the transition-moment-rotation hypothesis (see below).

The forgotten theoretical results of the 1978 paper are contained in a discussion

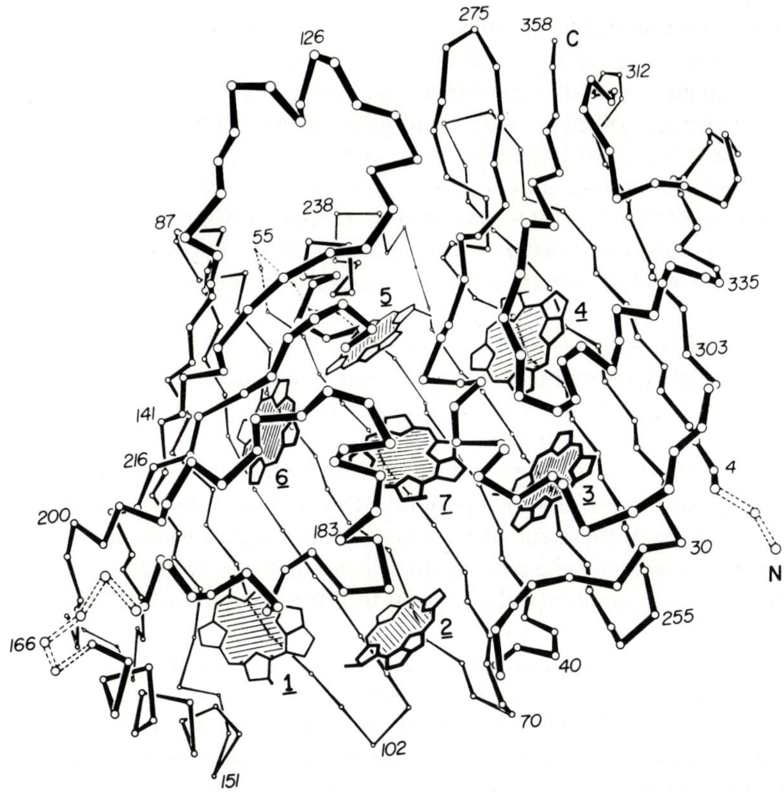

Fig. 1. Structure of one subunit of the water-soluble BChl *a*-protein trimer from the green photosynthetic bacterium *Prosthecochloris aestuarii* [48].

of possible alternatives to the rotation hypothesis. Pearlstein and Hemenger noted *eleven* such alternatives, and discussed one, differential environmental shifts of the BChl Q_y transition energies, in some detail. In this alternative proposal, the seven inequivalent BChls would have their Q_y transition wavelengths individually red-shifted (from 770 nm, the value in an organic solvent such as diethyl ether) by different amounts as a result of differing local, nonexcitonic interactions with the protein environment. However, extensive computer searches involving these seven red-shifts as independent search parameters yielded no calculated CD spectrum that resembles the observed one [16]. This point is overlooked in the more recent literature [4,47].

Of the other ten alternatives on the Pearlstein-Hemenger list, three deserve further discussion. One of these, interactions involving higher electronic excited states, has been invoked recently as important for understanding purple bacterial antenna and RC spectra [18,20,22]. It has also been resurrected recently as a possible explanation for the spectral anomalies of the Fenna-Matthews structure [4]. However, both from the 1978 calculations and again from very recent ones (R.M. Pearlstein, unpublished results), it is quite clear that simultaneous inclusion in the exciton interaction framework of all four-orbital states (Q_y, Q_x, B_x, B_y) for all seven BChls has only slight effects on calculated spectra. There is no contradiction in this finding. The higher states do contribute significantly to intensity borrowing and rotational strength redistribution in RCs and in the Scherz-Parson antenna model, in each of which at least two BChls are close enough ($\approx$7 Å, center-to-center) to induce substantial *off-resonance* exciton interactions (Q_y of one BChl with Q_x, B_x or B_y of another). Center-to-center distances in the Fenna-Matthews structure are too large ($\gtrsim$12 Å) for this to occur in any significant way. Off-resonance exciton interactions cannot explain the spectral anomalies of the Fenna-Matthews structure.

Another alternative worthy of further mention is residual errors in the BChl atomic coordinates as reported by Fenna et al. [44]. These are not likely to be large, or to explain the spectral anomalies, but they should be taken into account. In this regard, it is noteworthy that a refinement of the 2.8-Å Fenna-Matthews structure to 1.9 Å has now been reported by Tronrud et al. [48]. Subsequently, the complete amino acid sequence of the polypeptide was established [49]. A new refinement, making full use of the sequence data, will thus be necessary, although the present 1.9-Å coordinates are certainly more reliable than the original ones. In any event, the refined coordinates should be used in future spectral calculations.

The third noteworthy alternative is BChl–BChl charge transfer. This has recently been implicated as a significant factor in the interpretation of RC spectra [22], and the question has been raised anew whether it may play a role in antenna complexes as well (W.W. Parson, personal communication). In the familiar neutral (resonant) exciton interaction, only energy is transferred from one molecule to another: an electron in an excited state of one BChl is demoted to the ground state of that same BChl while an electron in the ground state of an interacting BChl is simultaneously promoted to the corresponding (resonant) state of that interacting BChl. On the other hand, with the charge-transfer exciton interaction both energy and electrons are exchanged: an electron in an excited state of one BChl is demoted to the ground state of an interacting BChl while, simultaneously, an electron in the ground state of that interacting BChl is promoted to the excited state of the first BChl. For the charge-transfer exciton it is not center-to-center distance that matters so much as favorably close approaches of the π-electron systems of two BChls. Pearlstein and Hemenger [16] argued against any significant contribution of charge transfer in the Fenna-Matthews structure because, on the basis of the coordinates of Fenna et al. [44], there are no approaches closer than 5 Å (and only one or two of these), which is substantially more than van der Waals contact ($\approx$3.4 Å). The two BChls of the special pair in the *Rps. viridis* RC, however, *are* in van der Waals contact [9], and the occurrence of charge transfer in this case would therefore not be surprising. In the light of the recently published refinement, this issue should be re-examined for the Fenna-Matthews structure, although it is not likely that errors in the original coordinates masked van der Waals contacts of BChls. The possibility of resonance interaction, neutral and charge-transfer, between BChl and aromatic amino acid side-chains must also be considered in view of the recent finding [48] that all seven BChls have contacts with at least one such sidechain.

Finally, let us return to the transition-moment-rotation hypothesis. Standard molecular orbital descriptions of porphyrins, chlorins and opposite bacteriochlorins include approximate D_{2h} or D_{4h} symmetry of the π-electron system, and therefore find all Q or B transitions to be polarized along (or very nearly along) the molecular symmetry (x or y) axes [27,50,51]. In organic solvents there are no perturbing interactions which can be expected to significantly break this symmetry. But in the Fenna-Matthews structure there is the possibility of symmetry-breaking interactions with nearby aromatic residues or with hydrogen-bond donors to the keto and/or acetyl oxygens (both within the π system). Such interactions may occur in other BChl-protein complexes as well (H. Zuber, personal communication). Although such symmetry breaking is unlikely to produce 90° rotation of the longest-wavelength electronic transition, rotations large enough to seriously affect distributions of intensities and rotational strengths cannot be ruled out.

6. *Purple bacterial reaction centers*

Historically, the second pigment-protein complex of photosynthetic interest for which an X-ray-diffraction structural model has been established is the reaction center of the BChl *b*-containing purple bacterium *Rhodopseudomonas viridis* [9–12]. Although its functions, as far as is known, are energy trapping and charge separation, and not excitation energy transfer, its 6 pigment moleclues (4 BChl and 2 BPh) are held in close proximity by its 3 polypeptides (this does not include the bound cytochrome) and therefore it can exhibit exciton effects.

Long before the X-ray structure was reported, Sauer et al. [52] were the first to propose that specific features in the Q_y region of absorption and CD spectra of purple bacterial RCs should be interpreted excitonically. After it had been suggested that a noncovalently linked dimer of BChl is the primary electron donor in purple bacteria [53], RC spectra were re-examined for evidence of dimeric exciton behavior. It was held that the longest-wavelength Q_y absorption band (at $\approx$870 nm in BChl *a* organisms, 960 nm in the BChl *b* case; both 300 K wavelengths) is the lower-energy transition of the exciton dimer, with that transition having 90% or more of the combined Q_y dipole strength of the dimer (see Refs. 1 and 54 for reviews). The upper-energy transition would then be a very weak one at a much shorter wavelength ($\lesssim$810 nm in the BChl *a* case). By this reasoning, the longest-wavelength (positive) CD band is also attributed to the exciton dimer. There is, however, a serious drawback to this interpretation. The observed strength of this CD band is unusually large, 3–4 DBm (Debye-Bohr magneton) [55,56]. For BChl 'special pair' models (prior to the X-ray results) the rotational strength from a simple exciton calculation is $\approx$1 DBm (R.M. Pearlstein, unpublished results), too small by a factor of $\lesssim$3.

Before the X-ray structure was known, few suggestions had been made to resolve this dilemma. One, to abandon the exciton interpretation of the CD band altogether [1], seems improbable because of the dearth of other possible mechanisms to produce from a dipole-allowed transition such a large rotational strength in the patent absence of either long-range order or differential light scattering. Another relies for success on freely adjusting intra-dimer geometry into a configuration which resembles neither the old special-pair models nor the X-ray structural model [57]. This paper [57] also implies that a differential environmental shift (confusingly called an 'asymmetric' exciton) can help to explain the observed rotational strength. Indeed, this is so for the unrealistic geometries considered. However, for any of the special-pair models, including the X-ray structure, this cannot help because, all other factors remaining constant, the rotational strength is *maximal* for zero differential shift ('symmetric' exciton) [1].

The advent of the X-ray model per se has also not resolved the rotational-strength discrepancy. Although the geometry of the BChl dimer revealed by the X-ray analysis (Ref. 9; and Chapter 3) differs significantly from those proposed earlier [58–60], from the point of view of the exciton interaction the situation is quite similar. From Tables 1 and 2 it follows that for values appropriate to the *Rps. viridis* dimer (R = 7 Å, μ^2 = 45 D^2, ν = 10400 cm^{-1}, κ_{rot} = 0.24) [17], the rotational strength of P-960 should be 1.3 DBm. This is still smaller by a factor of $\approx$3 than the experimental results [61]. (If P-870 (in BChl *a*-containing bacteria) were to have identical geometry, the corresponding calculated strength would also be 1.3 DBm because a decrease of μ^2 to 40 D^2 is offset by an increase of ν to 11500 cm^{-1}.)

Two independent calculations of absorption and CD spectra based on the (unrefined) atomic coordinates of the *Rps. viridis* RC have now been published. Although based on quite different sets of theoretical assumptions, each claims success in explaining major features of optical spectra, including CD. However, one of them apparently still fails to explain the rotational strength of the long-wavelength Q_y band, and the other is possibly too complicated to be completely satisfactory.

The calculation by Knapp et al. [17,62,63] uses extended rather than point transition dipoles, and includes overall and differential environmental shifts as adjustable parameters in the interaction matrix. The calculation is done separately for Q_y and Q_x bands (i.e., two separate 1-orbital calculations). Quantities calculated include exciton energy levels and eigenvectors, total absorbance dipole strengths, absorbance strengths polarized parallel and perpendicular to a crystal orientation axis, and rotational strengths. However, the various absolute strengths are not actually reported; rather, theoretical results are given as spectra of Gaussian shape and appropriate (asymmetric) widths, and presented comparatively with experimental spectra in graphs whose ordinates are scribed in 'arbitrary units'. (Explicit results for the spectra of oxidized reaction centers are not given.) No CD results are presented for the Q_x band. Theoretical curves shown agree well with experiments in wavelength, apparent amplitude and sign of peaks (or troughs), but it is not possible to determine from the figures whether absolute amplitudes agree. Although such agreement is stated in the text, this claim could not be confirmed for the rotational stength of the longest-wavelength CD band. This stength still seems to be too small by a factor of almost 3, just as in the simpler calculations discussed above.

The calculation presented by Parson et al. [22] includes simultaneous treatment of all 4-orbital states, and considers both neutral and charge-transfer exciton transitions of the BChl dimer. Overall red-shifts of Q_y and Q_x transitions are assumed, but differential environmental shifts are not. Charge transfer elements of the interaction matrix are calculated by the molecular-orbital method known as 'QCFF/PI', applied to the geometry of the dimer. Neutral interaction energies are calculated by either the point dipole or point monopole methods. A correction for doubly excited states is also included. Results appear in generally good agreement with experimental absorption and CD spectra. As shown, the theoretical rotational strength of the 960-nm band is again too small by the 'usual' factor of 3. However, more refined calculations (W.W. Parson, personal communication) give much closer agreement. Thus, the calculation by Parson et al. [22] should be viewed as most promising. On the other hand, their calculational scheme is quite complex, and therefore fraught with possibilities for error. For example, all spectral quantities are quite sensitive to the magnitudes of the charge-transfer energies, which, unlike the neutral transition energies, cannot be related to the transition energies of isolated BChl; charge-transfer bandwidths are adjustable parameters to which calculated CD spectra are particularly sensitive; and calculated rotational strengths are seemingly extremely variable depending on whether or not doubly excited states are included – a correction supposed to have only a small effect [21] on spectral quantities. Consequently, it is too early to conclude that the calculation of Parson et al. provides an accurate theoretical picture of pigment interactions in the reaction center.

It is worth noting that the failure of exciton theory thus far to explain the spectra of the *Prosthecochloris* BChl *a*-protein should be viewed as a problem also for exciton theories of *Rps. viridis* RC spectra. As discussed in Section 5, this failure cannot be attributed to neutral higher excited states, charge-transfer transitions, or differential environmental shifts, but may be due to the omission from theoretical treatments of possibly symmetry-breaking interactions with specific amino acid residues. Thus, possible effects of such interactions should also be considered in the RC. In particular, if the BChls of the dimer are affected, their long-wavelength transition dipoles could be significantly rotated away from their molecular *y*-axes. Depending on sign and magnitude of the rotation, the value of κ_{rot} could change substantially. Even in the simplest exciton theory, an increase of κ_{rot} by a factor of ≈ 3 (from 0.24 to 0.7, say) would explain the observed rotational strength of the 960-nm band.

Very recently, Vasmel et al. [64] have performed an exciton analysis of the RC of the filamentous green bacterium *Chloroflexus aurantiacus* by assuming its structure to be virtually isomorphic to that of the *Rps. viridis* RC. Since the *Chloroflexus* RC has 3 BChl and 3 BPh, whereas the *Rps. viridis* RC has 4 BChl and 2 BPh, Vasmel et al. assume that the only structural difference between the two RCs is that in *Chloroflexus* the third BPh replaces one of the two 'accessory' BChls found in *Rps. viridis.* The particular assumption that it is the M-branch accessory BChl that is so replaced leads to a very interesting set of results. Good agreement is found with experimental optical spectra, including, it is claimed, low-temperature absorption, CD, LD and fluorescence (excitation) polarization. Indeed in some ways the agreement between experiment and the predictions of this theoretical model appears to be better than in the case of the *Rps. viridis* RC, an impressive achievement given the lack of an explicit X-ray structure for the *Chloroflexus* RC. For example, it is stated that the theoretical rotational strength of the longest-wavelength CD band is only 30% less than its observed strength, a much closer agreement than that found for *Rps. viridis* under the same calculational conditions. However, those conditions are much simpler than those applied in published studies of the *Rps. viridis* RC. Only point dipoles are assumed, and only Q_y (1-orbital) transitions are included; charge-transfer and doubly excited states are left out. Thus, in spite of its impressiveness, the agreement may be merely fortuitous. Alternatively, it may signal a simpler situation in the *Chloroflexus* RC. For example, it may mean that (bacteriochlorin orbital) symmetry-breaking interactions are weaker in *Chloroflexus*. If so, this would reinforce the view that what has primarily been missing from photosynthetic exciton analyses is inclusion of orbital symmetry-breaking, not a variety of corrections to the 1-orbital, point-dipole exciton calculation itself.

7. *C-phycocyanin*

The subject of energy transfer in phycobilisomes and their sub-structures already has a large literature (see Ref. 65 for a review), mostly beyond the scope of this chapter. However, two of these sub-structures – trimeric C-phycocyanin from the thermophilic cyanobacterium *Mastigocladus laminosus* and hexameric C-phycocyanin from the cyanobacterium *Agmenellum quadruplicatum* – have very recently become respectively the third and fourth photosynthetic pigment-protein complexes for which structural models based on single-crystal X-ray diffraction near atomic resolution are now available (Refs. 66,67; and Chapter 11). Since these are presently the only such complexes, in addition to the two already discussed (Sections 5 and 6), it seems appropriate to conclude this review of exciton effects with some brief remarks on these C-phycocyanin structures.

The two structures are closely related. The *Mastigocladus* structure [66] consists of 3 monomers arranged in C3 symmetry. Each monomer is composed of two polypeptides, an A chain and a B chain, and associated covalently linked pigments. The pigments in this case are chemically all phycocyanobilin, an open-chain tetrapyrrole. Each A chain has one pigment, each B chain two. They are denoted by the residue numbers, in the linear sequence, of the amino acids to which they are linked. Thus, the 3 pigments, structurally inequivalent by location (though chemically identical), are denoted A84, B84 and B155. In the *Agmenellum* structure [67] two identical trimers, each closely resembling the *Mastigocladus* trimer, are stacked head-to-head to form the hexamer. These hexamers are presumed to be the basic phycocyanin stacking units in native phycobilisome rods [65]. It is assumed that pairs of *Mastigocladus* trimers form similar hexamers in that organism's phycobilisomes.

In the trimer, the most closely juxtaposed pigments are an A84 and a B84 on neighboring monomers, e.g., 1A84 and 2B84. The published separation of these pigments' centroids is $R = 24.4$. Å, but more recent analysis (H. Zuber, personal communication) suggests that actually $R < 20$ Å. According to Schirmer et al. [67] for this pair $|\kappa| = 0.4$. The squared dipole strength of the longest-wavelength transition ($\approx$620 nm), is $\approx$150 D^2/pigment in the trimer, and probably larger in the hexamer [68]. With these numbers (taking $R = 20$ Å), one sees from the formula of Table 2 that the exciton dipole coupling energy is $|J| = 38\ cm^{-1}$, corresponding to an exciton splitting $\approx$3 nm at these wavelengths. However, the uncertainties in all these quantities are quite large. Because the pigments, although basically in extended conformations, are somewhat bent and twisted, full molecular orbital calculations are necessary to determine the actual locations and orientations of the individual transition dipoles. Thus, the values of R and κ, in particular, must be viewed as provisional. The exciton splitting actually predicted by the structure could be substantially greater or less than 3 nm. In addition, the interpretation of CD spectra is much more difficult for these pigments than for BChls. This is because the intrinsic chiralities of the open-chain tetrapyrroles [68] can be much greater than those of the (more nearly planar) porphyrins, and thus intrinsic and excitonic sources of CD may be more difficult to separate. Again, molecular orbital calculations will be essential. One concludes that, in spite of the high-resolution X-ray structures, much remains to be done before it can be safely stated that exciton effects can or cannot be observed in the optical spectra of C-phycocyanin.

As just discussed, uncertainties in R and κ are currently the rule for all pigment pairs in C-phycocyanin. Because these quantities enter, respectively as R^6 and κ^2, the formula for the pairwise Förster energy transfer rates (see, e.g., Ref. 1), those rates are also still highly uncertain despite the structural information. It is therefore premature to attempt to analyse the kinetic properties of energy transfer in C-phycocyanin on the basis of the Förster formula. In the meantime, the structural data provide a useful guide for the analysis of experimental results relevant to energy transfer in C-phycocyanin [69].

References

1 Pearlstein, R.M. (1982) in Photosynthesis, Vol. 1 (Govindjee, ed.) pp. 293–330, Academic Press, New York.

2 Pearlstein, R.M. (1982) Photochem. Photobiol. 35, 835–844.

3 Pearlstein, R.M. (1984) in Advances in Photosynthesis Research, Vol. I (Sybesma, C., ed.) Part 1, pp. 13–20, Martinus Nijhoff/Dr. W. Junk Publishers, Dordrecht, The Netherlands.

4 Van Grondelle, R. (1985) Biochim. Biophys. Acta 811, 147–195.

5 Van Grondelle, R. and Amesz, J. (1986) in Light Emission by Plants and Bacteria (Govindjee et al., eds.) Academic Press, New York.

6 Beddard, G.S. and Cogdell, R.J. (1982) in Light Reaction Path of Photosynthesis (Fong, F.K., ed.) pp. 46–79, Springer-Verlag, Berlin.

7 Olson, J.M., Gerola, P.D., van Brakel, G.H., Meiburg, R.F. and Vasmel, H. (1985) in Antennas and Reaction Centers of Photosynthetic Bacteria (Michel-Beyerle, M.E., ed.) pp. 67–73, Springer-Verlag, Berlin.

8 Van Dorssen, R.J., Vasmel, H. and Amesz, J. (1986) Photosynth. Res. 9, 33–46.

9 Deisenhofer, J., Epp, O., Miki, K., Huber, R. and Michel, H. (1984) J. Mol. Biol. 180, 385–398.
10 Michel, H., Weyer, K.A., Gruenberg, H. and Lottspeich, F. (1985) EMBO J. 4, 1667–1672.
11 Deisenhofer, J., Epp, O., Miki, K., Huber, R. and Michel, H. (1985) Nature 318, 618–624.
12 Michel, H., Weyer, K.A., Gruenberg, H., Dunger, I., Oesterhelt, D. and Lottspeich, F. (1986) EMBO J. 5, 1149–1158.
13 Kasha, M. (1963) Radiat. Res. 20, 55–71.
14 Kasha, M., Rawls, H.R. and El-Bayoumi, M.A. (1965) Pure Appl. Chem. 11, 371–392.
15 Tinoco, I., Jr. (1963) Radiat. Res. 20, 133–139.
16 Pearlstein, R.M. and Hemenger, R.P. (1978) Proc. Natl. Acad. Sci. USA 75, 4920–4924.
17 Knapp, E.W., Fischer, S.F., Zinth, W., Sander, M., Kaiser, W., Deisenhofer, J. and Michel, H. (1985) Proc. Natl. Acad. Sci. USA 82, 8463–8467.
18 Scherz, A. and Parson, W.W. (1986) Photosynth. Res. 9, 21–32.
19 Gülen, D. and Knox, R.S. (1984) Photobiochem. Photobiophys. 7, 277–286.
20 Scherz, A. and Parson, W.W. (1984) Biochim. Biophys. Acta 766, 666–678.
21 Tinoco, I., Jr. (1962) Adv. Chem. Phys. 4, 113–157.
22 Parson, W.W., Scherz, A. and Warshel, A. (1985) in Antennas and Reaction Centers of Photosynthetic Bacteria (Michel-Beyerle, M.E., ed.) pp. 122–130, Springer-Verlag, Berlin.
23 Hemenger, R.P. (1978) J. Chem. Phys. 68, 1722–1728.
24 Hemenger, R.P. (1977) J. Chem. Phys. 67, 262–264.
25 Hemenger, R.P. (1977) J. Chem. Phys. 66, 1795–1801.
26 Lagos, R.E. and Friesner, R.A. (1984) J. Chem. Phys. 81, 5899–5905.
27 Weiss, C., Jr. (1972) J. Mol. Spectrosc. 44, 37–80.
28 Cogdell, R.J. and Scheer, H. (1985) Photochem. Photobiol. 42, 669–678.
29 Thornber, J.P. (1986) in Encyclopedia of Plant Physiology, New Series, Vol. 19 (Staehelin, L.A. and Arntzen, C.J., eds.) pp. 98–142, Springer-Verlag, Berlin.
30 Jay, F., Lambilotte, M., Stark, W. and Muhlethaler, K. (1984) EMBO J. 3, 773–776.
31 Zuber, H., Sidler, W., Füglistaller, P., Brunisholz, R. and Theiler, R. (1985) in Molecular Biology of the Photosynthetic Apparatus (Steinback, K.E., et al., eds.) pp. 183–195, Cold Spring Harbor Laboratory, Cold Spring Harbor, New York.
32 Philipson, K.D. and Sauer, K. (1972) Biochemistry 11, 1880–1885.
33 Loach, P.A., Parkes, P.S., Miller, J.F., Hinchigeri, S. and Callahan, P.M. (1985) in Molecular Biology of the Photosynthetic Apparatus (Steinback, K.E., et al., eds.) pp. 197–209, Cold Spring Harbor Laboratory, Cold Spring Harbor, New York.
34 Kramer, H.J.M., van Grondelle, R., Hunter, C.N., Westerhuis, W.H.J. and Amesz, J. (1984) Biochim. Biophys. Acta 765, 156–165.
35 Pearlstein, R.M. and Zuber, H. (1985) in Antennas and Reaction Centers of Photosynthetic Bacteria (Michel-Beyerle, M.E., ed.) pp. 53–61, Springer-Verlag, Berlin.
36 Van Metter, R.L. (1977) Biochim. Biophys. Acta 462, 642–658; Ph.D. Thesis, University of Rochester, New York.
37 Shepanski, J.F. and Knox, R.S. (1981) Isr. J. Chem. 21, 325–331.
38 Gülen, D., Knox, R.S. and Breton, J. (1986) Photosynth. Res. 9, 13–20.
39 Faludi-Dániel, Á. and Mustardy, L.A. (1983) Plant Physiol. 73, 16–19.
40 Holzwarth, G. and Holzwarth, N.A.W. (1973) J. Opt. Soc. Am. 63, 324–331.
41 Faludi-Dániel, Á., Bialek, G.E., Horvath, G., Szent-Rozsa, Z. and Gregory, R.P.F. (1978) Biochem. J. 174, 647–651.
42 Philipson, K.D. and Sauer, K. (1973) Biochemistry 12, 3454–3458.
43 Fenna, R.E. and Matthews, B.W. (1975) Nature 258, 573–577.
44 Fenna, R.E., ten Eyck, L.F. and Matthews, B.W. (1977) Biochem. Biophys. Res. Commun. 75, 751–756.
45 Matthews, B.W., Fenna, R.E., Bolognesi, M.C., Schmid, M.F. and Olson, J.M. (1979) J. Mol. Biol. 131, 259–285.
46 Mayne, L.C. and Hudson, B.S. (1983) Biophys. J. 41, 316a.
47 Lutz, M., Hoff, A.J. and Bréhamet, L. (1982) Biochim. Biophys. Acta 679, 331–341.
48 Tronrud, D.E., Schmid, M.F. and Matthews, B.W. (1986) J. Mol. Biol. 188, 443–454.

49 Daurat-Larroque, S.T., Brew, K. and Fenna, R.E. (1986) J. Biol. Chem. 261, 3607–3615.
50 Petke, J.D., Maggiora, G.M., Shipman, L.L. and Christoffersen, R.E. (1980) Photochem. Photobiol. 32, 399–414.
51 Scherz, A. and Levanon, H. (1985) Mol. Phys. 55, 923–937.
52 Sauer, K., Dratz, E.A. and Coyne, L. (1968) Proc. Natl. Acad. Sci. USA 61, 17–21.
53 Norris, J.R., Uphaus, R.A., Crespi, H.L. and Katz, J.J. (1971) Proc. Natl. Acad. Sci. USA 68, 625–629.
54 Parson, W.W. (1982) Annu. Rev. Biophys. Bioeng. 11, 57–80.
55 Reed, D.W. and Ke, B. (1973) J. Biol. Chem. 248, 3041–3045.
56 Sauer, K. and Austin, L.A. (1978) Biochemistry 17, 2011–2019.
57 Mar, T. and Gingras, G. (1984) Biochim. Biophys. Acta 764, 283–294.
58 Boxer, S.G. and Closs, G.L. (1976) J. Am. Chem. Soc. 98, 5406–5408.
59 Fong, F.K. and Koester, V.J. (1976) Biochim. Biophys. Acta 423, 52–64.
60 Shipman, L.L., Cotton, T.M., Norris, J.R. and Katz, J.J. (1976) Proc. Natl. Acad. Sci. USA 73, 1791–1794.
61 Shuvalov, V.A. and Asadov, A.A. (1979) Biochim. Biophys. Acta 545, 296–308.
62 Knapp, E.W. and Fischer, S.F. (1985) in Antennas and Reaction Centers of Photosynthetic Bacteria (Michel-Beyerle, M.E., ed.) pp. 103–108, Springer-Verlag, Berlin.
63 Zinth, W., Knapp, E.W., Fischer, S.F., Kaiser, W., Deisenhofer, J. and Michel, H. (1985) Chem. Phys. Lett. 119, 1–4.
64 Vasmel, H., Amesz, J. and Hoff, A.J. (1986) Biochim. Biophys. Acta 852, 159–168.
65 Glazer, A.N. (1984) Biochim. Biophys. Acta 768, 29–51.
66 Schirmer, T., Bode, W., Huber, R., Sidler, W. and Zuber, H. (1985) J. Mol. Biol. 184, 257–277.
67 Schirmer, T., Huber, R., Schneider, M., Bode, W., Miller, M. and Hackert, M.L. (1986) J. Mol. Biol. 188, 651–676.
68 Scheer, H. (1982) in Light Reaction Path of Photosynthesis (Fong, F.K., ed.) pp. 7–45, Springer-Verlag, Berlin.
69 Mimuro, M., Füglistaller, P., Rümbeli, R. and Zuber, H. (1986) Biochim. Biophys. Acta 848, 155–166.

J. Amesz (ed.) *Photosynthesis*

CHAPTER 14

Genetics and synthesis of chloroplast membrane proteins

JOHN C. GRAY

Botany School, University of Cambridge, Downing Street, Cambridge CB2 3EA, U.K.

1. Introduction

The thylakoid membrane system of chloroplasts is the most extensive membrane system in the leaves of higher plants, where it may account for up to 90% of cell membranes. The protein components of the thylakoid membrane are mainly involved in the light reactions of photosynthesis and include the supramolecular complexes of photosystems I and II, the cytochrome *b-f* complex and ATP synthase. These complexes have been extensively characterized in recent years and the subunit compositions of the individual complexes are becoming established. A major problem is to understand how the synthesis of the individual polypeptides of the membrane complexes is regulated to ensure that the thylakoid membrane is photosynthetically competent. It has become clear that the thylakoid membrane is the product of a complex interaction between the nucleocytoplasmic and chloroplast genetic systems, and that the expression of the genes for membrane components is regulated at a large number of steps. This chapter will consider our present knowledge of the genetics and synthesis of the photosynthetic membrane complexes of higher plants.

2. Photosystem II

Photosystem II appears to be the most complicated of the complexes of the thylakoid membrane both in structure and in regulation of the synthesis of the individual components. The complex may, superficially, be regarded as made up of three assemblies of polypeptides, a light-harvesting complex (LHC II), a core complex, containing primary electron donor P-680, and an oxygen-evolving complex. Oxygen-evolving PS II preparations may be prepared from thylakoid membranes by detergent fractionation or by mechanical fragmentation followed by partition in an aqueous polymer two-phase system. These preparations may be treated with detergent to separate the light-harvesting complex from the core complex with its associated oxygen-evolving complex. The polypeptides of the oxygen-evolving complex may be separated from the core complex by a variety of salt treatments.

A full discussion of the preparation of PS II complexes is given elsewhere [1]. The polypeptides associated with these complexes and their genetics and synthesis are discussed below.

2.1. Polypeptides of PS II

The polypeptides of the light-harvesting complex are most easily resolved by polyacrylamide gel electrophoresis of thylakoid membranes solubilized with sodium or lithium dodecyl sulphate. This produces the green chlorophyll-protein band, CP II, which on staining for protein reveals up to four polypeptides of 24–27 kDa [2], each of which is believed to bind Chl *a* and Chl *b*. In most plants two polypeptides predominate, but other minor polypeptides may be resolved [2].

The core complex is an intrinsic membrane complex containing P-680 and the electron acceptors Q_A and Q_B. Two polypeptides of 47 and 44 kDa appear to constitute the Chl-protein CPa resolved by polyacrylamide gel electrophoresis of thylakoid membranes in dodecyl sulphate [3]. These polypeptides bind Chl *a* and appear to be components of an antenna pigment system transferring light energy from LHC II to P-680. The polypeptide of 47 kDa has been suggested to be the site of P-680 [4,5], but this seems unlikely in view of the high degree of homology between the reaction centre polypeptides of *Rhodopseudomonas viridis* and two polypeptides of approximately 32 kDa of the higher plant core complex [6]. These two intrinsic membrane polypeptides have been called D1 and D2 because of their diffuse nature following SDS-polyacrylamide gel electrophoresis of *Chlamydomonas* thylakoid membranes [7]. D1 has been identified as the Q_B-binding polypeptide by labelling with the photoaffinity reagent azido-atrazine [8]. By analogy with the *Rps. viridis* reaction centre [9], it seems likely that D2 may form the binding site for Q_A, and the two polypeptides together form the binding site for P-680 [10]. The two polypeptides would also be expected to bind Chl *a* and pheophytin if this model is correct. The core complex also contains several smaller polypeptides whose functions are obscure. Cyt *b*-559 is composed of two polypeptides of 9 and 4 kDa, both of which provide histidine residues as ligands for the haem group [11]. The N-terminal amino acid sequence of both of these polypeptides has been determined [11,12]. Polypeptides of 7, 6.5, 5.5 and 5 kDa have been shown to be associated with a spinach PS II preparation [13]. The three larger polypeptides are hydrophobic, as shown by partitioning in Triton X-114, and are presumed to be intrinsic membrane proteins. The 5 kDa polypeptide is hydrophilic and therefore presumably peripheral to the core complex. This polypeptide has been purified and its amino acid composition determined [13].

A hydrophobic 10 kDa protein is also associated with PS II preparations [14], but its function is obscure. The protein is phosphorylated by a membrane-bound protein kinase [15] and the identity of this phosphoprotein has been the subject of much speculation [16]. It is clearly not the 9 kDa polypeptide of Cyt *b*-559 or the 8 kDa proteolipid subunit of ATP synthase as shown by N-terminal amino acid sequence [14].

The oxygen-evolving complex, responsible for the transfer of electrons from water to P-680, is composed of three well-characterized hydrophilic proteins of 33, 23 and 16 kDa [17]. Removal of these polypeptides from PS II preparations results in the loss of oxygen evolution, and activity can be reconstituted by their readdition to the core complex. Amino acid compositions of each of these polypeptides have been determined [18,19] and the N-terminal amino acid sequence of the 33 kDa polypeptide from spinach has been published by two laboratories [20,21]. However, with differences in 6 residues out of 16 residues at the N-terminus, it is not clear if this is a very variable protein or if the determined amino acid sequence is unreliable. Other polypeptides of 24, 22 and 10 kDa have been identified as associated with the oxygen-evolving complex by immunoprecipitation of detergent-solubilized PS II preparations with antibodies to the 33 and 23 kDa polypeptides [22]. The 24 and 22 kDa may be intrinsic membrane proteins but the 10 kDa polypeptide appears to be a peripheral protein, which may also be released from PS II preparations by salt treatment. This 10 kDa polypeptide is distinct from the 10 kDa hydrophobic phosphoprotein.

Approximately 20 polypeptides have therefore been identified as associated with PS II preparations. These are all presumably required in a particular stoichiometry for the optimal functioning of PS II. The stoichiometry of the subunits may be controlled by the synthesis and/or degradation of these subunits.

2.2. Genes for PS II components

The genes for the polypeptides of the PS II complexes are distributed between the nuclear and chloroplast genomes in higher plants. The only nuclear genes to be characterized to date are those for the polypeptides of LHC II. A nuclear location for these genes was first indicated by the Mendelian inheritance of the pattern of tryptic peptides of the Chl-protein CP II obtained from interspecific F_1 hybrids of *Nicotiana* [23]. Genes (*cab*) for LHC II polypeptides have been characterized from pea [24], wheat [25], *Lemna* [26,27], petunia [28,29] and tomato [30]. These genes have been identified by hybridization with cDNA produced from the mRNA from pea [28,31]. The genes make up a small gene family in all plants examined; in petunia [28], tomato [30] and *Lemna* [26], there are approximately 12–16 genes, and a similar number is suggested by Southern hybridization to pea nuclear DNA [31]. In wheat [25], there are at least 7 genes identified by S_1 nuclease analysis of transcription start sites, but the number of genes is probably much higher than this although the Southern blot analysis was difficult to interpret. The lowest number of genes is found in *Arabidopsis thaliana* [32], which has the smallest nuclear genome of any higher plant; three *cab* genes have been identified and characterized.

In *Petunia*, the 16 *cab* genes appear to fall into 5 different families, distinguished by the nucleotide sequence of the 3′ coding and non-coding regions of the genes [28,29]. Within a family the genes appear to be closely linked in nuclear DNA. Four *cab* genes are arranged in tandem at one gene locus in tomato [30], and at another locus three genes and a truncated gene are clustered [30]. In this latter case, two of the genes and the truncated gene are arranged in tandem, whereas the other gene is in the opposite orientation. The arrangement of genes in tandem, and in the inverted orientation, is also observed in *Petunia* [28] (see Fig. 1). The *cab* genes have been mapped to specific chromosomes in pea [33] and tomato [30]. In pea, all the variation in the length of restriction fragments hybridizing to a cDNA probe mapped to a single locus on chromosome 2. In tomato the two cloned clusters of *cab* genes mapped to separate chromosomes. The four genes in tandem mapped to chromosome 2, whereas the three genes with the truncated gene mapped to chromosome 3.

Nucleotide sequences of nuclear genes for LHC II polypeptides have been determined for petunia [28,29], tomato [31], *Lemna* [26,27], pea [24], wheat [25] and *Arabidopsis thaliana* [34]. One of the *Lemna* genes contains an intron of 84 bp within the coding sequence [26], but otherwise the genes do not contain intervening sequences. The *Lemna* intron has features characteristic of a transposable element, suggesting that it may not be an ancestral feature of *cab* genes but may have been introduced into the gene during subsequent evolution. The genes code for polypeptides of 264–269 amino acid residues which appear to be composed of a transit peptide of 33–36 residues and a mature protein of 229–233 residues. The exact site of cleavage of the transit peptide has not been determined, but it has been assumed to be on the N-terminal side of the sequence Met–Arg which is conserved in all species. Cleavage of a transit peptide on the N-terminal side of a Met residue is illustrated by the small subunit of ribulose-bisphosphate carboxylase [35]. Although the N-terminal amino acid of the mature pea polypeptide is probably Arg [36], the Met may have been removed within the chloroplast, in common with several other chloroplast proteins.

Although the mature LHC II polypeptides are relatively highly conserved, there is considerable variability in the amino acid sequence of the transit peptide. Only the N-terminal residues Met-Ala-Ala are conserved in the transit peptide deduced from all the sequenced *cab* genes. However, it has been suggested [34] that three blocks of residues are shared by transit peptides of other nuclear-encoded chloroplast proteins. There is also considerable variation at the N-terminal region of the mature polypeptide; deletions or insertions in this region may partly account for the difference in size of the LHC II polypeptides indicated by gel electrophoresis in the presence of SDS. Dunsmuir [29] has suggested that the different polypeptides are the products of different genes in *Petunia*. However, it has been demonstrated that a single gene-product can produce multiple bands in experiments examining uptake and assembly by chloroplasts in vitro [27]. Variation in the N-terminal 15 amino acid residues may well be related to the interaction of the protein with other LHC II polypeptides and other chloroplast components. This region of LHC II polypeptides has been identified as involved in grana stacking and excitation-energy utilization; phosphorylation of a threonine residue near the N-terminus of the protein leads to changes in these phenomena [2]. However, threonine residues are absent from this region of the polypeptides predicted from the nucleotide sequence of the tomato *cab* genes [30]. The remainder of the mature polypeptide ($\approx$220 amino acid residues) is well-conserved between the different plant species, presumably reflecting a conservation of functional residues involved in Chl binding and insertion into the thylakoid membrane. A model of the struc-

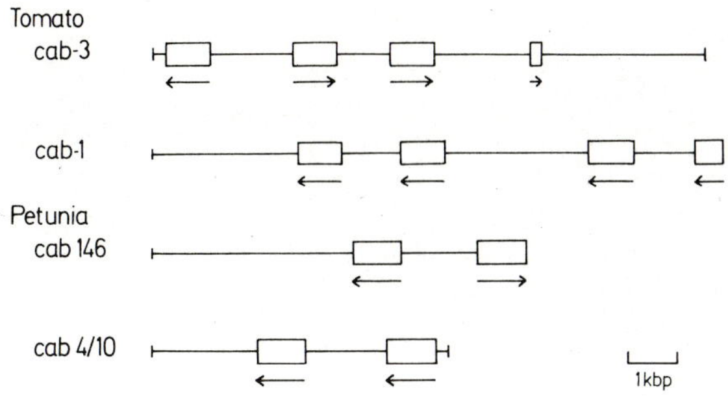

Fig. 1. Organization of *cab* genes in nuclear DNA from petunia and tomato, showing the tandem and head-to-head arrangement of the genes. Each region of DNA represents a restriction fragment of nuclear DNA cloned in a λ phage vector. Redrawn from Dunsmuir et al. [28] and Pichersky et al. [30].

ture of the polypeptide with three membrane-spanning segments has been proposed [26]. The sites of Chl-binding are unknown.

The genes for eight polypeptides of the PS II core complex have been located in chloroplast DNA from a number of higher plants. The location of these genes in the chloroplast genomes of pea and wheat is shown in Fig. 2. The genes are designated *psb*A-*psb*H and are tabulated with chloroplast genes for other photosynthetic membrane components in Table 1.

The gene *psb*A, for the 32 kDa Q_B protein (D1) was the first of the genes for PS II to be isolated [37] and has subsequently been characterized from a large number of plants, including spinach [38], tobacco [38,39], *Amaranthus* [40], *Solanum nigrum* [41,42], soybean [43], mustard [44] and pea [45]. Open reading frames of 353 codons have been detected in chloroplast DNA from each of these plants, but there is still some doubt as to the identity of the translation initiation codon. Initiation at the first AUG codon would produce a polypeptide of 39 kDa consisting of 353 amino acid residues. In the absence of any determined N-terminal amino acid sequence for the Q_B protein it is not possible to identify the translation start site unequivocally. The deduced amino acid sequence indicates a highly hydrophobic protein with the potential to form several transmembrane spans. The deduced amino acid sequence is homologous to the sequences of the L and M subunits of the reaction centre of *Rhodopseudomonas viridis* [6], which have been shown to fold with five membrane-spanning segments [9]. The *psb*A gene has been shown to be the locus of mutations producing resistance to the triazine group of herbicides in a number of plants. Atrazine has been shown to bind to the Q_B site and prevent electron transfer out of photosystem II [8,46]. Characterization of the mutant genes has indicated that a single point mutation is responsible for the change in herbicide susceptibility. In all atrazine-resistant higher plants examined, including *Amaranthus hybridus* [40], *Solanum nigrum* [41,42] and *Senecio vulgaris* [47] an A→G transition results in the replacement of a serine residue with a glycine residue at position 264. As would be expected from its location in chloroplast DNA, this mutation shows maternal inheritance in crosses with the wild-type atrazine-susceptible plants [48].

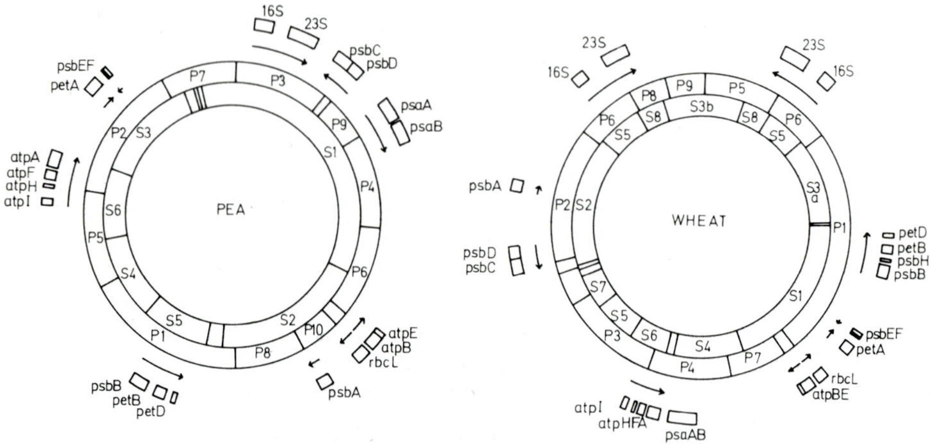

Fig. 2. Location of genes for photosynthetic membrane components in chloroplast DNA from pea and wheat. Restriction sites for the enzymes *Pst*I (P) and *Sal*GI (S) are shown. Genes are designated as in Table 1. The direction of transcription of the genes is indicated with arrows.

TABLE 1

Chloroplast genes for photosynthetic membrane components

Complex	Gene designation	Gene product	Amino acid residues
PS II	*psb*A	32 kDa Q_B protein	353
	*psb*B	47 kDa Chl *a*-protein	508
	*psb*C	44 kDa Chl *a*-protein	473
	*psb*D	D2	353
	*psb*E	9 kDa Cyt *b*-559	83
	*psb*F	4 kDa Cyt *b*-559	39
	*psb*G	24 kDa polypeptide	248
	*psb*H	10 kDa phosphoprotein	73
Cyt *b*-*f* complex	*pet*A	Cyt *f*	320
	*pet*B	Cyt *b*-563	211
	*pet*D	17 kDa polypeptide	139
PS I	*psa*A	P700-Chl *a*-protein	751
	*psa*B	P700-Chl *a*-protein	735
ATP synthase	*atp*A	CF_1 α subunit	504
	*atp*B	CF_1 β subunit	498
	*atp*E	CF_1 ε subunit	137
	*atp*F	CF_0 subunit I	183
	*atp*H	CF_0 subunit III	81
	*atp*I	CF_0 subunit IV (or *a*)	247

D2	Asp Gln Pro His Glu Asn Leu Ile Phe Pro Glu Glu Val Leu Pro Arg Gly Asn Ala Leu
Wheat	GAT CAG CCT CAT GAA AAT CTT ATA TTC CCT GAG GAG GTT CTA CCA CGT GGA AAC GCT CTT TAA TGG ACT
Pea	GAT CAG CCT CAT GAA AAC CTT ATA TTC CCT GAG GAG GTT CTA CCA CGT GGA AAC GCT CTT TAA TGG ACT
44kDa	Met Lys Ile Leu Tyr Ser Leu Arg Arg Phe Tyr His Val Glu Thr Leu Phe Asn Gly Thr
	Thr

Fig. 3. Overlap of the *psb*D and *psb*C genes in chloroplast DNA from pea and wheat.

The gene for the D2 polypeptide (*psb*D) was originally located in chloroplast DNA from *Chlamydomonas* [49] and has subsequently been located in pea [50], spinach [51,52] and wheat [53]. The gene codes for a protein that is homologous to the 32 kDa Q_B protein. The translation initiation codon has not been identified unequivocally and it is possible that the D2 polypeptide is composed of 353 amino acid residues. The protein is predicted to show the same folding pattern in the thylakoid membrane as the 32 kDa Q_B protein. The homology of the D1 and D2 proteins with the L and M subunits of the bacterial reaction centre has led to the suggestions that in PS II the D1 and D2 proteins are responsible for binding P-680, Q_A and Q_B [10]. However, experimental evidence in favour of this suggestion remains to be established.*

In all higher plants examined the gene (*psb*D) for the D2 polypeptide overlaps by 50 bp the gene (*psb*C) for the 44 kDa polypeptide of the antenna Chl *a*-protein. This overlapping region where the nucleotide sequence is translated in two different reading frames has been found in spinach [51,52], pea [50] and wheat (S.M. Hird et al., unpublished). The overlapping region is shown in Fig. 3. The *psb*C gene predicts a polypeptide of 473 amino acid residues which would be expected to span the thylakoid membrane several times. The residues involved in binding Chl *a* have not been identified.

The gene (*psb*B) for the 47 kDa polypeptide was originally located by the translation of RNA hybrid-selected by restriction fragments of spinach chloroplast DNA [54]. This method was also used for the localization of the *psb*C gene [54]. The *psb*B gene has subsequently been located in pea and wheat chloroplast DNA using probes derived from the spinach gene [53]. The nucleotide sequence of the spinach gene [55], indicates a polypeptide of 508 amino acid residues which is homologous to the 44 kDa polypeptide. Both polypeptides are predicted to fold into similar structures in the thylakoid membrane, suggesting that they may have related functions as antenna Chl-proteins.

Two genes (*psb*E and *psb*F) for polypeptides of Cyt *b*-559 have been located and characterized from spinach [56], wheat [57], tobacco [58], *Oenothera hookeri* [58] and pea (D.L. Willey, unpublished). The gene for the 9 kDa polypeptide known to be associated with Cyt *b*-559 preparations was first of all located in spinach [59] and wheat [57] by hybrid-release translation or by coupled transcription-translation of fragments of chloroplast DNA. Nucleotide sequencing [56,57] revealed an open reading frame of 83 codons, encoding a protein with the known N-

* Such evidence was recently presented by K. Satoh at the 7th Int. Congress of Photosynthesis (Providence, RI, August 1986).

TABLE 2

Translation products of nuclear genes for photosynthetic membrane components

Complex	Gene product	Translation product	
		Primary	Processed
PS II	LHC polypeptides	29–32 kDa	24–27 kDa
	33 kDa polypeptide	40 kDa	34 kDa
	23 kDa polypeptide	33 kDa	23 kDa
	16 kDa polypeptide	26 kDa	16 kDa
Cyt *b-f* complex	Rieske Fe-S protein	26 kDa	19 kDa
PS I	22 kDa polypeptide	26 kDa	22 kDa
ATP synthase	CF_1 γ subunit	46 kDa	35 kDa
	CF_1 δ subunit	27 kDa	21 kDa
	CF_0 subunit II	26 kDa	18 kDa

terminal sequence of the 9 kDa polypeptide, closely followed by an open reading frame of 39 codons. The deduced amino acid sequence of this gene product was homologous to that of the 9 kDa polypeptide, and it was suggested that this polypeptide may also be associated with Cyt *b*-559 [56]. Both the polypeptides contained single histidine residues located in putative membrane-spanning regions. Two histidine residues are required as the axial ligands of the haem iron in Cyt *b*-559 [60]. Subsequently the N-terminal amino acid sequence of a small polypeptide associated with the Cyt *b*-559 preparation confirmed that the gene *psb*F was the structural gene for the 4 kDa polypeptide of Cyt *b*-559 [11]. The arrangement of the two genes for the 9 kDa and 4 kDa polypeptides of Cyt *b*-559 is conserved in all higher plants examined.

Two other open reading frames in chloroplast DNA have recently been identified as genes for components of PS II [61] (S.M. Hird et al., unpublished). A synthetic peptide deduced from the nucleotide sequence of an open reading frame in maize chloroplast DNA was used to prepare antibodies in rabbits. The anti-peptide antibodies reacted with a 24 kDa polypeptide in a PS II preparation. It is not clear if this polypeptide is the 24 kDa polypeptide associated with the polypeptides of the water-splitting complex described by Ljungberg et al. [22]. The maize gene (*psb*G) codes for a polypeptide of 248 or 255 amino acid residues which may be predicted to form several membrane-spanning regions. The gene (*psb*H) for the 10 kDa phosphoprotein has recently been identified in wheat chloroplast DNA, close to the gene for the 47 kDa polypeptide (S.M. Hird et al., unpublished). An open reading frame of 73 codons shows homology to the determined N-terminal amino acid sequence of the spinach protein [14], and there are considerable similarities in the determined and deduced amino acid compositions of the spinach and wheat proteins.

Several other open reading frames in the vicinity of these PS II genes have been located by nucleotide sequencing. Although the products of these open reading frames have not been identified, their transcription with other PS II genes suggests they may be genes for small polypeptides associated with PS II preparations. An open reading frame of 62 codons is located downstream of the *psb*D and *psb*C genes, in spinach [51], pea [62], barley [63] and wheat [64]. A polypeptide with two putative membrane-spanning regions is predicted from the nucleotide sequence. Open reading frames of 38 and 40 codons are present downstream from the *psb*E and *psb*F genes in pea and wheat, and would be expected to give polypeptides with single membrane-spanning regions (D.L. Willey et al., unpublished). However, it has not been established whether polypeptides are produced from these open reading frames in vivo or whether the predicted polypeptides are associated with PS II.

The genes for the hydrophilic polypeptides of the water-splitting complex have not been isolated, but their location in nuclear DNA is predicted from the synthesis of precursor forms of these polypeptides from poly(A) RNA isolated from spinach [65]. There is no evidence for the location of the genes for the other polypeptides associated with PS II.

2.3. Synthesis of PS II

The regulation of the synthesis of the polypeptides of PS II appears to be particularly complex, with evidence for regulation at transcriptional, translational and post-translational levels. The synthesis of the individual polypeptides does not appear to be tightly coordinated, with certain polypeptides accumulating in the absence of other PS II polypeptides under a variety of experimental conditions. The synthesis of a functional PS II complex is strongly dependent on light, both for its effect on the transcription of PS II genes and for its absolute requirement for Chl synthesis.

Light stimulates the transcription of the nuclear genes for LHC II polypeptides, as shown by increases in translatable [66–68] and hybridizable [69] mRNA and by increased incorporation of [^{32}P]UTP into run-off transcripts by isolated leaf nuclei [70,71]. The light receptor in these experiments was phytochrome, as indicated by red–far-red reversibility of the light effects [72,73]. Light-regulated expression of a pea *cab* gene is dependent on the presence of a 400 bp sequence immediately upstream from the coding sequence [74]. This region is also sufficient for correct tissue-specific expression. The mechanisms of light-regulation and tissue-specific expression appear to be conserved between monocotyledonous and dicotyledonous plants. A wheat *cab* gene shows light-regulated expression in leaves of transgenic petunia and tobacco plants [75].

Transcripts of *cab* genes are translated on cytoplasmic polysomes to produce soluble precursor forms of the polypeptides approximately 3–4 kDa larger than the mature form [76]. The precursors bind to the outer envelope membrane of the chloroplasts [77] before being imported into the chloroplasts by an energy-dependent process [78]. The mechanism by which the soluble precursors cross the two envelope membranes is not known, but translocation may occur in regions where the two envelope membranes are appressed. After uptake, the transit peptide is cleaved from the N-terminus of the precursor polypeptide. It has recently been shown that this processing of the precursor is not necessary for the insertion of the polypeptide into the thylakoid membrane nor for its assembly into the light-harvesting complex [79]. The processing protease responsible for the cleavage of the transit peptide of the LHC II precursor has not been identified.

Accumulation of LHC II in the thylakoid membrane is dependent on a supply of Chl *a* and *b*. In the absence of chlorophyll synthesis, either in the dark [80] or in mutant plants [81], the LHC II polypeptides are rapidly degraded. Bennett [80] has shown that much of the newly synthesized LHC II in pea seedlings is unstable in the dark, with the degradation of both LHC II polypeptides and Chl *a* and *b*. It was suggested that this is part of the normal physiological mechanism for co-ordinating the accumulation of the pigment and protein components of LHC II. Carotenoid is also required for a functional LHC II complex, and in carotenoid-deficient mutants, or in plants treated with norflurazon, a herbicide which blocks carotenoid synthesis, LHC II synthesis is prevented [82,83]. These plants fail to accumulate LHC II mRNA due to an inability to increase the rate of transcription of *cab* genes on illumination. It has been suggested that plastid-dependent factors are required for the continuous light-dependent transcription of nuclear *cab* genes.

The synthesis of the polypeptides of the core complex takes place exclusively in the chloroplasts, as indicated by synthesis in isolated chloroplasts or by translation of chloroplast RNA in vitro. Regulation of the synthesis appears to operate at transcriptional, translational and post-translational levels, with light playing a major role.

The most extensively studied of the core complex polypeptides is the 32 kDa Q_B protein, which is the main membrane-bound product of protein synthesis in chloroplasts in vivo [84] or in vitro [85]. The protein is synthesized initially in a precursor form of 34.5 kDa, which is slowly processed by an unidentified thylakoid protease. Processing appears to remove 12–16 amino acid residues at the C-terminus of the precursor [86]. This processing event may take place in the stromal lamellae, with the mature 32 kDa polypeptide accumulating in the granal membranes. The mature protein appears to turn-over extremely rapidly and this may be related to its role in electron transfer [87]. Turnover is prevented by diuron (DCMU) and atrazine, which inhibit electron transfer by binding at the Q_B site [87]. Light-regulation of these processes is complex. Although originally described as a 'photogene' [37], indicating that light affected the levels of transcripts of the *psb*A gene in maize seedlings, it is clear that light is not absolutely required for transcription. Indeed, in *Spirodela* fronds it has been shown that light has no effect on the levels of transcripts for the 32 kDa polypeptide [88], although it profoundly affects the synthesis of the polypeptide itself. It appears that translation is light-dependent, and this may be due to a specific inhibition of the translation of the mRNA for the 32 kDa polypeptide in the dark, as other chloroplast RNAs are translated in the dark [88]. A specific translational block may be related to the availability of chlorophyll, which would be light-dependent, and would be required for the functioning of the 32 kDa polypeptide if it forms part of the reaction centre. The degradation of the 32 kDa polypeptide is also light-dependent, with rapid degradation in the light but not in the dark [87]. Degradation in the light is not dependent on ATP formed by photophosphorylation, and it has been suggested that damage to the Q_B protein, as an inherent consequence of its electron transfer properties, changes the conformation of the protein, rendering it susceptible to proteolytic degradation [87].

The regulation of the synthesis of the 47 kDa and 44 kDa Chl *a*-proteins and of the D2 polypeptide has not been studied so thoroughly. The 47 and 44 kDa Chl *a*-proteins are not present in dark-grown plants, and this appears to be due to the absence of the apoproteins [89]. However, transcripts of the *psb*B and *psb*C genes are present in dark-grown plants [90], indicating that regulation of the appearance of the gene products occurs post-transcriptionally. Regulation may occur by a specific block on translation in the dark, or by rapid degradation of newly synthesized polypeptides. Degradation of newly synthesized 44 kDa polypeptides is observed in the nuclear mutant *hcf*-3 of maize, where the polypeptides of the core complex fail to accumulate [91]. The regulation of the synthesis of the D2 polypeptide has not been studied, but the *psb*D gene is co-transcribed with the gene (*psb*C) for the 44 kDa polypeptide [51]. These two genes overlap by 50 bp and it is to be expected that any transcriptional controls affecting one polypeptide will also affect the other. This does not, however, preclude the possibility that separate post-transcriptional controls operate on the expression of these genes. Separate post-transcriptional controls appear to be necessary to explain the differential expression of the gene (*psb*B) for the 47 kDa polypeptide compared to the two genes, *pet*B and *pet*D, for components of the Cyt *b-f* complex, which are co-transcribed with *psb*B [55]. The 47 kDa polypeptide is synthesized only in the light, whereas the products of the *pet*B and *pet*D genes accumulate in the dark [92].

Cyt *b*-559, detected spectroscopically, also accumulates in dark-grown plants [93] and its regulation therefore appears to be different from that of the other core polypeptides. The genes, *psb*E and *psb*F, for the 9 and 4 kDa polypeptides of Cyt *b*-559 are co-transcribed [56] and, due to the close spacing of the genes, may also be translationally coupled. This may ensure the synthesis of equal amounts of the two polypeptides, but as yet the stoichiometry of these two polypeptides in PS II preparations has not been established.

The polypeptides of the water-splitting complex accumulate in dark-grown plants [89,94] and there is no absolute requirement for light for the expression of the nuclear genes. The polypeptides are synthesized on cytoplasmic ribosomes as precursor forms, with an additional 4–7 kDa extension [65]. This additional sequence is necessary for uptake by chloroplasts and presumably for targetting to the lumenal space of the thylakoid membrane. The primary sequences of the 33, 23 and 16 kDa polypeptides and their transit peptides should soon be forthcoming from analysis of cDNA clones.

3. *Cytochrome b-f complex*

The Cyt *b-f* complex appears to be the least complicated of the complexes of the thylakoid membrane, at least in structure. Only four polypeptides appear to be necessary to constitute a Cyt complex showing plastoquinol-plastocyanin oxido-reductase activity [95]. The complex may be extracted from thylakoid membranes with octylglucoside and cholate and subsequently purified by ammonium sulphate fractionation and then a variety of centrifugation or chromatographic steps [95–97].

3.1. Polypeptides of the cytochrome b-f complex

The Cyt *b-f* complex contains the redox components Cyt *f*, Cyt *b*-563 and the Rieske Fe-S protein, which in spinach have been identified as polypeptides of 33, 23 and 20 kDa respectively [95,98]. The spinach complex contains in addition a polypeptide of 17 kDa with no known redox function. The reported sizes of these polypeptides estimated by SDS-gel electrophoresis vary somewhat between different laboratories, presumably because of slightly different electrophoretic procedures. The estimated size of the Cyt *f* polypeptide varies between different plants, even when analysed in the same electrophoresis system, although the gene sequences predict polypeptides of very similar relative molecular mass.

Additional polypeptides ascribed to the Cyt *b-f* complex are a bound form of ferredoxin-$NADP^+$ reductase (FNR) [99] and one or more smaller polypeptides [100]. An association of the complex with ferredoxin-$NADP^+$ reductase may be expected in view of the reported role of FNR in cyclic electron flow from PS I to the Cyt complex [101]. FNR remains associated with the complex during the early stages of the purification of the complex but there is no evidence that it is an intrinsic component of the complex necessary for plastoquinol-plastocyanin oxido-reductase. The presence of small polypeptides in the complex requires further investigation. Polypeptides of about 5 kDa have been reported to be associated with the spinach complex [100].

3.2. Genes for components of the cytochrome b-f complex

As with all the other photosynthetic membrane complexes, the genes for the components of the cytochrome complex are distributed between the nuclear and chloroplast genomes of higher plants. The chloroplast genes for the Cyt *f*, Cyt *b*-563 and 17 kDa polypeptides have been extensively characterized, but the nuclear gene(s) for the Rieske Fe-S protein have not yet been isolated.

A chloroplast location of the gene for Cyt *f* was suggested by two plastome mutants of *Oenothera* which were deficient in spectrally detected Cyt *f* [102], and by the maternal mode of inheritance of Cyt *f* in interspecific F_1 hybrids of *Nicotiana* [103]. The structural gene (*pet*A) for the Cyt *f* polypeptide was initially localized and characterized from pea chloroplast DNA [104,105], and subsequently from wheat [106], spinach [107] and *Oenothera* [108]. Open reading frames of 320 codons have been detected in chloroplast DNA from each of these plants. Comparison with the determined N-terminal sequence of Cyt *f* indicates the presence of a 35 amino acid residue pre-sequence and a mature polypeptide of 285 amino acid residues. A putative haem-binding site, Cys-Ala-Asn-Cys-His, is located near the

N-terminus of the mature polypeptide and a single hydrophobic sequence, which may be involved in anchoring Cyt *f* in the thylakoid membrane, is located near the C-terminus. A model of Cyt *f* with a single membrane-spanning α-helix and the bulk of the polypeptide in the intrathylakoid space has been proposed and tested by partial proteolytic cleavage [105].

The genes, *pet*B and *pet*D, for the Cyt *b*-563 and 17 kDa polypeptides are located close together in spinach [109] and in several other plants, including pea [53], wheat [53] and tomato [110]. Nucleotide sequences of these genes indicate polypeptides of 223 and 139 amino acid residues [111,112], which are homologous to the N-terminal and C-terminal regions of mitochondrial Cyt *b*, respectively [113]. Comparison with Cyt *b* has indicated several conserved histidine residues, which have been implicated in haem binding. The Cyt *b*-563 is proposed to fold with five membrane-spanning segments arranged so as to bind two haem molecules within the hydrophobic interior of the protein [113]. The 17 kDa polypeptide is suggested to form three membrane-spanning segments [112,113] and presumably interacts with the Cyt *b*-563 polypeptide to form binding sites for other proteins and possibly for electron transfer components such as plastoquinol.

The gene(s) for the Rieske Fe-S protein has not been isolated, but its location in nuclear DNA is predicted from the synthesis of a precursor form from poly(A) RNA isolated from spinach [109]. cDNA clones containing the coding sequence of the Rieske Fe-S protein should be forthcoming shortly.

3.3. Synthesis of the cytochrome b-f complex

The accumulation of the polypeptides of the cytochrome complex is not strictly dependent on light, although there appear to be differences between plants in the magnitude of light effects. In cereals, such as barley and wheat, Cyt *f*, Cyt *b*-563 and the Rieske Fe-S centre are present in dark-grown plants in amounts comparable to light-grown plants, and there is little change on illumination [114,115]. In contrast, in bean and pea, Cyt *f* and *b*-563 are present in low amounts in etiolated leaves and the amounts increase dramatically on illumination [93]. These studies have recently been repeated using immunochemical detection of all four polypeptides of the complex; this confirms that the components of the complex are synthesized in the dark [92].

The synthesis of Cyt *f*, Cyt *b*-563 and the 17 kDa polypeptide takes place in the chloroplasts. All three polypeptides are synthesized and assembled into the Cyt *b*-*f* complex in isolated pea chloroplasts, although assembly of the Cyt *b*-563 polypeptide required the addition of 5 mM Mg-ATP [116]. The step in the assembly of Cyt *b*-563 requiring this additional Mg-ATP has not been identified. The Cyt *f* polypeptide is synthesized initially with a 35 amino acid residue N-terminal extension which appears to act as a signal sequence directing the polypeptide to the thylakoid membrane [104]. This N-terminal sequence acts as a signal sequence in *Escherichia coli*, where it is capable of directing β-galactosidase to the inner membrane [117]. A signal peptidase for the removal of this N-terminal extension is presumably present in chloroplast membranes but it has not been characterized. Nothing is known of the mechanism of haem addition.

The Cyt *b*-563 and 17 kDa polypeptides do not appear to be synthesized with recognisable N-terminal extensions, although the N-terminal amino acid sequences of the mature polypeptides have not been determined. The mechanisms of membrane insertion and haem addition to the Cyt *b*-563 polypeptide have not been investigated. The genes *pet*B and *pet*D, for the Cyt *b*-563 and 17 kDa polypeptides, are co-transcribed [111], together with the gene *psb*B, for the 47 kDa polypeptide of PS II [55], and this may be a means of ensuring stoichiometric production of the two polypeptides. However, the primary transcript undergoes considerable processing and it is not clear what effect this will have on the translation of the RNA. The mechanisms ensuring stoichiometric production of the three chloroplast-coded polypeptides are not clear, and it may be that this is achieved by degradation of excess unassembled polypeptides.

The Rieske Fe-S protein is synthesized on cytoplasmic ribosomes to give a 27 kDa precursor form [109]. This precursor is presumably required for targetting to the chloroplasts. There is no information on the regulation of expression of the gene for the Rieske Fe-S protein, nor on the assembly of the protein into the Cyt *b-f* complex.

4. Photosystem I

PS I is the least-characterized of the photosynthetic membrane complexes with regard to its polypeptide composition and synthesis. This stems largely from considerable differences in the number and size of smaller polypeptides associated with the complex reported from different plants. In addition, a light-harvesting complex (LHC I) specifically associated with PS I has been characterized only recently [118]. PS I preparations containing both the light-harvesting complex and the reaction-centre core complex may be obtained by careful Triton X-100 solubilization of chloroplast membranes followed by sucrose density gradient centrifugation in the presence of low Triton X-100 concentrations [119]. The PS I complex may be dissociated by increasing the concentration of Triton X-100, and the LHC I and core complexes purified separately [118,119]. Preparations of the core complex may also be obtained following solubilization of chloroplast membranes in a number of other detergents [120–122].

4.1. Polypeptides of PS I

The polypeptides of the reaction centre core complex are most easily resolved by electrophoresis in the presence of SDS. This allows the separation of two polypeptides of approximately 66 kDa together with six or more polypeptides of lower relative molecular mass. In pea, these have been reported to be 21, 17, 16.5, 11.5, 11 and 10.5 kDa [119], but the electrophoretic mobility of some of these polypeptides is highly dependent on the electrophoresis conditions, making comparisons between preparations from different laboratories difficult. A preparation containing only two smaller polypeptides of 18 and 15 kDa has been described from barley chloroplast membranes [122].

The two polypeptides of 66 kDa have been identified as the site of the primary electron donor P-700 and the electron acceptors A_0 and A_1, the first of which may be specialized chlorophyll molecule (see Chapter 4). These redox centres may be identified in preparations of CP I, a 130 kDa complex of the 66 kDa polypeptides together with chlorophyll and other pigments, separated by mild SDS-gel electrophoresis of chloroplast thylakoid membranes [123–125]. The CP I complex may also be obtained by mild SDS treatment of PS I preparations [126]. The identity of the polypeptides carrying the bound Fe-S centres A and B, which act as the stable electron acceptors in PS I, is not established. Polypeptides of 18 and 15 kDa in the barley PS I preparation have been suggested to carry Fe-S centres on the basis of labelling with ^{59}Fe [122], and a polypeptide of 8 kDa in the spinach complex has been suggested on the basis of a high cysteine content [127]. The function of none of the smaller polypeptides has been established unequivocally. A polypeptide of 20 kDa in the spinach complex has been implicated in electron transfer from plastocyanin to P-700 [128], but it is not clear whether the polypeptide provides a binding site for plastocyanin or has some other role.

The LHC I complex consists of Chl *a* and *b* bound to three or four polypeptides of 22–25 kDa in pea chloroplasts [118]. Variation in the number and size of the LHC I polypeptides in different plants and in different laboratories may be due to differences in electrophoresis procedures. The LHC I polypeptides migrate in SDS-gel electrophoresis in a position similar to that of the LHC II polypeptides, and the reported antigenic similarity between LHC I and LHC II polypeptides [129] may be due to cross-contamination of the LHC complexes.

4.2. Genes for PS I components

Only two genes for PS I components have been characterized to date; these are the chloroplast genes, *psa*A and *psa*B, for the 66 kDa reaction centre polypeptides [130]. These two genes were localized in chloroplast DNA from spinach by hybrid-selection of chloroplast RNA coding for the polypeptides by cloned restriction fragments of chloroplast DNA [131]. Nucleotide sequence analysis of a comparable region of maize chloroplast DNA revealed the presence of two large open reading frames of 751 and 735 codons separated by only 25 bp [130]. These two genes have been shown to code for the two polypeptides of the reaction centre of PS I. The deduced amino acid sequences of the polypeptides indicate that they are homologous, with about 45% identical amino acid residues. A similar arrangement of genes coding for the two reaction centre polypeptides has been found in spinach (W. Kirsch et al., unpublished) and pea chloroplast DNA (J. Lehmbeck et al., unpublished). A separate region of pea chloroplast DNA that was reported to contain the gene for a P-700-Chl *a*-protein [132] has now been sequenced (A.G. Smith et al., unpublished) and shows no homology with the *psa*A and *psa*B genes. This region contains a large open reading frame which is homologous to sequences in tobacco and liverwort chloroplast DNA. However, the predicted polypeptide does not appear to correspond to any of the identified polypeptides of PS I. Further characterization of this gene and its product are required.

Two or three of the smaller polypeptides of the PS I complex may be encoded in chloroplast DNA, on the basis of their reported synthesis on chloroplast ribosomes, but the genes have not yet been located. The remaining polypeptides of the complex, including the LHC I polypeptides, are presumed to be encoded in nuclear DNA, but none of the genes for these components has been isolated.

4.3. Synthesis of PS I

The appearance of a functional PS I in the thylakoids of higher plants is dependent on light. However, there is some controversy over the presence of PS I polypeptides in dark-grown plants. Significant amounts of the 66 kDa polypeptides were reported to be present in dark-grown plants of oats, bean and spinach [133], whereas these polypeptides have been reported to be absent from etiolated barley, wheat and pea seedlings [92,134]. The difference may be due to the specificity of the antisera used to detect the polypeptides after transfer from the polyacrylamide gel to nitrocellulose paper. The smaller polypeptides of the core complex are absent from dark-grown plants and accumulate on illumination of the plants [133], as do the 66 kDa polypeptides [92,133]. LHC I polypeptides also accumulate on illumination of dark-grown pea plants [135].

The regulation of the appearance of the 66 kDa polypeptides may take place at a post-transcriptional step. In maize, transcripts of the two genes, *psa*A and *psa*B, are present in dark-grown plants [90], and therefore, if maize does not accumulate 66 kDa polypeptides in the dark, there must be some mechanism to prevent translation of the mRNA or to degrade newly synthesized polypeptides. There is no information on the transcripts for other polypeptides of PS I in dark-grown plants.

The sites of synthesis of the smaller polypeptides of the core complex are not clearly established. Several methods, including the use of specific inhibitors of protein synthesis and synthesis in isolated chloroplasts, have suggested that two of the polypeptides are synthesized on chloroplast ribosomes. Experiments examining the effect of chloramphenicol and cycloheximide on the labelling of PS I polypeptides have suggested that in *Spirodela* two polypeptides of 12 and 8 kDa are synthesized on chloroplast ribosomes [121], whereas in pea a polypeptide of about 6 kDa was labelled in the presence of cycloheximide but not chloramphenicol [136]. In both plants the 66 kDa polypeptides were synthesized in the presence of cycloheximide but not chloramphenicol, indicating their synthesis on chloroplast ribosomes [121,136]. Incorporation of labelled amino acids into PS I polypeptides in isolated chloroplasts has indicated that one or two polypeptides, in addition to the 66 kDa polypeptides [137], are synthesized on chloroplast ribosomes. In pea chloroplasts polypeptides of 15 kDa [136] and 17 and 11 kDa [138] have been reported to be labelled, and in wheat a polypeptide of 15 kDa has been reported to be labelled in isolated etiochloroplasts [139]. The different electrophoresis systems used preclude any direct comparison of the sizes of these polypeptides, but the conclusion must be that two, or more, of the smaller polypeptides of the PS I complex are synthesized on chloroplast ribosomes.

The remaining polypeptides are assumed to be synthesized on cytoplasmic ribosomes, and this is indicated by the synthesis in the presence of chloramphenicol of a 16 kDa polypeptide in *Spirodela* [121] and of polypeptides of 20, 11.5, 11 and 8 kDa in pea [136]. In addition, the LHC I polypeptides were labelled in the presence of chloramphenicol in pea [136]. Translation of spinach poly(A) RNA in vitro indicates that the 22 kDa polypeptide of the core complex is synthesized as a larger precursor of 26 kDa [131]. In pea the corresponding 20 kDa polypeptide has been shown to be synthesized on poly(A) RNA, post-translationally imported into isolated chloroplasts and inserted into the thylakoid membrane [136]. Similar post-translational import into isolated chloroplasts and insertion into the thylakoid membranes has been observed with two polypeptides of the pea LHC I complex [136].

Further study is needed to define the polypeptide composition of PS I and in particular to identify equivalent polypeptides in PS I preparations from different plants. Only then will it be possible to compare results from different laboratories and produce a comprehensive view of the synthesis of PS I in higher plants.

5. *ATP synthase*

Chloroplast ATP synthase is a well-defined complex which may be solubilized from thylakoid membranes by treatment with octylglucoside and cholate and purified by ammonium sulphate fractionation and sucrose density gradient centrifugation [140]. The complex is composed of two assemblies of polypeptides: CF_1, a peripheral membrane complex, which may be washed from thylakoid membranes with EDTA and which shows latent ATPase activity, and CF_0, the intrinsic membrane sector, which translocates protons across the thylakoid membrane.

5.1. Polypeptides of ATP synthase

The CF_1 complex is composed of five different polypeptides in all plants examined so far. There is some variation in the reported M_r values of the polypeptides from different plants, and, although some of the variation may be due to the use of different electrophoretic techniques, there appear to be real differences in the sizes of some of the individual polypeptides. The sizes of the polypeptides of wheat CF_1 are: α, 58 kDa; β, 57kDa; γ, 38 kDa; δ, 25 kDa; and ϵ, 14 kDa [141]. There appears to be considerable variation in the size of the δ subunit, which is reported to be 19.5 kDa in spinach [142]. The sizes of the polypeptides of CF_1 are similar to those of the *E. coli* F_1, and it appears that the chloroplast and bacterial polypeptides are structurally and functionally homologous [143]. The stoichiometry of the polypeptides of CF_1 is believed to be 3α:3β:1γ:1δ:1ϵ, as in *E. coli*.

The polypeptide composition of CF_0 is deduced from the polypeptide composition of the purified ATP synthase preparations. CF_0 appears to be composed of four different polypeptides [140], although these may not all be resolved as separate bands by SDS-polyacrylamide gel electrophoresis. This led to the belief that there were only three polypeptides in CF_0, as in the *E. coli* F_0 complex, and these were called subunit I (18 kDa), subunit II (16 kDa) and subunit III (8 kDa). Subunit III is the dicyclohexylcarbodiimide (DCCD)-binding proteolipid, whose amino acid sequence has been determined [144]. Two-dimensional gel electrophoresis [140] or electrophoresis in the presence of urea [145,146] is able to resolve a fourth CF_0 polypeptide of approximately 19 kDa which has been called subunit IV [146]. Gene sequences have indicated that subunit IV is homologous to the F_0 subunit *a* of *E. coli* [147], whereas subunit I is equivalent to F_0 subunit *b* [148] and subunit III is equivalent to F_0 subunit *c* [149]. There is apparently no counterpart to CF_0 subunit II in the *E. coli* F_0 complex. CF_0 subunit II may correspond to one of the additional polypeptides associated with the mitochondrial ATP synthase, or it may be similar to the diverged form of the subunit *b* found in *Synechococcus* 6301 [147].

5.2. Genes for ATP synthase polypeptides

Six genes for ATP synthase polypeptides have been located and characterized in chloroplast DNA from a number of plants. These genes are located in two clusters and the genes within each cluster are co-transcribed, suggesting an operon structure. The genes, *atp*B and *atp*E, for the β and ϵ subunits of CF_1 are located close together in all higher plants, and nucleotide sequence analysis reveals that the genes overlap by 4 bp in maize [150], spinach [151], tobacco [152], barley [153] and wheat [154]. In pea, however, these genes do not overlap because a small duplication at the end of *atp*B results in an alteration of the reading frame and a 25 bp separation between the genes [155]. The amino acid sequences deduced from the nucleotide sequences of these genes show great similarities between different plants. The higher plant β subunit sequences contain approximately 95% identical residues, and show approximately 70% identical residues to bacterial and mitochondrial β subunit sequences. This presumably reflects the role of the β subunit in the catalytic activity of the ATP synthase. The amino acid sequence of the ϵ subunit is much less conserved, showing approximately 65% identical residues between higher plants and approximately 25% identical residues between plant and bacterial sequences.

The structural gene (*atp*A) for the α subunit is located 20–40 kbp away from the genes for β and ϵ subunits in chloroplast DNA [142,156]. Nucleotide sequences of the α subunit genes from tobacco [157], wheat [154] and spinach [158] have been determined and indicate a high degree of conservation of the amino acid sequence. Nearly 90% of amino acid residues are identical in all higher plants examined and approximately 50% of the residues are identical to bacterial and mitochondrial α subunit sequences. The α subunit sequences also show homology to the β subunit sequences, particularly in regions implicated in adenine nucleotide binding [154]. This suggests that the α and β subunit genes have evolved from a single ancestral gene.

The genes for three subunits of CF_0 are clustered near the gene for the α subunit in chloroplast DNA. The gene (*atp*H) for CF_0 subunit III was originally located approximately 2 kbp upstream from *atp*A in wheat chloroplast DNA [149,159] and has subsequently been located in a similar position in other higher plants [156,160,161]. Nucleotide sequence determination indicates that the primary structure of the polypeptide is highly conserved, showing approximately 90% identical residues between the plant polypeptides but only 20–25% identical residues compared to mitochondrial and most bacterial polypeptides. However, comparison with cyanobacterial polypeptides shows 84% identical residues [144], perhaps lending support for an endosymbiotic origin of chloroplasts. The gene (*atp*F) for CF_0 subunit I was originally located between *atp*H and *atp*A in wheat chloroplast DNA and was shown to contain a large intron [148]. The presence of an intron in the *atp*F gene has subsequently been shown for tobacco [162], spinach [158] and pea (A.K. Huttly, unpublished). Comparison of the amino acid sequence deduced from the nucleotide sequence of the two exons of the gene with the determined N-terminal sequence of the spinach CF_0 subunit I suggests that the primary translation product is processed to remove 17 amino acid residues in wheat, spinach and tobacco and 6 amino acid residues in pea. Secondary structure predictions of the mature CF_0 subunit I suggest that the protein is homologous to F_0 subunit *b* of *E. coli*. The third gene (*atp*I) for a CF_0 subunit was identified upstream from the *atp*H gene in pea [147], spinach [158] and wheat (A.S. Höglund, unpublished). An open reading frame coding for a hydrophobic protein of 247 amino acid residues homologous to the F_0 subunit *a* of *E. coli* was indicated by the nucleotide sequence. The deduced amino acid sequences show less than 20% residues identical to the polypeptide from *E. coli* and mitochondria, but 70% of residues are identical to the F_0 subunit *a* from *Synechococcus* 6301 [147]. It seems probable that the product of this chloroplast gene corresponds to CF_0 subunit IV separated by electrophoresis in the presence of urea [146].

The genes for the γ and δ subunits of CF_1 and for CF_0 subunit II are most likely located in nuclear DNA, but they have not yet been isolated. Each of these polypeptides has been shown to be synthesized as a larger precursor form on translation of poly(A) RNA in vitro [142,146]. The isolation and characterization of the nuclear genes for these three subunits of ATP synthase is urgently required.

5.3. *Synthesis of ATP synthase*

Chloroplast ATP synthase is present in dark-grown plants, indicating that light is not essential for the synthesis of any of the subunits of this complex. However, the regulatory features necessary to coordinate the expression of chloroplast and nuclear genes are not yet fully understood. One feature that needs to be investigated in more detail is the provision of sufficient subunits for assembly in the stoichiometry demanded by the ATP synthase complex. The necessary subunit stoichiometry cannot be provided solely by transcriptional control mechanisms, and regulation at translational or post-translational steps will also be required.

The chloroplast genes for the CF_0 subunits IV, III and I and for the α subunit of CF_1 are co-transcribed to give a primary transcript which is then processed to remove the intron in *atp*F and to generate a number of smaller transcripts [148]. Nothing is known of the mechanisms of these processing events. The chloroplast genes for the β and ϵ subunits of CF_1 are also co-transcribed [151,152]. The transcripts are translated on chloroplast ribosomes, as indicated by the sensitivity of the synthesis of the subunits to chloramphenicol [121,163] and by the incorporation of labelled amino acids into the individual subunits by isolated chloroplasts [164,166].

The γ and δ subunits of CF_1 and subunit II of CF_0 are synthesized on cytoplasmic ribosomes, as indicated by the sensitivity of their synthesis to cycloheximide in pea [163] and *Spirodela* [121]. Each of these subunits has been shown to be synthesized as a higher-M_r form on translation of poly(A) RNA in vitro [142,146]. The γ subunit is synthesized initially as a 46 kDa form, the δ subunit as a 27 kDa form and subunit II as a 24 kDa form. These precursors are 7–9 kDa larger than the mature polypeptides, and probably contain N-terminal transit peptide sequences necessary for the import of the polypeptides into the chloroplasts. However, this has not been established for any of the nuclear-encoded subunits of ATP synthase.

The assembly of newly synthesized chloroplast-encoded polypeptides into ATP synthase complexes by isolated chloroplasts has been observed [166], suggesting that the chloroplasts contain pools of the cytoplasmically synthesized subunits γ, δ and II. It has been shown that in the absence of the chloroplast-synthesized polypeptides of ATP synthase, newly synthesized γ and δ subunits are rapidly degraded [167]. The apparent half-lives of the unassembled γ and δ subunits in rye seedlings grown at 32°C were 2 h and 4 h respectively. This suggests that rapid degradation of excess unassembled subunits is a regulatory feature ensuring the correct stoichiometry of the ATP synthase subunits.

6. *Conclusions and future directions*

Although the majority of the chloroplast genes for photosynthetic membrane proteins have been located and characterized, much work remains on the isolation and characterization of the nuclear genes. However, even with all the genes characterized, a large amount of research will be necessary to understand the mechanisms regulating the expression of these genes and the assembly of the polypeptides into the functional membrane complex. It is clear from the investigations carried out so far that the expression of genes for photosynthetic membrane components is regulated at transcriptional, translational and post-translational events. It seems probable that the expression of chloroplast genes is regulated primarily at post-transcriptional stages, whereas nuclear genes are regulated at the level of transcription, but much work remains to substantiate this view.

Acknowledgements

I am most grateful to Dave Willey, Paul Dunn and Richard Blyden for helpful comments while writing this review, and to the SERC and AFRC for supporting the research carried out in my laboratory.

References

1 Andersson, B. and Anderson, J.M. (1985) in Modern Methods of Plant Analysis, New Series Vol. 1 (Linskens, H.F. and Jackson, J.F., eds.) pp. 231–258, Springer-Verlag, Berlin.

2 Bennett, J. (1983) Biochem. J. 212, 1–13.

3 Thornber, J.P. (1986) In Encyclopedia of Plant Physiology, New Series Vol. 19 (Staehelin, L.A. and Arntzen, C.J., eds.) pp. 98–142, Springer-Verlag, Berlin.

4 Camm, E.L. and Green, B.R. (1983) J. Cell. Biochem. 23, 171–179.

5 Nakatani, H.Y., Ke, B., Dolan, E. and Arntzen, C.J. (1984) Biochim. Biophys. Acta 765, 347–352.

6 Michel, H., Weyer, K.A., Gruenberg, H., Dunger, I., Oesterhelt, D. and Lottspeich, F. (1986) EMBO J. 5, 1149–1158.

7 Chua, N.-H. and Gillham, N.W. (1977) J. Cell Biol. 77, 441–452.

8 Steinback, K.E., McIntosh, L., Bogorad, L. and Arntzen, C.J. (1981) Proc. Natl. Acad. Sci. USA 78, 7463–7467.

9 Deisenhofer, J., Epp, O., Miki, K., Huber, R. and Michel, H. (1985) Nature 318, 618–624.

10 Trebst, A. (1986) Z. Naturforsch. 41c, 240–246.

11 Widger, W.R., Cramer, W.A., Hermodson, M. and Herrmann, R.G. (1985) FEBS Lett. 191, 186–190.

12 Widger, W.R., Cramer, W.A., Hermodson, M., Meyer, D. and Gullifor, M. (1984) J. Biol. Chem. 259, 3870–3876.

13 Ljungberg, U., Henrysson, T., Rochester, C.P., Åkerlund, H.E. and Andersson, B. (1986) Biochim. Biophys. Acta 849, 112–120.

14 Farchaus, J. and Dilley, R.A. (1986) Arch. Biochem. Biophys. 244, 94–101.

15 Bennett, J. (1979) FEBS Lett. 103, 342–344.

16 Alfonzo, R., Nelson, N. and Racker, E. (1980) Plant Physiol. 65, 730–734.

17 Andersson, B. (1986) in Encyclopedia of Plant Physiology, New Series Vol. 19 (Staehelin, L.A. and Arntzen, C.J. eds.) pp. 447–456, Springer-Verlag, Berlin.

18 Kuwabara, T. and Murata, N. (1984) in Advances in Photosynthesis Research, Vol. 1 (Sybesma, C., ed.) pp. 375–378, Martinus Nijhoff/Dr. W. Junk, The Hague.

19 Jansson, C. (1984) in Advances in Photosynthesis Research, Vol. 1. (Sybesma C., ed.) pp. 375–378, Martinus Nijhoff/Dr. W. Junk, The Hague.

20 Wolter, F.P., Schmitt, J.M., Bohnert, H.J. and Tsugita, A. (1984) Plant Sci. Lett. 34, 323–334.

21 Yamamoto, Y., Hermodson, M.A. and Krogmann, D.W. (1986) FEBS Lett. 195, 155–158.

22 Ljungberg, U., Åkerlund, H.E., Larsson, C. and Andersson, B. (1984) Biochim. Biophys. Acta 767, 145–152.

23 Kung, S.D., Thornber, J.P. and Wildman, S.G. (1972) FEBS Lett. 24, 185–188.

24 Cashmore, A.R. (1984) Proc. Natl. Acad. Sci. USA 81, 2960–2964.

25 Lamppa, G.K., Morelli, G. and Chua, N.-H. (1985) Mol. Cell. Biol. 5, 1370–1378.

26 Karlin-Neumann, G.A., Kohorn, B.D., Thornber, J.P. and Tobin, E.M. (1985) J. Mol. Appl. Genet. 3, 45–61.

27 Kohorn, B.D., Harel, E., Chitnis, P.R., Thornber, J.P. and Tobin, E.M. (1986) J. Cell Biol. 102, 972–981.

28 Dunsmuir, P., Smith, S.M. and Bedbrook, J. (1983) J. Mol. Appl. Genet. 2, 285–300.

29 Dunsmuir, P. (1985) Nucleic Acids Res. 13, 2503–2518.

30 Pichersky, E., Bernatzky, R., Tanksley, S.D., Briedenbach, R.B., Kausch, A.P. and Cashmore, A.R. (1985) Gene 40, 247–258.

31 Corruzi, G., Broglie, R., Cashmore, A. and Chua, N.-H. (1983) J. Biol. Chem. 258, 1399–1402.

32 Meyerowitz, E.M. and Pruitt, R.E. (1985) Science 229, 1214–1218.

33 Polans, N.O., Weeden, N.F. and Thompson, W.F. (1985) Proc. Natl. Acad. Sci. USA 82, 5083–5087.

34 Karlin-Neumann, G.A. and Tobin, E.M. (1986) EMBO J. 6, 9–13.

35 Schmidt, G.W., Devilliers-Thiery, A., Desmisseaux, H., Blobel, G. and Chua, N.-H. (1979) J. Cell Biol. 83, 615–622.

36 Mullet, J.E. (1983) J. Biol. Chem. 258, 9941–9948.

37 Bedbrook, J.R., Link G., Coen, D.M., Bogorad, L. and Rich, A. (1978) Proc. Natl. Acad. Sci. USA 75, 3060–3064.

38 Zurawski, G., Bohnert, H.J., Whitfeld, P.R. and Bottomley, W. (1982) Proc. Natl. Acad. Sci. USA 79, 7699–7703.
39 Sugita, M. and Sugiura, M. (1984) Mol. Gen. Genet. 195, 308–313.
40 Hirschberg, J. and McIntosh, L. (1983) Science 222, 1346–1349.
41 Goloubinoff, P., Edelman, M. and Hallick R.B. (1984) Nucleic Acids Res. 12, 9489–9496.
42 Hirschberg, J., Bleecker, A., Kyle, D.J., McIntosh, L. and Arntzen, C.J. (1984) Z. Naturforsch. 39c, 412–420.
43 Spielmann, A. and Stutz, E. (1983) Nucleic Acids Res. 11, 7157–7167.
44 Link, G. and Langridge, U. (1984) Nucleic Acids Res. 12, 945–958.
45 Oishi, K.K., Shapiro, D.R. and Tewari, K.K. (1984) Mol. Cell. Biol. 4, 2556–2563.
46 Pfister, K., Steinback, K.E., Gardner, G. and Arntzen, C.J. (1981) Proc. Natl. Acad. Sci. USA 78, 981–985.
47 Blyden, E.R. and Gray, J.C. (1986) Biochem. Soc. Trans. 14, 62.
48 Souza-Machado, V., Bandeen, J.D., Stephenson, G.R. and Lavigne, P. (1978) Can. J. Plant. Sci. 58, 977–981.
49 Rochaix, J.-D., Dron, M., Rahire, M. and Malnoe, P. (1984) Plant Mol. Biol. 3, 363–370.
50 Rasmussen, O.F., Bookjans, G., Stummann, B.M. and Henningsen, K.W. (1984) Plant Mol. Biol. 3, 191–199.
51 Holschuh, K., Bottomley, W. and Whitfeld, P.R. (1984) Nucleic Acids Res. 12, 8819–8834.
52 Alt, J., Morris, J., Westhoff, P. and Herrmann, R.G. (1984) Curr. Genet. 8, 597–806.
53 Courtice, G.R.M., Bowman, C.M., Dyer, T.A. and Gray, J.C. (1985) Curr. Genet. 10, 329–333.
54 Westhoff, P., Alt, J. and Herrmann, R.G. (1983) EMBO J. 2, 2229–2237.
55 Morris, J. and Herrmann, R.G. (1984) Nucleic Acids Res. 12, 2837–2850.
56 Herrmann, R.G., Alt. J., Schiller, B., Widger, W.R. and Cramer, W.A. (1984) FEBS Lett. 176, 239–244.
57 Hird, S.M., Willey, D.L., Dyer, T.A. and Gray, J.C. (1986) Mol. Gen. Genet. 203, 95–100.
58 Carillo, N., Seyer, P., Tyagi, A. and Herrmann, R.G. (1986) Curr. Genet. 10, 619–624.
59 Westhoff, P., Alt, J., Widger, W.R. Cramer, W.A. and Herrmann, R.G. (1985) Plant Mol. Biol. 4, 103–110.
60 Babcock, G.T., Widger, W.R., Cramer, W.A., Oertling, W.A. and Metz, J.G. (1985) Biochemistry 24, 3638–3645.
61 Steinmetz, A.A., Castroviejo, M., Sayre, R.T. and Bogorad, L. (1986) J. Biol. Chem. 261, 2485–2488.
62 Bookjans, G., Stummann, B.M., Rasmussen, O.F. and Henningsen, K.W. (1986) Plant Mol. Biol. 6, 359–366.
63 Oliver, R.P. and Poulsen, C. (1984) Carlsberg Res. Commun. 49, 647–673.
64 Quigley, F. and Weil, J.H. (1985) Curr. Genet. 9, 495–503.
65 Westhoff, P., Jansson, C., Klein-Hitpass, L., Berzborn, R., Larsson, C. and Bartlett, S.G. (1985) Plant Mol. Biol. 4, 135–146.
66 Apel, K. and Kloppstech, K. (1978) Eur. J. Biochem. 85, 581–588.
67 Tobin, E.M. (1981) Plant. Physiol. 67, 1078–1083.
68 Cuming, A.C. and Bennett, J. (1981) Eur. J. Biochem. 118, 71–80.
69 Thompson, W.F., Everett, M., Polans, N.O., Jorgensen, R.A. and Palmer, J.D. (1983) Planta 158, 487–500.
70 Gallagher, T.F. and Ellis, R.J. (1982) EMBO J. 1, 1493–1498.
71 Silverthorne, J. and Tobin, E.M. (1984) Proc. Natl. Acad. Sci. USA 81, 1112–1116.
72 Apel, K. (1979) Eur. J. Biochem. 97, 183–188.
73 Tobin, E.M. (1981) Plant Mol. Biol. 1, 35–51.
74 Simpson, J., Timko, M.P., Cashmore, A.R., Schell, J., van Montagu, M. and Herrera-Estrella, L. (1985) EMBO J. 4, 2723–2729.
75 Lamppa, G.K., Nagy, F. and Chua, N.-H. (1985) Nature 316, 750–752.
76 Schmidt, G.W., Bartlett, S.G., Grossman, A.R., Cashmore A.R. and Chua, N.-H. (1982) J. Cell Biol. 9, 468–478.
77 Cline, K., Werner-Washbourne, M., Andrews, J. and Keegstra, K. (1985) J. Biol. Chem. 260, 3691–3696.

78 Grossman, A.R., Bartlett, S.G. and Chua, N.-H. (1980) Nature 285, 625–628.
79 Chitnis, P.R., Harel, E., Kohorn, B.D., Tobin, E.M. and Thornber, J.P. (1986) J. Cell Biol. 102, 982–988.
80 Bennett, J. (1981) Eur. J. Biochem. 118, 61–70.
81 Bellemare, G., Bartlett, S.G. and Chua, N.-H. (1982) J. Biol. Chem. 257, 7762–7767.
82 Mayfield, S.P. and Taylor, W.C. (1984) Eur. J. Biochem. 144, 79–84.
83 Batchauer, A., Mosinger, E., Kreuz, K., Dorr, I. and Apel, K. (1986) Eur. J. Biochem. 154, 625–634.
84 Reisfeld, A., Gressel, J., Jakob, K.M. and Edelman, M. (1978) Photochem. Photobiol. 27, 161–165.
85 Eaglesham, A.R.J. and Ellis, R.J. (1974) Biochim. Biophys. Acta 335, 396–407.
86 Marder, J.B., Goloubinoff, P. and Edelman, M. (1984) J. Biol. Chem. 259, 3900–3908.
87 Mattoo, A.K., Hoffman-Falk, H., Marder, J.B. and Edelman, M. (1984) Proc. Natl. Acad. Sci. USA 81, 1380–1384.
88 Fromm, H., Devic, M., Fluhr, R. and Edelman, M. (1985) EMBO J. 4, 291–295.
89 Liveanu, V., Yocum, C.F. and Nelson, N. (1986) J. Biol. Chem. 261, 5296–5300.
90 Zhu, Y.S., Kung, S.-D. and Bogorad, L. (1985) Plant Physiol. 79, 371–376.
91 Leto, K.J., Bell, E. and McIntosh, L. (1985) EMBO J. 4, 1645–1653.
92 Takabe, T., Takabe, T. and Akazawa, T. (1986) Plant Physiol. 81, 60–66.
93 Gregory, P. and Bradbeer, J.W. (1973) Planta 109, 317–326.
94 Ryrie, I.J., Young, S. and Andersson, B. (1984) FEBS Lett. 177, 269–273.
95 Hurt, E.C. and Hauska, G. (1981) Eur. J. Biochem. 117, 591–599.
96 Phillips, A.L. and Gray, J.C. (1983) Eur. J. Biochem. 137, 553–560.
97 Clark, R.D. and Hind, G. (1983) J. Biol. Chem. 258, 10348-10354.
98 Hurt, E., Hauska, G. and Malkin, R. (1981) FEBS Lett. 134, 1–5.
99 Clark, R.D., Hawkesford, M.J., Coughlan, S.J., Bennett, J. and Hind, G. (1984) FEBS Lett. 174, 137–142.
100 Hurt, E. and Hauska, G. (1982) J. Bioenerg. Biomembranes 14, 405–424.
101 Shahak, Y., Crowther, D. and Hind, G. (1981) Biochim. Biophys. Acta 636, 234–243.
102 Hallier, U.W. and Heber, U.W. (1977) In Photosynthetic Organelles (Miyachi, S., Katoh, S., Fujita, Y. and Shibata, K., eds.) pp. 257–273, Jap. Soc. Plant Physiol., Japan.
103 Gray, J.C. (1980) Eur. J. Biochem. 112, 39–46.
104 Willey, D.L., Huttly, A.K., Phillips, A.L. and Gray, J.C. (1983) Mol. Gen. Genet. 189, 85–89.
105 Willey, D.L., Auffret, A.D. and Gray, J.C. (1984) Cell 36, 555–562.
106 Willey, D.L., Howe, C.J., Auffret, A.D., Bowman, C.M., Dyer, T.A. and Gray, J.C. (1984) Mol. Gen. Genet. 194, 416–422.
107 Alt, J. and Herrmann, R.G. (1984) Curr. Genet. 8, 551–557.
108 Tyagi, A.K. and Herrmann, R.G. (1986) Curr. Genet. 10, 481–486.
109 Alt, J. Westhoff, P., Sears, B.B., Nelson, N., Hurt, E., Hauska, G. and Herrmann, R.G. (1983) EMBO J. 2, 979–986.
110 Phillips, A.L. (1985) Curr. Genet. 10, 153–161.
111 Heinemeyer, W., Alt. J. and Herrmann, R.G. (1984) Curr. Genet. 8, 543–549.
112 Phillips, A.L. and Gray, J.C. (1984) Mol. Gen. Genet. 194, 477–484.
113 Widger, W.R., Cramer, W.A., Herrmann, R.G. and Trebst, A. (1984) Proc. Natl. Acad. Sci. USA 81, 674–678.
114 Plesnicar, M. and Bendall, D.S. (1973) Biochem. J. 136, 803–812.
115 Baltimore, B.G. and Malkin, R. (1977) Plant Physiol. 60, 76–80.
116 Phillips, A.L. and Gray, J.C. (1984) Eur. J. Biochem. 138, 591–595.
117 Rothstein, S.J., Gatenby, A.A., Willey, D.L. and Gray, J.C. (1985) Proc. Natl. Acad. Sci. USA 82, 7955–7959.
118 Haworth, P., Watson, J.L. and Arntzen, C.J. (1983) Biochim. Biophys. Acta 724, 151–158.
119 Mullet, J.E., Burke, J.J. and Arntzen, C.J. (1980) Plant Physiol. 65, 814–822.
120 Bengis, C. and Nelson, N. (1975) J. Biol. Chem. 250, 2783–2788.
121 Nechushtai, R., Nelson, N., Mattoo, A.K. and Edelman, M. (1981) FEBS Lett. 125, 115–119.
122 Moller, B.L., Hoyer-Hansen, G. and Hiller, R.G. (1981) in Photosynthesis, Vol. 3 (Akoyunoglou, G., ed.) pp. 245–256, Balaban Int. Sci. Services, Philadelphia, PA.

123 Mathis, P. , Sauer, K. and Remy, R. (1978) FEBS Lett. 88, 2575–2578.
124 Anderson, J.M. (1980) Biochim. Biophys. Acta 591, 113–126.
125 Vierling, E. and Alberte, R.S. (1983) Plant Physiol. 72, 625–633.
126 Anderson, J.M., Brown, J.S., Lam, E. and Malkin, R. (1983) Photochem. Photobiol. 38, 205–210.
127 Lagoutte, B., Setif, P. and Duranton, J. (1984) FEBS Lett. 174, 24–29.
128 Bengis, C. and Nelson, N. (1977) J. Biol. Chem. 252, 4564–4569.
129 Evans, P.K. and Anderson, J.M. (1986) FEBS Lett. 199, 227–233.
130 Fish, L.E., Kuck, U. and Bogorad, L. (1985) J. Biol. Chem. 260, 1413–1421.
131 Westhoff, P., Alt, J., Nelson, N., Bottomley, W., Bunemann, H. and Herrmann, R.G. (1983) Plant Mol. Biol. 2, 95–107.
132 Smith, A.G. and Gray, J.C. (1984) Mol. Gen. Genet. 194, 471–476.
133 Nechushtai, R. and Nelson, N. (1985) Plant Mol. Biol. 4, 377–384.
134 Vierling, E. and Alberte, R.S. (1983) J. Cell Biol. 97, 1806–1814.
135 Mullet, J.E., Burke, J.J. and Arntzen, C.J. (1980) Plant Physiol. 65, 823–827.
136 Mullet, J.E., Grossman, A.R. and Chua, N.-H. (1981) Cold Spring Harbor Symp. Quant. Biol. 46, 979–984.
137 Zielinski, R.E. and Price, C.A. (1980) J. Cell Biol. 85, 435–445.
138 Smith, A.G. and Gray, J.C. (1984) in Advances in Photosynthesis Research, Vol. 4 (Sybesma, C., ed.) pp. 513–516, Martinus Nijhoff/Dr. W. Junk, The Hague.
139 Obokata, J. (1986) Plant Physiol. 81, 7805–7807.
140 Pick, U. and Racker, E. (1979) J. Biol. Chem. 254, 2793–2799.
141 Moase, E.H. and Green, B.R. (1981) Eur. J. Biochem. 119, 145–150.
142 Westhoff, P., Nelson, N., Bunemann, H. and Herrmann, R.G. (1981) Curr. Genet. 4, 109–120.
143 Walker, J.E. and Tybulewicz, V.L.J. (1986) in Molecular Biology of the Photosynthetic Apparatus (Arntzen, C.J., Bogorad, L., Bonitz, S. and Steinback, K., eds.) pp. 141–153, Cold Spring Harbor, New York.
144 Sebald, W. and Hoppe, J. (1981) Curr. Top. Bioenerg. 12, 1–64.
145 Suss, K.-H. (1980) FEBS Lett. 112, 255–259.
146 Westhoff, P., Alt, J., Nelson, N. and Herrmann, R.G. (1985) Mol. Gen. Genet. 199, 290–299.
147 Cozens, A.L., Walker, J.E., Phillipps, A.L., Huttly, A.K. and Gray, J.C. (1986) EMBO, J. 5, 217–222.
148 Bird, C.R., Koller, B., Auffret, A.D., Huttly, A.K., Howe, C.J., Dyer, T.A. and Gray, J.C. (1985) EMBO J. 4, 1381–1388.
149 Howe, C.J., Auffret, A.D., Doherty, A., Bowman, C.M., Dyer, T.A. and Gray, J.C. (1982) Proc. Natl. Acad. Sci. USA 79, 6903–6907.
150 Krebbers, E.T., Larrinua, I.M., MacIntosh, L. and Bogorad, L. (1982) Nucleic Acids Res. 10, 4985–5002.
151 Zurawski, G., Bottomley, W. and Whitfeld, P.R. (1982) Proc. Natl. Acad. Sci. USA 79, 6260–6264.
152 Shinozaki, K., Deno, H., Kato, A. and Sugiura, M. (1983) Gene 24, 147–155.
153 Zurawski, G. and Clegg, M. (1984) Nucleic Acids Res. 12, 2549–2559.
154 Howe, C.J., Fearnley, I.M., Walker, J.E., Dyer, T.A. and Gray, J.C. (1985) Plant Mol. Biol. 4, 333–345.
155 Zurawski, G., Bottomley, W. and Whitfeld, P.R. (1986) Nucleic Acids Res. 14, 3974.
156 Huttly, A.K. and Gray, J.C. (1984) Mol. Gen. Genet. 194, 402–409.
157 Deno, H., Shinozaki, K. and Sugiura, M. (1983) Nucleic Acids Res. 11, 2185–2191.
158 Hennig, J. and Herrmann, R.G. (1986) Mol. Gen. Genet. 203, 117–128.
159 Howe, C.J., Bowman, C.M., Dyer, T.A. and Gray, J.C. (1983) Mol. Gen. Genet. 190, 51–55.
160 Alt, J., Winter, P., Sebald, W., Moser, J.G., Schedel, R., Westhoff, P. and Herrmann, R.G. (1983) Curr. Genet. 7, 129–138.
161 Deno, H., Shinozaki, K. and Sugiura, M. (1984) Gene 32, 195–201.
162 Shinozaki, K., Deno, H., Wakasagi, T. and Sugiura, M. (1986) Curr. Genet. 10, 421–423.
163 Bouthyette, P.-Y. and Jagendorf, A. (1978) Plant Cell Physiol. 19, 1169–1174.
164 Mendiola-Morgenthaler, L.R., Morgenthaler, J.J. and Price, C.A. (1976) FEBS Lett. 62, 96–100.
165 Doherty, A. and Gray, J.C. (1980) Eur. J. Biochem. 108, 131–136.
166 Nelson, N., Nelson, H. and Schatz, G. (1980) Proc. Natl. Acad. Sci. USA 77, 1361–1364.
167 Biekmann, S. and Feierabend, J. (1985) Eur. J. Biochem. 152, 529–535.

J. Amesz (ed.) *Photosynthesis*

CHAPTER 15

Evolution of photosynthesis

H.J. VAN GORKOM

Department of Biophysics, Huygens Laboratory of the State University, P.O.Box 9504, 2300 RA Leiden, The Netherlands

1. Introduction

Like all complicated biological structures, the photosynthetic apparatus may be expected to bear traces of its history, which may range in importance from the fundamental design of the whole thing to a now meaningless scar that has not had enough time yet to heal. A full understanding of the structure and function of the photosynthetic apparatus requires that these traces are recognized as such. Although in most, if not all, cases only tentative answers can be expected, evolutionary considerations often do answer otherwise enigmatic questions in a very satisfactory and illuminating way. On the other hand, to distinguish fact and fiction in evolution theory is not a trivial problem. To illustrate the point: now 12 years of intensive research have passed since E. Broda in his monograph on 'the evolution of the bioenergetic processes' [1] accumulated over two thousand references, most of them relevant to the subject of this chapter. The average specialist in one of the many fields of photosynthesis research is at best superficially acquainted with the relevant literature and will usually not resort to speculations on evolution except as an afterthought, to polish the last paragraph of his research paper, instead of using them as a source of inspiration for new experiments. Written by such a specialist, the present chapter does no justice to the rich literature on the evolution of photosynthesis. Its primary purpose is to illustrate the significance of evolutionary concepts in understanding the photosynthetic machinery as it operates today.

2. The origin of chloroplasts

Although hardly relevant to the evolution of photosynthesis as such, the evolutionary origin of the chloroplast will be discussed first, because it provides a striking example of an evolutionary concept that explains a large number of observations which otherwise seem to make no sense at all.

Globally speaking, photosynthesis takes place in the chloroplast and respiration takes place in the mitochondrion. Both chloroplasts and mitochondria are organelles of eukaryotic cells. They look in many respects like cells within the cell and it is now generally accepted that these organelles are descendants of once free-living eubacteria. Comprehensive accounts of this theory may be found in Refs. 2 and 3. The protozoa are thought to originate from some ‘protoeukaryote’ (perhaps similar to the anaerobic amoeba *Pelomyxa palustris* [2]) which acquired an efficient energy metabolism by the uptake and tolerance in its cytoplasm of some aerobic respiring eubacterium and the subsequent establisment of a stable symbiotic association with this bacterium by synchronization of their reproductive cycles. It appears that all mitochondria may be descendants of the same endosymbiont, related phylogenetically to *Agrobacterium tumefaciens* [4]. Later, and probably continuing to this day, various symbiotic associations between protozoa and eubacteria with oxygenic photosynthesis were established, some of which led to the different groups of algae now in existence. Some features of the chloroplasts of Cryptomonads even suggest that these are derived from a eukaryotic endosymbiont, most likely a red alga: endosymbiosis and integration as an organelle took place twice in succession in this case [5]. The red algae have chloroplasts which may be derived from cyanobacteria, since they contain similar antenna pigment-protein complexes; all phycobiliproteins are closely related [6]. One oxygen-evolving prokaryote with Chl *b* as an accessory pigment is known (*Prochloron*), but shows no special affinity to the chloroplasts of green algae [7]. No Chl *c*-containing prokaryotes have been described. Presumably a wider variety, especially with regard to antenna pigments, of prokaryotic oxygen-evolving organisms existed before the eukaryotes took their place as the main primary producers in the biosphere.

Chloroplasts and mitochondria have their own enclosing membrane, and the lack of protoplasmic continuity between the chloroplast stroma or mitochondrial matrix and the cytosol of the eukaryote almost defines these organelles as separate cells. They have their own DNA, RNA and protein synthesis machinery and in all these fundamental cellular properties they resemble the eubacteria much more than the eukaryote host cell. Many chloroplast proteins which must have been endogenous to the endosymbiont, because homologous proteins are found in Cyanobacteria, are now encoded by nuclear genes and synthesized in the cytoplasm [8]. The genetic integration of the mitochondria is even more extensive. Perhaps chloroplasts are in the process of losing all their genetic material to the nucleus, but some genes may be much more resistant to the process than others. As a general rule, the intrinsic thylakoid proteins are synthesized in the chloroplast. The transport across the chloroplast envelope and subsequent insertion in the thylakoid membrane requires a complex processing machinery and, although it does exist for LHC II, the main light-harvesting Chl *a*/*b* protein [9], one can imagine that such a system takes time to evolve.

In view of the endosymbiotic theory no further explanation is needed for the detailed similarity between mitochondrial respiration and that of some eubacteria, nor for the striking resemblance of chloroplast photosynthesis to that of cyanobacteria.

3. *The origin of photosynthesis*

How old is photosynthesis? The present atmospheric oxygen level is determined by the balance between photosynthesis and respiration. Geochemical evidence indicates that a stable aerobic environment was established some 1700 million years ago. Most likely, this marked the ultimate saturation of a vast supply of oxygen sinks (banded iron formations) after a long period of increasing photosynthetic oxygen evolution, starting more than 2500 million years ago [10]. The widespread occurence of photosynthetic bacteria as early as 3500 million years ago is strongly indicated by fossil evidence (stromatolites) [11]. Since the oldest sediments known are not much older (3800 million years) and highly modified by geochemical processes [12], there is little hope of obtaining hard evidence on the origins of photosynthesis other than that preserved in the genomes of extant organisms.

The photosynthesizing organisms all belong to the eubacteria, if we consider chloroplasts as eubacteria and disregard the light-dependent proton pump of the Halobacteria (which is totally unrelated to what is normally called 'photosynthesis' (Ref. 13; and Chapter 2). The different types of photosystem known today belong to different eubacterial 'phyla' which are too distantly related to determine their relative phylogenetic positions by the current methods [14]. Thus, bacterial phylogeny does not yet help to decide which type of photosystem may be most similar to the ancestral type and does not even support the general assumption that all photosystems are derived from the same ancestral type. On the basis of this assumption it now seems likely that the common ancestor of all eubacteria already had a well-developed photosynthetic apparatus. On the other hand, to postulate that photosynthesis is as old as life itself [13] may be taking the point a little bit too far, in view of the absence of photosynthetic eukaryotes without chloroplasts and of photosynthetic archaebacteria. It seems simpler to assume that these organisms have not lost photosynthesis but are true relics of pre-photosynthetic 'progenotes', which by some specialization managed to survive the appearance of photosynthetic cells: the surviving archaebacteria may stem from progenotes which escaped in some dark corner; the specialization of the 'urkaryote' [15] may have been that by that time it had already begun to feed on other cells or their remains. In this picture, photosynthesis was invented by one of the progenotes, which thereby became the ancestor of a successful and rapidly expanding family: the eubacteria.

At present neither fossil evidence nor the phylogeny of extant bacteria can tell what the ancestral photosystem looked like and what properties predisposed its possessor among the progenotes to the acquisition of photosynthesis. Another approach to these questions is to search the photosynthetic apparatus itself for possible traces of its history.

4. *Reaction center structure*

Of one purple bacterium, *Rhodopseudomonas viridis*, the structure of the reaction

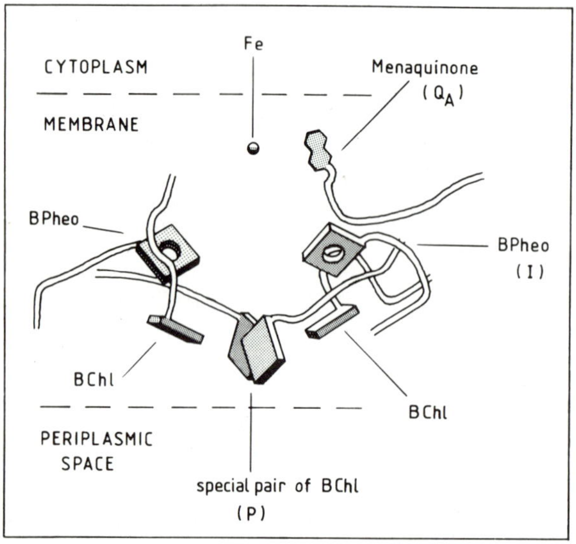

Fig. 1. Arrangement of the pigments in the reaction center of *Rhodopseudomonas viridis*, based on the crystallographic data in Ref. 16. Upon excitation an electron is transferred from the special pair P to the menaquinone Q_A. (From Ref. 18).

center has been resolved by X-ray crystallography [16,17]. Fig. 1 shows the arrangement of the six porphyrins and the permanently bound quinone Q_A in this reaction center. The most striking feature is its almost perfect two-fold rotational symmetry, which is not limited to the arrangement of the chromophores shown in Fig. 1, but applies to the apoproteins (L and M subunits) and the binding site for the exchangeable quinone Q_B relative to that of Q_A as well. Such a symmetry is expected when identical, asymmetrical subunits form dimers. Probably the present, structurally minor, differences between the two subunits reflect a specialization afterwards, and the formation of dimers had some advantage when the subunits were still identical. One possible advantage might be in the dimeric structure of the primary electron donor, but even if this proves to be essential, it is not clear why a similar 'special pair' could not arise in a monomeric reaction center protein. In fact the presence of two BChl molecules per subunit may perhaps indicate that a 'special pair' was already present when the subunits were still independent reaction centers. A more obvious reason for the pairing of reaction centers may have been that the reaction center did not yet have a permanently bound quinone as a secondary electron acceptor. The reaction center dimer could, upon excitation, transfer an electron to either side, thereby reducing the risk of wasting a charge separation by lack of an electron acceptor to stabilize it. This hypothesis presupposes electronic coupling between the pigments acting as the primary electron donors in each of the two monomeric reaction centers and suggests that the present 'special pair' may be regarded as the ultimate result of a tendency to enhance this coupling.

At some stage gene duplication occurred and differentiation of the two branches of electron transfer in the dimeric reaction center could begin. The most obvious change is that on one side the quinone binding site has become permanently occupied by a quinone molecule, Q_A, which upon reduction is not replaced, but in turn reduces the quinone at the other binding site, Q_B, when present. Thus, the differentiation of the two reaction center subunits, just like their association, was probably selected primarily because it reduced the risk of losing a charge separation by lack of a bound quinone to stabilize it.

Given the permanent presence of a bound quinone at one branch, it makes sense that the intermediary porphyrins in the other branch have lost their electron transport function. Their continued presence, apparently without gross conformational changes, may merely indicate a structural role. It may not have been worth while to replace them by a modified protein structure, or such a modification may require an improbably large number of simultaneous mutations in order to maintain activity.

It should be mentioned that electron transfer to the quinone pool, both by PS II and by the reaction centers of purple bacteria, now proceeds via a two-electron gating mechanism: after one electron has arrived at the temporarily bound quinone Q_B, the semiquinone remains in its unprotonated, negatively charged form and tightly bound to the reaction center; only after a second photoreaction does its full reduction, protonation and release as a quinol take place [19]. This procedure may also have played a role in the selection of the dimeric reaction center structure, but its importance most likely has to do with the reactivity of semiquinones with molecular oxygen and in that case it probably appeared much later.

Like Q_A, other stabilizing secondary electron donors and acceptors in reaction centers may have originated as substrate molecules which have become permanently bound. The ferredoxin-reducing reaction centers of green sulfur bacteria and of PS I have iron-sulfur centers as secondary electron acceptors. With the exception of PS II, all reaction centers seem to oxidize a Cyt *c* and in those that have a permanently bound secondary electron donor, this donor is a *c*-type cytochrome. It seems reasonable to assume that the reaction centers acquired their secondary, stabilizing redox components by increasing the affinity of their binding site for the – mobile – substrate (quinone, ferredoxin or Cyt *c*) and by development of electron transfer between the bound substrate molecule and its mobile colleagues. It should be mentioned that such an electron transfer would automatically become thermodynamically favored it the reaction center succeeded in increasing its affinity for the substrate more than for the product (reduced quinone or ferredoxin, or oxidized Cyt *c*), since the midpoint potential of the bound form would thereby be shifted relative to that of the free form.

If this hypothesis is correct, some interesting inferences may be made: no iron-sulfur centers have been found in the quinone-reducing systems (PS II and the reaction centers of purple bacteria), but there is evidence for the presence of a low-potential quinone in PS I [20] and in *Heliobacterium chlorum* [21], and no evidence against its presence in green sulfur bacteria. This suggests that ferredoxin-reducing systems evolved from quinone-reducing systems, in line with the general view in the evolution literature that linear photosynthetic electron transport came later than photosynthetic energy conversion by cyclic electron flow, which is based on the assumption that a shortage of energy preceded a shortage of reduced carbon compounds.

5. *A minimal model*

The simplest cyclic photosystem might require a polypeptide spanning the cytoplasmic membrane, carrying two different chlorophyll-like chromophores (perhaps a special pair plus an electron acceptor) arranged in such a way that excitation would result in electron transfer towards the cytoplasmic side, and the presence of lipophilic quinones, probably menaquinone, in the membrane. Quinones are quite unreactive in aprotic media [22], and in the absence of specialized binding sites would be expected to react only near the cytoplasmic or near the outside surface of the membrane, where protonation and deprotonation reactions could stabilize the product. The directionality of the charge separation must be postulated to ensure that, if an electron transfer to quinone took place, it would be followed by proton uptake from the cytoplasm. The positive charge left behind on the pigments should be localized preferentially near the outside surface, where it could lead to quinol oxidation upon proton release towards the outside. This reaction sequence is the most likely one, in view of the electrochemical properties of quinones: the oxidation of quinol is normally energetically feasible only after dissociation of a proton and is therefore relatively slow [22]. The quantum yield of the light-driven proton pump thus obtained would be dependent largely on the probability that a quinone molecule happened to be in the right position to accept an electron at the moment a photon was absorbed, and subsequent developments to enhance that probability are precisely what the present reaction center structure seems to show.

A possible objection against the above model is that no contemporary photosystem oxidizes the mobile quinol in the membrane, but in PS II two quinol molecules seem to be bound close to the reaction center chlorophyll, P-680, and one of those acts as the secondary electron donor Z [23,24]. In this case no deprotonation is observed upon oxidation, but that may be regarded as a specialization of the reaction center protein which took place much later, when the need to oxidize very high-potential exogenous electron donors arose. At an earlier stage, the 'pre-PS II' reaction center most likely had some other reason to bind and oxidize quinol and a cyclic electron transport via the quinone pool, pumping protons out of the cell, seems the most simple explanation. With the advent of the Cyt *b/c* complex and the Q-cycle [25], quinol oxidation directly by the reaction center became energetically wasteful and therefore selected against. Since it was too slow to contribute to the stabilization of the charge separation, it could disappear without trace.

The history of the Cyt *b/c* complex itself may be closely related to that of the reaction centers. The occurrence of such a complex in photosynthetic and respiratory electron transport chains of chloroplasts, mitochondria and eubacteria suggests that all these chains have a common evolutionary origin. It appears that the Cyt *b* of chloroplasts and that of mitochondria show considerable homology [26], in spite of their phylogenetic distance. The bound Cyt *c* and the Rieske Fe-S protein from chloroplasts and those from mitochondria show little homology, but these proteins may have had more freedom and more reason to change; they are largely external to the chloroplast stroma and to the mitochondrial matrix, they have a covalently bound prosthetic group and only one, nearly terminal membrane-spanning sequence, and they are presumably inserted in to the membrane from opposite sides in the two types of organelle (the mitochondrial proteins are synthesized in the cytoplasm) [27]. The Cyt *b*, on the other hand, is endogenous in both organelles, is highly intrinsic and spans the membrane five times [26], just like the L and M subunits of the *Rhodopseudomonas viridis* reaction center. The cytochrome complex has its two Cyt *b* hemes arranged in such a way as to facilitate electron transfer across the membrane [26], it oxidizes quinol with concomitant proton release on the outside, and it can reduce quinone with concomitant proton uptake from the cytoplasm. These properties suggest an evolutionary relationship to the primitive reaction center postulated above.

6. *Photosynthesis*

By the simple reaction center–quinone cycle described above a photochemical proton pump would be obtained, initially with little requirement for specificity of the polypeptide involved and without any further proteins or redox components. Its components might be of abiotic origin, accumulated in the membrane due to their hydrophobicity. The selective advantage of the proton pump, e.g. in the uptake of amino acids, would be clear even for the first membrane-enclosed 'cell'. Only a proton-gradient-driven ATP synthase would be needed to establish a photosynthetic energy metabolism. Proton-translocating ATPases are ubiquitous and probably date back to the progenotes [28], and the same applies to cytochromes and iron-sulfur proteins, important building blocks of contemporary photosynthetic electron transport. Olson and Pierson [13], emphasizing the possible simplicity of ancestral photosynthesis and the possible abiotic origin of its components, argue that fermentative energy metabolism is in fact much more complicated and unlikely to have preceded photosynthesis as an energy source for the first cells. They note that the Embden-Meyerhof-Parnas pathway, shared by most fermentations, is absent in archaebacteria and requires a large number of highly specific enzymes. If photosynthesis was invented by organisms with a fermentative energy metabolism, a more simple fermentation mechanism would indeed be expected. While this consideration may argue against the idea that photosynthesis originated after the establishment of accessory oxidant-dependent fermentation as we know it now [29], it does not remove the necessity of assuming a central role of carbohydrate metabolism at the earliest stages of evolution [30], perhaps even earlier than the development of a membrane-enclosed cell as required for photophosphorylation. Substrate-linked phosphorylations carried out by structures as simple and as available as those required for the most primitive type of photosynthetic energy conversion may be hard to conceive, but, without equally demanding biosyntheses coupled to pyrophosphate bond hydrolysis, photophosphorylation serves no purpose.

References

1 Broda, E. (1975) The Evolution of the Bioenergetic Processes, Pergamon Press, New York.
2 Margulis, L. (1981) Symbiosis in Cell Evolution, Freeman and Co., San Francisco.
3 Cavalier-Smith, T. (1981) in Molecular and Cellular Aspects of Microbial Evolution (Carlisle, M.J., Collins, J.F. and Moseley, B.E.B., eds.) pp. 33–84, Cambridge University Press, Cambridge.
4 Yang, Y., Oyaizu, H., Olsen, G.J. and Woese, C.R. (1985) Proc. Natl. Acad. Sci. USA 82, 4443–4447.
5 Gillott, M.A. and Gibbs, S.P. (1980) J. Phycol. 16, 558–568.
6 Glazer, A.N. (1984) Biochim. Biophys. Acta 768, 29–51.
7 Lewin, R.A. (1984) Phycologia 23, 203–208.
8 Ellis, R.J. (1981) Annu. Rev. Plant. Physiol. 32, 111–137.
9 Schmidt, G.W., Bartlett, S.G., Grossman, A.R., Cashmore, A.R. and Chua, N.-H. (1981) J. Cell Biol. 91, 468–478.
10 Walker, J.C.G., Klein, C., Schidlowski, M., Schopf, J.W., Stevenson, D.J. and Walter, M.R. (1983) in Earth's Earliest Biosphere; its Origin and Evolution (Schopf, J.W., ed.) pp. 260–290, Princeton University Press, Princeton.
11 Schopf, J.W. and Walter, M.R. (1983) ibid., pp. 214–239.
12 Hayes, J.M., Kaplan, I.R. and Wedeking, K.W. (1983) ibid, pp. 93–134.
13 Olson, J.M. and Pierson, B.K. (1986) Int. Rev. Cytol., in press.
14 Woese, C.R. (1985) in Evolution of Prokaryotes (Schleifer, K.H. and Stackebrandt, E., eds.) pp. 1–30, Academic Press, London.
15 Woese, C.R. and Fox, G.E. (1977) Proc. Natl. Acad. Sci. USA 74, 5088–5090.
16 Deisenhofer, J., Epp, O., Miki, K., Huber, R. and Michel, H. (1984) J. Mol. Biol. 180, 385–398.
17 Deisenhofer, J., Epp. O., Miki, K., Huber, R. and Michel, H. (1985) Nature (London) 318, 618–624.
18 Van Gorkom, H.J. (1986) Bioelectrochem. Bioenerg. 16, 77–87.
19 Crofts, A.R. and Wraight, C.A. (1983) Biochim. Biophys. Acta 726, 149–185.
20 Rutherford, A.W. and Heathcote, P. (1985) Photosynth. Res. 6, 295–316.
21 Brok, M., Vasmel, H., Horikx, J.T.G. and Hoff, A.J. (1986) FEBS Lett. 194, 322–326.
22 Rich, P.R. (1981) Biochim. Biophys. Acta 637, 28–33.
23 O'Malley, P.J. and Babcock, G.T. (1984) Biochim. Biophys. Acta 765, 370–379.
24 Dekker, J.P., Van Gorkom, H.J., Brok, M. and Ouwehand, L. (1984) Biochim. Biophys. Acta 764, 301–309.
25 Slater, E.C. (1983) Trends Biochim. Sci. 8, 239–242.
26 Widger, W.R., Cramer, W.A., Herrmann, R.G. and Trebst, A. (1984) Proc. Natl. Acad. Sci. USA 81, 674–678.
27 Hauska, G. (1985) in Molecular Biology of the Photosynthetic Apparatus, (Steinback, K.E., Bonitz, S., Arntzen, C.J. and Bogorad, L., eds.) pp. 79–87, Cold Spring Harbor Laboratory, Cold Spring Harbor, NY.
28 Wilson, T.H. and Lin, E.C.C. (1980) J. Supramol. Struct. 13, 421–446.
29 Gest, H. (1980) FEMS Microbiol. Lett. 7, 73–77.
30 Quayle, J.R. and Ferenci, T. (1978) Microbiol. Rev. 42, 251–273.

Subject index

Adenylsulfate role in sulfide oxidation, 204
ADRY reagents, 142
Agmenellum quadruplicatum, 257, 314
Allophycocyanin, 249–261
Amaranthus, 323
Amine binding to oxygen-evolving complex, 147, 148
 effect on oxygen evolution, 145–148
Ammonia
 as uncoupler, 162
 binding to oxygen-evolving complex, 147, 148
 effect on oxygen evolution, 145–147
Amytal, 202
A_0, 70–72, 112
A_1, 70, 72, 112, 113
Arabidopsis thaliana, 321, 322
Aspartate transport, 180, 181
ATP
 in carbon dioxide assimilation, 176, 177, 182, 184
 in NAD^+ reduction, 201, 202
 synthesis, see photophosporylation
ATP synthase, 8, 33, 162, 167–169, 283
 biosynthesis, 217, 218, 324, 337, 338
 evolution of, 349, 350
 genes for, 336, 337
 in *Chlorobium*, 28
 in purple bacteria, 33
 polypeptides, 168, 169, 216, 218, 335–338
 reconstitution, 167–169
 structure, 216–218, 324, 335, 336
ATP-P_i exchange, 162
ATPase, 162, 163
ATP/e_2^- ratio photophosphorylation, 11, 160, 166
Atrazine binding, 323
Atriplex spongiosa, 180, 181

Bacteriochlorophyll
 distribution in photosynthetic bacteria, 36, 37
 spectroscopic properties, 299–301
Bacteriochlorophyll *a*
 protein complex, 24, 32, 33, 237–247, 301–306, 308–311
 structure, 23
Bacteriochlorophyll *b*
 protein complex, 237, 240
 structure, 23
Bacteriochlorophyll *c*
 as primary electron acceptor, 27, 29, 46, 112
 in chlorosomes, 24, 27, 245, 246
 protein complex, 24, 27, 245, 246
 structure, 23
Bacteriochlorophyll *d* structure, 23
Bacteriochlorophyll *e* structure, 23
Bacteriochlorophyll *g*
 in *Heliobacterium chlorum*, 28, 29
 structure, 38
Bacteriopheophytin
 as electron acceptor, 33, 44–47, 54–56, 111, 112
 EPR, 111, 112
Bacteriorhodopsin, 38
Bohr magneton, 100
Bundle sheath cells, 178–181, 192

Calvin-Benson cycle, see reductive pentose phosphate cycle
Carbon dioxide assimilation, see reductive pentose phosphate cycle
Carotenoid
 as electron donor, 87
 distribution in photosynthetic bacteria, 37
 in reaction center, 53
 in light-harvesting complex, 238, 242, 243, 276, 294
 in photosystem I, 219, 220
 in photosystem II, 223, 224, 263, 280
Ca^{2+} effect on oxygen evolution, 131, 138–139, 144–146
CF_0, see ATP synthase
CF_1, see ATP synthase
Charge transfer, 56, 301, 310
Chlamydomonas, 222, 223, 320, 325
 reinhardtii, 262, 287
Chlorella, 176, 183
Chloride
 binding to oxygen-evolving complex, 145, 146
 effect on oxygen evolution, 4, 130, 131, 138–139, 142, 144–146

Chlorobium, 26–28
  limicola, 110, 203, 206, 207
Chloroflexus aurantiacus, 23–26, 45, 48, 52, 110, 200, 201, 203, 236, 240, 245, 246, 313, 314
Chlorophyll
  protein complex, 219, 220, 223–226, 261–266, 275–277, 306, 307
  spectroscopic properties, 299–301
Chlorophyll *a*
  as electron acceptor, 70, 112
  structure, 3
  synthesis, 222
Chlorophyll *b* structure, 3
Chloroplast
  DNA, 215–218, 222, 226, 227, 323–336, 344
  membrane carriers, 187, 188
  origin, 343–345
  ribosomes, 215
  RNA, 226, 328, 333–335, 344
  see also thylakoid
Chlorosomes
  of *Chloroflexus aurantiacus*, 24, 245, 246
  of green sulfur bacteria, 27
Chromatium vinosum, 33, 34, 43, 47, 56, 111, 202–208, 239, 244
Chroomonas, 250, 259–261
CIDEP, 71
Circular dichroism, 300, 301, 303–309, 311–315
Crassulacean acid metabolism, 180–183
  regulation, 193–194
Cytochromes
  in green sulfur bacteria, 27
  in purple bacteria, 33
  of *Chloroflexus aurantiacus*, 25, 26
Cytochrome *b*-559, 4, 77, 87, 88, 90, 130, 133, 145, 285
  biosynthesis, 255, 329
  polypeptides, 280, 320, 326
Cytochrome b_6, 5, 226, 277–279, 285
  biosynthesis, 215, 216, 331–333
  polypeptides, 277, 278, 324, 330–333
Cytochrome bc_1 complex
  Chloroflexus aurantiacus, 25
  mitochondria, 204, 214
  purple bacteria, 45, 204
Cytochrome b_6-*f* complex
  biosynthesis, 215, 216, 226, 285, 324, 329–332
  evolution of, 348
  function, 5, 6, 214
  genes for, 330–331
  polypeptides, 214, 215, 277, 278, 330–332
  structure, 214, 215, 277–279, 324, 329–332
Cytochrome c_2, 49
Cytochrome *c*-551, 207
Cytochrome *c*-553, 72, 74
Cytochrome *c*-555, 33, 43, 207
Cytochrome *f*, 6, 7, 64, 277–279, 285
  biosynthesis, 214, 215, 331–333
  polypeptides, 214, 215, 277, 278, 324, 330–333
Cytochrome *o*, 34
Cytoplasmic membrane bacteria, 24, 27–29
C_3 cycle, see reductive pentose phosphate cycle
C_4 pathway, 178–180
  regulation, 191–193
C_4 plants, 178–180, 185, 192

DCCD, 162
DCMU, 79, 82, 83, 127, 286, 328
Dihydroxyacetone phosphate
  formation, 177, 183
  regulatory function, 190, 191
  transport, 187, 188
Dio-9, 162
DNA
  chloroplast, 215–218, 222, 226, 227, 323–336, 344
  nuclear, 222, 321, 323, 327, 331, 334

Ectothiorhodospira halochloris, 23, 33, 242
EDTA, 218, 335
Electrochromism, 54
Electron spin polarization, 113, 116–117
ENDOR, 105, 107
EPR Signal II, 86, 115, 134
Erythrobacter longus, 35
ESE, 117–119
Excitation energy distribution, 14–17, 293–294
Exciton effects, 299–302

FAD, 204, 206
FCCP, 162
Ferredoxin, 6, 67
  in photosystem I, 219–221, 330
  in purple bacteria, 33
  thioredoxin system, 185–187, 192
Ferredoxin-NADP$^+$ reductase, 6, 67, 330
Flavocytochrome *c*, 205–207
Fluorescence, see photosystem II
Fluorescence polarization, 306, 313
Free energy electron transport, 44, 53
Fremyella diplosiphon, 250–258
Fructose 1,6-bisphosphate
  formation, 177
  regulatory function, 189–191, 195

Galactolipids, 274, 281, 282
Glyceraldehyde 3-phosphate formation, 176, 178
Grana, see thylakoid

Halobacteria, 21, 36
Heliobacterium chlorum, 28, 29, 39, 110, 112, 347
Hole burning, 56
Hydroxylamine
  binding to oxygen-evolving complex, 147, 148
  effect on oxygen evolution, 137, 147

Inside-out vesicles, 13, 275, 284
Interaction
  dipolar, 100, 101, 109
  exchange, 101, 111, 112, 117, 118
  exciton, 299–302
  hyperfine, 100, 101
  Zeeman, 105
Iron non-heme in reaction center, 52, 56, 77, 80, 109
Iron-sulfur centers
  EPR, 67, 104, 111
  evolution of as electron acceptors, 347, 348
  in *Chloroflexus aurantiacus*, 26, 203
  in green sulfur bacteria, 27, 46, 110
  in *Heliobacterium chlorum*, 29, 110
  photosystem I, 6, 7, 67–70, 104, 111
  Rieske, 6, 26, 215, 278, 279, 331, 332

Lemna, 321, 322
  gibba, 263, 264
Light-harvesting complex
  biosynthesis plants, 223–227, 334, 335
  B1015 structure, 237, 240
  B800-850 structure, 237, 240, 243–244, 301–306
  B880 structure, 238–242
  cryptomonads, 247
  cyanobacteria, 247
  energy transfer, 14–17, 24, 233–235
  exciton calculations, 303–311, 314–315
  genes for, 223–227, 321–327, 333, 334
  green bacteria, 27, 245–247, 308–311
  phosphorylation of, 16, 288–290
  photosystem I, 74, 262, 263
  photosystem II, 14, 223–225, 263–266, 276, 277, 279–281, 288–290, 306, 307, 320–328
  polypeptides plants, 223–227, 332–335
  purple bacteria, 32, 33, 238–245, 314–315
  red algae, 247
Light-harvesting protein *Prosthecochloris* crystal structure, 308–311

Linear dichroism, 300, 301, 313
Linker polypeptides, 255, 256
Luminescence, 83, 141, 170

Magnetic quantum number, 100
Malate
  formation, 181, 185
  transport, 180, 181
Manganese
  complex model, 150, 151
  in oxygen evolution, 4, 115, 131, 134–138
  in reaction center purple bacteria, 51
  multiline EPR signal, 115, 135, 136, 142, 146, 147
  release, 136, 137
  UV absorption changes, 135, 141, 149
  X-ray absorption spectroscopy, 135
Mastigocladus laminosus, 249–258, 314
Menaquinone, 25, 28, 41, 45, 52, 110, 111
  see also Q_A, Q_B
Mesophyll cells, 178–181, 192
Mitchell's hypothesis, 8–12, 165–167
Mitochondrion origin, 344

NAD^+ reduction, 201–203, 207
NADH dehydrogenase, 202, 203
NADP-G3P dehydrogenase, 176, 177
$NADP^+$ reduction, 7, 201–203, 207
NADPH in carbon dioxide assimilation, 176, 177, 182, 184
Nicotiana, 321, 325
Nigericin, 162

Oenothera hookeri, 325, 330
Oxaloacetate
  metabolism, 179, 181
  transamination, 180
Oxygen-evolving complex, 4, 131–151
  biosynthesis, 225, 329
  EPR, 115, 135
  inhibition, 143–146
  molecular structure, 150, 151
  polypeptides, 128–132, 225, 329
  see also S states

P-680, 4, 84–86, 88, 90, 108, 109, 132, 139–141, 144, 223–225, 288, 319–321, 476
  electron donors to, 4, 83–87, 132, 139–143
P-700, 14, 64–67, 70, 73–75, 108, 219–221, 281, 283, 288, 290
P-700-protein biosynthesis, 222, 333–335
P-798, 29
P-840, 27

P-865, 25
P-870, 33, 43–57, 107, 108
P-960, 33, 54, 67
Panicum maximum, 180, 181
Petunia, 321, 322
Pheophytin
  as electron acceptor, 4, 81, 111, 113
  EPR, 113
Phlorizin, 162
Phospho*enol*pyruvate
  carboxylation, 179, 181, 192–194
  in mesophyll cells, 192
  production, 182, 192, 194
3-Phosphoglycerate
  formation, 176, 177
  phosphorylation, 176, 177
  regulatory function, 190
  transport, 187, 188
Photogene, 226, 328
Photophosphorylation
  acid-base, 164, 169
  ATP/2e ratio, 11, 160
  coupling sites, 161
  cyclic, 7, 27
  electric field, 165, 169
  energy transfer in relation to, 162
  evolution of, 349, 350
  H^+/ATP ratio, 160
  H^+/e^- ratio, 11, 160, 165
  in single turnover flashes, 10
  ΔpH in relation to, 165–167, 169
  post-illumination, 164
  $\Delta\Psi$ in relation to, 8–12, 163–167, 170
  uncouplers, 162
  see also ATP synthase
Photorespiration in C_4 plants, 179
Photosynthetic bacteria
  comparative biochemistry, 21–39
  ecology, 24, 26, 28, 34, 35
  phylogeny, 21, 22, 35–37
  taxonomy, 29–32
Phyotosystem I
  biosynthesis, 222, 334–335
  electron acceptors, 6, 64–73, 112, 113, 219–221
  genes for, 262, 263, 333–334
  location in thylakoid, 283–288
  polypeptides, 73, 74, 219–223, 262, 263, 279, 280, 332–335
  reaction center, 6, 65, 73–75
  see also P-700
Photosystem II
  alternative electron acceptors and donors, 80, 82, 84, 87, 140
  biosynthesis, 225–227, 327–329
  core complex, 127, 128, 130
  electron acceptors, 5, 75–88, 113
  electron donors, 4, 83–87, 132, 139–143
  energy transfer, 14–17, 265, 266
  fluorescence, 5, 12, 14–17, 75, 76, 80–83
  genes for, 223–225, 321–327
  heterogeneity, 285, 286
  location in thylakoid, 283–288
  luminescence, 83, 141, 142, 144, 170
  polypeptides, 5, 89–91, 128–132, 143–146, 223–225, 263–266, 280, 320, 321
  preparations, 126–129
  reaction center, 5, 89–91, 129, 130
  see also P-680
Phycobiliproteins, 247–261
  energy transfer, 315
  structure, 248–261, 314, 315
Phycobilisomes
  distribution, 249–255
  structure, 248–261
Phycocyanin, 249–261
Phycoerythrin, 249–261
Phycoerythrocyanin, 249–261
Piericidin A, 202
Plastocyanin, 2, 64, 72, 281, 290, 291
Plastoquinone, 5–7, 16, 77, 111, 141, 223, 288, 290
  as electron donor to P-680, 86, 134
  see also Q_A, Q_B
Porphyridium cruentum, 253
Prochloron, 344
Prosthecochloris, 26, 27
  aestuarii, 46, 47, 110, 236, 246, 247, 308, 309
Purple bacteria
  primary charge separation, 53–56
  taxonomy, 29–31
Pyruvate metabolism, 179, 181

Q_A, 5, 44–57, 76, 80, 109, 111, 140, 169, 170, 275, 320
Q_B, 5, 44, 45, 52, 53, 56, 77–80, 111, 275, 286, 320, 328
Quinones
  distribution in photosynthetic bacteria, 37
  evolution of as electron acceptors, 346, 347

Reaction center
  Chloroflexus aurantiacus, 24, 25, 48, 52, 313, 314
  exciton calculations, 311–314
  see also photosystem I, photosystem II

Reaction center purple bacteria, 21, 32, 33, 43–57
absorption spectra, 53–55
crystal structure, 49–51
electron transport, 32, 33, 43, 53–56, 109–112
evolution, 345–348
isolation, 46
pigment arrangement, 51, 52
polypeptides, 33, 47–50
structure, 47–51, 311–314
X-ray analysis, 47–51
Reductive pentose phosphate cycle, 176–178
free energy changes, 183
in CAM plants, 181–183
in C_4 plants, 178–180
regulation, 183–187
Reverse electron transport, 169, 170, 204
Rhodobacter capsulatus, 34, 47, 49, 200, 203, 236–240, 243, 244
Rhodobacter sphaeroides, 31, 33, 34, 43–57, 107, 108, 110–112, 118, 119, 203, 236–240, 243, 244
Rhodocyclus gelatinosus, 31, 47, 238, 239
Rhodopseudomonas, see also *Rhodobacter, Rhodocyclus*
Rhodopseudomonas acidophila, 238, 239, 244, 245
Rhodopseudomonas palustris, 238, 239, 244
Rhodopseudomonas viridis, 23, 33, 45, 47–52, 54–57, 67, 107, 112, 114, 115, 129, 237–242, 299, 303, 305, 311, 313, 345, 349
Rhodospirillum rubrum, 32–34, 43, 47, 67, 77, 107, 201, 203, 236–242, 244
Ribosomes chloroplast, 215
Ribulose 5-phosphate formation, 177
Ribulose-1,5-bisphosphate carboxylase, see rubisco
Rieske iron-sulfur protein, see iron-sulfur centers
RNA chloroplast, 226, 328, 333–335, 344
Rotenone, 202
Rubisco
activation, 184
carboxylation, 178
oxygenase activity, 178, 184, 185
3-PGA formation, 178
RYDMAR, 117–119

S states, 3, 85, 88, 125, 140–151
manganese involvement in, 4, 115, 134–138
proton release, 4, 148, 149
UV difference spectra, 135, 141, 149
see also oxygen-evolving complex
Scenedesmus, 176
Signal II, 86, 115, 134
Solanum nigrum, 323
Spirodela, 328, 334, 335, 338
Stark effect, 54
Stromabolites, 345
Succinate oxidation, 201, 203, 204
Sucrose synthesis, 188–191
Sulfide oxidation, 204
Synechococcus, 256, 337
Synechocystis, 222

Tentoxin, 162
Thermoluminescence, 141, 142, 144
Thioredoxin, 185–187, 192
Thiosulfate oxidation, 207
Three-light reaction scheme for photosynthesis, 7
Thylakoid
appressed and non-appressed regions of, 12, 13, 281–294
heterogeneity, 283–288
permeability, 10, 168
stacking, 292–294
structure, 12, 274–277
Triphenyltin, 162
Triplet state
ADMR, 114
EPR, 113–115

Ubiquinone, 33, 44, 45
see also Q_A, Q_B

Vitamin K_1 as electron acceptor, 72
see also A_1

X-ray absorption spectroscopy manganese, 135

Zea mays, 179, 181
Zeeman energy, 98, 101